The Cognitive Neuroscience of Metacognition

Stephen M. Fleming · Christopher D. Frith
Editors

The Cognitive Neuroscience of Metacognition

Springer

Editors
Stephen M. Fleming
Center for Neural Science
New York University
New York, NY
USA

Christopher D. Frith
Aarhus University Hospital
Aarhus
Denmark

ISBN 978-3-642-45189-8 ISBN 978-3-642-45190-4 (eBook)
DOI 10.1007/978-3-642-45190-4
Springer Heidelberg New York Dordrecht London

Library of Congress Control Number: 2014930091

Printed on acid-free paper

Springer is part of Springer Science+Business Media (www.springer.com)

Acknowledgments

The authors gratefully acknowledge the support of All Souls College, Oxford (CDF), and the Wellcome Trust (SMF) during the preparation of this book.

Contents

Chapter 1
Metacognitive Neuroscience: An Introduction

Stephen M. Fleming and Christopher D. Frith

Abstract The past two decades have witnessed the birth of the cognitive neurosciences, spurred in large part by the advent of brain scanning technology. From this discipline our understanding of psychological constructs ranging from perception to memory to emotion have been enriched by knowledge of their neural underpinnings. The same is now true of metacognition. This volume represents a first attempt to take stock of the rapidly developing field of the neuroscience of metacognition in humans and non-human animals, and in turn examine the implications of neuroscience data for psychological accounts of metacognitive processes.

In the introduction to a recent volume on metacognition, Michael Beran and colleagues wrote, "The very idea of publishing another book on metacognition needs a word of justification as there is already a number of collections available in this rapidly growing field" [1]. As the book you are holding follows their excellent volume, it is even more pressing for us to address this question. Fortunately, it is relatively straightforward to do so. The past two decades have witnessed the birth of the cognitive neurosciences, spurred in large part by the advent of brain scanning technology. From this discipline our understanding of psychological constructs ranging from perception to memory to emotion have been enriched by knowledge of their neural underpinnings. The same is now true of metacognition.

S. M. Fleming (✉)
Department of Experimental Psychology, University of Oxford, Oxford, UK
e-mail: sf102@nyu.edu

S. M. Fleming
Center for Neural Science, New York University, New York, USA

C. D. Frith (✉)
Wellcome Trust Centre for Neuroimaging, University College London, London, UK
e-mail: c.frith@ucl.ac.uk

C. D. Frith
Interacting Minds Centre, Aarhus University, Aarhus, Denmark

S. M. Fleming and C. D. Frith (eds.), *The Cognitive Neuroscience of Metacognition*,
DOI: 10.1007/978-3-642-45190-4_1,

This volume represents a first attempt to take stock of the rapidly developing field of the neuroscience of metacognition in humans and non-human animals, and in turn examine the implications of neuroscience data for psychological accounts of metacognitive processes.

1.1 Defining Metacognition

Before previewing the chapters in this book, let us start with a definition of metacognition. There are at least two reasons why the term metacognition sometimes leads to confusion. The first is that it evokes different domain-specific associations. For example, metacognition may take on different connotations within education research, in memory research and in perception research. A second, more subtle reason is that metacognition is sometimes associated with conscious (and by implication, human) reflective awareness. We think this latter reason presents a barrier to a satisfactory computational and biological explanation of metacognition, so we take some time here to outline what such an explanation might look like.

The simplest definition of metacognition is cognition about cognition. A metacognitive process is meta-level with respect to an object-level cognitive process. This framework was originated by Flavell [2], and later Nelson and Narens [6], in the study of learning and memory. Memory still gives us our most subjectively vivid examples of metacognition: the decision to stop revising for an exam and the feeling of a tip-of-the-tongue experience are both quintessential metacognitive experiences. Importantly however, these examples of metacognition happen to be associated with explicit conscious awareness. While metacognition may be accompanied by conscious awareness in humans, this need not be the case, suggesting a division between "explicit", conscious metacognition and "implicit" metacognition [3, 4, 8]. To take a concrete example rooted in neuroscience, consider that a visual image of a face leads to activity in the fusiform cortex. We can think of the fusiform response to the face as an "object-level" process. A meta-level process (say in the prefrontal cortex) may represent confidence that fusiform activity is signaling a face is present. This meta-level process may or may not be associated with reflective awareness, but it is nevertheless metacognitive.

Appreciating this point helps admit a broader range of evidence in the study of the neuroscience of metacognition. For example, finding neurons in non-human animal brains that covary with confidence in a previous decision would reveal a plausible neural substrate of metacognition, despite the difficulty of assessing whether metacognitive judgments are explicit or implicit in non-human animals. It also makes clear that an important question for future study is the difference in neural implementation between implicit and explicit metacognition in humans, and the degree to which this neural circuitry is shared with non-human animals. Crucially neuroscience may be able to provide a window on the representational architecture of a metacognitive system, a point to which we turn next.

1.2 Why Neuroscience?

Constraints on neurobiological implementation serve to shape psychological theory, and these constraints might prove particularly important in the study of metacognition. It is helpful to draw an analogy with the well-established field of memory neuroscience. The discovery of intact implicit memory despite impaired episodic memory in patient HM has provided a strong constraint on every systems-level theory of memory since the 1950s. More recently brain imaging technology has revealed links between components of psychological models and their putative divisions of labour at an implementational level [5].[1] Psychological models of metacognition are particularly ripe for such an analysis. Consider again the example of face perception. Imagine that a subject's task is to rate their confidence in having seen a series of blurry faces while in a brain scanner. There are at least two possible neural and psychological accounts of how this second-order confidence judgment is made. First, it might be achieved by a direct readout of properties of an object-level representation in the fusiform cortex (a non-metacognitive implementation of a second-order behavior). Second, the judgment might rely on a meta-level representation of a subset of properties of the fusiform activity, say in prefrontal cortex. These inevitably over-simplified hypotheses make neuroscientific predictions: in the former case, lesions to putative meta-level representations in prefrontal cortex should not affect confidence judgments; in the latter case, confidence judgments may be selectively affected by lesions while leaving first-order behavioural responses (such as a forced-choice judgment) to the face intact.

This schematic example makes clear that cognitive neuroscience has much to offer a psychological-level understanding of metacognition. We might find that some behaviours traditionally thought of as metacognitive are implemented in a manner that does not require meta-level representations; in turn, a detailed understanding of those systems that do permit meta-level representation will refine psychological-level models. Finally, we note that by rooting our psychological-level models in cognitive neuroscience the division between meta- and object-level becomes less sharp and more nuanced, reflecting the intricate interplay between higher-order and primary sensory and mnemonic brain areas.

It should be clear from previous paragraphs that neuroscience does not stand apart from behavioural measurement or theoretical models. In this spirit, we have included an opening section entitled "Quantifying metacognition for the neurosciences" which reviews types of metacognitive judgments and theoretical and computational frameworks within which to understand these judgments. We hope that these chapters form a self-contained section while not retreading ground that has already been covered in excellent previous collections.

[1] The next trend will be to understand how individual functional specialisations predicted by psychological-level models are integrated via analysis of functional and structural connectivity between brain regions.

1.3 An Outline of the Book

The first section, "Quantifying metacognition for the neurosciences", outlines behavioural and analytic techniques important for the development of metacognitive neuroscience. Bennett and Schwartz (Chap. 2) emphasise that human metacognitive judgments are likely to arise from multiple component psychological processes, using the tip-of-the-tongue experience as a case in point. A pressing issue in the quantification of metacognition is distilling a metacognitive component of behavior from other confounding factors. Lau and Maniscalco (Chap. 3) present an overview of their recently developed computational measure of metacognitive efficiency that achieves this control within a signal detection theoretic framework. Overgaard and Sandberg (Chap. 4) review different types of metacognitive report about perception, and discuss how different types of report may map onto different kinds of metacognitive access. Finally, Smith, Couchman and Beran (Chap. 5) outline progress on the quantification of metacognition in non-verbal animal species, and consider the various theoretical interpretations of these data.

The second section, "Computational approaches", focuses on the utility of computational models for bridging behavioural and neural data. Computational models have proven very useful for revealing neural correlates of "hidden" internal states that would not otherwise be apparent in analysis of behaviour alone (e.g. [7]). Kepecs and Mainen (Chap. 6) present a signal detection theoretic model of decision confidence that can be powerfully applied to understanding confidence signals in neural data across different species. Yeung and Summerfield (Chap. 7) outline evidence accumulation models as a common framework in which to understand studies of error detection and confidence judgments. Fleming and De Martino (Chap. 8) present a case study of the application of an evidence accumulation model to understand the neural basis of confidence and metacognition during human value-based decision-making. Bang, Mahmoodi, Olsen, Roepstorff, Rees, Frith and Bahrami (Chap. 9) outline recent developments in modeling metacognitive judgments during social decision-making.

A third section reviews the cognitive neuroscience of metacognition across several inter-related areas of study. Middlebrooks, Abzug and Sommer (Chap. 10) review three recent studies examining metacognition-related activity in single neurons recorded from macaque monkeys and rats. Fleming and Dolan (Chap. 11) review the psychological and neural basis of metacognitive accuracy in humans, drawing on data from studies of perception, decision-making and memory. Chua, Weintraub and Pergolizzi (Chap. 12) present a comprehensive review of cognitive neuroscience studies on metacognition of human memory, covering the neural basis of subjective confidence and metacognitive accuracy. Metacognition shares an intriguing relationship with studies of human mind-wandering, which at first glance seems to be the opposite of deliberate metacognitive monitoring and control. However scholarly analysis of this link has been lacking: in their chapter, Fox and Christoff (Chap. 13) outline how metacognition and mind-wandering may share more than just an antagonistic relationship, with metacognition actively guiding the wandering mind. Finally, Chambon, Filevich and Haggard (Chap. 14) consider whether the human sense of agency should be considered a metacognitive object, and review recent work from their laboratory on understanding the neural basis of agency.

In a final section we turn to the interface between neuropsychiatric disorders and metacognition. A neuroscience of metacognition has great promise for understanding metacognitive deficits observed in neuropsychiatric disorders such as Alzheimer's disease and schizophrenia. In turn, studies of metacognition in neuropsychiatric patients can provide a novel window onto the mechanisms of metacognition. The chapter by David, Bedford, Wiffen and Gilleen (Chap. 15) describes the link between metacognitive failures and lack of insight in psychosis, noting that while insight appears separable from primary symptomology, the relationship between cognitive and clinical insight remains poorly understood. Metcalfe, Van Snellenberg, DeRosse, Balsam and Malhotra (Chap. 16) describe a study of judgments of agency in schizophrenic subjects, revealing impairment in self-related, or "autonoetic" metacognition. Finally, Cosentino (Chap. 17) reviews studies aimed at understanding the impairments of metacognition that often occur in Alzheimer's disease.

1.4 Conclusions

Neuroscience has had a dramatic impact on our understanding of individual domains of cognition, from vision to memory. We hope that a cognitive neuroscience of metacognition will bear similar fruits. This is an exciting time to be a metacognition researcher: cognitive neuroscience is maturing as a field, and has available a wealth of tools with which to investigate the biological basis of mind. These tools, combined with advanced behavioural techniques and computational modeling, have great promise to advance our nascent understanding of metacognition.

References

1. Beran MJ, Brandl J, Perner J, Proust J (2012) Foundations of metacognition. Oxford University Press, Oxford
2. Flavell J (1979) Metacognition and cognitive monitoring: a new area of cognitive–developmental inquiry. Am Psychol 34(10):906
3. Fleming SM, Dolan RJ, Frith CD (2012) Metacognition: computation, biology and function. Philos Trans R Soc Lond B Biol Sci 367(1594):1280–1286
4. Frith CD (2012) The role of metacognition in human social interactions. Philos Trans R Soc Lond B Biol Sci 367(1599):2213–2223

5. Henson R (2005) What can functional neuroimaging tell the experimental psychologist? Q J Exp Psychol 58(2):193–233
6. Nelson TO, Narens L (1990) Metamemory: a theoretical framework and new findings. Psychol Learn Motiv: Adv Res Theory 26:125–173
7. O'Doherty JP, Hampton A, Kim H (2007) Model-based fMRI and its application to reward learning and decision making. Ann N Y Acad Sci 1104:35–53
8. Reder LM, Schunn CD (1996) Metacognition does not imply awareness: strategy choice is governed by implicit learning and memory. In: Reder LM (ed) Implicit memory and metacognition. Erlbaum, Mahwah

Part I
Quantifying Metacognition for the Neurosciences

Chapter 2
Quantifying Human Metacognition for the Neurosciences

Bennett L. Schwartz and Fernando Díaz

Abstract The study of metacognition examines the relation between internal cognitive processes and mental experience. To investigate metacognition researchers ask participants to make confidence judgments about the efficacy of some aspect of their cognition or memory. We are concerned that, in our haste to understand metacognition, we mistakenly equate the judgments we elicit from participants with the processes that underlie them. We assert here that multiple processes may determine any metacognitive judgment. In our own research, we explore the tip-of-the-tongue phenomenon (TOT). Both behavioral and neuroscience evidence suggest that a number of processes contribute to the TOT. The fMRI, electroencephalography (EEG), and magnetoencephalography (MEG) data find that retrieval failure and TOT experience map onto different areas of the brain and at different times following the presentation of a stimuli. Behavioral data suggest that there are multiple cognitive processes that contribute to the TOT, including cue familiarity and the retrieval of related information. We assert that TOTs occur when retrieval processes fail and a separate set of processes monitor the retrieval failure to determine if the target can eventually be recovered. Thus, the TOT data support a model in which different underlying processes are responsible for the cognition and the metacognition that monitors it. Thus, understanding any metacognitive judgment must involve understanding the cognition it measures and the multiple processes that contribute to the judgment.

Although this chapter concerns metacognition, we start with psychophysics. The earliest psychological science was that of psychophysics, which was (and still is) the study of the relation between external energy and internal experience [14].

B. L. Schwartz (✉)
Department of Psychology, Florida International University, University Park, Miami, FL 33199, USA
e-mail: bennett.schwartz@fiu.edu

F. Díaz
University of Santiago de Compostela, Santiago de Compostela, Spain

S. M. Fleming and C. D. Frith (eds.), *The Cognitive Neuroscience of Metacognition*,
DOI: 10.1007/978-3-642-45190-4_2,

In psychophysics, we measure the wavelength of light and correlate it with the experience of color. Or we measure the frequency of sound and correlate it with the perception of pitch. Time and time again, such correlations yield replicable patterns within and across people. We argue here that, at its heart, metacognition aims to achieve something similar to the goals of psychophysics. However, metacognition's goal is to study the relation between internal cognitive processes and mental experience. For example, we study the strength of a memory and its relation to a subjective judgment of learning. Or we study the accessibility of an item and its correlation with tip-of-the-tongue (TOT) experiences. Such a goal would have likely been impossible to achieve 150 years ago, but now, with our understanding of cognitive psychology and neuroscience, it is possible to study internal mental experiences, such as metacognition, in a robust scientific fashion.

Cognitive processes are internal processes that carry out a particular function. These cognitive processes are of course, based on physical networks in the brain. Some cognitive processes may be open to introspection, and others may not. Thus, retrieval processes produce conscious memories, though the process itself is difficult if not impossible to introspect on. We define mental experiences as our subjective states that rise into consciousness. For example, our cognitive processes retrieve a memory of snorkeling on a coral reef, but the recollected experience of color, excitement, and warm water is our mental experience. The final part of this equation is behavior. Scientific psychology is rooted in the study of behavior. This applies to metacognition as well. As good experimental psychologists, we assess both cognitive processes and mental experience by observing and eliciting behavior.

2.1 The Doctrine of Concordance

We turn to the relation between cognition, behavior, and experience and Tulving's [40] doctrine of concordance. Tulving argued that there was a traditional bias in psychology, including cognitive psychology, to assume a strong correlation between cognitive processes, behavior, and experience. That is, a particular cognitive process, such as retrieval, is associated with a particular behavior, a verbal description of an earlier episode, and that this behavior is always associated with a particular conscious experience, in this case mental time travel. Tulving [40] claimed that this model no longer worked—there were too many demonstrations of conscious experience not accompanying a particular behavior to warrant its challenge. He cited studies on implicit memory, in particular, in which memory processes create a change in behavior, but without the accompanying mental experience. More recently, we can point to research in which mental experiences of memory arise from cognitive processes not tied to the retrieval process. For example, Cleary et al. [9] showed that déjà vu experiences arise when a familiarity experience occurs without corresponding retrieval of event details.

Tulving's [40] challenge to the assumptions of the doctrine of concordance underlies the basis of a great deal of research in metacognition (see [31, 32, 35]).

Metacognitive experiences arise from cognitive processes and correspond to particular behaviors. For example, an object is recognized as having been seen before (cognitive process), accompanied by an experience of confidence, and the person then says that they know the answer (behavior). However, the challenge to the doctrine arises from the repeated observations that the cognitive processes that give rise to metacognition are not the same cognitive processes that produce the behavior. One set of processes drives the recall of information, but another set of processes drives our awareness of it. In the case of retrieval, most metacognition research shows that the process that produces the metacognitive experience of confidence is dissociable from the process that elicits the retrieval. Retrieval success is determined by the strength of the target, but feeling-of-knowing judgments are determined by the ease of access to partial information and the strength of the cue [2, 34, 39]. In the aforementioned déjà vu experience, recollective processes convince the participant that the event is new, but familiarity processes drive the déjà vu experience. Thus, the processes that produce metacognition are not identical to the processes that produce the cognition they reflect.

Based on the data and reasoning above, some theorists view metacognition as heuristic in nature, that is, that metacognitive processes are not the same as the cognitive processes they monitor. Our metacognitive processes accurately predict memory performance because the processes that produce metacognition are correlated with the processes that produce cognition. Thus, as cue familiarity is correlated with target retrieval, using cue familiarity to predict recall leads to accurate feeling-of-knowing judgments. Such a heuristic model, therefore, explains why metacognition is generally accurate at predicting performance, but also why it sometimes does not predict performance; it depends on whether there is a strong positive correlation between the processes that lead to the behavior and those that lead to the internal mental experience. Thus, metamemory fails to predict performance when the metacognitive processes are not correlated with the cognitive processes used in the base process. For example, Benjamin et al. [3] found that memory strength predicted recall, but that ease of earlier processing predicted participants' judgments of learning (henceforth, JOLs), thus leading to a negative correlation between judgment and performance. To summarize, metacognition is a heuristic—it capitalizes on processes that correlate with cognitive processes and allow the organism to predict ongoing processes. Metacognitive judgments measure metacognitive experience and, in turn, are based upon underlying cognitive processes that produce them. These cognitive processes are correlated with the cognitive processes they are monitoring, but seldom identical. Thus, the cognitive processes that produce feeling of knowing may be partially based on cue familiarity, but the processes they monitor are based on retrieval strength. A generation of research has documented such dissociation in process.

2.2 Metacognition: An Introduction

Metacognition's chief empirical tool is to ask participants to make confidence judgments about the efficacy of some aspect of their cognition. The mainstay of metacognitive research, for largely historical reasons, has been judgments concerning memory [13]. Within this domain, one finds a plethora of judgments related to different aspects of the learning and retrieval processes. Ease-of-learning judgments are assessments of perceived difficulty of items in advance of study. Judgments of learning (JOLs) are assessment of whether an item being studied now will be recalled later. Turning to retrieval processes, feeling-of-knowing judgments (FOKs) are an assessment that a currently unrecalled item will be recognized later. TOT states refer to the strong feeling that a currently unrecalled item will be recalled shortly. Finally, confidence judgments can assess the feeling that a retrieved answer is actually correct. These are the main judgments used in metamemory research, although a variety of other judgments have been employed to assess specific aspects of memory (see [13], for a review).

2.3 Multiple Processes Underlie Judgments

Although it is largely accepted that cognition and metacognition are dissociable, there is less consensus on the relation between metacognitive processes and metacognitive judgments. We are concerned that, in our haste to understand metacognition, we mistakenly equate the judgments we elicit from participants with the processes that underlie them. For example, there have been disagreements as to whether FOKs are caused by cue familiarity, partial information, retrieval, or unconscious access to the target [13, 20, 26]. Clearly, all three may contribute to the judgments but in order to assert this we need to see that judgments and process are not identical.

In this chapter we challenge the assumption that there is a 1:1 correspondence between the processes that drive metacognition and the specific judgments that we make concerning metacognition. We assert here that multiple processes may determine any metacognitive judgment, and thus the judgments we measure are not pure indicators of the metacognitive processes that we are interested in. To be more concrete, consider the feeling of the TOT state [6, 35]. It is likely that the experience of the TOT is determined by the familiarity of the cue, the amount and intensity of related information retrieved, the amount and intensity of partial information retrieved, and the activation strength of the item itself. As such, the TOT experience cannot serve as a stand-in for any single one of these processes. Any consideration of the TOT requires consideration of all of these cognitive processes. We will consider the TOT and its etiology at length later in the chapter.

We propose that multiple cognitive processes may underlie any particular metacognitive judgment, be it TOT, JOL, or a confidence judgment. Although this is not a controversial statement, its implications are that each judgment itself does not perfectly reflect one metacognitive process, as they are often thought to do. This becomes important in discussing neuroimaging studies of metacognition, in which one looks at the neural correlates of a particular metacognitive judgment. It may be hard—via one study—to determine which neural area is associated with which cognitive process or which neural area is associated with which mental experience because each is multiply determined. This further complicates the issue of the relation between process, task, and subjective experience.

Does this leave an awful intractable mess? Behavior need not be correlated with subjective experience (metacognition), subjective experience may be correlated but not caused by the same processes that drive the behavior, and all of these may be influenced by multiple cognitive factors and driven by diverse mechanisms neurally. Not necessarily; just as psychophysics established principles that governed the relation between physical energy and subjective experience, we are committed to the view that studies of metacognition can develop principles that govern the relation between internal cognitive processes and subjective experience.

It is with these issues and concerns in mind that each author of this chapter began investigating the TOT phenomenon. The TOT offers a number of features that make it an excellent case study in the relation among process, behavior, and experience. TOTs are a universal experience, they are relatively frequent in everyday life, and they are easy to induce in the laboratory. More importantly, TOTs are closely linked to a particular set of cognitive processes, namely those of retrieval, and TOTs engage a specific experience linked to a specific referent, namely a particular word. These characteristics make TOTs a good candidate for a case study in the scientific examination of human phenomenology, in particular the relation between subjective experience, cognition, and behavior [35].

2.4 Tip-of-the-Tongue States

A TOT state is the feeling that we will be able to recall a currently unrecalled word. In short, a TOT is a feeling of temporary inaccessibility. We argue that there is a strong correlation between the feeling of temporary inaccessibility (the phenomenological TOT) and actual temporary inaccessibility (sometimes, called the cognitive TOT, [1]). In general, the TOT experience is predictive of resolution of temporary inaccessibility (but see [30]). This positive correlation means that TOTs are adaptive, in a functional sense, as they alert us to correctible retrieval failures. However, they also provide us with a manner of understanding the relation of process, experience, and behavior [5, 31]. The first author has argued at length elsewhere for the reasons why it is necessary to consider that the cognitive processes that produce the phenomenological TOT are different from the processes that result in temporary inaccessibility (see [27, 31, 32, 35, 36]).

Applying the logic of the challenge to the Doctrine of Concordance, what should we expect to see when we examine TOTs? TOTs are subjective experiences, which monitor unretrieved target memories. The process by which TOTs are produced should, therefore, be separable from but related to the processes that actually engage in retrieval. Moreover, it should be possible to find multiple neural components associated with retrieval and the TOT experience. These processes should overlap, but it should also be possible to see some brain regions involved with and responsible for retrieval but not the TOT and vice versa. We will now examine the neuroscience literature with these ideas in mind.

2.5 The Neuroscience of TOTs

fMRI studies Two studies have directly applied the logic above to an fMRI analysis of the neural correlates of the TOT experience [24, 25]. The first study compared TOTs, correct responses, and don't know responses, and the second study did a similar analysis, but also included feeling-of-knowing judgments. Participants were presented with definitions of words or general-information questions (e.g., "Carmen composer") and were asked to retrieve the word that matched them (e.g., "Bizet"). Participants made one of three responses while being monitored by fMRI, indicating that they (1) recalled the answer, (2) did not know the answer, or (3) were in a TOT for the answer. Because the participants were in the scanner, these responses were made via finger presses. Follow-up questions showed a relatively low rate of commission errors in the "recalled" condition. Maril et al. [24, 25] compared brain activity across these three responses. Results from fMRI studies are often complex, but there were clearly areas of higher activity in TOTs than in either the "recalled" or "don't know" condition. These areas of the brain more activated during TOTs were mostly in the frontal cortex, including right inferior frontal and right medial frontal, right dorsolateral frontal, bilateral anterior frontal, and anterior cingulate cortices (also see [19]).

The prefrontal lobe neural regions are intriguing because they have been associated with metacognition in other studies (see Chua et al., this volume; Fleming and Dolan, this volume). These areas have been previously associated with a number of monitoring and supervisory functions, including executive control (see [33, 38]). Some of the areas above are associated with monitoring and control in other tasks. For example, dorsolateral prefrontal cortex is implicated in judgments of learning and feeling of knowing. The anterior cingulate cortex is associated with surprise monitoring across a number of domains from metacognition to emotional regulation [4]. Thus, the areas of the brain activated during TOTs support the idea that a TOT is a metacognitive signal, as they are functionally related to the processes that produce the phenomenological TOT as well as the processes that cause the temporary inaccessibility.

Other fMRI studies have directed analysis toward the processes by which words become temporarily inaccessible. These studies find that temporary inaccessibility is a function of other areas in the brain. For instance, Shafto et al. [37] used a celebrity-naming task, in which participants were asked to identify the name of celebrity when a photograph of the person's face was presented. This study was mainly interested in age differences and whether these correlated with changes in temporary inaccessibility. When focusing on temporary inaccessibility, these authors found a relation between the insula and phonological processing, and that the degree of atrophy of this region in older people could contribute to the age-related increase in temporary inaccessibility. Furthermore, similar to Maril et al. [24, 25] findings, they found higher activation in the anterior cingulate and inferior frontal cortex (among other areas) in the TOT condition than in the know condition, indicating that these regions are correlated with the experience of the TOT. Thus, the pattern that emerges from the fMRI data is that temporary inaccessibility seems to be correlated with processes related to language processing, associated with the insula, but that the feeling of temporary inaccessibility might be associated with the anterior cingulate and prefrontal areas.

Although the temporal resolution of fMRI is improving, fMRI is still too slow to capture the rapid changes that occur in neural processes as active cognition unfolds over time. In order to look at rapid changes in the brain over time, it is still necessary to employ electroencephalography (EEG) or magnetoencephalography (MEG) techniques that allow study of the direct electromagnetic activity of neuron populations with a temporal resolution in the order of milliseconds. EEG and MEG can not only isolate areas of the brain (although with a spatial resolution lesser than fMRI), but can also observe changes over time with an optimal temporal resolution. Thus, the EEG and MEG data provide an excellent way to evaluate the model of how retrieval and metacognition interact.

EEG and MEG studies of TOTs Díaz and his colleagues have extensively studied both the retrieval process and the TOT using EEG and MEG techniques. We review this work in this section. In an initial study, Díaz et al. [10] examined face naming and TOTs for unrecalled names while the EEG was monitoring participants. In the task, participants were presented with the face of a famous person and required to press a button to indicate whether they were sure that they knew the person's name. They were also required to name the person. If they felt they knew the name but could not recall it, they were asked to indicate a TOT. If a person indicated a TOT, they were given a phonological cue and an opportunity to retrieve the name again (i.e., the same face was presented again). In this study, Díaz et al. were able to compare EEG patterns for successful retrieval (Know), unsuccessful retrieval (Don't Know), and unsuccessful retrieval accompanied by TOTs (see Fig. 2.1 for the general design of these studies).

The logic of the procedure was to look for systematic event-related potential (ERP) differences between task categories time-locked to the onset of the stimulus. Thus, a face of the basketball player Pau Gasol is presented at time 0. Then the EEG can register changes in the brain-wave patterns locked from the stimulus onset. Using this technique, Díaz et al. [10] were able to look at differences in the

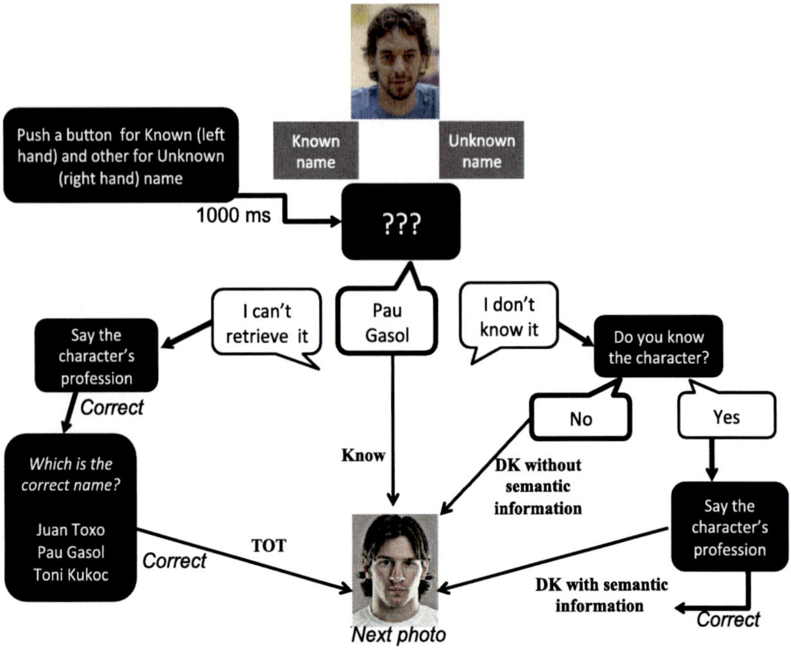

Fig. 2.1 Procedure used in Buján et al. [7], but indicative of the general procedure in these experiments. After pressing the corresponding *left* or *right* button, the participant has to say aloud **a** the correct name (Know), after which the next photograph appeared, **b** "I can't retrieve it" after which a series of questions appeared (TOT) and **c** "I don't know it" after which a series of questions appeared (Don't Know)

event-related potential between items that were correctly remembered (Know) and those that induced TOTs. Interestingly, the data from this study showed differences in the patterns between these two internal states.

During the initial presentation of the faces, there were no differences between the Know and the TOT categories in the ERP components for the first 550 ms (milliseconds) after presentation [10]. That is, the ERP correlates of perceptual processing, face recognition, and access to semantic and lexical information, as indicated by fame domain and name recall, did not differ between Know and TOT categories. This is consistent with a model in which retrieval is initiated, and the person engages in a recall attempt. Thus, for the first approximately half-second after presentation, we cannot yet distinguish retrieval and metacognitive processes. However, between 550 and 750 ms after presentation, a wave known as the late P3, which is associated with the response categorization (Know, TOT or Don't Know) was significantly larger for retrieved items than for TOT items. This result was attributed to a division of processing resources in TOTs between the categorization of the stimulus and the continuous search for complete phonological information about the name.

A second EEG component that differed among output condition was the late negative wave occurring about 1,300 ms after presentation of the face stimulus. In the Know and Don't Know response categories, late negative waves appeared after the motor response, whereas in TOTs, it appeared after the stimulus classification but before the motor response. Late negative waves were largest in Don't Know items, intermediate in Know items, and smallest in TOT items, perhaps associated with the level of uncertainty about the categorization of the stimulus and with the release of processing resources with the response. The later dissociation among retrieved, unretrieved, and TOT items supports the idea that processes diverge once the recall attempt has failed during TOTs. In a subsequent study [8], in which the ERPs were averaged in relation to a manual response, it was shown that the preparation and the execution of the responses (manual+verbal) differently modulated the stimulus-related ERP components in each response category, explaining in part the differences between categories in the amplitude of the late negative wave. Galdo-Álvarez et al. [15, 16] replicated the Díaz et al. [10] findings and found no differences in time course of ERP between older and younger participants for the Know and TOT responses, although there were age-related differences in ERP amplitudes and their scalp distribution.

In a subsequent replication, Lindín and Díaz [21] replaced the manual plus verbal response (for Know, Don't Know, TOT responses) with a manual response that was separated 1 s from the verbal response (three question marks were presented authorizing the participant to perform the corresponding verbal response). Using this methodology, there was a longer latency of the N450 wave for TOTs. This means that for TOTs the N450 wave occurred slightly later than it did for Know and Don't Know responses, probably indicating the slowing in the retrieval of semantic and lexical-semantic information during a TOT. Again, the late P3 distinguished TOTs and retrieved items. However, in this study, there were no differences at the late 1,300 ms stage between TOTs and other states. Though the form of response brought out one feature and suppressed another, we still see that TOTs and successful recall are dissociable by their EEG patterns.

Lindín et al. [22] used a similar behavioral methodology but with MEG technology in addition to EEG. The use of MEG technology allowed the researchers to pinpoint more accurately the spatial correlates of the behavioral measures with the same temporal resolution provided by the EEG. The goal in this study was to characterize the spatiotemporal course of brain activation in both successful recall and during the TOT. Consistent with the earlier findings, there were no differences in the MEG data for the first 210 ms after presentation of a face. However, during the interval from 210 to 520 ms, there was greater activation for Know responses than for TOT responses in a variety of brain regions, mostly in the left hemisphere, including left anterior medial prefrontal cortex, left orbitofrontal cortex, the left superior temporal pole, and the left inferior, middle and superior temporal gyri, as well as bilateral parahipocampal gyrus, right fusiform gyrus, and Broca's area. These are consistent with the processes involved in successfully retrieving a stored memory. They also found that at the later interval, 580–820 ms, there was greater activation for TOTs in the bilateral inferior and middle occipital gyri as well as left temporal and right frontal and parietal regions, consistent with the role of monitoring in TOT experiences (because of the right frontal activity) and with the active but fruitless search of the name.

In sum, whereas the ERP data showed that the amplitude differences between Know responses and TOTs were observed between 550 and 750 ms post-stimulus, coinciding with the categorization of stimulus, the MEG data showed that the differences are already apparent from 210 ms and, consistently, from 310 ms. MEG data also showed that there are two distinct phases: the first, between 310 and 510 ms corresponding with the successful access to the phonology of the name (greater activation of a brain network related with name recall in the case of Know responses) and with the genesis of the TOT (hypoactivation of the network in the TOT responses); and the second between 580 and 820 ms, with greater activation for TOT than for Know. The 310–510 activity may correspond to the active search of information in memory about the name. The 580–820 activation may correspond to the metacognitive monitoring in TOT experiences because this activity may be responsible for partial retrieval, such as the retrieval of visual information and partial lexical information [10].

With the aim of determining the timing of the phonological retrieval, Buján et al. [7] carried out another ERP study using a face-naming task. In this study, the early components of the ERP were again equivalent in TOTs and Know responses. However, there were differences among response options later in processing. Again, after 550 ms, there were smaller positive amplitudes in the TOTs than in Know responses, and after 1,100 ms there were higher negative amplitudes in the TOT than in Know and Don't know responses. This may correspond to the metacognitive control of retrieval, as resources may be diverted to conflict management and a continued search for the missing word, consistent with a metacognitive component to TOTs (see Fig. 2.2).

Probably the most interesting result of Buján et al.'s [7] study is the difference between the Know category and the TOT in the Lateralized Readiness Potential (LRP). The onset latency of the stimulus-related LRP was 360 ms for the Know category, whereas the TOT (and also the Don't Know category) showed a significant delay. The onset latency of the stimulus-related LRP is a correlate of response selection; consequently, when the response selection starts, access to the phonological output lexicon (lexemes) has already taken place. The response-related LRP also showed earlier onset latency in Know than in TOT, that is, the start of the actual preparation of the response was slower for TOT than for Know responses. These LRP data are consistent with the MEG data and indicate that around 360 ms, the phonological information of the name was retrieved in Know (which allow the corresponding selection and preparation of the response), but in TOTs the delay in the selection and in the preparation of the response indicated the failure in retrieving the complete phonology of the name.

To summarize, it has become clear that the TOT can be thought of as both a problem with retrieval and as successful metacognition (see [36]). The retrieval failure occurs because of any number of problems with the retrieval process, whereas successful metacognition monitors that failure. We argue here that the

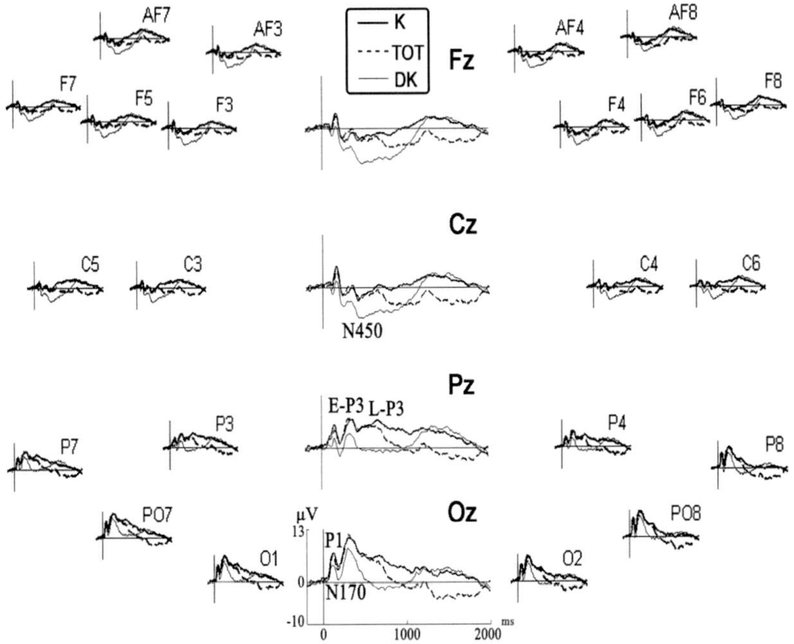

Fig. 2.2 Grand-averaged ERP waveforms for the Know (K, *thick line*), TOT (*dashed line*) and Don't Know (DK, *thin line*) response categories at *midline* electrodes and at several of the lateralized electrodes used

neural data presented here support this view. Both the fMRI, EEG and MEG data find that retrieval failure and TOT experience map onto different areas of the brain and at different times following the presentation of a stimuli. That is, there is one set of processes that are responsible for retrieval. When these processes fail, a separate set of processes monitor the retrieval failure to determine if the target can eventually be recovered. When these processes indicate a possibility of this, the TOT is experienced. Thus, the TOT data support a model in which different underlying processes are responsible for the cognition and the metacognition that monitors it.

2.6 Are Metacognitive Judgments Process-Pure?

We now return to the question raised earlier in the chapter. Do metacognitive judgments reflect multiple underlying metacognitive processes? We think our discussion of TOTs suggests that even straightforward metacognitive judgments like TOTs may involve numerous factors that influence its occurrence. We present here our reasoning that this conclusion likely extends to other metacognitive judgments.

Consider a neuroimaging study that examines the neural correlates of a particular metacognitive judgment, say a judgment of learning (JOL). A participant must determine if a particular item, say the Icelandic-English translation pair, "fiðlu—violin" has been learned. The participant must indicate his or her JOL on a scale of 0–100, translating the experience of the JOL into an outputted number on the scale. Multiple processes are surely at work as one does this task. Is the Icelandic word familiar, easy to pronounce, studied before, and does it resemble an English word? How long in the future will be the memory test, will it be a cued recall or a forced-choice recognition test, how many other pairs might also be required for the test? Can an easy linkword (e.g., "fiddle") be generated to help encode the pair? We contend that there are many processes used to determine a single number given in the JOL. When we look at the pattern of activity in the brain, it is difficult to correlate any activated area of the brain with one and only one of these potential sources of information for the JOL.

When one looks at the neuroscience literature, one finds that JOLs are correlated with different regions in the brain in different studies. For example, Kao et al. [18] found that JOLs were associated with activity in left ventromedial prefrontal cortex. However, Do Lam et al. [12] found that JOLs were associated with activation in medial PFC, orbital frontal, and anterior cingulate cortices. Thus, ventromedial cortex appears common to both studies, but additional areas were activated during JOLs in one study that were not in the other. Why does one judgment correlate with such different brain regions across studies? It likely has to do with the procedural differences between the two studies. Kao et al. had participants make JOLs on photographs of visual scenes (e.g., a mountain sunset) for eventual recognition whereas Do Lam et al. had participants make JOLs on photographs of faces for eventual cued recall of names. Thus, because of the different tasks, the JOLs were based on different sources of information and thus different areas of the brain were recruited. Because the processes underlying JOLs are sensitive to differences in tasks, the JOL is not a process-pure measure of metacognition.

Feeling-of-knowing judgments (FOKs) are generally defined as a feeling that an unrecalled item will eventually be recognized [32]. We also find that different studies find different regions of the brain are associated with FOK. For example, although Maril et al. [23] and Jing et al. [17] found that FOK was uniquely associated with activity in left inferior prefrontal cortex, Schnyer et al. [29] found that FOK was uniquely associated with ventromedial PFC. Moreover, Reggev et al. [28] found different areas of the prefrontal cortex were uniquely associated with FOK for episodic and semantic memory. The point is not these studies should all be the same, but that small changes in procedure elicit different processes that draw on different areas of the brain. Thus, the FOK task is not a process-pure measure of metacognition either.

This is not a pessimistic argument. We are not arguing against the use of neuroimaging or against the use of particular tasks to investigate metacognition or the association between brain regions and particular judgments. Our point is simple: metacognitive judgments draw on multiple underlying cognitive processes that likely draw on different underlying brain processes. Small changes in procedure can therefore shift the relevance of different processes for the same judgment, leading to the pattern of data observed in JOLs and FOKs, in which small shifts in procedure yield different unique activity associated with a particular area. So to repeat, we must be cautious because metacognitive judgments are not necessarily process-pure.

2.7 Future Directions

We welcome and embrace metacognitive neuroscience. Already neuroscience research has contributed to our understanding of the underlying cognitive mechanisms involved in some metacognition paradigms (e.g., [24]). And we certainly agree that it is important to know which brain regions correlate with which cognitive processes. Our point here is a simple one—that any metacognitive judgment may map onto multiple cognitive processes and each of these processes may correlate with different neural networks. This leads to the conclusion that any attempt to map the neurocognition of metacognition will require us to start disentangling the processes that underlie a particular judgment. Thus, TOTs may be determined by cue familiarity, partial retrieval, and perhaps the fluency of the broken retrieval process. The TOT experience is an amalgamation of these different underlying processes. In looking at the brain, one must try to look for the regions and time course of activity associated with these different neural elements and networks (see [11]).

We close by returning to the issue of Tulving's challenge to the notion of the doctrine of concordance. Tulving challenged the view that experience, behavior, and cognitive process were always perfectly correlated. Schwartz [31] modified this view to challenge the view that metacognitive experiences are based on the processes they are supposed to monitor. Indeed, much research now suggests that metacognitive judgments are largely a set of heuristics we use to infer what our cognitive processes are doing [20, 26]. We think it is clear from the arguments and data advanced here that it is not tenable to speak of metacognitive experience as being identical to the cognitive processes these experiences track. We think our data show that the TOT experience, for example, may be partially but not completely based on the processes that lead to retrieval failure and partially based on heuristics such as the amount of related information, with more retrieved related information leading to a greater likelihood of a TOT [35]. Thus, as neuroscience explores the nature of mental experience and metacognition, researchers must bear in mind the importance of distinguishing object-level processes from meta-level processes. We look forward to seeing continued work linking metacognition to brain function.

References

1. Bacon E, Schwartz BL, Paire-Ficout L, Izaute M (2007) Dissociation between the cognitive process and the phenomenological experience of the TOT: effect of the anxiolytic drug lorazepam on TOT states. Cogn Conscious 16:360–373
2. Benjamin AS (2005) Response speeding mediates the contribution of cue familiarity and target retrievability to metamnemonic judgments. Psychon Bull Rev 12:874–879
3. Benjamin AS, Bjork RA, Schwartz BL (1998) The mismeasure of memory: when retrieval fluency is misleading as a metamnemonic index. J Exp Psychol Gen 127:55–68
4. Botvinick M (2007) Conflict monitoring and decision making: reconciling two perspectives on anterior cingulate function. Cogn Affect Behav Neurosci 7:356–366
5. Brown AS (1991) A review of the tip-of-the-tongue experience. Psychol Bull 109:204–223. doi:10.1037/0033-2909.109.2.204
6. Brown AS (2012) Tip of the tongue state. Psychology Press, New York, Jan 2011
7. Buján A, Galdo-Álvarez S, Lindín M, Díaz F (2012) An event-related potentials study of face naming: evidence of phonological retrieval deficit in the tip-of-the-tongue state. Psychophysiology 49:980–990. doi:10.1111/j.1469-8986.2012.01374.x
8. Buján A, Lindín M, Díaz F (2009) Movement related cortical potentials in a face naming task: influence of the tip-of-the-tongue state. Int J Psychophysiol 72:235–245. doi:10.1016/j.ijpsycho.2008.12.012
9. Cleary AM, Brown AS, Sawyer BD, Nomi JS, Ajoku AC, Ryals AJ (2012) Familiarity from the configuration of objects in 3-dimensional space and its relation to déjà vu: a virtual reality investigation. Conscious Cogn 21:969–975
10. Díaz F, Lindín M, Galdo-Álvarez S, Facal D, Juncos-Rabadán O (2007) An event-related potentials study of face identification and naming: the tip-of-the tongue state. Psychophysiology 44(1):50–68. doi:10.1111/j.1469-8986.2006.00483.x
11. Díaz F, Lindín M, Galdo-Álvarez S, Buján A (in press) Neurofunctional correlates of the tip-of-the-tongue state. In: Schwartz BL, Brown AS (eds) The tip-of-the-tongue and related phenomena. Cambridge University Press, Cambridge
12. Do Lam ATA, Axmacher N, Fell J, Staresina BP, Gauggel S et al (2012) Monitoring the mind: the neurocognitive correlates of metamemory. PLoS one 7(1):e30009. doi:10.1371/journal.pone.0030009
13. Dunlosky J, Metcalfe J (2009) Metacognition. Sage Publications Inc, Thousand Oaks
14. Fechner GT (1860/1966) Elements of psychophysics. Holt, Rinehart, and Winston, New York
15. Galdo-Álvarez S, Lindín M, Díaz F (2009) Age-related prefrontal over-recruitment in semantic memory retrieval: evidence from successful face naming and the tip-of-the-tongue state. Biol Psychol 82(1):89–96. doi:10.1016/j.biopsycho.2009.06.003
16. Galdo-Álvarez S, Lindín M, Díaz F (2009) The effect of age on event-related potentials (ERP) associated with face naming and with the tip-of-the-tongue (TOT) state. Biol Psychol 81:14–23. doi:10.1016/j.biopsycho.2009.01.002
17. Jing L, Niki K, Xiaoping Y, Yue-jia L (2004) Knowing that you know and knowing that you don't know: a fMRI study on feeling-of-knowing (FOK). Acta Psychol Sin 36:426–433
18. Kao Y-C, Davis ES, Gabrieli JDE (2005) Neural correlates of actual and predicted memory formation. Nat Neurosci 8:1776–1783
19. Kikyo H, Ohki K, Sekihara K (2001) Temporal characterization of memory retrieval processes: an fMRI study of the 'tip of the tongue' phenomenon. Eur J Neurosci 14(5):887–892. doi:10.1046/j.0953-816x.2001.01711.x
20. Koriat A (1993) How doe we know that we know? The accessibility account of the feeling of knowing. Psychol Rev 100:609–639
21. Lindín M, Díaz F (2010) Event-related potentials in face naming and tip-of-the-tongue state: further results. Int J Psychophysiol 77(1):53–58. doi:10.1016/j.ijpsycho.2010.04.002
22. Lindín M, Díaz F, Capilla A, Ortiz T, Maestú F (2010) On the characterization of the spatio-temporal profiles of brain activity associated with face naming and the tip-of-the-tongue

state: a magnetoencephalographic (MEG) study. Neuropsychologia 48(6):1757–1766. doi:10.1016/j.neuropsychologia.2010.02.025
23. Maril A, Simon JS, Mitchell JP, Schwartz BL, Schacter DL (2003) Feeling-of-knowing in episodic memory: an event-related fMRI study. NeuroImage 18:827–836
24. Maril A, Simons JS, Weaver JJ, Schacter DL (2005) Graded recall success: an event-related fMRI comparison of tip of the tongue and feeling of knowing. NeuroImage 24:1130–1138
25. Maril A, Wagner AD, Schacter DL (2001) On the tip of the tongue: an event-related fMRI study of semantic retrieval failure and cognitive conflict. Neuron 31:653–660
26. Metcalfe J (1993) Novelty monitoring, metacognition, and control in a composite holographic associative recall model: interpretations for Korsakoff amnesia. Psychol Rev 100:3–22
27. Metcalfe J, Schwartz BL, Joaquim SG (1993) The cue familiarity heuristic in metacognition. J Exp Psychol Learn Mem Cogn 19:851–861. doi:10.1037/0278-7393.19.4.851
28. Reggev N, Zuckerman M, Maril A (2011) Are all judgments created equal? An fMRI study of semantic and episodic metamnemonic predictions. Neuropsychologia 49:1332–1343
29. Schnyer DM, Nicholls L, Verfaellie M (2005) The role of VMPC in metamemorial judgments of content retrievability. J Cogn Neurosci 17:832–846
30. Schwartz BL (1998) Illusory tip-of-the-tongue states. Memory 6:623–642
31. Schwartz BL (1999) Sparkling at the end of the tongue: the etiology of tip-of-the-tongue phenomenology. Psychon Bull Rev 6:379–393
32. Schwartz BL (2006) Tip-of-the-tongue states as metacognition. Metacogn Learn 1:149–158
33. Schwartz BL, Bacon E (2008) Metacogn neurosci. In: Dunlosky J, Bjork R (eds) Handbook of metamemory and memory. Psychology Press, New York, pp 355–371
34. Schwartz BL, Metcalfe J (1992) Cue familiarity but not target retrievability enhances feeling-of-knowing judgments. J Exp Psychol Learn Mem Cogn 18:1074–1083
35. Schwartz BL, Metcalfe J (2011) Tip-of-the-tongue (TOT) states: retrieval, behavior, and experience. Mem Cogn 39(5):737–749. doi:10.3758/s13421-010-0066-8
36. Schwartz BL, Metcalfe J (in press) Tip-of-the-tongue (TOT) states: mechanisms and metacognitive control. In: Schwartz BL, Brown AS (eds) The tip-of-the-tongue and related phenomena. Cambridge University Press, Cambridge
37. Shafto M, Stamatakis E, Tam P, Tyler L (2010) Word retrieval failures in old age: the relationship between structure and function. J Cogn Neurosci 22(7):1530–1540. doi:10.1162/jocn.2009.21321
38. Shimamura AP (2008) A neurocognitive approach to metacognitive monitoring and control. In: Dunlosky J, Bjork RA (eds) Handbook of memory and metamemory: essays in honor of Thomas O. Nelson. Psychology Press, New York, pp 373–390
39. Thomas AK, Bulevich JB, Dubois S (2011) The role of contextual information in episodic feeling of knowing. J Exp Psychol Learn Mem Cogn 38:96–108
40. Tulving E (1989) Memory: performance, knowledge, and experience. Eur J Cogn Psychol 1:3–26

Chapter 3
Signal Detection Theory Analysis of Type 1 and Type 2 Data: Meta-d', Response-Specific Meta-d', and the Unequal Variance SDT Model

Brian Maniscalco and Hakwan Lau

Abstract Previously we have proposed a signal detection theory (SDT) methodology for measuring metacognitive sensitivity (Maniscalco and Lau, Conscious Cogn 21:422–430, 2012). Our SDT measure, meta-d', provides a response-bias free measure of how well confidence ratings track task accuracy. Here we provide an overview of standard SDT and an extended formal treatment of meta-d'. However, whereas meta-d' characterizes an observer's sensitivity in tracking overall accuracy, it may sometimes be of interest to assess metacognition for a particular kind of behavioral response. For instance, in a perceptual detection task, we may wish to characterize metacognition separately for reports of stimulus presence and absence. Here we discuss the methodology for computing such a "response-specific" meta-d' and provide corresponding Matlab code. This approach potentially offers an alternative explanation for data that are typically taken to support the unequal variance SDT (UV-SDT) model. We demonstrate that simulated data generated from UV-SDT can be well fit by an equal variance SDT model positing different metacognitive ability for each kind of behavioral response, and likewise that data generated by the latter model can be captured by UV-SDT. This ambiguity entails that caution is needed in interpreting the processes underlying relative operating characteristic (ROC) curve properties. Type 1 ROC curves generated by combining type 1 and type 2 judgments, traditionally interpreted in

B. Maniscalco (✉)
National Institute of Neurological Disorders and Stroke, National Institutes of Health,
10 Center Drive, Building 10, Room B1D728, MSC 1065, Bethesda, MD 20892-1065, USA
e-mail: bmaniscalco@gmail.com

B. Maniscalco · H. Lau
Department of Psychology, Columbia University, 406 Schermerhorn Hall,
1190 Amsterdam Avenue MC 5501, New York, NY 10027, USA
e-mail: hakwan@gmail.com

H. Lau
Department of Psychology, UCLA, 1285 Franz Hall, Box 951563 Los Angeles,
CA 90095-1563, USA

S. M. Fleming and C. D. Frith (eds.), *The Cognitive Neuroscience of Metacognition*,
DOI: 10.1007/978-3-642-45190-4_3,

terms of low-level processes (UV), can potentially be interpreted in terms of high-level processes instead (response-specific metacognition). Similarly, differences in area under response-specific type 2 ROC curves may reflect the influence of low-level processes (UV) rather than high-level metacognitive processes.

3.1 Introduction

Signal detection theory (SDT; [10, 12]) has provided a simple yet powerful methodology for distinguishing between *sensitivity* (an observer's ability to discriminate stimuli) and *response bias* (an observer's standards for producing different behavioral responses) in stimulus discrimination tasks. In tasks where an observer rates his confidence that his stimulus classification was correct, it may also be of interest to characterize how well the observer performs in placing these confidence ratings. For convenience, we can refer to the task of classifying stimuli as the type 1 task, and the task of rating confidence in classification accuracy as the type 2 task [2]. As with the type 1 task, SDT treatments of the type 2 task are concerned with independently characterizing an observer's type 2 sensitivity (how well confidence ratings discriminate between an observer's own correct and incorrect stimulus classifications) and type 2 response bias (the observer's standards for reporting different levels of confidence).

Traditional analyses of type 2 performance investigate how well confidence ratings discriminate between all correct trials versus all incorrect trials. In addition to characterizing an observer's overall type 2 performance in this way, it may also be of interest to characterize how well confidence ratings discriminate between correct and incorrect trials corresponding to a particular kind of type 1 response. For instance, in a visual detection task, the observer may classify the stimulus as "signal present" or "signal absent." An overall type 2 analysis would investigate how well confidence ratings discriminate between correct and incorrect trials, regardless of whether those trials corresponded to classifications of "signal present" or "signal absent." However, it is possible that perceptual and/or metacognitive processing qualitatively differs for "signal present" and "signal absent" trials. In light of this possibility, we may be interested to know how well confidence characterizes correct and incorrect trials *only* for "signal present" responses, or *only* for "signal absent" responses (e.g. [11]). Other factors, such as experimental manipulations that target one response type or another (e.g. [7]) may also provide impetus for such an analysis. We will refer to the analysis of type 2 performance for correct and incorrect trials corresponding to a particular type 1 response as the analysis of *response-specific*[1] type 2 performance.

---

[1] We have previously used the phrase "response-conditional" rather than "response-specific" [13]. However, [2] used the terms "stimulus-conditional" and "response-conditional" to refer to

In this article, we present an overview of the SDT analysis of type 1 and type 2 performance and introduce a new SDT-based methodology for analyzing response-specific type 2 performance, building on a previously introduced method for analyzing overall type 2 performance [13]. We first provide a brief overview of type 1 SDT. We then demonstrate how the analysis of type 1 data can be extended to the type 2 task, with a discussion of how our approach compares to that of Galvin et al. [9]. We provide a more comprehensive methodological treatment of our SDT measure of type 2 sensitivity, meta-d' [13], than has previously been published. With this foundation in place, we show how the analysis can be extended to characterize response-specific type 2 performance.

After discussing these methodological points, we provide a cautionary note on the interpretation of type 1 and type 2 relative operating characteristic (ROC) curves. We demonstrate that differences in type 2 performance for different response types can generate patterns of data that have typically been taken to support the unequal variance SDT (UV-SDT) model. Likewise, we show that the UV-SDT model can generate patterns of data that have been taken to reflect processes of a metacognitive origin. We provide a theoretical rationale for this in terms of the mathematical relationship between type 2 ROC curves and type 1 ROC curves constructed from confidence ratings, and discuss possible solutions for these difficulties in inferring psychological processes from patterns in the type 1 and type 2 ROC curves.

3.2 The SDT Model and Type 1 and Type 2 ROC Curves

3.2.1 Type 1 SDT

Suppose an observer is performing a task in which one of two possible stimulus classes (*S*1 or *S*2)[2] is presented on each trial, and that following each stimulus presentation, the observer must classify that stimulus as "*S*1" or "*S*2."[3] We may define four possible outcomes for each trial depending on the stimulus and the observer's response: hits, misses, false alarms, and correct rejections (Table 3.1).

(Footnote 1 continued)
the type 1 and type 2 tasks. Thus, to avoid confusion, we now use "response-specific" to refer to type 2 performance for a given response type. We will use the analogous phrase "stimulus-specific" to refer to type 2 performance for correct and incorrect trials corresponding to a particular stimulus.

[2] Traditionally, *S*1 is taken to be the "signal absent" stimulus and *S*2 the "signal present" stimulus. Here we follow [12] in using the more neutral terms *S*1 and *S*2 for the sake of generality.

[3] We will adopt the convention of placing "*S*1" and "*S*2" in quotation marks whenever they denote an observer's classification of a stimulus, and omitting quotation marks when these denote the objective stimulus identity.

Table 3.1 Possible outcomes for the type 1 task

Stimulus	Response	
	"*S*1"	"*S*2"
*S*1	Correct rejection (CR)	False alarm (FA)
*S*2	Miss	Hit

When an *S*2 stimulus is shown, the observer's response can be either a hit (a correct classification as "*S*2") or a miss (an incorrect classification as "*S*1"). Similarly, when *S*1 is shown, the observer's response can be either a correct rejection (correct classification as "*S*1") or a false alarm (incorrect classification as "*S*2").[4]

A summary of the observer's performance is provided by hit rate and false alarm rate[5]:

$$\text{Hit Rate} = \text{HR} = p(\text{resp} = \text{“}S2\text{”} \mid \text{stim} = S2) = \frac{n(\text{resp} = \text{“}S2\text{”}, \text{stim} = S2)}{n(\text{stim} = S2)}$$

$$\text{False Alarm Rate} = \text{FAR} = p(\text{resp} = \text{“}S2\text{”} \mid \text{stim} = S1) = \frac{n(\text{resp} = \text{“}S2\text{”}, \text{stim} = S1)}{n(\text{stim} = S1)}$$

where $n(C)$ denotes a count of the total number of trials satisfying the condition C.

ROC curves define how changes in hit rate and false alarm rate are related. For instance, an observer may become more reluctant to produce "*S*2" responses if he is informed that *S*2 stimuli will rarely be presented, or if he is instructed that incorrect "*S*2" responses will be penalized more heavily than incorrect "*S*1" responses (e.g. [12, 22]); such manipulations would tend to lower the observer's probability of responding "*S*2," and thus reduce false alarm rate and hit rate. By producing multiple such manipulations that alter the observer's propensity to respond "*S*2," multiple (FAR, HR) pairs can be collected and used to construct the ROC curve, which plots hit rate against false alarm rate (Fig. 3.1b[6]).

On the presumption that such manipulations affect only the observer's *standards* for responding "*S*2," and not his underlying ability to discriminate *S*1 stimuli from *S*2 stimuli, the properties of the ROC curve as a whole should be

---

4 These category names are more intuitive when thinking of *S*1 and *S*2 as "signal absent" and "signal present." Then a hit is a successful detection of the signal, a miss is a failure to detect the signal, a correct rejection is an accurate assessment that no signal was presented, and a false alarm is a detection of a signal where none existed.

5 Since hit rate and miss rate sum to 1, miss rate does not provide any extra information beyond that provided by hit rate and can be ignored; similarly for false alarm rate and correct rejection rate.

6 Note that the example ROC curve in Fig. 3.1b is depicted as having been constructed from confidence data (Fig. 3.1a), rather than from direct experimental manipulations on the observer's criterion for responding "*S*2". See the section titled *Constructing pseudo type 1 ROC curves from type 2 data* below.

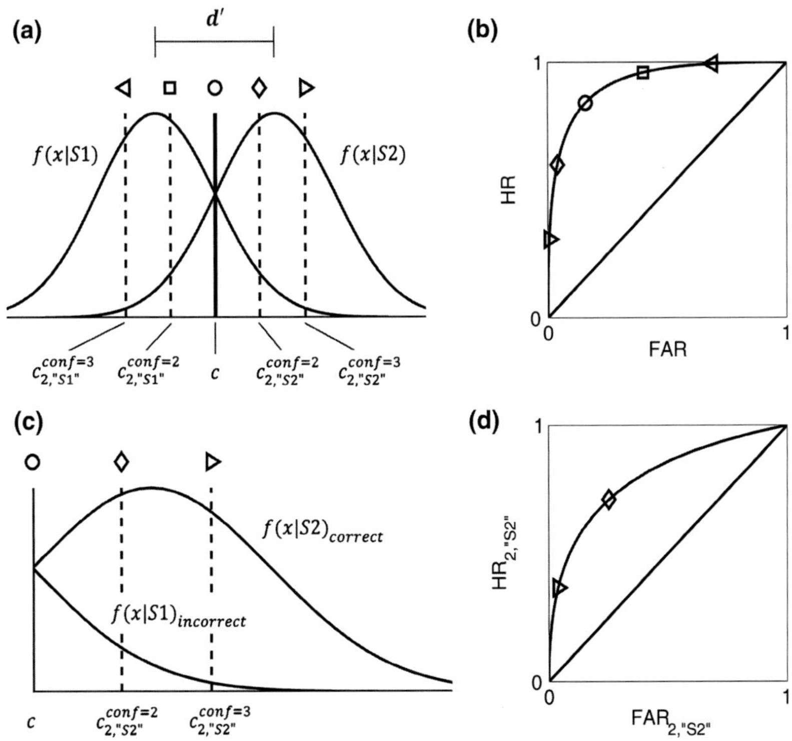

Fig. 3.1 Signal detection theory models of type 1 and type 2 ROC curves. **a** *Type 1 SDT model.* On each trial, a stimulus generates an internal response *x* within an observer, who must use *x* to decide whether the stimulus was *S*1 or *S*2. For each stimulus type, *x* is drawn from a normal distribution. The distance between these distributions is *d*′, which measures the observer's ability to discriminate *S*1 from *S*2. The stimulus is classified as "*S*2" if *x* exceeds a decision criterion *c*, and "*S*1" otherwise. In this example, the observer also rates decision confidence on a scale of 1–3 by comparing *x* to the additional response-specific type 2 criteria (*dashed vertical lines*). **b** *Type 1 ROC curve. d*′ and *c* determine false alarm rate (*FAR*) and hit rate (*HR*). By holding *d*′ constant and changing *c*, a characteristic set of (FAR, HR) points—the ROC curve—can be generated. In this example, shapes on the ROC curve mark the (FAR, HR) generated when using the corresponding criterion in panel **a** to classify the stimulus. (Note that, because this type 1 ROC curve is generated in part by the type 2 criteria in panel 1**a**, it is actually a pseudo type 1 ROC curve, as discussed later in this paper.) **c** *Type 2 task for "S2" responses.* Consider only the trials where the observer classifies the stimulus as "*S*2," i.e. only the portion of the graph in panel a exceeding *c*. Then the *S*2 stimulus distribution corresponds to correct trials, and the *S*1 distribution to incorrect trials. The placement of the type 2 criteria determines the probability of high confidence for correct and incorrect trials—type 2 HR and type 2 FAR. *d*′ and *c* jointly determine to what extent correct and incorrect trials for each response type are distinguishable. **d** *Type 2 ROC curve for "S2" responses.* The distributions in panel c can be used to derive type 2 FAR and HR for "*S*2" responses. By holding *d*′ and *c* constant and changing $c_{2,\text{"}S2\text{"}}$, a set of type 2 (FAR, HR) points for "*S*2" responses—a response-specific type 2 ROC curve—can be generated. In this example, shapes on the ROC curve mark the ($FAR_{2,\text{"}S2\text{"}}$, $HR_{2,\text{"}S2\text{"}}$) generated when using the corresponding criterion in panel **c** to rate confidence

informative regarding the observer's *sensitivity* in discriminating $S1$ from $S2$, independent of the observer's overall *response bias* for producing "$S2$" responses. The observer's sensitivity thus determines the set of possible (FAR, HR) pairs the observer can produce (i.e. the ROC curve), whereas the observer's response bias determines which amongst those possible pairs is actually exhibited, depending on whether the observer is conservative or liberal in responding "$S2$." Higher sensitivity is associated with greater area underneath the ROC curve, whereas more conservative response bias is associated with (FAR, HR) points falling more towards the lower-left portion of the ROC curve.

Measures of task performance have implied ROC curves [12, 19]. An implied ROC curve for a given measure of performance is a set of (FAR, HR) pairs that yield the same value for the measure. Thus, to the extent that empirical ROC curves dissociate sensitivity from bias, they provide an empirical target for theoretical measures of performance to emulate. If a proposed measure of sensitivity does not have implied ROC curves that match the properties of empirical ROC curves, then this measure cannot be said to provide a bias-free measure of sensitivity.

A core empirical strength of SDT ([10, 12]; Fig. 3.1a) is that it provides a simple computational model that provides close fits to empirical ROC curves [10, 20]. According to SDT, the observer performs the task of discriminating $S1$ from $S2$ by evaluating internal responses along a decision axis. Every time an $S1$ stimulus is shown, it produces in the mind of the observer an internal response drawn from a Gaussian probability density function. $S2$ stimulus presentations also generate such normally distributed internal responses. For the sake of simplicity, in the following we will assume that the probability density functions for $S1$ and $S2$ have an equal standard deviation σ.

The observer is able to discriminate $S1$ from $S2$ just to the extent that the internal responses produced by these stimuli are distinguishable, such that better sensitivity for discriminating $S1$ from $S2$ is associated with larger separation between the $S1$ and $S2$ internal response distributions. The SDT measure of sensitivity, d', is thus the distance between the means of the $S1$ and $S2$ distributions, measured in units of their common standard deviation:

$$d' = \frac{\mu_{S2} - \mu_{S1}}{\sigma}$$

By convention, the internal response where the $S1$ and $S2$ distributions intersect is defined to have the value of zero, so that $\mu_{S2} = \sigma\, d'/2$ and $\mu_{S1} = -\sigma\, d'/2$. For simplicity, and without loss of generality, we can set $\sigma = 1$.

In order to classify an internal response x on a given trial as originating from an $S1$ or $S2$ stimulus, the observer compares the internal response to a *decision criterion*, c, and only produces "$S2$" classifications for internal responses that surpass the criterion.

$$\text{response} = \begin{cases} \text{"}S1\text{"}, & x \le c \\ \text{"}S2\text{"}, & x > c \end{cases}$$

Since hit rate is the probability of responding "*S*2" when an *S*2 stimulus is shown, it can be calculated on the SDT model as the area underneath the portion of the *S*2 probability density function that exceeds c. Since the cumulative distribution function for the normal distribution with mean μ and standard deviation σ evaluated at x is

$$\Phi(x, \mu, \sigma) = \int_{-\infty}^{x} \frac{1}{\sigma\sqrt{2\pi}} e^{\frac{-(x-\mu)^2}{2\sigma^2}}$$

then hit rate can be derived from the parameters of the SDT model as

$$\mathrm{HR} = 1 - \Phi(c, \mu_{S2}) = 1 - \Phi\left(c, \frac{d'}{2}\right)$$

And similarly,

$$\mathrm{FAR} = 1 - \Phi(c, \mu_{S1}) = 1 - \Phi\left(c, -\frac{d'}{2}\right)$$

where omitting the σ parameter in ϕ is understood to be equivalent to setting $\sigma = 1$.

By systematically altering the value of c while holding d' constant, a set of (FAR, HR) pairs ranging between (0, 0) and (1, 1) can be generated, tracing out the shape of the ROC curve (Fig. 3.1b). The family of ROC curves predicted by SDT matches well with empirical ROC curves across a range of experimental tasks and conditions [10, 20].

The parameters of the SDT model can be recovered from a given (FAR, HR) pair as

$$d' = z(\mathrm{HR}) - z(\mathrm{FAR})$$
$$c = -0.5 \times [z(\mathrm{HR}) + z(\mathrm{FAR})]$$

where z is the inverse of the normal cumulative distribution function. Thus, SDT analysis allows us to separately characterize an observer's sensitivity (d') and response bias (c) on the basis of a single (FAR, HR) pair, obviating the need to collect an entire empirical ROC curve in order to separately characterize sensitivity and bias—provided that the assumptions of the SDT model hold.

3.2.2 Type 2 SDT

Suppose we extend the empirical task described above, such that after classifying the stimulus as *S*1 or *S*2, the observer must provide a confidence rating that characterizes the likelihood of the stimulus classification being correct. This confidence rating task can be viewed as a secondary discrimination task. Just as the observer first had to discriminate whether the stimulus was *S*1 or *S*2 by means of providing a stimulus classification response, the observer now must discriminate whether that stimulus classification response itself was correct or incorrect by means of providing a confidence rating.[7] Following convention, we will refer to the task of classifying the stimulus as the "type 1" task, and the task of classifying the accuracy of the stimulus classification as the "type 2" task [2, 9].

3.2.2.1 Type 2 Hit Rates and False Alarm Rates

A similar set of principles for the analysis of the type 1 task may be applied to the type 2 task. Consider the simple case where the observer rates confidence as either "high" or "low." We can then distinguish 4 possible outcomes in the type 2 task: high confidence correct trials, low confidence correct trials, low confidence incorrect trials, and high confidence incorrect trials. By direct analogy with the type 1 analysis, we may refer to these outcomes as type 2 hits, type 2 misses, type 2 correct rejections, and type 2 false alarms, respectively (Table 3.2).[8]

Type 2 hit rate and type 2 false alarm rate summarize an observer's type 2 performance and may be calculated as

$$\text{type 2 HR} = \text{HR}_2 = p(\text{high conf} \mid \text{stim} = \text{resp}) = \frac{n(\text{high conf correct})}{n(\text{correct})}$$

$$\text{type 2 FAR} = \text{FAR}_2 = p(\text{high conf} \mid \text{stim} \neq \text{resp}) = \frac{n(\text{high conf incorrect})}{n(\text{incorrect})}$$

Since the binary classification task we have been discussing has two kinds of correct trials (hits and correct rejections) and two kinds of incorrect trials (misses and false alarms), the classification of type 2 performance can be further subdivided into a *response-specific* analysis, where we consider type 2 performance only for trials where the type 1 stimulus classification response was "*S*1" or "*S*2" (Table 3.3).[9]

---

[7] In principle, since the observer should always choose the stimulus classification response that is deemed most likely to be correct, then in a two-alternative task he should always judge that the chosen response is more likely to be correct than it is to be incorrect. Intuitively, then, the type 2 decision actually consists in deciding whether the type 1 response is *likely* to be correct or not, where the standard for what level of confidence merits being labeled as "likely to be correct" is determined by a subjective criterion than can be either conservative or liberal. Nonetheless, viewing the type 2 task as a discrimination between correct and incorrect stimulus classifications facilitates comparison with the type 1 task.

[8] The analogy is more intuitive when thinking of *S*1 as "signal absent" and *S*2 as "signal present". Then the type 2 analogue of "signal absent" is an incorrect stimulus classification, whereas the analogue of "signal present" is a correct stimulus classification. The type 2 task can then be thought of as involving the detection of this type 2 "signal."

[9] It is also possible to conduct a stimulus-specific analysis and construct stimulus-specific type 2 ROC curves. For *S*1 stimuli, this would consist in a plot of p(high conflcorrect rejection) vs p(high conflfalse alarm). Likewise for *S*2 stimuli—p(high conflhit) vs p(high conflmiss). However, as will be made clear later in the text, the present approach to analyzing type 2 ROC

Table 3.2 Possible outcomes for the type 2 task

Accuracy	Confidence	
	Low	High
Incorrect	Type 2 correct rejection	Type 2 false alarm
Correct	Type 2 miss	Type 2 hit

Table 3.3 Possible outcomes for the type 2 task, contingent on type 1 response (i.e., response-specific type 2 outcomes)

Response			Confidence	
			Low	High
"*S*1"	Accuracy	Incorrect (Type 1 miss)	$\text{CR}_{2,\text{"S1"}}$	$\text{FA}_{2,\text{"S1"}}$
		Correct (Type 1 correct rejection)	$\text{Miss}_{2,\text{"S1"}}$	$\text{Hit}_{2,\text{"S1"}}$
"*S*2"	Accuracy	Incorrect (Type 1 false alarm)	$\text{CR}_{2,\text{"S2"}}$	$\text{FA}_{2,\text{"S2"}}$
		Correct (Type 1 hit)	$\text{Miss}_{2,\text{"S2"}}$	$\text{Hit}_{2,\text{"S2"}}$

Thus, when considering type 2 performance only for "*S*1" responses,

$$\text{HR}_{2,\text{"}S1\text{"}} = p(\text{high conf} \mid \text{stim} = S1, \text{resp} = \text{"}S1\text{"}) = \frac{n(\text{high conf correct rejection})}{n(\text{correct rejection})}$$

$$\text{FAR}_{2,\text{"}S1\text{"}} = p(\text{high conf} \mid \text{stim} = S2, \text{resp} = \text{"}S1\text{"}) = \frac{n(\text{high conf miss})}{n(\text{miss})}$$

where the subscript "*S*1" indicates that these are type 2 data for type 1 "*S*1" responses.

Similarly for "*S*2" responses,

(Footnote 9 continued)
curves in terms of the type 1 SDT model requires each type 2 (FAR, HR) pair to be generated by the application of a type 2 criterion to two overlapping distributions. For stimulus-specific type 2 data, the corresponding type 1 model consists of only one stimulus distribution, with separate type 2 criteria for "*S*1" and "*S*2" responses generating the type 2 FAR and type 2 HR. (e.g. for the *S*2 stimulus, a type 2 criterion for "*S*1" responses rates confidence for type 1 misses, and a separate type 2 criterion for "*S*2" responses rates confidence for type 1 hits.) Thus there is no analogue of meta-d' for stimulus-specific type 2 data, since d' is only defined with respect to the relationship between two stimulus distributions, whereas stimulus-specific analysis is restricted to only one stimulus distribution. It is possible that an analysis of stimulus-specific type 2 ROC curves could be conducted by positing how the type 2 criteria on either side of the type 1 criterion are coordinated, or similarly by supposing that the observer rates confidence according to an overall type 2 decision variable. For more elaboration, see the section below titled "Comparison of the current approach to that of [9]."

$$\mathrm{HR}_{2,\text{``}S2\text{''}} = p(\text{high conf} \mid \text{stim} = S2,\ \text{resp} = \text{``}S2\text{''}) = \frac{n(\text{high conf hit})}{n(\text{hit})}$$

$$\mathrm{FAR}_{2,\text{``}S2\text{''}} = p(\text{high conf} \mid \text{stim} = S1,\ \text{resp} = \text{``}S2\text{''}) = \frac{n(\text{high conf false alarm})}{n(\text{false alarm})}$$

From the above definitions, it follows that overall type 2 FAR and HR are weighted averages of the response-specific type 2 FARs and HRs, where the weights are determined by the proportion of correct and incorrect trials originating from each response type:

$$\begin{aligned}\mathrm{HR}_2 &= \frac{n(\text{high conf correct})}{n(\text{correct})} = \frac{n(\text{high conf hit}) + n(\text{high conf CR})}{n(\text{hit}) + n(\text{CR})} \\ &= \frac{n(\text{hit}) \times \mathrm{HR}_{2,\text{``}S2\text{''}} + n(\text{CR}) \times \mathrm{HR}_{2,\text{``}S1\text{''}}}{n(\text{hit}) + n(\text{CR})} \\ &= p(\text{hit} \mid \text{correct}) \times \mathrm{HR}_{2,\text{``}S2\text{''}} + [1 - p(\text{hit} \mid \text{correct})] \times \mathrm{HR}_{2,\text{``}S1\text{''}}\end{aligned}$$

And similarly,

$$\mathrm{FAR}_2 = p(\text{FA} \mid \text{incorrect}) \times \mathrm{FAR}_{2,\text{``}S2\text{''}} + [1 - p(\text{FA} \mid \text{incorrect})] \times \mathrm{FAR}_{2,\text{``}S1\text{''}}$$

Confidence rating data may be richer than a mere binary classification. In the general case, the observer may rate confidence on either a discrete or continuous scale ranging from 1 to H. In this case, we can arbitrarily select a value h, $1 < h \leq H$, such that all confidence ratings greater than or equal to h are classified as "high confidence" and all others, "low confidence." We can denote this choice of imposing a binary classification upon the confidence data by writing e.g. $H_2^{\text{conf}=h}$, where the superscript conf $= h$ indicates that this type 2 hit rate was calculated using a classification scheme where h was the smallest confidence rating considered to be "high." Thus, for instance,

$$\mathrm{HR}_{2,\text{``}S2\text{''}}^{\text{conf}=h} = p(\text{high conf} \mid \text{stim} = S2,\ \text{resp} = \text{``}S2\text{''}) = p(\text{conf} \geq h \mid \text{hit})$$

Each choice of h generates a type 2 (FAR, HR) pair, and so calculating these for multiple values of h allows for the construction of a type 2 ROC curve with multiple points. When using a discrete confidence rating scale ranging from 1 to H, there are $H - 1$ ways of selecting h, allowing for the construction of a type 2 ROC curve with $H - 1$ points.

3.2.2.2 Adding Response-Specific Type 2 Criteria to the Type 1 SDT Model to Capture Type 2 Data

As with the type 1 task, type 2 ROC curves allow us to separately assess an observer's sensitivity (how well confidence ratings discriminate correct from incorrect trials) and response bias (the overall propensity for reporting high confidence) in the type 2 task. However, fitting a computational model to type 2 ROC curves is somewhat more complicated than in the type 1 case. It is not appropriate to assume that correct and incorrect trials are associated with normal probability density functions in a direct analogy to the *S*1 and *S*2 distributions of type 1 SDT. The reason for this is that specifying the parameters of the type 1 SDT model—d' and c—places strong constraints on the probability density functions for correct and incorrect trials, and these derived distributions are not normally distributed [9]. In addition to this theoretical consideration, it has also been empirically demonstrated that conducting a type 2 SDT analysis that assumes normal distributions for correct and incorrect trials does not give a good fit to data [6].

Thus, the structure of the SDT model for type 2 performance must take into account the structure of the SDT model for type 1 performance. Galvin et al. [9] presented an approach for the SDT analysis of type 2 data based on analytically deriving formulae for the type 2 probability density functions under a suitable transformation of the type 1 decision axis. Here we present a simpler alternative approach on the basis of which response-specific type 2 ROC curves can be derived directly from the type 1 model.

In order for the type 1 SDT model to characterize type 2 data, we first need an added mechanism whereby confidence ratings can be generated. This can be accomplished by supposing that the observer simply uses additional decision criteria, analogous to the type 1 criterion c, to generate a confidence rating on the basis of the internal response x on a given trial. In the simplest case, the observer makes a binary confidence rating—high or low—and thus needs to use two additional decision criteria to rate confidence for each kind of type 1 response. Call these response-specific type 2 criteria $c_{2,\text{``}S1\text{''}}$ and $c_{2,\text{``}S2\text{''}}$, where $c_{2,\text{``}S1\text{''}} < c$ and $c_{2,\text{``}S2\text{''}} > c$. Intuitively, confidence increases as the internal response x becomes more distant from c, i.e. as the internal response becomes more likely to have been generated by one of the two stimulus distributions.[10] More formally,

$$\text{confidence}_{\text{resp}=\text{``}S1\text{''}} = \begin{cases} \text{low}, & x \geq c_{2,\text{``}S1\text{''}} \\ \text{high}, & x < c_{2,\text{``}S1\text{''}} \end{cases}$$

$$\text{confidence}_{\text{resp}=\text{``}S2\text{''}} = \begin{cases} \text{low}, & x \leq c_{2,\text{``}S2\text{''}} \\ \text{high}, & x > c_{2,\text{``}S2\text{''}} \end{cases}$$

In the more general case of a discrete confidence scale ranging from 1 to H, then $H - 1$ type 2 criteria are required to rate confidence for each response type. (See e.g. Fig. 3.1a, where two type 2 criteria on left/right of the type 1 criterion allow for confidence for "*S*1"/"*S*2" responses to be rated on a scale of 1–3.) We may define

[10] See "Comparison of the current approach to that of Galvin et al. [9]" and footnote 12 for a more detailed consideration of the type 2 decision axis.

$$\underline{\boldsymbol{c}}_{2,\text{“}S1\text{”}} = \left(c_{2,\text{“}S1\text{”}}^{\text{conf}=2}, c_{2,\text{“}S1\text{”}}^{\text{conf}=3}, \ldots, c_{2,\text{“}S1\text{”}}^{\text{conf}=H} \right)$$

$$\underline{\boldsymbol{c}}_{2,\text{“}S2\text{”}} = \left(c_{2,\text{“}S2\text{”}}^{\text{conf}=2}, c_{2,\text{“}S2\text{”}}^{\text{conf}=3}, \ldots, c_{2,\text{“}S2\text{”}}^{\text{conf}=H} \right)$$

where e.g. $\underline{\boldsymbol{c}}_{2,\text{“}S1\text{”}}$ is a tuple containing the $H - 1$ type 2 criteria for "$S1$" responses. Each $c_{2,\text{“}S1\text{”}}^{\text{conf}=y}$ denotes the type 2 criterion such that internal responses more extreme (i.e. more distant from the type 1 criterion) than $c_{2,\text{“}S1\text{”}}^{\text{conf}=y}$ are associated with confidence ratings of at least y. More specifically,

$$\text{confidence}_{\text{resp}=\text{“}S1\text{”}} = \begin{cases} 1, & x \geq c_{2,\text{“}S1\text{”}}^{\text{conf}=2} \\ y, & c_{2,\text{“}S1\text{”}}^{\text{conf}=y+1} \leq x < c_{2,\text{“}S1\text{”}}^{\text{conf}=y}, \quad 1 < y < H \\ H, & x < c_{2,\text{“}S1\text{”}}^{\text{conf}=H} \end{cases}$$

$$\text{confidence}_{\text{resp}=\text{“}S2\text{”}} = \begin{cases} 1, & x \leq c_{2,\text{“}S2\text{”}}^{\text{conf}=2} \\ y, & c_{2,\text{“}S2\text{”}}^{\text{conf}=y} < x \leq c_{2,\text{“}S2\text{”}}^{\text{conf}=y+1}, \quad 1 < y < H \\ H, & x > c_{2,\text{“}S2\text{”}}^{\text{conf}=H} \end{cases}$$

The type 1 and type 2 decision criteria must have a certain ordering in order for the SDT model to be meaningful. Response-specific type 2 criteria corresponding to higher confidence ratings must be more distant from c than type 2 criteria corresponding to lower confidence ratings. Additionally, c must be larger than all type 2 criteria for "$S1$" responses but smaller than all type 2 criteria for "$S2$" responses. For convenience, we may define

$$\underline{\boldsymbol{c}}_{\text{ascending}} = \left(c_{2,\text{“}S1\text{”}}^{\text{conf}=H}, c_{2,\text{“}S1\text{”}}^{\text{conf}=H-1}, \ldots, c_{2,\text{“}S1\text{”}}^{\text{conf}=1}, c, c_{2,\text{“}S2\text{”}}^{\text{conf}=1}, c_{2,\text{“}S2\text{”}}^{\text{conf}=2}, \ldots, c_{2,\text{“}S2\text{”}}^{\text{conf}=H} \right)$$

The ordering of decision criteria in $\underline{\boldsymbol{c}}_{\text{ascending}}$ from first to last is the same as the ordering of the criteria from left to right when displayed on an SDT graph (e.g. Fig. 3.1a). These decision criteria are properly ordered only if each element of $\underline{\boldsymbol{c}}_{\text{ascending}}$ is at least as large as the previous element, i.e. only if the Boolean function $\gamma\left(\underline{\boldsymbol{c}}_{\text{ascending}}\right)$ defined below is true:

$$\gamma\left(\underline{\boldsymbol{c}}_{\text{ascending}}\right) = \bigcap_{i=1}^{2H-2} \underline{\boldsymbol{c}}_{\text{ascending}}(i+1) \geq \underline{\boldsymbol{c}}_{\text{ascending}}(i)$$

It will be necessary to use this function later on when discussing how to fit SDT models to type 2 data.

3.2.2.3 Calculating Response-Specific Type 2 (FAR, HR) from the Type 1 SDT Model with Response-Specific Type 2 Criteria

Now let us consider how to calculate response-specific type 2 HR and type 2 FAR from the type 1 SDT model. Recall that

$$\mathrm{HR}_{2,\text{``}S2\text{''}}^{\mathrm{conf}=h} = p(\mathrm{conf} \geq h \mid \mathrm{stim} = S2,\ \mathrm{resp} = \text{``}S2\text{''}) = \frac{p(\mathrm{conf} \geq h,\ \mathrm{hit})}{p(\mathrm{hit})}$$

As discussed above, p(hit), the hit rate, is the probability that an $S2$ stimulus generates an internal response that exceeds the type 1 criterion c. Similarly, $p(\mathrm{conf} \geq h$, hit), the probability of a hit endorsed with high confidence, is just the probability that an $S2$ stimulus generates an internal response that exceeds the high-confidence type 2 criterion for "$S2$" responses,$c_{2,\text{``}S2\text{''}}^{\mathrm{conf}=h}$. Thus, we can straightforwardly characterize the probabilities in the numerator and denominator of $\mathrm{HR}_{2,\text{``}S2\text{''}}^{\mathrm{conf}=h}$ in terms of the type 1 SDT parameters, as follows:

$$\mathrm{HR}_{2,\text{``}S2\text{''}}^{\mathrm{conf}=h} = \frac{p(\mathrm{conf} \geq h,\ \mathrm{hit})}{p(\mathrm{hit})} = \frac{1 - \Phi\left(c_{2,\text{``}S2\text{''}}^{\mathrm{conf}=h}, \frac{d'}{2}\right)}{1 - \Phi\left(c, \frac{d'}{2}\right)}$$

By similar reasoning,

$$\mathrm{FAR}_{2,\text{``}S2\text{''}}^{\mathrm{conf}=h} = \frac{1 - \Phi\left(c_{2,\text{``}S2\text{''}}^{\mathrm{conf}=h}, -\frac{d'}{2}\right)}{1 - \Phi\left(c, -\frac{d'}{2}\right)}$$

And likewise for "$S1$" responses,

$$\mathrm{HR}_{2,\text{``}S1\text{''}}^{\mathrm{conf}=h} = \frac{\Phi\left(c_{2,\text{``}S1\text{''}}^{\mathrm{conf}=h}, -\frac{d'}{2}\right)}{\Phi\left(c, -\frac{d'}{2}\right)}$$

$$\mathrm{FAR}_{2,\text{``}S1\text{''}}^{\mathrm{conf}=h} = \frac{\Phi\left(c_{2,\text{``}S1\text{''}}^{\mathrm{conf}=h}, \frac{d'}{2}\right)}{\Phi\left(c, \frac{d'}{2}\right)}$$

Figure 3.1c illustrates how type 2 (FAR, HR) arise from type 1 d' and c along with a type 2 criterion. For instance, suppose $h = 3$. Then the type 2 hit rate for "$S2$" responses, $\mathrm{HR}_{2,\text{``}S2\text{''}}^{\mathrm{conf}=3}$, is the probability of a high confidence hit (the area in the $S2$ distribution beyond $c_{2,\text{``}S2\text{''}}^{\mathrm{conf}=3}$) divided by the probability of a hit (the area in the $S2$ distribution beyond c).

By systematically altering the value of the type 2 criteria while holding d' and c constant, a set of (FAR_2, HR_2) pairs ranging between (0, 0) and (1, 1) can be generated, tracing out a curvilinear prediction for the shape of the type 2 ROC curve (Fig. 3.1d). Thus, according to this SDT account, specifying type 1 sensitivity (d') and response bias (c) is already sufficient to determine response-specific type 2 sensitivity (i.e. the family of response-specific type 2 ROC curves).

3.2.3 Comparison of the Current Approach to that of Galvin et al. [9]

Before continuing with our treatment of SDT analysis of type 2 data, we will make some comparisons between this approach and the one described in Galvin et al. [9].

3.2.3.1 SDT Approaches to Type 2 Performance

Galvin et al. were concerned with characterizing the *overall* type 2 ROC curve, rather than response-specific type 2 ROC curves. On their modeling approach, an (FAR_2, HR_2) pair can be generated by setting a single type 2 criterion on a type 2 decision axis. All internal responses that exceed this type 2 criterion are labeled "high confidence," and all others "low confidence." By systematically changing the location of this type 2 criterion on the decision axis, the entire overall type 2 ROC curve can be traced out.

However, if the internal response x is used to make the binary confidence decision in this way, the ensuing type 2 ROC curve behaves oddly, typically containing regions where it extends below the line of chance performance [9]. This suboptimal behavior is not surprising, in that comparing the raw value of x to a single criterion value essentially recapitulates the decision rule used in the type 1 task and does not take into account the relationship between x and the observer's type 1 criterion, which is crucial for evaluating type 1 performance. The solution is that some *transformation* of x must be used as the type 2 decision variable, ideally one that depends upon both x and c.

For instance, consider the transformation $t(x) = |x - c|$. This converts the initial raw value of the internal response, x, into the distance of x from the type 1 criterion. This transformed value can then plausibly be compared to a single type 2 criterion to rate confidence, e.g. an observer might rate confidence as high whenever $t(x) > 1$. Other transformations for the type 2 decision variable are possible, and the choice is not arbitrary, since different choices for type 2 decision variables can lead to different predictions for the type 2 ROC curve [9]. The optimal type 2 ROC curve (i.e. the one that maximizes area under the curve) is derived by using the likelihood ratio of the type 2 probability density functions as the type 2 decision variable [9, 10].

We have adopted a different approach thus far. Rather than characterizing an overall (FAR_2, HR_2) pair as arising from the comparison of a single type 2 decision variable to a single type 2 criterion, we have focused on response-specific (FAR_2, HR_2) data arising from comparisons of the type 1 internal response x to separate type 2 decision criteria for "$S1$" and "$S2$" responses (e.g. Fig. 3.1a). Thus, our approach would characterize the overall (FAR_2, HR_2) as arising from a pair of response-specific type 2 criteria set on either side of the type 1 criterion on the type 1 decision axis, rather than from a single type 2 criterion set on a type 2 decision axis. We have posited no constraints on the setting of these type 2 criteria other than that they stand in appropriate ordinal relationships to eachother. For the sake of brevity in comparing these two approaches, in the following we will refer to Galvin et al.'s approach as G and the current approach as C.

3.2.3.2 Type 2 Decision Rules and Response-Specific Type 2 Criterion Setting

Notice that choosing a reasonable type 2 decision variable for G is equivalent to setting constraints on the relationship between type 2 criteria for "$S1$" and "$S2$" responses on C. For instance, on G suppose that the type 2 decision variable is defined as $t(x) = |x - c|$ and confidence is high if $t(x) > 1$. On C, this is equivalent to setting response-specific type 2 criteria symmetrically about the type 1 criterion, i.e. $t(c_{2,\text{“}S1\text{”}}) = t(c_{2,\text{“}S2\text{”}}) = |c_{2,\text{“}S1\text{”}} - c| = |c_{2,\text{“}S2\text{”}} - c| = 1$. In other words, assuming (on G) the general rule that confidence is high whenever the distance between x and c exceeds 1 requires (on C) that the type 2 criteria for each response type both satisfy this property of being 1 unit away from c. Any other way of setting the type 2 criteria for C would yield outcomes inconsistent with the decision rule posited by G. Similarly, if the type 2 decision rule is that confidence is high when type 2 likelihood ratio $LR_2(x) > c_{LR2}$, this same rule on C would require $LR_2(c_{2,\text{“}S1\text{”}}) = LR_2(c_{2,\text{“}S2\text{”}}) = c_{LR2}$, i.e. that type 2 criteria for both response types be set at the locations of x on either side of c corresponding to a type 2 likelihood ratio of c_{LR2}.

On G, choosing a suboptimal type 2 decision variable can lead to decreased area under the overall type 2 ROC curve. This can be understood on C as being related to the influence of response-specific type 2 criterion placement on the response-specific type 2 (FAR, HR) points, which in turn affect the overall type 2 (FAR, HR) points. As shown above, overall type 2 FAR and HR are weighted averages of the corresponding response-specific type 2 FARs and HRs. But computing a weighted average for two (FAR, HR) pairs on a concave down ROC curve will yield a new (FAR, HR) pair that lies below the original ROC curve. As a consequence, more exaggerated differences in the response-specific type 2 FAR and HR due to more exaggerated difference in response-specific type 2 criterion placement will tend to drive down the area below the overall type 2 ROC curve. Thus, the overall type 2 ROC curve may decrease even while the response-specific curves stay constant, depending on how criterion setting for each response type is coordinated. This reduced area under the overall type 2 ROC curve on C due to response-specific type 2 criterion placement is closely related to reduced area under the overall type 2 ROC curve on G due to choosing a suboptimal type 2 decision variable.

For example, consider the SDT model where $d' = 2$, $c = 0$, $c_{2,\text{“}S1\text{”}} = -1$, and $c_{2,\text{“}S2\text{”}} = 1$. This model yields $\text{FAR}_{2,\text{“}S1\text{”}} = \text{FAR}_{2,\text{“}S2\text{”}} = \text{FAR}_2 = 0.14$ and $\text{HR}_{2,\text{“}S1\text{”}} = \text{HR}_{2,\text{“}S2\text{”}} = \text{HR}_2 = 0.59$. The type 1 criterion is optimally placed and the type 2 criteria are symmetrically placed around it. This arrangement of criteria on C turns out to be equivalent to using the type 2 likelihood ratio on G, and thus yields an optimal type 2 performance. Now consider the SDT model where $d' = 2$, $c = 0$, $c_{2,\text{“}S1\text{”}} = -1.5$, and $c_{2,\text{“}S2\text{”}} = 0.76$. This model yields $\text{FAR}_{2,\text{“}S1\text{”}} = 0.04$, $\text{HR}_{2,\text{“}S1\text{”}} = 0.37$, $\text{FAR}_{2,\text{“}S2\text{”}} = 0.25$, $\text{HR}_{2,\text{“}S2\text{”}} = 0.71$, and overall $\text{FAR}_2 = 0.14$, $\text{HR}_2 = 0.54$. Although d' and c are the same as in the previous example, now the type 2 criteria are set asymmetrically about c, yielding different outcomes for the type 2 FAR and HR for “$S1$” and “$S2$” responses. This has the effect of yielding a lower overall HR_2 (0.54 vs. 0.59) in spite of happening to yield the same FAR_2 (0.14). Thus, this asymmetric arrangement of response-specific type 2 criteria yields worse performance on the overall type 2 ROC curve than the symmetric case for the same values of d' and c. On G, this can be understood as being the result of choosing a suboptimal type 2 decision variable in the second example (i.e. a decision variable that is consistent with the way the response-specific type 2 criteria have been defined on C). In this case, the asymmetric placement of the response-specific type 2 criteria is inconsistent with a type 2 decision variable based on the type 2 likelihood ratio.

3.2.3.3 A Method for Assessing Overall Type 2 Sensitivity Based on the Approach of Galvin et al.

In the upcoming section, we will discuss our methodology for quantifying type 2 sensitivity with meta-d'. Meta-d' essentially provides a single measure that jointly characterizes the areas under the response-specific type 2 ROC curves for both “$S1$” and “$S2$” responses, and in this way provides a measure of overall type 2 sensitivity. However, in doing so, it treats the relationships of type 2 criteria across response types as purely a matter of criterion setting. However, as we have discussed, coordination of type 2 criterion setting could also be seen as arising from the construction of a type 2 decision variable, where the choice of decision variable influences area under the overall type 2 ROC curve. We take it to be a substantive conceptual, and perhaps empirical, question as to whether it is preferable to characterize these effects as a matter of criterion setting (coordinating response-specific type 2 criteria) or sensitivity (constructing a type 2 decision variable). However, if one were to decide that for some purpose it were better to view this as a sensitivity effect, then the characterization of type 2 performance provided by Galvin et al. may be preferable to that of the current approach.

In the interest of recognizing this, we provide free Matlab code available online (see note at the end of the manuscript) that implements one way of using Galvin et al.'s approach to evaluate an observer's overall type 2 performance. Given the parameters of an SDT model, this code outputs the theoretically optimal[11] overall type 2 ROC curve—i.e. the overall type 2 ROC curve based on type 2 likelihood ratio, which has the maximum possible area under the curve. Maniscalco and Lau [13], building on the suggestions of Galvin et al. [9], proposed that one way of evaluating an observer's type 2 performance is to compare her empirical type 2 ROC curve with the theoretical type 2 ROC curve, given her type 1 performance. By comparing an observer's empirical overall type 2 ROC curve with the theoretically optimal overall type 2 ROC curve based on type 2 likelihood ratios, the observer's overall type 2 sensitivity can be assessed with respect to the SDT-optimal level. This approach will capture potential variation in area under the overall type 2 ROC curve that is ignored (treated as a response-specific criterion effect) by the meta-d' approach.

3.2.3.4 Advantages of the Current Approach

Our SDT treatment of type 2 performance has certain advantages over that of Galvin et al. One advantage is that it does not require making an explicit assumption regarding what overall type 2 decision variable an observer uses, or even that the observer constructs such an overall type 2 decision variable to begin with.[12] This is because our approach allows the type 2 criteria for each response to vary independently, rather than positing a fixed relationship between their locations. Thus, if an observer does construct an overall type 2 decision variable, our treatment will capture this implicitly by means of the relationship between the response-specific type 2 criteria; and if an observer does not use an overall type 2 decision variable to begin with, our treatment can accommodate this behavior. The question of what overall type 2 decision variables, if any, observers tend to use is a substantive empirical question, and so it is preferable to avoid making assumptions on this matter if possible.

A second, related advantage is that our approach is potentially more flexible than Galvin et al.'s in capturing the behavior of response-specific type 2 ROC curves, without loss of flexibility in capturing the overall type 2 ROC curve. (Since overall type 2 ROC curves depend on the response-specific curves, as shown above, our focus on characterizing the response-specific curves does not entail a deficit in capturing the overall curve.) A third advantage is that our approach provides a simple way to derive response-specific type 2 ROC curves from the

---

[11] Provided the assumptions of the SDT model are correct.

[12] Of course, our approach must at least implicitly assume a type 2 decision variable *within* each response type. In our treatment, the implicit type 2 decision variable for each response type is just the distance of x from c. However, for the analysis of response-specific type 2 performance for the equal variance SDT model, distance from criterion and type 2 likelihood ratio are equivalent decision variables. This is because they vary monotonically with each other [9], and so produce the same type 2 ROC curve [5, 21].

type 1 SDT model, whereas deriving the overall type 2 ROC curve is more complex under Galvin et al.'s approach and depends upon the type 2 decision variable being assumed.

3.3 Characterizing Type 2 Sensitivity in Terms of Type 1 SDT: Meta-d'

Since response-specific type 2 ROC curves can be derived directly from d' and c on the SDT model, this entails a tight theoretical relationship between type 1 and type 2 performance. One practical consequence is that type 2 sensitivity—the empirical type 2 ROC curves—can be quantified in terms of the type 1 SDT parameters d' and c [13]. However, it is necessary to explicitly differentiate instances when d' is meant to characterize type 1 performance from those instances when d' (along with c) is meant to characterize type 2 performance. Here we adopt the convention of using the variable names meta-d' and meta-c to refer to type 1 SDT parameters when used to characterize type 2 performance. We will refer to the type 1 SDT model as a whole, when used to characterize type 2 performance, as the meta-SDT model. Essentially, d' and c describe the type 1 SDT model fit to the type 1 ROC curve,[13] whereas meta-d' and meta-c—the meta-SDT model—quantify the type 1 SDT model when used exclusively to fit type 2 ROC curves.

How do we go about using the type 1 SDT model to quantify type 2 performance? There are several choices to make before a concrete method can be proposed. In the course of discussing these issues, we will put forth the methodological approach originally proposed by Maniscalco and Lau [13].

3.3.1 Which Type 2 ROC Curves?

As discussed in the preceding section "Comparison of the current approach to that of Galvin et al. [9]," we find the meta-SDT fit that provides the best simultaneous fit to the response-specific type 2 ROC curves for "$S1$" and "$S2$" responses, rather than finding a model that directly fits the overall type 2 ROC curve. As explained in more detail in that prior discussion, we make this selection primarily because (1) it allows more flexibility and accuracy in fitting the overall data set, and (2) it does not require making an explicit assumption regarding what type 2 decision variable the observer might use for confidence rating.

[13] When the multiple points on the type 1 ROC curve are obtained using confidence rating data, it is arguably preferable to calculate d' and c only from the (FAR, HR) pair generated purely by the observer's type 1 response. The remaining type 1 ROC points incorporate confidence rating data and depend on type 2 sensitivity, and so estimating d' on the basis of these ROC points may confound type 1 and type 2 sensitivity. See the section below titled "Response-specific meta-d' and the unequal variance SDT model".

3.3.2 Which Way of Using Meta-d′ and Meta-c to Derive Response-Specific Type 2 ROC Curves?

A second consideration is how to characterize the response-specific type 2 ROC curves using meta-d' and meta-c. For the sake of simplifying the analysis, and for the sake of facilitating comparison between d' and meta-d', an appealing option is to *a priori* fix the value of meta-c so as to be similar to the empirically observed type 1 response bias c, thus effectively allowing meta-d' to be the sole free parameter that characterizes type 2 sensitivity. However, since there are multiple ways of measuring type 1 response bias [12], there are also multiple ways of fixing the value of meta-c on the basis of c. In addition to the already-introduced c, type 1 response bias can be measured with the relative criterion, c':

$$c' = c/d'$$

This measure takes into account how extreme the criterion is, *relative to* the stimulus distributions.

Bias can also be measured as β, the ratio of the probability density function for $S2$ stimuli to that of $S1$ stimuli at the location of the decision criterion:

$$\beta = e^{cd'}$$

Figure 3.2 shows an example of how c, c', and β relate to the stimulus distributions when bias is fixed and d' varies. Panel a shows an SDT diagram for $d' = 3$ and $c = 1$. In panel b, $d' = 1$ and the three decision criteria are generated by setting c, c', and β to the equivalent values of those exhibited by these measures in panel a. Arguably, c' performs best in terms of achieving a similar "cut" between the stimulus distributions in panels a and b. This is an intuitive result given that c' essentially adjusts the location of c according to d'. Thus, holding c' constant ensures that, as d' changes, the location of the decision criterion remains in a similar location with respect to the means of the two stimulus distributions.

By choosing c' as the measure of response bias that will be held constant in the estimation of meta-d', we can say that when the SDT and meta-SDT models are fit to the same data set, they will have similar type 1 response bias, in the sense that they have the same c' value. This in turn allows us to interpret a subject's meta-d' in the following way: "Suppose there is an ideal subject whose behavior is perfectly described by SDT, and who performs this task with a similar level of response bias (i.e. same c') as the actual subject. Then in order for our ideal subject to produce the actual subject's response-specific type 2 ROC curves, she would need her d' to be equal to meta-d'."

Thus, meta-d' can be found by fitting the type 1 SDT model to response-specific type 2 ROC curves, with the constraint that meta-$c' = c'$. (Note that in the below we list meta-c, rather than meta-c', as a parameter of the meta-SDT model. The constraint meta-$c' = c'$ can thus be satisfied by ensuring meta-c = meta-$d' \times c'$.)

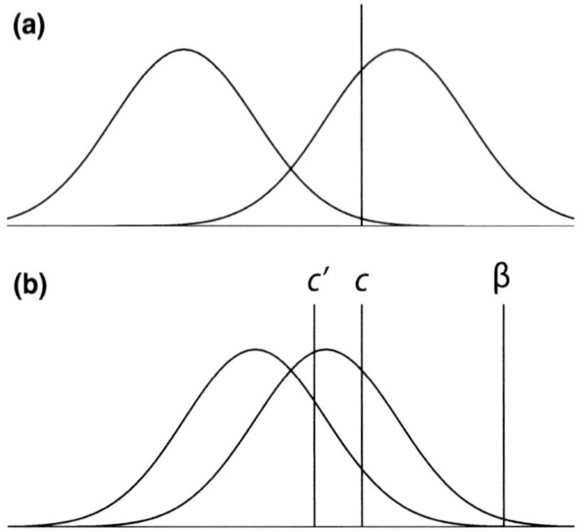

Fig. 3.2 Example behavior of holding response bias constant as d' changes for c, c', and β. **a** An SDT graph where $d' = 3$ and $c = 1$. The criterion location can also be quantified as $c' = c/d' = 1/3$ and $\log \beta = c \times d' = 3$. **b** An SDT graph where $d' = 1$. The three decision criteria plotted here represent the locations of the criteria that preserve the value of the corresponding response bias exhibited in panel a. So e.g. the criterion marked c' in panel b has the same value of c' as the criterion in panel a (=1/3), and likewise for c (constant value of 1) and β (constant value of 3)

3.3.3 What Computational Method of Fitting?

If the response-specific type 2 ROC curves contain more than one empirical (FAR_2, HR_2) pair, then in general an exact fit of the model to the data is not possible. In this case, fitting the model to the data requires minimizing some loss function, or maximizing some metric of goodness of fit.

Here we consider the procedure for finding the parameters of the type 1 SDT model that maximize the likelihood of the response-specific type 2 data. Maximum likelihood approaches for fitting SDT models to type 1 ROC curves with multiple data points have been established [4, 16]. Here we adapt these existing type 1 approaches to the type 2 case. The likelihood of the type 2 data can be characterized using the multinomial model as

$$L_{\text{type 2}}(\theta \mid \text{data}) \propto \prod_{y,s,r} \text{Prob}_\theta(\text{conf} = y \mid \text{stim} = s,\ \text{resp} = r)^{n_{\text{data}}(\text{conf}=y \mid \text{stim}=s,\ \text{resp}=r)}$$

Maximizing likelihood is equivalent to maximizing log-likelihood, and in practice it is typically more convenient to work with log-likelihoods. The log-likelihood for type 2 data is given by

$$\log L_{\text{type 2}}(\theta \mid \text{data}) \propto \sum_{y,s,r} n_{\text{data}} \log \text{Prob}_\theta$$

θ is the set of parameters for the meta-SDT model:

$$\theta = \left(\text{meta-}d', \text{meta-}c, \text{meta-}\underline{\boldsymbol{c}}_{2,\text{“}S1\text{”}}, \text{meta-}\underline{\boldsymbol{c}}_{2,\text{“}S2\text{”}}\right)$$

$n_{\text{data}}(\text{conf} = y \mid \text{stim} = s,\ \text{resp} = r)$ is a count of the number of times in the data a confidence rating of y was provided when the stimulus was s and response was r. y, s, and r are indices ranging over all possible confidence ratings, stimulus classes, and stimulus classification responses, respectively.
$\text{prob}_\theta(\text{conf} = y \mid \text{stim} = s,\ \text{resp} = r)$ is the model-predicted probability of generating confidence rating y for trials where the stimulus and response were s and r, given the parameter values specified in θ.

Calculation of these type 2 probabilities from the type 1 SDT model is similar to the procedure used to calculate the response-specific type 2 FAR and HR. For notational convenience, below we express these probabilities in terms of the standard SDT model parameters, omitting the “meta” prefix.

For convenience, define

$$\dot{\underline{\boldsymbol{c}}}_{2,\text{“}S1\text{”}} = \left(c, c_{2,\text{“}S1\text{”}}^{\text{conf}=2}, c_{2,\text{“}S1\text{”}}^{\text{conf}=3}, \ldots, c_{2,\text{“}S1\text{”}}^{\text{conf}=H}, -\infty\right)$$

$$\dot{\underline{\boldsymbol{c}}}_{2,\text{“}S2\text{”}} = \left(c, c_{2,\text{“}S2\text{”}}^{\text{conf}=2}, c_{2,\text{“}S2\text{”}}^{\text{conf}=3}, \ldots, c_{2,\text{“}S2\text{”}}^{\text{conf}=H}, \infty\right)$$

Then

$$\text{Prob}(\text{conf} = y \mid \text{stim} = S1,\ \text{resp} = \text{“}S1\text{”}) = \frac{\Phi\left(\dot{\underline{\boldsymbol{c}}}_{2,\text{“}S1\text{”}}(y), -\frac{d'}{2}\right) - \Phi\left(\dot{\underline{\boldsymbol{c}}}_{2,\text{“}S1\text{”}}(y+1), -\frac{d'}{2}\right)}{\Phi\left(c, -\frac{d'}{2}\right)}$$

$$\text{Prob}(\text{conf} = y \mid \text{stim} = S1,\ \text{resp} = \text{“}S2\text{”}) = \frac{\Phi\left(\dot{\underline{\boldsymbol{c}}}_{2,\text{“}S2\text{”}}(y+1), -\frac{d'}{2}\right) - \Phi\left(\dot{\underline{\boldsymbol{c}}}_{2,\text{“}S2\text{”}}(y), -\frac{d'}{2}\right)}{1 - \Phi\left(c, -\frac{d'}{2}\right)}$$

$$\text{Prob}(\text{conf} = y \mid \text{stim} = S2,\ \text{resp} = \text{“}S2\text{”}) = \frac{\Phi\left(\dot{\underline{\boldsymbol{c}}}_{2,\text{“}S2\text{”}}(y+1), \frac{d'}{2}\right) - \Phi\left(\dot{\underline{\boldsymbol{c}}}_{2,\text{“}S2\text{”}}(y), \frac{d'}{2}\right)}{1 - \Phi\left(c, \frac{d'}{2}\right)}$$

An illustration of how these type 2 probabilities are derived from the type 1 SDT model is provided in Fig. 3.3.

The multinomial model used as the basis for calculating likelihood treats each discrete type 2 outcome ($\text{conf} = y \mid \text{stim} = s,\ \text{resp} = r$) as an event with a fixed probability that occurred a certain number of times in the data set, where outcomes across trials are assumed to be statistically independent. The probability of the entire set of type 2 outcomes across all trials is then proportional to the product of the probability of each individual type 2 outcome, just as e.g. the probability of

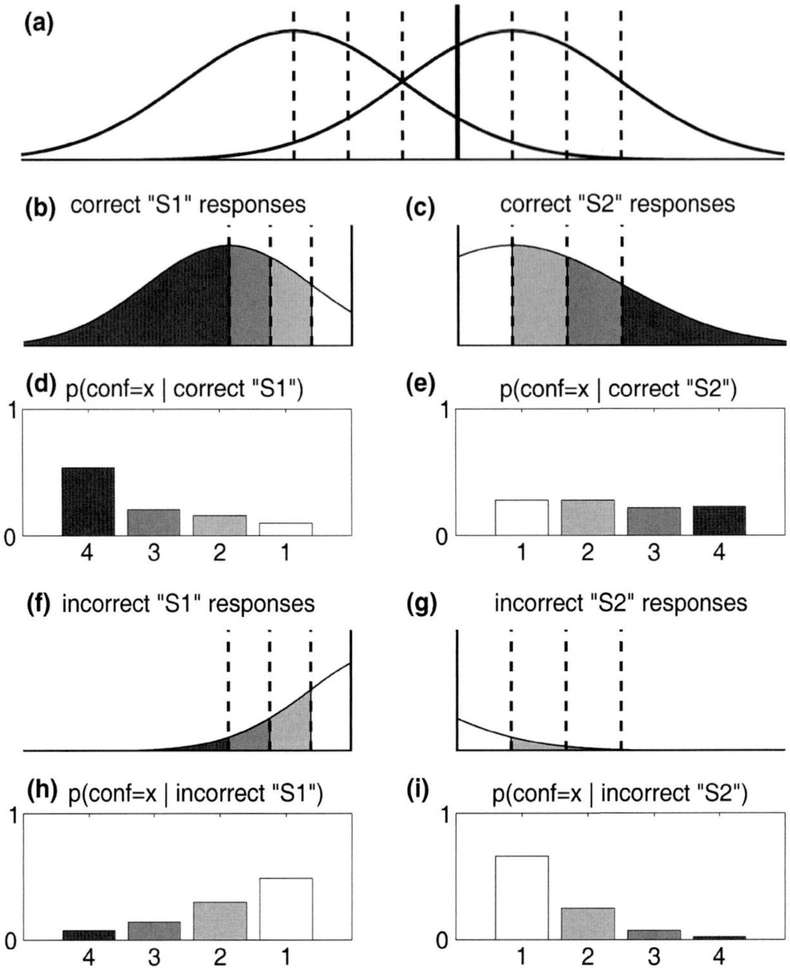

Fig. 3.3 Type 2 response probabilities from the SDT model. **a** An SDT graph with $d' = 2$ and decision criteria $c = 0.5$, $\boldsymbol{c}_{2,\text{“}S1\text{”}} = (0, -0.5, -1)$, and $\boldsymbol{c}_{2,\text{“}S2\text{”}} = (1, 1.5, 2)$. The type 1 criterion (solid vertical line) is set to the value of 0.5, corresponding to a conservative bias for providing "*S*2" responses, in order to create an asymmetry between "*S*1" and "*S*2" responses for the sake of illustration. Seven decision criteria are used in all, segmenting the decision axis into 8 regions. Each region corresponds to one of the possible permutations of type 1 and type 2 responses, as there are two possible stimulus classifications and four possible confidence ratings. **b–i** Deriving probability of confidence rating contingent on type 1 response and accuracy. How would the SDT model depicted in panel (**a**) predict the probability of each confidence rating for correct "*S*1" responses? Since we wish to characterize "*S*1" responses, we need consider only the portion of the SDT graph falling to the left of the type 1 criterion. Since "*S*1" responses are only correct when the *S*1 stimulus was actually presented, we can further limit our consideration to internal responses generated by *S*1 stimuli. This is depicted in panel (**b**). This distribution is further subdivided into 4 levels of confidence by the 3 type 2 criteria (*dashed vertical lines*), where darker regions correspond to higher confidence. The area under the *S*1 curve in each of these

◀ regions, divided by the total area under the *S*1 curve that falls below the type 1 criterion, yields the probability of reporting each confidence level, given that the observer provided a correct "*S*1" response. Panel (**d**) shows these probabilities as derived from areas under the curve in panel (**b**). The remaining panels display the analogous logic for deriving confidence probabilities for incorrect "*S*1" responses (**f**, **h**), correct "*S*2" responses (**c**, **e**), and incorrect "*S*2" responses (**g**, **i**)

throwing 4 heads and 6 tails for a fair coin is proportional to $0.5^4 \times 0.5^6$. (Calculation of the exact probability depends on a combinatorial term which is invariant with respect to θ and can therefore be ignored for the purposes of maximum likelihood fitting.)

Likelihood, $L(\theta)$, can be thought of as measuring how probable the empirical data is, according to the model parameterized with θ. A very low $L(\theta)$ indicates that the model with θ would be very unlikely to generate a pattern like that observed in the data. A higher $L(\theta)$ indicates that the data are more in line with the typical behavior of data produced by the model with θ. Mathematical optimization techniques can be used to find the values of θ that maximize the likelihood, i.e. that create maximal concordance between the empirical distribution of outcomes and the model-expected distribution of outcomes.

The preceding approach for quantifying type 2 sensitivity with the type 1 SDT model—i.e. for fitting the meta-SDT model—can be summarized as a mathematical optimization problem:

$$\theta^* = \arg\max_{\theta} L_{\text{type 2}}(\theta \mid \text{data}), \quad \text{subject to:} \quad \text{meta-}c' = c', \ \gamma\left(\text{meta-}\underline{\boldsymbol{c}}_{\text{ascending}}\right)$$

where type 2 sensitivity is quantified by meta-$d' \in \theta^*$.

$\gamma\left(\text{meta-}\underline{\boldsymbol{c}}_{\text{ascending}}\right)$ is the Boolean function described previously, which returns a value of "true" only if the type 1 and type 2 criteria stand in appropriate ordinal relationships.

We provide free Matlab code, available online, for implementing this maximum likelihood procedure for fitting the meta-SDT model to a data set (see note at the end of the manuscript).

*3.3.4 Toy Example of Meta-*d′ *Fitting*

An illustration of the meta-*d′* fitting procedure is demonstrated in Fig. 3.4 using simulated data. In this simulation, we make the usual SDT assumption that on each trial, presentation of stimulus S generates an internal response x that is drawn from the probability density function of S, and that a type 1 response is made by comparing x to the decision criterion c. However, we now add an extra mechanism to the model to allow for the possibility of added noise in the type 2 task. Let us call the internal response used to rate confidence x_2. The type 1 SDT model we

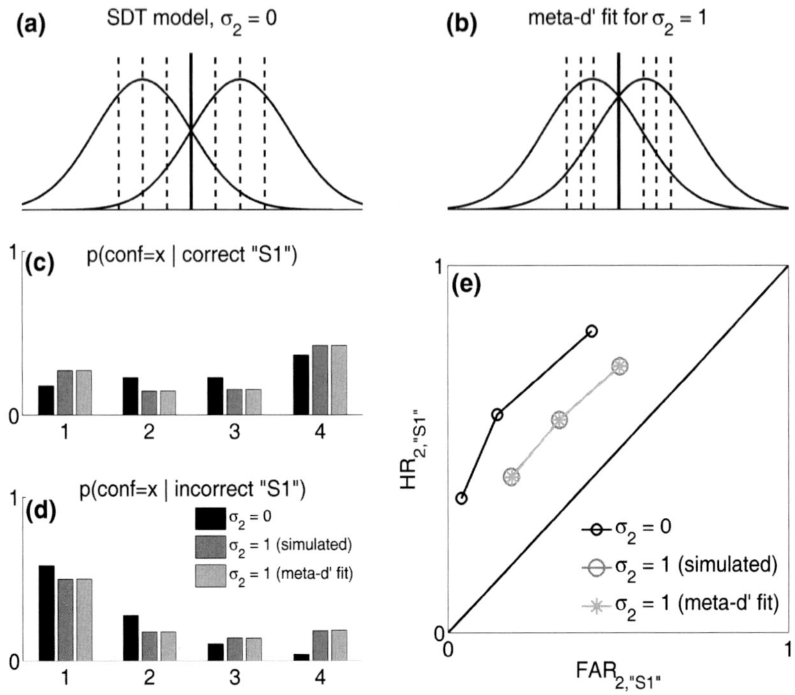

Fig. 3.4 Fitting meta-d' to response-specific type 2 data. **a** Graph for the SDT model where $d' = 2$ and $\sigma_2 = 0$ (see text for details). **b** A model identical to that in panel a, with the exception that $\sigma_2 = 1$, was used to create simulated data. This panel displays the SDT graph of the parameters for the meta-d' fit to the $\sigma_2 = 1$ data. **c**, **d** Response-specific type 2 probabilities. The maximum likelihood method of fitting meta-d' to type 2 data uses response-specific type 2 probabilities as the fundamental unit of analysis. The type 1 SDT parameters that maximize the likelihood of the type 2 data yield distributions of response-specific type 2 probabilities closely approximating the empirical (here, simulated) distributions. Here we only show the probabilities for "$S1$" responses; because of the symmetry of the generating model, "$S2$" responses follow identical distributions. **e** Response-specific type 2 ROC curves. ROC curves provide a more informative visualization of the type 2 data than the raw probabilities. Here it is evident that there is considerably less area under the type 2 ROC curve for the $\sigma_2 = 1$ simulation than is predicted by the $\sigma_2 = 0$ model. The meta-d' fit provides a close match to the simulated data

have thus far considered assumes $x_2 = x$. In this example, we suppose that x_2 is a noisier facsimile of x. Formally,

$$x_2 = x + \xi, \quad \xi \sim N(0, \sigma_2)$$

where $N(0, \sigma_2)$ is the normal distribution with mean 0 and standard deviation σ_2. The parameter σ_2 thus determines how much noisier x_2 is than x. For $\sigma_2 = 0$ we expect meta-$d' = d'$, and for $\sigma_2 > 0$ we expect meta-$d' < d'$.

The simulated observer rates confidence on a 4-point scale by comparing x_2 to response-specific type 2 criteria, using the previously defined decision rules for confidence in the type 1 SDT model.[14]

We first considered the SDT model with $d' = 2$, $c = 0$, $\underline{c}_{2,\text{“}S1\text{”}} = (-0.5, -1, -1.5)$, $\underline{c}_{2,\text{“}S2\text{”}} = (0.5, 1, 1.5)$ and $\sigma_2 = 0$. Because $\sigma_2 = 0$, this is equivalent to the standard type 1 SDT model. The SDT graph for these parameter values is plotted in Fig. 3.4a. Using these parameter settings, we computed the theoretical probability of each confidence rating for each permutation of stimulus and response. These probabilities for “$S1$” responses are shown in panels c and d, and the corresponding type 2 ROC curve is shown in panel e. (Because the type 1 criterion c is unbiased and the type 2 criteria are set symmetrically about c, confidence data for “$S2$” responses follow an identical distribution to that of “$S1$” responses and are not shown.)

Next we simulated 10,000,000 trials using the same parameter values as the previously considered model, with the exception that $\sigma_2 = 1$. With this additional noise in the type 2 task, type 2 sensitivity should decrease. This decrease in type 2 sensitivity can be seen in the type 2 ROC curve in panel e. There is more area underneath the type 2 ROC curve when $\sigma_2 = 0$ than when $\sigma_2 = 1$.

We performed a maximum likelihood fit of meta-d' to the simulated type 2 data using the fmincon function in the optimization toolbox for Matlab (MathWorks, Natick, MA), yielding a fit with parameter values meta-$d' = 1.07$, meta-$c = 0$, meta-$\underline{c}_{2,\text{“}S1\text{”}} = (-0.51, -0.77, -1.06)$, and meta-$\underline{c}_{2,\text{“}S2\text{”}} = (0.51, 0.77, 1.06)$. The SDT graph for these parameter values is plotted in Fig. 3.4b.

Panels c and d demonstrate the component type 2 probabilities used for computing the type 2 likelihood. The response-specific type 2 probabilities for $\sigma_2 = 0$ are not distributed the same way as those for $\sigma_2 = 1$, reflecting the influence of adding noise to the internal response for the type 2 task. Computing meta-d' for the $\sigma_2 = 1$ data consists in finding the parameter values of the ordinary type 1 SDT model that maximize the likelihood of the $\sigma_2 = 1$ response-specific type 2 data. This results in a type 1 SDT model whose theoretical type 2 probabilities closely

[14] Note that for this model, it is possible for x and x_2 to be on opposite sides of the type 1 decision criterion c (see, e.g. Fig. 3.5a, b). This is not problematic, since only x is used to provide the type 1 stimulus classification. It is also possible for x_2 to surpass some of the type 2 criteria on the opposite side of c. For instance, suppose that $x = -0.5$, $x_2 = +0.6$, $c = 0$, and $c_{2,\text{“}S2\text{”}}^{\text{conf}=h} = +0.5$. Then x is classified as an $S1$ stimulus, and yet x_2 surpasses the criterion for rating “$S2$” responses with a confidence of h. Thus, there is potential for the paradoxical result whereby the type 1 response is “$S1$” and yet the type 2 confidence rating is rated highly due to the relatively strong “$S2$”-ness of x_2. In this example, the paradox is resolved by the definition of the type 2 decision rules stated above, which stipulate that internal responses are only evaluated with respect to the response-specific type 2 criteria that are congruent with the type 1 response. Thus, in this case, the decision rule would not compare x_2 with the type 2 criteria for “$S2$” responses to begin with. Instead, it would find that x_2 does not surpass the minimal confidence criterion for “$S1$” responses (i.e., $x_2 > c > c_{2,\text{“}S1\text{”}}^{\text{conf}=2}$) and would therefore assign x_2 a confidence of 1. Thus, in this case, the paradoxical outcome is averted. But such potentially paradoxical results need to be taken into account for any SDT model that posits a potential dissociation between x and x_2.

match the empirical type 2 probabilities for the simulated $\sigma_2 = 1$ data (Fig. 3.4c, d). Because type 2 ROC curves are closely related to these type 2 probabilities, the meta-d' fit also produces a type 2 ROC curve closely resembling the simulated curve, as shown in panel e.

3.3.5 *Interpretation of Meta-*d′

Notice that because meta-d' characterizes type 2 sensitivity purely in terms of the type 1 SDT model, it does not explicitly posit any mechanisms by means of which type 2 sensitivity varies. Although the meta-d' fitting procedure gave a good fit to data simulated by the toy σ_2 model discussed above, it could also produce similarly good fits to data generated by different models that posit completely different mechanisms for variation in type 2 performance. In this sense, meta-d' is descriptive but not explanatory. It describes how an ideal SDT observer with similar type 1 response bias as the actual subject would have achieved the observed type 2 performance, rather than explain how the actual subject achieved their type 2 performance.

The primary virtue of using meta-d' is that it allows us to quantify type 2 sensitivity in a principled SDT framework, and compare this against SDT expectations of what type 2 performance *should have been*, given performance on the type 1 task, all while remaining agnostic about the underlying processes. For instance, if we find that a subject has $d' = 2$ and meta-$d' = 1$, then (1) we have taken appropriate SDT-inspired measures to factor out the influence of response bias in our measure of type 2 sensitivity; (2) we have discovered a violation of the SDT expectation that meta-$d' = d' = 2$, giving us a point of reference in interpreting the subject's metacognitive performance in relation to their primary task performance and suggesting that the subject's metacognition is suboptimal (provided the assumptions of the SDT model hold); and (3) we have done so while making minimal assumptions and commitments regarding the underlying processes.

Another important point for interpretation concerns the raw meta-d' value, as opposed to its value in relation to d'. Suppose observers A and B both have meta-$d' = 1$, but $d'_{\mathrm{A}} = 1$ and $d'_{\mathrm{B}} = 2$. Then there is a sense in which they have equivalent metacognition, as their confidence ratings are equally sensitive in discerning correct from incorrect trials. But there is also a sense in which A has superior metacognition, since A was able to achieve the same level of meta-d' as B in spite of a lower d'. In a sense, A is more metacognitively ideal, according to SDT. We can refer to the first kind of metacognition, which depends only on meta-d', as "absolute type 2 sensitivity," and the second kind, which depends on the relationship between meta-d' and d', as "relative type 2 sensitivity." Absolute and relative type 2 sensitivity are distinct constructs that inform us about distinct aspects of metacognitive performance.

For more on the interpretation of meta-d', see Maniscalco and Lau [13].

3.4 Response-Specific Meta-d'

Thus far we have considered how to characterize an observer's overall type 2 sensitivity using meta-d', expounding upon the method originally introduced in Maniscalco and Lau [13]. Here we show how to extend this analysis and characterize response-specific type 2 sensitivity in terms of the type 1 SDT model.

In the below we focus on "$S1$" responses, but similar considerations apply for "$S2$" responses.

We wish to find the type 1 SDT parameters θ that provide the best fit to the type 2 ROC curve for "$S1$" responses, i.e. the set of empirical $\left(\mathrm{FAR}_{2,\text{“}S1\text{”}}^{\mathrm{conf}=h}, \mathrm{HR}_{2,\text{“}S1\text{”}}^{\mathrm{conf}=h}\right)$ for all h satisfying $2 \le h \le \mathrm{H}$. Thus, we wish to find the θ that maximizes the likelihood of the type 2 probabilities for "$S1$" responses, using the usual meta-d' fitting approach. This is essentially equivalent to applying the original meta-d' procedure described above to the subset of the model and data pertaining to "$S1$" responses.

Thus, we wish to solve the optimization problem

$$\theta^*_{\text{“}S1\text{”}} = \underset{\theta_{\text{“}S1\text{”}}}{\arg\max}\, L_{2,\text{“}S1\text{”}}(\theta_{\text{“}S1\text{”}} \mid \text{data}),$$

$$\text{subject to:} \quad \text{meta-}c'_{\text{“}S1\text{”}} = c', \quad \gamma\left(\text{meta-}\underline{\mathbf{c}}_{\text{ascending}}\right)$$

where

$$\theta_{\text{“}S1\text{”}} = \left(\text{meta-}d'_{\text{“}S1\text{”}}, \text{meta-}c_{\text{“}S1\text{”}}, \text{meta-}\underline{\mathbf{c}}_{2,\text{“}S1\text{”}}\right)$$

$$L_{2,\text{“}S1\text{”}}(\theta_{\text{“}S1\text{”}} \mid \text{data}) \propto \prod_{y,s} \text{Prob}_\theta(\text{conf} = y \mid \text{stim} = s,\ \text{resp} = \text{“}S1\text{”})^{n_{\text{data}}(\text{conf}=y \mid \text{stim}=s,\ \text{resp}=\text{“}S1\text{”})}$$

meta-$d'_{\text{“}S1\text{”}} \in \theta^*_{\text{“}S1\text{”}}$ measures type 2 sensitivity for "$S1$" responses.

The differences between this approach and the "overall" meta-d' fit are straightforward. The same likelihood function is used, but with the index r fixed to the value "$S1$". $\theta_{\text{“}S1\text{”}}$ is equivalent to θ except for its omission of meta$\underline{c}_{2,\text{“}S2\text{”}}$, since type 2 criteria for "$S2$" responses are irrelevant for fitting "$S1$" type 2 ROC curves. The type 1 criterion meta-$c_{\text{“}S1\text{”}}$ is listed with a "$S1$" subscript to distinguish it from meta-$c_{\text{“}S2\text{”}}$, the type 1 criterion value from the maximum likelihood fit to "$S2$" type 2 data. Since the maximum likelihood fitting procedure enforces the constraint meta-$c'_{\text{“}S1\text{”}} = c'$, it follows that meta-$c_{\text{“}S1\text{”}}$ = meta-$d'_{\text{“}S1\text{”}} \times c'$. Thus, in the general case where meta-$d'_{\text{“}S1\text{”}} \neq$ meta-$d'_{\text{“}S2\text{”}}$ and $c' \neq 0$, it is also true that meta-$c_{\text{“}S1\text{”}} \neq$ meta-$c_{\text{“}S2\text{”}}$.

We provide free Matlab code, available online, for implementing this maximum likelihood procedure for fitting the response-specific meta-SDT model to a data set (see note at the end of the manuscript).

3.4.1 Toy Example of Response-Specific Meta-d′ Fitting

An illustration of the response-specific meta-d' fitting procedure is demonstrated in Fig. 3.5 using simulated data. We use a similar model as that used in the previous toy example of meta-d' fitting. That is, we use the usual type 1 SDT model, except we suppose that the internal response used to produce the type 2 judgment, x_2, may be a noisier version of its type 1 counterpart, x. This time, we additionally allow the degree of added noisiness in x_2 to differ for "$S1$" and "$S2$" responses. Formally,

$$x_2 = \begin{cases} x + \xi_{\text{“}S1\text{”}}, & \xi_{\text{“}S1\text{”}} \sim N(0, \sigma_{2\text{“}S1\text{”}}) \quad \text{if } x \le c \\ x + \xi_{\text{“}S2\text{”}}, & \xi_{\text{“}S2\text{”}} \sim N(0, \sigma_{2,\text{“}S2\text{”}}) \quad \text{if } x > c \end{cases}$$

Different levels of type 2 noisiness for each response type allows for the possibility that response-specific type 2 sensitivity can differ for "$S1$" and "$S2$" responses.

We first considered the SDT model with $d' = 2, c = 0, \underline{c}_{2,\text{“}S1\text{”}} = (-0.5, -1, -1.5)$, $\underline{c}_{2,\text{“}S2\text{”}} = (0.5, 1, 1.5)$ and $\sigma_{2,\text{“}S1\text{”}} = \sigma_{2,\text{“}S2\text{”}} = 0$. Because $\sigma_{2,\text{“}S1\text{”}} = \sigma_{2,\text{“}S2\text{”}} = 0$, this is equivalent to the standard type 1 SDT model. The SDT graph for these parameter values were used in the previous example, as illustrated in Fig. 3.4a. Using these parameter settings, we constructed theoretical response-specific type 2 ROC curves, as shown in Fig. 3.5c, d.

Next we simulated 10,000,000 trials using the same parameter values as the previously considered model, with the exception that $\sigma_{2,\text{“}S1\text{”}} = 2/3$ and $\sigma_{2,\text{“}S2\text{”}} = 1/3$. Since $\sigma_{2,\text{“}S2\text{”}} < \sigma_{2,\text{“}S1\text{”}}$, for these simulated data there is more area underneath the type 2 ROC curve for "$S2$" than for "$S1$" responses (Fig. 3.5c, d). The simulated distributions of x_2 values for correct and incorrect "$S1$" and "$S2$" responses is shown in the top halves of Fig. 3.5a, b. Note that this model generates some x_2 values that lie on the opposite side of the type 1 criterion as the corresponding x value (which determines the type 1 response). For all such trials, the type 1 response was determined only by x and confidence was set to 1. See footnote 14 above for more details.

We first performed a maximum likelihood fit of overall meta-d' to the simulated data, yielding a fit with parameter values meta-$d' = 1.17$, meta-$c = 0$, meta-$\underline{c}_{2,\text{“}S1\text{”}} = (-0.59, -0.79, -1.01)$, and meta-$\underline{c}_{2,\text{“}S2\text{”}} = (0.43, 0.80, 1.2)$.The theoretical type 2 ROC curves predicted by the SDT model with these parameter values is displayed in Fig. 3.5c, d alongside the simulated data. Inspection of these graphs suggests that the meta-d' fit was able to account for differences in overall levels of confidence for "$S1$" and "$S2$" responses, as reflected by the fact that the response-specific curves are scaled in such a way as to mirror the scaling of the empirical type 2 ROC curves. However, the meta-d' fit cannot account for the difference in type 2 sensitivity for "$S1$" and "$S2$" responses. Instead, the fit produces overlapping type 2 ROC curves located midway between the empirical "$S1$" and "$S2$" curves, as if capturing something analogous to the average type 2 sensitivity for each response type. (See the inset of Fig. 3.5d for a plot of the meta-d' type 2 ROC curves for both response types.)

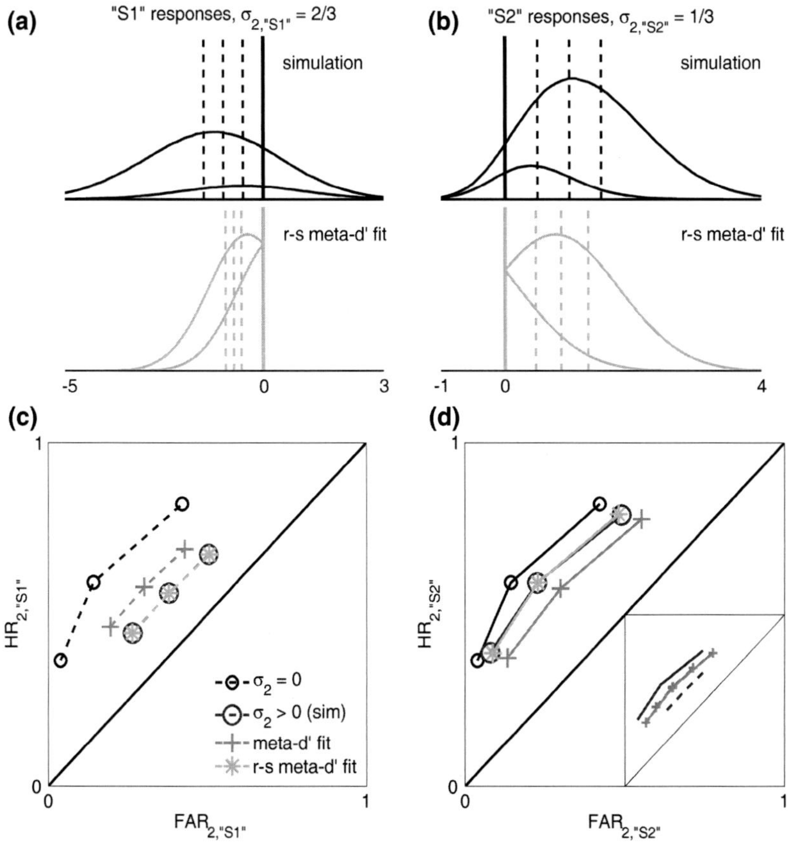

Fig. 3.5 Response-specific meta-d' fitting. **a** Simulated data and meta-d' fit for "S1" responses. *Top* Simulated distribution of x_2 values for correct and incorrect "$S1$" responses for simulated data with $\sigma_{2,\text{“}S1\text{”}} = 2/3$. (See main text for details.) Note that many x_2 values initially labeled "$S1$" cross over to the other side of the type 1 criterion after having type 2 noise added. These are considered to be "$S1$" responses with confidence =1. See footnote 14 in main text for further discussion. *Bottom* SDT parameters of meta-$d'_{\text{“}S1\text{”}}$ fit. **b** Same as A, but for "$S2$" responses. **c** Type 2 ROC curves for "S1" responses. Setting $\sigma_{2,\text{“}S1\text{”}} = 2/3$ substantially reduces type 2 sensitivity, as revealed by the comparison of area under the ROC curves for $\sigma_{2,\text{“}S1\text{”}} = 2/3$ and $\sigma_{2,\text{“}S1\text{”}} = 0$. Response-specific meta-d' fits the data well, but meta-d' provides an overestimate. **d** Type 2 ROC curves for "S2" responses. Response-specific meta-d' fits the "$S2$" data well, but meta-d' provides an underestimate. *Inset* Type 2 ROC curves for both "$S1$" and "$S2$" responses, shown for the simulated data (*black*) and the meta-d' fit (*gray*). The meta-d' fit generates type 2 ROC curves intermediate between the empirical (simulated) "$S1$" and "$S2$" curves

Next we performed a maximum likelihood fit for response-specific meta-d' to the simulated data. This yielded a fit with parameter values meta-$d'_{\text{“}S1\text{”}} = 0.77$, meta-$c_{\text{“}S1\text{”}} = 0$, meta-$\underline{\mathbf{c}}_{2,\text{“}S1\text{”}} = (-0.54, -0.73, -0.94)$ for “$S1$” responses, and meta-$d'_{\text{“}S2\text{”}} = 1.56$, meta-$c_{\text{“}S2\text{”}} = 0$, meta-$\underline{\mathbf{c}}_{2,\text{“}S2\text{”}} = (0.48, 0.87, 1.30)$ for “$S2$” responses. The SDT graph for these parameter values is plotted in the bottom halves of Fig. 3.5a, b. The theoretical type 2 ROC curves corresponding to these fits are displayed in Fig. 3.5c, d alongside the simulated data. It is evident that the response-specific meta-d' approach provides a close fit to the simulated data.

3.5 Response-Specific Meta-d' and the Unequal Variance SDT Model

3.5.1 *Inferring Unequal Variance from the z-ROC Curve Slope*

Thus far we have discussed SDT models assuming that the variance of the internal response distributions for $S1$ and $S2$ stimuli have equal variance. However, it is also possible to relax this assumption and allow the variances to differ. In conventional notation, we can define an additional parameter to the type 1 SDT model, s:

$$s = \frac{\sigma_{S1}}{\sigma_{S2}}$$

We may refer to the SDT model parameterized with s as the unequal variance SDT model, or UV-SDT. We may similarly refer to the more basic SDT model we have discussed thus far as the equal variance SDT model or EV-SDT.

UV-SDT has been shown to have advantages over EV-SDT in capturing certain data sets. The primary motivation for UV-SDT arises from the analysis of type 1 z-ROC curves. Given a set of type 1 (FAR, HR) points, a z-ROC curve may be constructed by plotting z(HR) against z(FAR), where z denotes the inverse of the cumulative distribution function for the normal distribution. That is,

$$z(p) = x, \quad \text{such that } \Phi(x, 0, 1) = p$$

According to SDT, since FAR and HR are generated by the normal cumulative distribution function evaluated at some location on the decision axis X, it should follow that z(FAR) and z(HR) correspond to locations on X. More specifically, it can be shown that z(FAR) quantifies the distance between the mean of the $S1$ distribution and the criterion used to generate that FAR, as measured in units of the standard deviation of the $S1$ distribution [and similarly for z(HR)]. That is,

$$z(\mathrm{FAR}_c) = \frac{\mu_{S1} - c}{\sigma_{S1}}, \quad \mathrm{FAR}_c = 1 - \Phi(c, \mu_{S1}, \sigma_{S1})$$
$$z(\mathrm{HR}_c) = \frac{\mu_{S2} - c}{\sigma_{S2}}, \quad \mathrm{HR}_c = 1 - \Phi(c, \mu_{S2}, \sigma_{S2})$$

The slope of the z-ROC curve for a set of (FAR_c, HR_c) represents how changes in $z(\mathrm{HR}_c)$ relate to changes in $z(\mathrm{FAR}_c)$. According to SDT, this is equivalent to how changes in the criterion c, as measured in σ_{S2} units, are related to changes in the same quantity c as measured in σ_{S1} units, since

$$z\text{-ROC slope} = \frac{\Delta z(\mathrm{HR})}{\Delta z(\mathrm{FAR})} = \frac{\Delta c/\sigma_{S2}}{\Delta c/\sigma_{S1}} = \frac{\sigma_{S1}}{\sigma_{S2}} = s$$

3.5.2 Constructing Pseudo Type 1 ROC Curves from Type 2 Data

Under EV-SDT, where $\sigma_{S1} = \sigma_{S2}$, the z-ROC curve should therefore be linear with a slope of 1, since changing c by δ units of σ_{S2} is equivalent to changing c by δ units of σ_{S1}. Under UV-SDT, the z-ROC curve should be linear with a slope of s, since changing c by δ units of σ_{S1} is equivalent to changing c by $s \times \delta$ units of σ_{S2}. Thus, empirical instances of linear z-ROC curves with non-unit slope have been taken to constitute empirical support for the UV-SDT model (e.g. [20]).

Constructing empirical type 1 ROC curves requires manipulating response bias in order to collect multiple type 1 (FAR, HR) points at the same level of sensitivity. One method of accomplishing this is to place the subject in multiple experimental conditions that tend to induce different response biases, e.g. due to different base rates of stimulus presentation or payoff structures [12, 22]. However, this method is somewhat resource intensive.

A popular alternative strategy for constructing empirical type 1 ROC curves is to use the conjunction of type 1 and type 2 judgments in order to emulate distinct type 1 judgments. For instance, suppose the observer classifies a stimulus as *S*1 or *S*2 and then rates confidence as high or low. FAR and HR are determined by how often the observer responds "*S*2." But we can also imagine that, had the subject been very conservative in responding "*S*2," he might have only done so for those trials in which he endorsed the "*S*2" response with high confidence. Thus, we can compute a second (FAR, HR) pair by provisionally treating only "high confidence *S*2" trials as "*S*2" responses. Similarly, we can emulate a liberal type 1 response bias by provisionally treating anything other than a "high confidence *S*1" response as an "*S*2" response. This procedure would thus allow us to derive 3 points on the type 1 ROC curve from a single experimental session.

Following the naming convention introduced by Galvin et al. [9], we will refer to the type 1 ROC curve constructed in this way as the pseudo type 1 ROC curve, and the extra (FAR, HR) points generated from confidence ratings as pseudo type 1 (FAR, HR). For a discrete H-point rating scale, we can derive 2H − 1 points on the pseudo type 1 ROC curve. In addition to the usual (FAR, HR) pair as determined by the observer's stimulus classification, we can compute new pseudo (FAR, HR) pairs from "$S1$" and "$S2$" responses at each level of confidence $h > 1$, as

$$\mathrm{HR}_{1\sim 2,\text{“}S1\text{”}}^{\mathrm{conf}=h} = 1 - p(\mathrm{resp} = \text{“}S1\text{”},\ \mathrm{conf} \geq h \mid \mathrm{stim} = S2)$$
$$\mathrm{FAR}_{1\sim 2,\text{“}S1\text{”}}^{\mathrm{conf}=h} = 1 - p(\mathrm{resp} = \text{“}S1\text{”},\ \mathrm{conf} \geq h \mid \mathrm{stim} = S1)$$

$$\mathrm{HR}_{1\sim 2,\text{“}S2\text{”}}^{\mathrm{conf}=h} = p(\mathrm{resp} = \text{“}S2\text{”},\ \mathrm{conf} \geq h \mid \mathrm{stim} = S2)$$
$$\mathrm{FAR}_{1\sim 2,\text{“}S2\text{”}}^{\mathrm{conf}=h} = p(\mathrm{resp} = \text{“}S2\text{”},\ \mathrm{conf} \geq h \mid \mathrm{stim} = S1)$$

The subscript "1 ~ 2" denotes that these pseudo type 1 (FAR, HR) pairs are being treated as type 1 data in spite of having been partially constructed from type 2 decisions.

The pseudo type 1 ROC curve has a straightforward interpretation on the SDT graph. Each pseudo type 1 (FAR, HR) pair can be computed from the SDT model by using the corresponding response-specific type 2 criterion in place of the type 1 criterion in the formula for FAR and HR:

$$\mathrm{HR}_{1\sim 2,\text{“}SX\text{”}}^{\mathrm{conf}=h} = 1 - \Phi\left(c_{2,\text{“}SX\text{”}}^{\mathrm{conf}=h}, \mu_{S2}, \sigma_{S2}\right)$$
$$\mathrm{FAR}_{1\sim 2,\text{“}SX\text{”}}^{\mathrm{conf}=h} = 1 - \Phi\left(c_{2,\text{“}SX\text{”}}^{\mathrm{conf}=h}, \mu_{S1}, \sigma_{S1}\right)$$

where "SX" denotes either "$S1$" or "$S2$." Figure 3.1a, b illustrates this principle.

3.5.3 Dependence of Pseudo Type 1 ROC Curves on Response-Specific Type 2 ROC Curves

However, because the pseudo type 1 (FAR, HR) points depend on both type 1 and type 2 judgments, they risk confounding type 1 and type 2 sensitivity. Indeed, we will now demonstrate that pseudo type 1 (FAR, HR) points directly depend upon type 1 and type 2 ROC data. For instance, consider the pseudo type 1 (FAR, HR) for "$S2$" responses. It follows from the definition of these that

$$\begin{aligned}\mathrm{HR}_{1\sim 2,\text{“}S2\text{”}}^{\mathrm{conf}=h} &= p(\mathrm{resp} = \text{“}S2\text{”},\ \mathrm{conf} \geq h \mid \mathrm{stim} = S2)\\ &= p(\mathrm{conf} \geq h \mid \mathrm{resp} = \text{“}S2\text{”},\ \mathrm{stim} = S2) \times p(\mathrm{resp} = \text{“}S2\text{”} \mid \mathrm{stim} = S2)\\ &= \mathrm{HR}_{2,\text{“}S2\text{”}}^{\mathrm{conf}=h} \times \mathrm{HR}_1\end{aligned}$$

$$
\begin{aligned}
\mathrm{FAR}_{1\sim 2,\text{“}S2\text{”}}^{\mathrm{conf}=h} &= p(\mathrm{resp} = \text{“}S2\text{”},\ \mathrm{conf} \geq h \mid \mathrm{stim} = S1) \\
&= p(\mathrm{conf} \geq h \mid \mathrm{resp} = \text{“}S2\text{”},\ \mathrm{stim} = S1) \times p(\mathrm{resp} = \text{“}S2\text{”} \mid \mathrm{stim} = S1) \\
&= \mathrm{FAR}_{2,\text{“}S2\text{”}}^{\mathrm{conf}=h} \times \mathrm{FAR}_1
\end{aligned}
$$

Similarly for “$S1$” responses,

$$
\begin{aligned}
\mathrm{HR}_{1\sim 2,\text{“}S1\text{”}}^{\mathrm{conf}=h} &= 1 - p(\mathrm{resp} = \text{“}S1\text{”},\ \mathrm{conf} \geq h \mid \mathrm{stim} = S2) \\
&= 1 - [p(\mathrm{conf} \geq h \mid \mathrm{resp} = \text{“}S1\text{”},\ \mathrm{stim} = S2) \times p(\mathrm{resp} = \text{“}S1\text{”} \mid \mathrm{stim} = S2)] \\
&= 1 - \left[\mathrm{FAR}_{2,\text{“}S1\text{”}}^{\mathrm{conf}=h} \times (1 - \mathrm{HR}_1)\right]
\end{aligned}
$$

$$
\begin{aligned}
\mathrm{FAR}_{1\sim 2,\text{“}S1\text{”}}^{\mathrm{conf}=h} &= 1 - p(\mathrm{resp} = \text{“}S1\text{”},\ \mathrm{conf} \geq h \mid \mathrm{stim} = S1) \\
&= 1 - [p(\mathrm{conf} \geq h \mid \mathrm{resp} = \text{“}S1\text{”},\ \mathrm{stim} = S1) \times p(\mathrm{resp} = \text{“}S1\text{”} \mid \mathrm{stim} = S1)] \\
&= 1 - \left[\mathrm{HR}_{2,\text{“}S1\text{”}}^{\mathrm{conf}=h} \times (1 - \mathrm{FAR}_1)\right]
\end{aligned}
$$

Thus, if separate cognitive mechanisms govern type 1 and type 2 judgments, then it is possible that patterns in the pseudo type 1 ROC curve reflect aspects of cognitive processing pertaining to type 2, rather than type 1, judgments. One such theoretical pattern is revealed in the case of chance type 2 responding, as discussed in Clifford et al. [3]. If an observer has chance levels of type 2 sensitivity, then confidence ratings do not differentiate between correct and incorrect trials, and so $\mathrm{HR}_2 = \mathrm{FAR}_2$. The pseudo type 1 ROC points constructed from such data would consist in a linear scaling of the “true” (FAR_1, HR_1) pair by some constant $k = \mathrm{HR}_2 = \mathrm{FAR}_2$. Thus, the pseudo type 1 ROC curve would consist of two line segments, one connecting (0, 0) to (FAR_1, HR_1) (corresponding to chance type 2 performance for “$S2$” responses), the other connecting (FAR_1, HR_1) to (1, 1) (corresponding to chance type 2 performance for “$S1$” responses); see Clifford et al.’s Fig. 3.2c.

Here we make the observation that pseudo type 1 z-ROC curves with non-unit slope can be generated by an EV-SDT model with differences in response-specific meta-d' (hereafter, RSM-SDT). By the same token, we observe that differences in the area under response-specific type 2 ROC curves can be generated purely as a consequence of the type 1 properties of the UV-SDT model. Thus, considerable caution is warranted in making inferences about the cognitive processes that underlie patterns in type 1 and type 2 ROC curves because of the possibility of confusing the effects of different variance for type 1 distributions and different suboptimalities for response-specific metacognitive sensitivity.

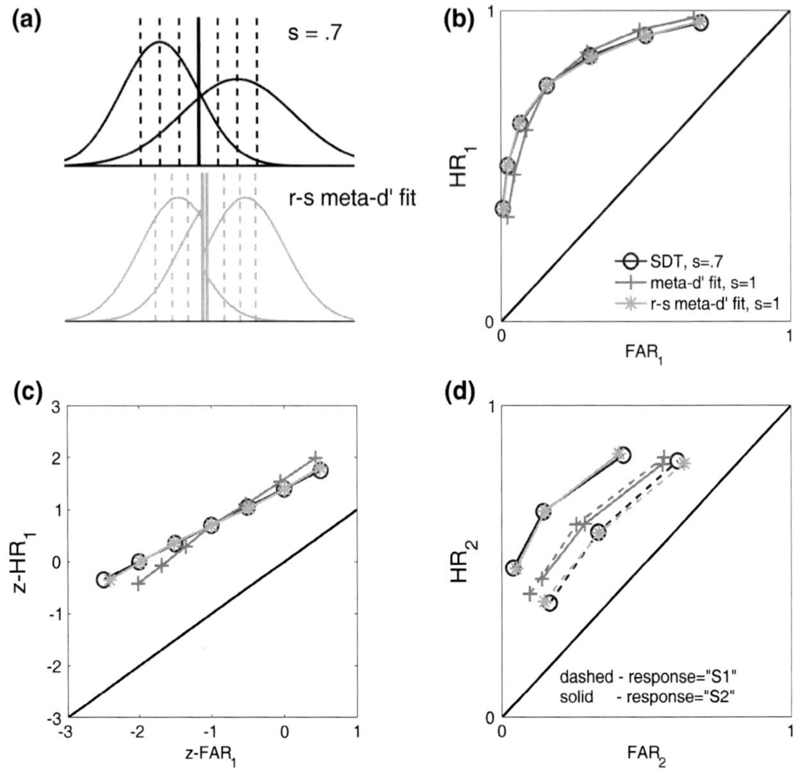

Fig. 3.6 Response-specific meta-d' model can fit patterns generated by the unequal variance SDT model. **a** UV-SDT model and response-specific meta-d' fit using EV-SDT. We used simulated trials from a UV-SDT model with $s = 0.7$ to generate type 1 and type 2 ROC curves. The response-specific meta-d' fit was able to emulate the differences in the degree of distribution overlap for "$S1$" and "$S2$" responses exhibited by the UV-SDT model (compare distribution overlaps on either side of the type 1 criterion in the top and bottom panels). **b** Type 1 ROC curve. We constructed pseudo type 1 ROC curves from the type 2 (FAR, HR) data produced by the meta-d' fits and the type 1 (FAR, HR) computed from the simulated data according to EV-SDT. Differences between UV-SDT and the meta-d' fits are difficult to discern on the pseudo type 1 ROC. **c** Type 1 z-ROC curve. On the pseudo type 1 z-ROC curve it is apparent that UV-SDT produces a curve with a non-unit slope, and that the curve based on response-specific meta-d' under EV-SDT produced a close match. By contrast, the curve based on the meta-d' fit under EV-SDT produced a unit slope. **d** Response-specific type 2 ROC curves. Under the UV-SDT model, there is more area under the type 2 ROC curve for "$S2$" responses than there is for "$S1$" responses. This pattern is closely connected to the non-unit slope on the type 1 z-ROC curve. As expected, response-specific meta-d' but not overall meta-d' produced a good fit to this type 2 data

3.5.4 RSM-SDT Fit to Data Generated by UV-SDT

We will illustrate the ability of differences in response-specific meta-d' to produce a non-unit slope on the pseudo type 1 z-ROC curve by simulating data from the UV-SDT model and fitting it with RSM-SDT. We used the UV-SDT model with $d'_1 = 2$, $c_1 = 0$, $\underline{c}_{2,\text{"}S1\text{"}} = (-0.5, -1, -1.5)$, $\underline{c}_{2,\text{"}S2\text{"}} = (0.5, 1, 1.5)$, and $s = 0.7$, where the "1" subscript for d' and c denotes that these are measured in σ_{S1} units. The SDT graph for these parameter values is plotted in Fig. 3.6a. We simulated 10,000,000 trials and constructed the pseudo type 1 ROC curve, pseudo type 1 z-ROC curve, and response-specific type 2 ROC curves, as plotted in Fig. 3.6b–d.

Next, we performed both an overall meta-d' fit and a response-specific meta-d' fit to the data, both times using the EV-SDT model as a basis. Performing the meta-d' fit requires first calculating d' and c for the simulated data. Performing the calculations for d' and c under the EV-SDT model yielded $d' = 1.7$ and $c = 0.15$.[15] The overall meta-d' fit resulted in parameter values of meta-$d' = 1.47$, meta-$c = 0.13$, $\underline{c}_{2,\text{"}S1\text{"}} = (-0.29, -0.74, -1.20)$, and $\underline{c}_{2,\text{"}S2\text{"}} = (0.51, 0.86, 1.20)$. The response-specific meta-d' fit resulted in parameter values of meta-$d'_{\text{"}S1\text{"}} = 1.05$, meta-$c_{\text{"}S1\text{"}} = 0.09$, meta-$\underline{c}_{2,\text{"}S1\text{"}} = (-0.28, -0.69, -1.13)$ for "$S1$" responses, and meta-$d'_{\text{"}S2\text{"}} = 2.40$,[16] meta-$c_{\text{"}S2\text{"}} = 0.21$, meta-$\underline{c}_{2,\text{"}S1\text{"}} = (0.65, 1.06, 1.45)$ for "$S2$" responses. From these parameter values, we computed the theoretical response-specific type 2 ROC curves (Fig. 3.6d). We also constructed the theoretical pseudo type 1 ROC curves (Fig. 3.6b, c) for the meta-d' fits. It was not possible to do this directly, since the meta-d' fits are meant to describe type 2 performance rather than type 1 outcomes. Thus, we performed the following procedure. From the meta-d' fits, we obtained a set of response-specific (FAR_2, HR_2) pairs. From the simulated data, we computed the "true" (FAR_1, HR_1) pair. Then we computed a set of pseudo type 1 ROC points, ($FAR_{1\sim2}$, $HR_{1\sim2}$), using the equations above that describe how to derive pseudo type 1 ROC points from (FAR_1, HR_1) and a set of response-specific (FAR_2, HR_2).

Figure 3.6c shows that the UV-SDT model produced a linear z-ROC curve with a slope lower than 1. It also demonstrates that the RSM-SDT fit produced a close approximation to the UV-SDT data, whereas the overall meta-d' fit did not. To quantify these observations, we performed maximum likelihood fits of the UV-SDT model onto (1) the simulated data originally generated by the UV-SDT model, and (2) a new set of 10,000,000 simulated trials that followed a distribution

[15] Note that the values for d' and c recovered by EV-SDT analysis are slightly different from those used in the generating UV-SDT model due to their differing assumptions about the distribution variances.

[16] The value of meta-$d'_{\text{"}S2\text{"}}$ at 2.4 was substantially larger than the value of d' at 1.7, an unusual result as we would typically expect meta-$d' \le d'$ [13]. However, constraining the RSM-SDT fit such that meta-$d'_{\text{"}S2\text{"}} \le d'$ still produced data that gave a reasonable approximation to the z-ROC curve. Fitting the UV-SDT model to the data distributed according to this RSM-SDT fit yielded $s = 0.83$, demonstrating that even with the constraint that meta-$d'_{\text{"}S2\text{"}} \le d'$, RSM-SDT still produced a z-ROC curve with non-unit slope.

of outcomes following the theoretical pseudo type 1 ROC curve generated by the RSM-SDT fit, and (3) similarly for the overall meta-d' fit. The UV-SDT fit to the UV-SDT generated data yielded $s = 0.7$, successfully recovering the true value of s in the generating model. The UV-SDT fit to the data distributed according to RSM-SDT yielded a closely matching $s = 0.72$. The UV-SDT fit to the data distributed according to the overall meta-d' fit yielded $s = 0.98$ since this model has no mechanism with which to produce non-unit slopes on the z-ROC curve.

The relationship between the slope of the pseudo type 1 z-ROC curve and area under the response-specific type 2 ROC curves is made evident in Fig. 3.6d. The data generated by the UV-SDT model produced a type 2 ROC curve for "$S2$" responses that has substantially more area underneath it than does the type 2 ROC curve for "$S1$" responses. Intuitively, this is due to the fact that when $s < 1$, the $S1$ and $S2$ distributions overlap less for "$S2$" responses than they do for "$S1$" responses (see Fig. 3.6a). As expected, the response-specific meta-d' fit is able to accommodate this pattern in the response-specific type 2 ROC curves, whereas the overall meta-d' fit is not.

3.5.5 UV-SDT Fit to Data Generated by RSM-SDT

Just as RSM-SDT can closely fit data generated by UV-SDT, here we show that UV-SDT can produce patterns of data similar to those generated by an RSM-SDT model. For this example, we once again use the model described in the section "Toy example of response-specific meta-d' fitting." This model has two parameters, $\sigma_{2,\text{“}S1\text{”}}$ and $\sigma_{2,\text{“}S2\text{”}}$, that control the level of noisiness in type 2 sensitivity for "$S1$" and "$S2$" responses. We simulated 10,000,000 trials using parameter values $d' = 2$, $c = 0$, $\underline{c}_{2,\text{“}S1\text{”}} = (-0.5, -1, -1.5)$, $\underline{c}_{2,\text{“}S1\text{”}} = (0.5, 1, 1.5)$, $\sigma_{2,\text{“}S1\text{”}} = 1.5$, and $\sigma_{2,\text{“}S2\text{”}} = 0$.

We fit the RSM-SDT model to this data set, yielding a fit with meta-$d'_{\text{“}S1\text{”}} = 0.78$, meta-$c_{\text{“}S1\text{”}} = 0$, meta-$\underline{c}_{2,\text{“}S1\text{”}} = (-0.54, -0.73, -0.94)$ for "$S1$" responses, and meta-$d'_{\text{“}S2\text{”}} = 2.00$, meta-$c_{\text{“}S2\text{”}} = 0$, and meta-$\underline{c}_{2,\text{“}S2\text{”}} = (0.50, 1.00, 1.50)$ for "$S2$" responses. The SDT graphs for these fits are plotted in the top half of Fig. 3.7a.

[17] Note that the nature of the UV-SDT model inherently places constraints upon the set of type 1 and type 2 ROC curves that can be exhibited at the same time, whereas the method for fitting meta-d' minimizes constraints of type 1 performance upon the type 2 fit. Additionally, the likelihood function for the UV-SDT model is built from pseudo type 1 probabilities of the form p(resp = r, conf = y|stim = s). This is different from the likelihood function for fitting meta-d', which is built from type 2 probabilities of the form p(conf = y|stim = s, resp = r). Thus, whereas the meta-d' algorithm is specialized for fitting type 2 data, the fit for the UV-SDT model must account for variance in both type 1 and type 2 responses, entailing potential tradeoffs in the fit. Fitting UV-SDT to the data with a type 2 likelihood function achieves a near perfect fit to the type 2 ROC curves, albeit with a very poor fit to the type 1 ROC curve (data not shown).

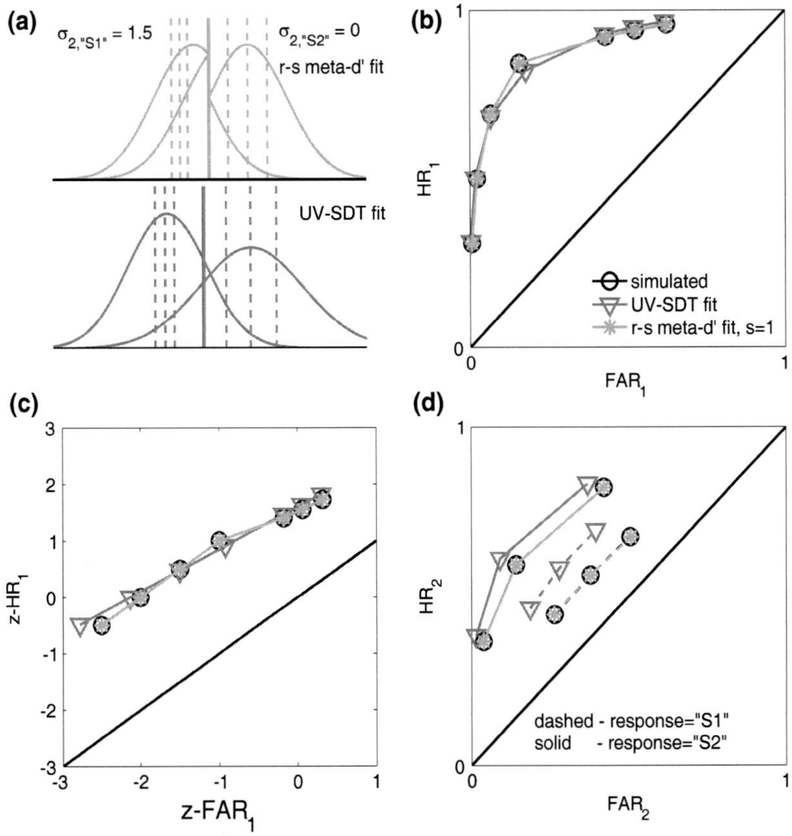

Fig. 3.7 The unequal variance SDT model can fit patterns generated by asymmetries in response-specific metacognitive sensitivity. **a** Response-specific meta-*d′* and UV-SDT fits to simulated data. We returned to the model depicted in Fig. 3.5, simulating trials with $\sigma_{2,\text{"}S1\text{"}} = 1.5$ and $\sigma_{2,\text{"}S2\text{"}} = 0$. The *top* half of this panel depicts the response-specific meta-*d′* fit for the simulated data. The *bottom* half depicts the UV-SDT fit. **b** Type 1 ROC curves. **c** Type 1 z-ROC curves. We produced type 1 ROC curves from the meta-*d′* fits using the same procedure as in Fig. 3.6. Both the response-specific meta-*d′* fit and the UV-SDT fit provided a close match to the type 1 ROC curves of the generating model. **d** Response-specific type 2 ROC curves. The UV-SDT model slightly overestimated area under the response-specific type 2 ROC curves, but still captured the fact that there is more area under the curve for "*S*2" responses than for "*S*1" responses

Next, we found the maximum likelihood fit of the UV-SDT model to this data set. This yielded a fit with $d'_1 = 2.14$, $c_1 = -0.15$, $\underline{c}_{2,\text{“}S1\text{”}} = (-0.89, -1.12, -1.37)$, $\underline{c}_{2,\text{“}S2\text{”}} = (0.43, 1.06, 1.72)$, and $s = 0.75$. The SDT graph for this fit is plotted in the bottom half of Fig. 3.7a.

As shown in Fig. 3.7c, the simulated data and meta-d' fit produce a pseudo type 1 z-ROC curve with a small deviation from linearity due to an upward-going kink in the curve corresponding to the “true” (FAR, HR) point. Nonetheless, this curve is closely approximated by the linear z-ROC curve with slope = 0.75 produced by the UV-SDT model fit. The deviation between the UV-SDT fit and the generating model is more clearly pronounced on the response-specific type 2 ROC curves. Although the UV-SDT model overestimates the area under both curves, it nonetheless captures the qualitative pattern that there is more area under the curve for “$S2$” responses than for “$S1$.”[17]

3.6 Discussion

3.6.1 *Implications for Interpretation and Methodology of SDT Analysis of Type 1 and Type 2 Processes*

The foregoing analyses suggest that extra caution should be exercised when interpreting ROC curves. Constructing z-ROC curves using confidence rating data risks conflating the contributions of type 1 and type 2 performance. Non-unit slopes on these pseudo type 1 z-ROC curves can occur due to response-specific differences in type 2 processing even when the underlying type 1 stimulus distributions have equal variance. Thus, inferences about the nature of type 1 processing based on the pseudo type 1 z-ROC curve slope may not always be justified.

This is especially a concern in light of empirical demonstrations that type 1 and type 2 performance can dissociate; e.g., Rounis et al. [17] found that applying transcranial magnetic stimulation to dorsolateral prefrontal cortex selectively diminishes type 2, but not type 1, sensitivity, and Fleming et al. [8] found that between-subject anatomical variation in frontal cortex correlates with variability in type 2 sensitivity even when type 1 sensitivity is held constant across subjects. This suggests that type 2 sensitivity is subject to sources of variation that do not affect type 1 processing. In turn, this suggests that estimates of the relative variances in type 1 stimulus distributions based on the pseudo type 1 ROC curve may be unduly affected by factors that cause variation in type 2, but not type 1, processing.

By the same token, however, differences in response-specific type 2 ROC curves do not necessarily entail differences specifically at the level of type 2 or “metacognitive” processing. Instead, such differences are potentially attributable to differences in basic attributes of type 1 processing, such as type 1 sensitivity, criterion placement, and/or the variability of the type 1 stimulus distributions. For instance, Kanai et al. [11] observed that area under the type 2 ROC curve for "signal absent" responses were poorer for manipulations that target perceptual, rather than attentional, processes. They inferred that perceptual, but not attentional, manipulations disrupted processing at early levels of processing, such that subjects lacked introspective awareness regarding the source of their failure to detect the target. However, an alternative explanation might be that the type 2 ROC curves differed purely as a consequence of differences in d', c, and s. Reducing the values of d', c, and s can all potentially lead to reductions in area under the type 2 ROC curve for "$S1$" responses. Thus, it is possible that the differences in the type 2 ROC curves for the perceptual and attentional manipulations might be explicable purely in terms of differences in low-level processing, rather than in terms of differences across levels of processing. This is an example of the more general principle upon which our whole approach to type 2 analysis is founded, the principle which necessitates the need for a measure like meta-d': Since type 2 ROC curves depend on the parameters of the type 1 SDT model, it is crucial to interpret type 2 data in the context of the empirical type 1 data, and to consider the extent to which the relationship between the type 1 and type 2 data conforms to or violates SDT expectation [9, 13].

Galvin et al. [9] similarly cautioned against the use of pseudo type 1 ROC curves to make inferences about type 1 processes. They suggested that so-called type 1 ratings (e.g. "rate your confidence that the stimulus was $S2$ on a scale of 1–8") may offer a window into type 1 processing that type 2 ratings (e.g. "rate your confidence that your "$S1$" or "$S2$" response was correct on a scale of 1–4") do not. However, it is not clear that the cognitive mechanisms required to generate such type 1 ratings would differ substantively from those needed for the type 2 ratings, and the informational content of type 1 and type 2 ratings may turn out to be identical, differing only in superficial aspects. In their discussion, Galvin et al. point out that it may be difficult to create a type 2 decision rule that captures the behavior of type 1 ratings. (Per our discussion in the section titled "Comparison of the current approach to that of Galvin et al. [9]", we might say that this is analogous to the problem regarding how to create a type 2 decision rule that adequately captures the empirically observed relationships between the placement of response-specific type 2 criteria.) However, we note that the potential difficulty of such a mapping may simply reflect the possibility that observers do not, in fact, explicitly compute an overall type 2 decision variable as such, or perhaps only do so in a heuristic or variable way.

It may be possible to use z-ROC data to estimate distribution variances without the confounding influence of response-specific type 2 processing by avoiding the use of pseudo type 1 z-ROC curves. Instead, type 1 ROC curves can be constructed by using experimental interventions that directly target type 1 decision processes, such as direct instruction, changes in stimulus base rates, or changes in the payoff matrix. On the presumption that such manipulations are not themselves targeting processes that depend on metacognitive or type 2 kind of processing, ROC curves constructed in this way might offer purer windows into the nature of type 1 processing, relatively uncontaminated by the influence of type 2 processing.

This suggestion is consistent with the observation that pseudo type 1 ROC curves do not always behave the same as "true" type 1 ROC curves generated by direct manipulation of the type 1 criterion. For instance, Markowitz and Swets [14] found that estimates of *s* in auditory detection tasks depend on signal strength for pseudo, but not true, type 1 ROC curves; Van Zandt [23] found that estimates of *s* based on pseudo type 1 ROC curves varied depending on the degree of bias in the true type 1 criterion (thus implying that not all pseudo type 1 ROC curves yield the same estimate of *s* as the "true" type 1 ROC curve); and Balakrishnan [1], replicated in Mueller and Weidemann [15], found that pseudo type 1 ROC points can fall below the true type 1 ROC curve constructed under the same experimental conditions. Empirical results like these suggest that pseudo and true-type 1 ROC curves may indeed tap into distinct cognitive processes, which is consistent with our observations that (1) the pseudo type 1 ROC curve has a direct mathematical relationship with response-specific type 2 ROC curves, and (2) type 2 ROC curves are subject to sources of variation that do not affect type 1 performance (e.g. [8, 17]).

These considerations also have implications for the methodology of estimating meta-d'. In the current work, and previously, we have considered estimation of meta-d' in the context of equal variance SDT. Only a simple extension of the methodology is needed to perform meta-d' analysis based on the UV-SDT model. Presumably the value of *s* would be set to a fixed value in the meta-SDT model based on the characteristics of the empirical data being characterized, analogous to the treatment of meta-c'. Then the interpretation of meta-d' based upon the UV-SDT model could be expanded to say e.g. "suppose there is an ideal SDT observer O who exhibits a response bias (c') and unequal type 1 variance (*s*) similar to those of subject X. In order for O to produce response-specific type 2 ROC curves like those of X, O would need a d' equal to so-and-so."

However, it is unclear how we could or should arrive at the value of *s* to be used for such an UV meta-SDT model. As we have seen, the pseudo type 1 ROC curve has a direct mathematical relationship with response-specific type 2 ROC curves, opening up the possibility that measures of *s* based on the pseudo type 1 ROC curve are confounded by independent sources of variation in type 2 sensitivity. It is not clear that deriving a value for *s* from pseudo type 1 data, and then using that value of *s* in a characterization of the type 2 sensitivity exhibited by the very same confidence ratings used to estimate the value of *s* in the previous step, would be desirable. One potential workaround, as discussed above, might be to independently estimate the type 1 ROC curve based upon direct manipulations of type 1 response bias across experimental conditions. The estimate of *s* derived from the "true" type 1 ROC curve could then be used to fit an UV meta-SDT model to the type 2 data.

Another option is to gracefully sidestep the problem of UV by utilizing experimental designs that tend to produce data that is adequately characterized by EV-SDT. For example, in 2-interval forced choice designs, the $S1$ stimulus may appear in one of two spatial or temporal intervals, while the $S2$ stimulus appears in the other. The observer must report whether the stimulus sequence on the current trial was <*S*1, *S*2> or <*S*2, *S*1> (e.g. spatially, "*S*1 was on the left and *S*2 was on the right" or temporally, "*S*2 came first and *S*1 came second"). Intuitively, internal responses should be equally variable for <*S*1, *S*2> and <*S*2, *S*1 > sequences, even if internal responses to *S*1 and *S*2 themselves are not equally variable. This result can be more formally derived from the SDT model [12] and has been observed in empirical data (e.g. [18]). Thus, the 2-inveral forced choice design may provide an experimental paradigm that circumvents concerns related to the UV-SDT model and facilitates usage of EV-SDT.

Another possibility is to create a variation of the SDT model that includes structures to account both for UV and variation in type 2 sensitivity (a simple example of the latter being the σ_2 model used earlier). It is possible that finding the best fit of such a model to a data set could arbitrate to some extent on the relative contributions of UV and response-specific metacognition to patterns in the data. Such an approach would constitute something of a departure from the meta-d' methodology discussed here. However, it seems likely that such a model-based approach would still need to be supplemented by experimental designs intended to produce data that specifically arbitrate between the mechanisms in question, and it is not clear that the standard form of the two-choice task with confidence ratings provides such a design. Ultimately, analysis of how computational models fit the data needs to be supplemented with other empirical and conceptual considerations in order to make strong inferences about the underlying cognitive processes.

3.7 Code for Implementing Overall and Response-Specific Meta-d' Analysis

We provide free Matlab scripts for conducting type 1 and type 2 SDT analysis, including functions to find the maximum likelihood fits of overall and response-specific meta-d' to a data set, at http://www.columbia.edu/~bsm2105/type2sdt

Acknowledgements This work was supported by Templeton Foundation Grant 21569 (H.L.). We thank Dobromir Rahnev and Guillermo Solovey for comments on an earlier version of the manuscript.

References

1. Balakrishnan JD (1998) Measures and Interpretations of vigilance performance: evidence against the detection criterion. Hum Factors: J Hum Factors Ergon Soc 40(4):601–623. doi:10.1518/001872098779649337
2. Clarke FR, Birdsall TG, Tanner J (1959) Two types of ROC curves and definitions of parameters. J Acoust Soc Am 31(5):629–630. doi:10.1121/1.1907764
3. Clifford CWG, Arabzadeh E, Harris JA (2008) Getting technical about awareness. Trends Cogn Sci 12(2):54–58. doi:10.1016/j.tics.2007.11.009

4. Dorfman DD, Alf E (1969) Maximum-likelihood estimation of parameters of signal-detection theory and determination of confidence intervals–Rating-method data. J Math Psychol 6(3):487–496. doi:10.1016/0022-2496(69)90019-4
5. Egan JP (1975) Signal detection theory and ROC analysis. Academic Press, New York
6. Evans S, Azzopardi P (2007) Evaluation of a "bias-free" measure of awareness. Spat Vis 20(1–2):61–77
7. Fleming SM, Maniscalco B, Amendi N, Ro T, Lau H (in review). Action-specific disruption of visual metacognition
8. Fleming SM, Weil RS, Nagy Z, Dolan RJ, Rees G (2010) Relating introspective accuracy to individual differences in brain structure. Science 329(5998):1541–1543. doi:10.1126/science.1191883
9. Galvin SJ, Podd JV, Drga V, Whitmore J (2003) Type 2 tasks in the theory of signal detectability: discrimination between correct and incorrect decisions. Psychon Bull Rev 10(4):843–876. doi:15000533
10. Green DM, Swets JA (1966) Signal detection theory and psychophysics. Wiley, New York
11. Kanai R, Walsh V, Tseng C-H (2010) Subjective discriminability of invisibility: a framework for distinguishing perceptual and attentional failures of awareness. Conscious Cogn. doi:10.1016/j.concog.2010.06.003
12. Macmillan NA, Creelman CD (2005) Detection theory: a user's guide, 2nd edn. Lawrence Erlbaum
13. Maniscalco B, Lau H (2012) A signal detection theoretic approach for estimating metacognitive sensitivity from confidence ratings. Conscious Cogn 21(1):422–430. doi:10.1016/j.concog.2011.09.021
14. Markowitz J, Swets JA (1967) Factors affecting the slope of empirical ROC curves: comparison of binary and rating responses. Percept Psychophysics 2(3):91–100. doi:10.3758/BF03210301
15. Mueller ST, Weidemann CT (2008) Decision noise: an explanation for observed violations of signal detection theory. Psychon Bull Rev 15(3):465–494. doi:18567246
16. Ogilvie JC, Creelman CD (1968) Maximum-likelihood estimation of receiver operating characteristic curve parameters. J Math Psychol 5(3):377–391. doi:10.1016/0022-2496(68)90083-7
17. Rounis E, Maniscalco B, Rothwell JC, Passingham RE, Lau H (2010) Theta-burst transcranial magnetic stimulation to the prefrontal cortex impairs metacognitive visual awareness. Cogn Neurosci 1(3):165–175. doi:10.1080/17588921003632529
18. Schulman AJ, Mitchell RR (1966) Operating characteristics from Yes-No and Forced-Choice procedures. J Acoust Soc Am 40(2):473–477. doi:10.1121/1.1910098
19. Swets JA (1986) Indices of discrimination or diagnostic accuracy: their ROCs and implied models. Psychol Bull 99(1):100–117. doi:3704032
20. Swets JA (1986) Form of empirical ROCs in discrimination and diagnostic tasks: implications for theory and measurement of performance. Psychol Bull 99(2):181–198
21. Swets JA, Tanner WP Jr, Birdsall TG (1961) Decision processes in perception. Psychol Rev 68(5):301–340. doi:10.1037/h0040547
22. Tanner WP Jr, Swets JA (1954) A decision-making theory of visual detection. Psychol Rev 61(6):401–409
23. Van Zandt T (2000) ROC curves and confidence judgements in recognition memory. J Exp Psychol Learn Mem Cogn 26(3):582–600

Chapter 4
Kinds of Access: Different Methods for Report Reveal Different Kinds of Metacognitive Access

Morten Overgaard and Kristian Sandberg

Abstract In experimental investigations of consciousness, participants are asked to reflect upon their own experiences by issuing reports about them in different ways. For this reason, a participant needs some access to the content of her own conscious experience in order to report. In such experiments, the reports typically consist of some variety of ratings of confidence or direct descriptions of one's own experiences. Whereas different methods of reporting are typically used interchangeably, recent experiments indicate that different results are obtained with different kinds of reporting. We argue that there is not only a theoretical, but also an empirical difference between different methods of reporting. We hypothesise that differences in the sensitivity of different scales may reveal that different types of access are used to issue direct reports about experiences and metacognitive reports about the classification process.

Keywords Metacognition · Consciousness · Awareness · Subliminal perception · Vision

---

This chapter is adapted from: Overgaard M, Sandberg K (2012) Kinds of access: Different methods for report reveal different kinds of metacognitive access. Phil Trans R Soc B 367:1287–1296

M. Overgaard (✉)
CCN, Department of Communication and Psychology, Aalborg University,
Kroghsttraede 3, 9220 Aalborg Oest, Aalborg, Denmark
e-mail: morten.storm.overgaard@ki.au.dk

M. Overgaard · K. Sandberg
CNRU, Hammel Neurorehabilitation and Research Center, CFIN, MindLab, Aarhus University, Noerrebrogade 44, Build. 10G 8000 Aarhus C, Denmark

S. M. Fleming and C. D. Frith (eds.), *The Cognitive Neuroscience of Metacognition*,
DOI: 10.1007/978-3-642-45190-4_4, © Springer-Verlag Berlin Heidelberg 2014

4.1 Introduction and Definitions

Although 'metacognition', 'verbal reports' and even 'introspection' have become legitimate concepts in cognitive neuroscience, they are rarely clearly defined, and their relations to cognition and consciousness are often even more elusive. Here, we attempt to show how these concepts may be understood, how they relate to each other, and present empirical arguments which support a hypothesis that introspection may be a unique type of metacognition, i.e. that introspective reports are empirically different from other kinds of metacognitive reports.

In consciousness research, authors typically refer to their behavioural measures of conscious experience as introspective or metacognitive (see for instance [1–3]), and the two terms are often used interchangeably although they differ from a theoretical perspective. In principle, metacognition is any cognitive process about a different cognitive process, whereas introspection is closely tied to conscious experience. Metacognition is thus a higher order process with cognition as its object, whereas introspection is a higher order process with conscious experience as its object.

Although consciousness and cognition by several accounts may be related to highly overlapping brain processes, the concepts are defined differently. Cognition, as described in Ulric Neisser's landmark publication *Cognitive Psychology* [4], is defined as the transformation and use of sensory input, and how these processes work even in the absence of input as in hallucinations or imagination. The 'transformations' or 'processes' are not themselves observable, but are inferred based on observations of behaviour. Cognition is thus some sort of processing, and it can usually be functionally defined—a cognitive scientist can examine the workings and purpose of the memory system, for instance. In cognitive science, it is not unusual to investigate certain cognitive states that are said to be about other cognitive states rather than external phenomena—so-called metacognitive states [5]. An important aspect of cognitive science theory is the idea that our knowledge of ourselves is basically inferential and not based on privileged access to one's own mental processes. Nevertheless, this inferential information, it is argued, is frequently causally related to our actions. It is because we have certain ideas about our own qualities or disadvantages that we decide to take a certain career path or give up on another. It is because we think we would like a particular kind of music that we chose to pay money to attend one particular concert. This way, a person can in principle perform a metacognitive judgment without trying to evaluate his mental processes or strategies directly, although at other times, a metacognitive judgment might involve some kind of evaluation of an internal process. This could for instance be the case when a participant in a psychological experiment is asked to rate his confidence in having performed a visual discrimination task successfully. In other words, metacognition does not require that the information used to evaluate a mental process is obtained by probing the system (by use of another, internal mental process), although in some cases it may.

There are several meanings of the word 'consciousness'. One use of 'conscious' is applied to a person's total or background state (what is sometimes called 'state consciousness'). A person is conscious, in this sense, if he or she is, say, alert and

Table 4.1 Structure of definitions

	'First order' mental state	'Second order' mental state
Functionally defined	Cognition	Metacognition
Subjectively defined	Consciousness	Introspection

awake rather than being asleep or even in a coma. However, most attempts to capture the meaning of the term consciousness seem to focus on a second aspect, the contents of consciousness. Some philosophers use terms such as 'qualia', while others prefer 'subjective experience', 'phenomenal consciousness', 'conscious awareness' and the like.

Consciousness has become a fast growing topic of interest in empirical research, and consequently scientists are methodologically dependent on more than the presence or absence of consciousness in order to study it. They need participants to be able to give some sort of report or externally observable indication that they experience something, thus leading to a renewed interest in how participants report and introspect. Introspection, however, is hardly a new method. At the beginning of experimental psychology William James, for one, believed that one method to study the mind should be the direct, inner observation of consciousness [6]. Common to classical as well as more recent discussions of introspection is the necessary link to conscious experience. A broadly discussed account, mentioned by William James among others, defines introspection as a self-reflective, retrospective process in which one's memory trace of the original target mental state is inspected. In more recent accounts, introspection is understood and investigated as an 'on-line' inspection of current and ongoing mental states. The available definitions of introspection agree, however, that it should be defined as some sort of observation of or attention to what is subjectively experienced [6–8], for which reason there cannot be introspection without experience as a matter of definition.

As we can see from the above, there are many intuitive similarities between the concepts of metacognition and introspection, yet the two are differently defined. Whereas metacognition is functionally defined as basically any cognitive state that is about another cognitive state, introspection can only be about a specifically conscious state. In this sense, both concepts can be said to be of 'second order' as they are about another (first order) state.

Based on the reasoning so far, the following distinctions can be made (Table 4.1).

Table 4.1 shows how introspection and metacognition are defined in different ways although their use overlaps greatly. Metacognition is a concept with a functional definition, whereas introspection is defined from a subjective perspective. This distinction has little to say about the relationship between metacognition and introspection. Clearly, the relationship need not be a matter of opposition. Rather, introspection might be a special kind of metacognition.

4.1.1 Methodological Consequences

Just as consciousness should not be confused with introspection, introspection should not be confused with a report. A report is, in this context, an intended communication by a participant and may be delivered verbally or by any other kind of controlled behaviour (signs, button presses or whatever). We may have full metacognitive or introspective knowledge of some mental event but choose not to report it, to lie about it or to report just one aspect of it, while not mentioning another. Thus, introspection and metacognition are fully dissociable from reports. However, the opposite is not the case: Reports about conscious or cognitive states are not dissociable from introspection or metacognition.

These conceptual distinctions lead logically to a number of methodological consequences. First of all, one methodological consequence is that any neuroscientific investigation of cognitive processes using metacognition or of consciousness using introspection will have difficulties sorting out which brain activations are related to the first order and to the second order states. This is, however, not to say that nothing can be done experimentally to tease the levels apart. For one thing, the second order states, as a matter of principle, cannot exist without the simultaneous presence of a first order state, which it is about. First order states, however, do not stand in any dependent relation to the second order states. One might assume a certain temporal relation between them, so that the first order mental states always occur before the second order state that is about the first order state.

The number of methodological arguments against the use of introspection to study consciousness is impressive. Most of them, however, circle around one central claim that introspection does not give reliable and valid information about conscious processes. Nisbett and Wilson [9] presented empirical evidence that subjects have little introspective access to the causes of their choices and behaviour. In one example, they showed that participants had a bias to prefer objects presented to the right, yet when asked, they never mentioned the location of the object as their reason for preferring it. However, participants giving an introspective report about liking objects presented to the right for some other reason than the object's location in space may be giving a perfectly good and scientifically usable report of what they experienced. Nisbett and Wilson correctly rejected introspection as a methodology to learn about (some aspects of) choice and decision-making, as the behavioural data suggested a very different explanation from the one that participants reported. Considering Table 4.1, it should be clear that an introspective report by definition is only about what is experienced and not about cognitive processes [9, 10].

The discussion rests on a philosophical debate about whether introspection is corrigible or not, or whether we have 'privileged access' to our own consciousness. In the view presented here, where consciousness is considered subjective and introspection is an attending to consciousness, introspection is incorrigible in the sense that no external measure can directly access a person's experiences.

The introspective report, of course, may be corrected by the subject himself on closer examination or if he was lying or otherwise somehow reporting suboptimally the first time. In this sense, you could be asked to calculate, say, 956 minus 394 and come up with the right answer, but when asked how you did it, produce a report describing a method that logically would lead to a wrong answer. This would make your report false as a description of how you actually performed the math. However, it may still be correct as an introspective report, telling what you experience when inspecting your memory of what happened.

Another argument against introspection has been that introspection is not exclusive, i.e. it is not specifically and solely about the relevant conscious state [11]. The flip side of the argument is that introspection is not exhaustive, i.e. there may be more to the conscious state than what is captured by introspection. A fast response to both sides of this argument is that it confuses introspection with the report. The report may not be exhaustive and exclusive, as one, obviously, may not report all aspects or report too many. One may, however, speculate that the argument runs deeper. For instance, it might be the case that we do not introspect all of our experiences at the same time, but, as it is the case with other functions of attention, only partial information is selected for further processing. If this were the case, one could ask participants to introspect on some aspects, but not all aspects, of an experience. This would in that case have great importance for how to study consciousness experimentally. The exact wording of instructions may have a great impact on the way a participant attends to her own experiences and which aspects that are introspectively accessed.

As the individuation of different conscious contents and cognitive states and processes is an empirical matter, these speculations seem at least in principle to be open to empirical investigation and resolution as well. Conscious contents are individuated though the process of introspection, whereas cognitive states and processes are individuated by inference from behaviour. The considerations so far predict that different types of instructions and different types of report will give rise to different results in experiments on consciousness due to a modulation of the 'kind of access' or 'target of introspective attention'. We will discuss these findings in the following.

4.2 Experiments Using Subjective Reports

Before discussing the most recent experimental findings, we provide a brief history of how such measures have been used in empirical science and how the quality of a measure might be evaluated.

4.2.1 Subjective Measures: Methods and Issues

Subjective measures have been used in psychological science for more than 100 years. Possibly the first study is reported by Sidis [12]. He presented participants with single letters or digits at various distances. The participants made (introspective) judgements as to whether or not they could see what was on the card followed by an attempt at identification. Sidis observed that even when participants claimed not to see the letter or digit, they performed above chance on the identification task. This kind of paradigm has also been referred to as the *subjective threshold* or *dissociation approach*, and has since been used in numerous studies, including very recent experiments [13]. Unconscious processing, in this case, is presumed to be responsible for any above-chance performance found when stimuli are below the so-called subjective criterion (i.e. when participants claim to have no experience of the identity of the stimulus) [14]. In a more recent variation of this, participants are asked instead to report their confidence in being correct. This method is referred to as establishing unconscious processing by 'the guessing criterion' [15].

Already, we see a difference between different scales of awareness. When using the scale introduced by Sidis, participants are asked only to introspectively report their experience, but not to draw any inferences about the accuracy of a classification process. On the other hand, when participants are asked to report their confidence in being correct (or in a slightly different paradigm described below, place a wager on being correct), they perform a metacognitive judgment ('how good was the classification?'), and they are presumed to use (only) their conscious experiences as the basis of this judgment [16]. One type of scale thus encourages introspection (judgment of the clarity of the experience), whereas the other type of scale encourages metacognition (judgment of the quality of the classification process). Various theoretical arguments have been made in favour of both types of scales [17], and in the following we focus mainly on empirical differences. Both types of scales, however, use the dissociation approach as noted above.

There are certain problems with the dissociation approach. As a participant, when you see something so vaguely that you have almost no confidence at all in what you see, you may be reluctant to claim that you are in fact seeing something. If anything, this tendency would be expected to increase when you know that an experimenter will analyse your data to see if you are actually correct when you claim to see something. In other words, it should in fact be expected that participants in experiments are holding back information about very vague experiences or about classifications in which they have very little confidence. However, it is required of a subjective measure that it detect all relevant conscious knowledge (or all experiences) [11, 18–20]. Technically, this has been referred to as *exhaustiveness* [11], and we might expect that different measures will differ in their degree of exhaustiveness.

Unfortunately, we cannot simply solve the problems associated with poor exhaustiveness by using the scale that shows the greatest sensitivity as some scales might misclassify unconscious processes as conscious and thus give 'opposite' results. If, for instance, we used a purely behavioural measure such as task accuracy to test for the presence of conscious experience—this is indeed sometimes done when experimenters want to make sure that no conscious experience is present [14, 21]—we would risk classifying some unconscious information as conscious. If we are to trust our scale, it should not classify any unconscious information as conscious—that is, it should be *exclusive* [11]. As with exhaustiveness, we might expect different scales to differ in the extent to which they are exclusive.

Subjective measures of consciousness should obviously be as exhaustive and as exclusive as possible, yet it is not entirely clear how to go about this. As we cannot simply use the most sensitive scale (which might not be exclusive), we would have to compare only those scales, which we have no reason to believe would mistake unconscious processing for conscious processing. If there is no a priori reason to suspect that a collection of scales are suboptimally exclusive, we can compare their sensitivity, and the scale that is the most sensitive will be the most exhaustive. This will thus be the preferred scale (all else being equal). Two common ways in which scales are compared are (1) The amount of unconscious processing they indicate, and (2) How well and how consistently ratings correlate with accuracy. Under usual circumstances, for a scale to be maximally exhaustive, it should indicate as little subliminal perception as possible (i.e. participants are holding back the smallest amount of information), and the correlation between accuracy and awareness should be as good as possible (i.e. there is a large difference in accuracy when participants claim to see nothing and when they claim to have a clear experience). A further test for the trustworthiness of a measure is its stability, e.g. it is not that 'seeing nothing' is associated with one accuracy level one moment, and a completely different level the next. One or more of these methods have been used to compare different awareness scales in a number of experiments. These experiments are summarised below.

4.2.2 The Impact of Scale Steps

Overall, the impact on exhaustiveness has been examined for two types of scale manipulations. First, modifications of the number and descriptions of scale steps have been examined, and second, changes in the type of scale (e.g. whether participants report consciousness, confidence or something else) has been examined. For the first manipulation, we are primarily aware of studies using purely introspective measures and thus not drawing upon other metacognitive skills (such as judgment of confidence in a classification process). For this reason, this section draws more heavily on purely introspective measures.

A number of experiments have found above-chance performance when participants claim to have no experience of the stimulus (i.e. when the stimulus is not perceived according to the subjective threshold as established by an introspective report). However, Ramsøy and Overgaard [2] drew attention to the fact that many such studies divide subjective experiences into 'clear, vivid experiences' and 'nothing at all', a division that might not capture the descriptions given by participants, and previous studies might thus have used inappropriate introspective measures. Participants often claim to experience stimuli in quite different ways from trial to trial, and for that reason Ramsøy and Overgaard examined if any subliminal perception was found if participants used a scale that followed their own descriptions of experiences rather than a dichotomous all-or-none scale. Both in a pilot study and in the actual experiment, participants performed a visual discrimination task [forced-choice of position (three possible locations), form (three geometric figures) and colour (three different colours] of stimulus) and subsequently reported their awareness. They were asked to construct their own awareness scale. Participants generally ended up using a 4-step scale with the following scale step labels 'No experience', 'Brief glimpse', 'Almost clear experience', and 'Clear experience' (this scale will be referred to as the perceptual awareness scale, or PAS, in future experiments). When participants used the 'Brief glimpse' category they reported no awareness of form, colour or shape (but instead only a general vague experience of having seen something). A few participants included two additional steps, but these had no separate descriptions and were rarely used.

The results demonstrated that each PAS rating was related to a different accuracy level, with accuracy increasing as a function of PAS rating. In other words, the individual subjective ratings corresponded to different levels of objective performance. Additionally, no statistically significant above-chance performance was found when participants used the scale step 'no experience'. However, if the scale step for which participants claimed not to see stimulus features or location (i.e. the 'Brief glimpse' category) was also included, above-chance performance was indeed found as in previous studies. The study can be criticised for not comparing the results to results obtained by an actual dichotomous scale, and partly for this reason a second study was performed by Overgaard et al. [22]. Another purpose of the study was to examine if awareness was gradual in a general sense—i.e. if any feature can be perceived more or less clearly, or whether partial awareness is simply full awareness of individual features as has been hypothesised by others [23].

Data from Ramsøy and Overgaard [2] seemed to support the notion that awareness is gradual in a general sense—i.e. that any stimulus feature can be perceived in a gradual way. Even a line segment or a dot might be perceived more or less clearly. Yet, based on the data, it was difficult to rule out that the reports of partial awareness were caused by participants perceiving, e.g. one line of a geometric figure. For this reason, very simple stimuli were used in the 2006 study: Participants were presented with small grey lines on a white background. Most grey lines were tilted 45° clockwise from vertical, but on each trial a small group of lines in one quadrant of the display were instead tilted 270°. This group of lines was the target, and the participants were asked to report in which quadrant the target appeared and subsequently rate their awareness of the target dichotomously or using the PAS.

The results replicated the earlier findings that accuracy increased as a function of PAS rating. Furthermore, accuracy was higher when participants rated a stimulus as unseen on the dichotomous scale than when they rated it as unseen on PAS (35 vs. 31 %)—PAS thus proving more exhaustive than a dichotomous scale—and accuracy was found to be a lot lower when participants rated a stimulus as seen on the dichotomous scale than on PAS (78 vs. 94 %). If consciousness is always all-or-none, there would be little or no reason that these differences should be observed, and Overgaard et al. [22] thus concluded that there is evidence that consciousness is a gradual phenomenon even when very simple stimuli are used.

The difference between using a dichotomous measure of awareness and a 4-step measure (PAS) has also been examined in blindsight. Blindsight patients arguably report no visual experiences in a part of their visual field corresponding to a neural injury to V1, and thus consider themselves to be (partially) blind. Nevertheless, in certain laboratory tests they perform well above chance when performing forced-choice tasks on visual stimulation in the blind field [24]. The common interpretation is that vision can occur in the complete absence of awareness, yet a few papers have reported the presence of weak visual experiences in blindsight [25, 26]. Having observed that PAS seemed more exhaustive than a dichotomous measure Overgaard et al. [27] examined whether different conclusions would be reached if a blindsight patient was tested with PAS rather than a dichotomous measure.

They examined a patient with damage to her left visual cortex and apparent corresponding blindness in her right visual field. In a first experiment, they presented her with letters in different parts of the visual field. She failed to report any letters displayed in the upper right quadrant in spite of successfully reporting seeing all letters presented elsewhere. In a second experiment, the patient was presented with geometric figures (in the healthy as well as the injured field) and asked to guess which one was presented on each trial and subsequently rate her awareness on a dichotomous scale. The typical blindsight findings were replicated; on the trials on which the patient reported a stimulus as 'not seen' in her injured field she nevertheless performed well above chance (46 % vs. a chance level of 33 %) and accuracy did not vary significantly as a function of awareness rating. However, when PAS was used in a third experiment, a strong relationship between awareness rating and accuracy was observed, and it seemed very similar to the relationship in the intact field. In other words, it was shown that at least for that particular blindsight patient, the observed above-chance performance was better explained by weak experiences than intact processing in the absence of awareness when PAS was used, thus indicating that the 4-point awareness scale with each step labelled by participants was more exhaustive than a dichotomous awareness scale.

We are aware of only a few studies examining the impact of different confidence scales on consciousness and these give somewhat mixed results. These studies all employ artificial grammar learning and test for conscious/unconscious knowledge.

In this paradigm, participants are typically first asked to memorise a large number of letter strings. The participants are subsequently told that each string obeyed one of two complex rules, and in the second part of the experiment they are presented with the strings again and for each string they are asked to indicate if they believe it obeyed rule a or b. After each classification, participants report their confidence, and the relationship between confidence and accuracy can be examined. Quite surprisingly, Tunney and Shanks [28] and Tunney [29] found that a binary 'high/low' scale was able to detect differences between conscious and unconscious processing, whereas a continuous scale from 50 to 100 (indicating the estimated accuracy from chance to complete certainty) was not. Both studies, however, used a very difficult task (accuracy ~55 %). Dienes [30] hypothesised that the results might reflect that our judgments of confidence are not numerical as such, and converting our feelings of confidence to a number might be a noisy process.

Dienes [30] repeated the experiment using six different scales (a 'high/low' scale, a 'guess/no guess' scale, a '50–100' with and without descriptions of what the numbers mean, a 'numerical categories 50–100' scale where it is only possible to report 50, 51–59, 60–69, …, 90–99, 100, and finally a 6-step scale with verbal categories). Using the same stimuli and Tunney and Shanks, he found that overall, there was no difference between scales—the only scale seemingly (but not significantly) performing slightly worse was the numerical categories scale. He concluded that the type of scale made little difference, at least in a very difficult task. Using easier stimuli, a comparison between the sensitivity of a 'high/low' and a '50–100' scale gave the opposite result of Tunney and Shanks, i.e. that a more fine-grained scale was more sensitive.

Taken together, the studies discussed above indicate that when introspecting, dichotomous scales are suboptimal in many cases and the subjective measure in visual awareness tasks seems to benefit from allowing the participants to define the number of scale steps and their description (or at least using a scale that is created by participants in similar studies rather than one created arbitrarily by the experimenter). For scales using metacognitive judgments of classification performance, only artificial grammar learning has been examined. Here, it seems that when task accuracy is very low, using a fine-grained scale makes no difference (or possibly it makes the results worse), whereas a fine-grained scale appears useful when the task is somewhat easier. In the following, we will discuss the research into scale comparison, i.e. whether a scale using purely introspection performs better or worse than the scales using metacognitive judgments about task accuracy.

4.2.3 The Impact of Scale Type

In different experiments, different measures of awareness have been used, yet very few studies have examined which of the scales in use is the most exhaustive. When examining conscious experience, the most intuitive thing to ask about is probably just that, conscious experience (i.e. asking participants to introspect on their experience). At least, this is the oldest method, and it is still used very frequently today (in the PAS experiments, for example). However, as mentioned above, introspection has often been criticised as inaccurate, and many scientists prefer measures that are functionally defined. Confidence ratings (CR) have been used as an alternative to ratings of experience in part because they do not ask participants to rate their experience as such, but instead ask participants about their insight into a cognitive process. Additionally, such measures can be used across many different paradigms (e.g. the same confidence scale can be used whether the participant is viewing geometric figures, viewing motion or even performing artificial grammar learning). CR have been used either with respect to perception itself, in which case participants report their confidence in having perceived something (note that this type of scale has some introspective qualities) [31, 32], or with respect to participants' performance, in which case they report their confidence in having provided a correct answer [15, 30, 33].

Asking participants to place a wager on their classification has been used as an alternative to ratings of experience or confidence. Originally, gambling was used in cases where the participants could not be expected to understand a confidence scale; Ruffman et al. [34] used it with 3-year olds, and Shields et al. [35], Kornell et al. [36], and Kiani and Shadlen [37] used it with rhesus monkeys. Recently, Persaud et al. [16] used post-decision wagering (PDW) with adult humans (normal participants and a blindsight patient), and they argued that it should be the preferred method because it required, according to them, no introspection (i.e. it was claimed to be an objective measure) and the prospect of making money motivates participants to reveal all knowledge. PDW was thus claimed to be the measure with the highest degree of exhaustiveness, yet no direct comparison was made regarding exhaustiveness in terms of performance at the subjective threshold and the correlation between accuracy and wagers/awareness rating. The claim that PDW is an objective measure was quickly challenged by Seth [38] who argued that PDW required metacognitive judgments just like any confidence scale, and Clifford et al. [39] argued that the unconscious processing indicated in the experiments of Persaud et al. was likely to be a consequence of whatever criterion the participants set for when to wager high (in other words, the results could be caused by suboptimal exhaustiveness). Additionally, Sandberg et al. [17] drew attention to the fact that the use of PDW with real money alters the performance of the objective classification task.

In order to empirically test the claims made in the PDW papers, Dienes and Seth [40] compared a PDW scale with another metacognitive measure that did not encourage participants to introspect directly (a confidence scale) in the context of an artificial grammar learning paradigm. Dienes and Seth wanted to examine if PDW is indeed a better or more objective measure of awareness than CR scales when participants belong to a population that is expected to be able to understand and use CR scales (in this case adult humans with well-developed linguistic abilities). In addition to simply comparing the scales, they administered a test of risk aversion in order to examine if PDW was more closely linked to risk aversion than CR.

Dienes and Seth performed two experiments. In their first experiment, they simply asked participants to assign letter strings to one of two categories and either rate their confidence in being correct or wager one or two sweets. Although CR performed numerically better, they found no statistically significant difference between groups for the amount of unconscious processing at the subjective threshold or for the correlation between confidence rating and accuracy. They also examined a different measure closely related to confidence-accuracy correlations, type 2 d', and found no difference here either. Interestingly, however, they did observe a negative correlation between risk aversion and type 2 d' and marginally so between risk aversion and confidence-accuracy correlation when participants used wagering, but not when they used CR. In other words, the more risk averse you are, the smaller the estimated amount of conscious knowledge as measured by a wagering scale, but not by a confidence scale.

In their second experiment, Dienes and Seth changed their wagering procedure so participants could no longer lose anything. After making their classification, participants could choose to stick to their decision and gain a sweet if they were correct or simply randomly determine if they would get a sweet or not by drawing a card (50 % probability). In this case, the lower step on the scale is clearly associated with complete chance performance, and in this way it can be said that participants are instructed to use this scale step only when they believe they are completely at chance—i.e. the scale now resembles a standard confidence scale. Not surprisingly, this manipulation resulted in participants using the scale very similarly to how the confidence scale was used in Experiment 1, and the correlation between risk aversion and conscious knowledge disappeared. The correlation between accuracy and awareness was also marginally higher than for PDW in Experiment 1 (but no different than for CR). The overall conclusion of Dienes and Seth's study is thus that PDW does not show superior performance in artificial grammar learning paradigms, and if scientists want to avoid risk aversion influencing the experimental outcome, they may have to alter the instructions to make the PDW scale very similar to a confidence scale. For this reason, a confidence scale seems preferable compared to a PDW scale whenever participants can be expected to be able to use such a scale.

In contrast to Dienes and Seth, Sandberg et al. [17] used a visual paradigm and they examined not only confidence about being correct and wagering, but also reports of the perceptual experience as revealed introspectively on the PAS. Participants were asked to discriminate briefly presented geometric figures (four choices), and subsequently rate their awareness on one of three scales (PDW, CR, PAS). All scales had four scale steps. Interestingly, Sandberg et al. [17] found that PAS indicated the lowest accuracy at the lowest scale step. An accuracy level of 27.9 % was found, only just significantly above chance, 25 %, when uncorrected for multiple comparisons. In comparison, accuracies of 36.6 % was found for CR, and 42 % was found for PDW. The correlation between accuracy and awareness rating was also examined for all scales. Again, PAS gave the best results; the best overall correlation was found and the ratings were used more consistently across stimulus durations (i.e. each awareness rating was more consistently related to a particular accuracy rating than in other scales). The experiment thus confirmed the findings of Dienes and Seth [40] that CR performs similar to or slightly better than PDW, but in a different paradigm. Additionally, the experiment showed that introspective reports of awareness performed better than either of the other two scales.

One possible reason for the results could be the response distribution. When participants use PAS, they use all scale steps roughly the same number of times, whereas participants used PDW and to some extent CR more dichotomously. This results in awareness ratings of 1 and 4 being used across many different accuracy levels, thus giving a worse correlation and poor exhaustiveness at awareness ratings of 1. The failure to increase wagers can be explained by risk aversion as shown by Dienes and Seth [40], yet this cannot explain why PAS performs better than CR. One straightforward, yet somewhat controversial (cf. Dienes and Seth [41] and Timmermans et al. [42]) explanation is that participants are quite able to report small differences in experience, but they have no knowledge that these small differences are significant enough to alter performance. In this context, it is interesting to note that CR and PDW requires metacognitive insight into the classification process, which is exactly what Nisbett and Wilson showed to be unreliable, whereas PAS does not. In other words, reporting confidence in the response might be considered a harder task than reporting awareness of the stimulus, and performing this task optimally requires some degree of introspection along with a successful additional metacognitive process of relating experience and accuracy, a skill that participants might not always possess. This hypothesis is not easily tested, but experimental designs might be proposed drawing on the fact that metacognition is a skill (individual differences and specific neural correlates are found [1, 43]). Thus, if CR and PDW reports tax metacognitive skills to a higher degree than introspection, they might be affected more by distracter tasks or pressure to report quickly.

An introspective measure has also been compared with PDW by Nieuwenhuis and de Kleijn [44]. They examined the attentional blink in four experiments using an awareness scale and wagering. During the attentional blink, the ability to detect a target stimulus, T2, is impaired by presenting it shortly after another target, T1, in a series of rapidly presented stimuli. Like Overgaard et al. [22], Nieuwenhuis and de Kleijn wanted to examine the claim that consciousness is an all-or-none phenomenon. Sergent and Dehaene [23] had found that this seemed to be the case at least during the attentional blink, and so far a continuous transition from unconscious to conscious processing had not been found in this paradigm. Sergent and Dehaene used a 21-point scale, which has been criticised for being confusing for participants [22]. Nieuwenhuis and de Kleijn reduced the number of scale steps to 7 (still an arbitrary, although lower number) in their first experiment, and replicated the findings of Sergent and Dehaene. When they used wagering in their second experiment, however, the dichotomous response pattern disappeared to some extent. This is somewhat surprising as wagering was found to be a poor measure in the experiments mentioned above. Inspection of the tasks as well as the PAS and the wagering scale, reveals some explanation for this.

The attentional blink is, in most cases, presented as a detection task, not an identification task, which is the type of task the PAS, for instance, is developed for. Nieuwenhuis and de Kleijn asked participants to perform a discrimination task for T1 while simply reporting the clarity of T2, which may or may not be present (when not present, a blank slide was shown instead). In this case, the awareness rating and the wagering response relates to a presence/absence judgment (a detection task), which is nevertheless not explicitly performed. For wagering, participants were allowed to wager three different amounts on the absence or presence of T2 and even not to wager, whereas awareness ratings were made on a single scale from 'not seen' to 'maximal visibility'. The lowest awareness rating 'not seen' thus covered anything from complete certainty that nothing was displayed (e.g. completely clear perception of the empty slide) to no awareness of anything. In other words, the lowest step on the PAS corresponded to a combination of 4 steps on the wagering scale. The two scales, in this sense, were not comparable, and it seems there is a need to construct a PAS which can be used in detection tasks. All this considered, Nieuwenhuis and de Kleijn nevertheless demonstrated a continuous transition between conscious and unconscious processing in their second experiment. However, as no attentional blink was found in their second experiment, they performed two additional experiments for which task difficulty was increased by changing both target single digit numbers and distractors to single letters. In these experiments, participants were also asked to identify both T1 and T2 after rating their awareness or placing their wager (still based on present/absent judgements). With increased task difficulty, a continuous transition pattern was found for both wagering and PASs. In this case, a PDW scale seemed to perform as well as or slightly better than an awareness scale with an arbitrary number of steps, which was not generated with a detection task in mind.

4.2.4 The Impact of Stimulus Intensity

Interestingly, all types of scales (introspective, confidence or wagering) indicate that unconscious processing occurs at very specific stimulus intensities. In the experiment by Sandberg et al. [17], they found that stimuli had to be presented for 50 ms for participants to be able to identify them at a rate above chance. When stimuli were presented for around 130 ms (or more), participants were able to identify them with an accuracy of almost 100 %. Unconscious perception, as indicated by subjective threshold analysis (task accuracy when participants claim to have no awareness), is plotted in Fig. 4.1a. Here it can be seen that all unconscious processing is found within this time window (i.e. unconscious perception starts when performance deviates from chance and disappears again shortly after peak performance is reached). However, one large problem for the analysis at the subjective threshold is that it is difficult to conclude much about unconscious processing at high stimulus intensities. The reason for this is that only

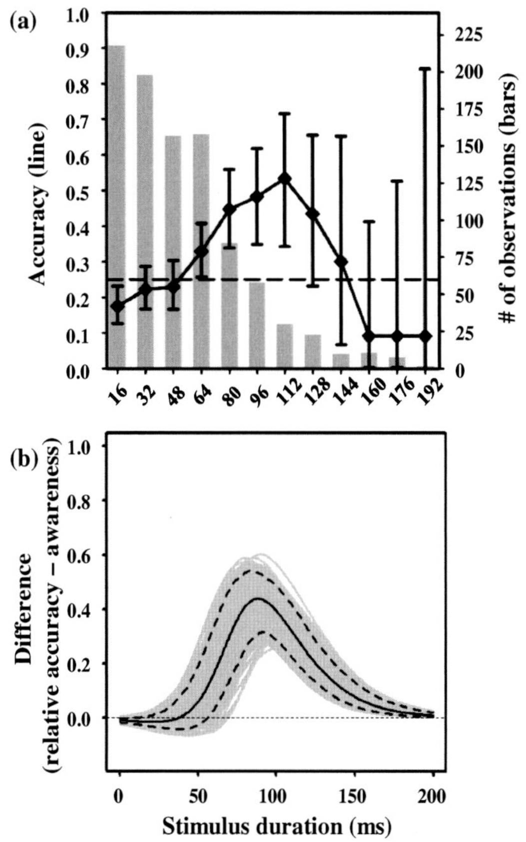

Fig. 4.1 A window of subliminal perception. Estimated subliminal perception across stimulus duration in a visual identification task across 12 participants using PAS using two different methods of analysis. Unconscious perception can be calculated using the subjective threshold. Here, unconscious perception accounts for any above-chance performance when the participant claims to have no experience of the stimulus. Unconscious perception can also be estimated from relative differences between average accuracy and awareness responses [45]. This method build on the fact that task accuracy and awareness can be described as sigmoid functions of stimulus duration (i.e. what has been called a psychometric and a conscious metric curve [46]. Using this method, unconscious perception is calculated by subtracting sigmoid accuracy functions from sigmoid awareness functions. **a** Subliminal perception calculated using the subjective threshold. Error bars indicate 95 % confidence intervals calculated from binomial distributions. **b** Subliminal perception calculated by subtracting the sigmoid awareness function from the sigmoid accuracy function. Note that both methods of analysis indicate unconscious perception to occur only for stimulus durations of around 50–150 ms

awareness ratings of 1 are used in the analysis, and when stimulus intensity is very high, participants rarely claim not to see anything (i.e. the number of observations dropped drastically when stimuli were presented for more than 100 ms as indicated by the bars in Fig. 4.1a). Yet, this possible 'window of subliminal perception' was confirmed by analysing the data in a different way.

Sandberg et al. [45] drew attention to the fact that accuracy and awareness in visual discrimination tasks can be described as sigmoid functions of stimulus intensity as had first been hypothesised by Koch and Preuschoff [46]. It has long been known that average task accuracy can be described as a sigmoid function (a standard psychometric curve), but Sandberg et al. [45] found this to be the case for awareness ratings as well. By fitting sigmoid functions to the average accuracy and awareness (taking into account variability between participants), group estimates of accuracy and awareness functions could be made. Sandberg et al. [45] found that in general, the awareness function lagged the accuracy function—which could be taken as an indication of unconscious perception. Their analysis confirmed that awareness was generally lagging accuracy, and that awareness increased more slowly than accuracy. In other words, accuracy and awareness functions start increasing from the bottom plateau at roughly the same time, yet the awareness function increases much more slowly.

Since every data point (i.e. accuracy and awareness rating for all trials, not just the ones where participants claimed no awareness) was used in the analysis, the overall confidence intervals with which the sigmoids were estimated were very small compared to the confidence intervals of subjective threshold approaches, and this allowed for additional analyses. In order to estimate the stimulus durations for which the awareness was lagging accuracy most clearly, the awareness function was subtracted from the accuracy function. The result of this subtraction is shown in Fig. 4.1, and it is clear that this method reveals a 'window of subliminal perception' that is quite similar to that found by use of the subjective threshold approach (although the confidence intervals are somewhat narrower for the curve estimations and the curve is smoother).

4.3 Concluding Discussion

Summing up the above, it appears that, at least for visual discrimination tasks, the best results are obtained using the measure drawing most heavily on introspection, i.e. the PAS, but not other metacognitive skills. In the one experiment [17] comparing PAS to CR and wagering, PAS indicated less subliminal perception as well as a better and more stable correlation between accuracy and awareness, while wagering performed the worst. This seems to indicate that PAS is unaffected by risk aversion (as are CR), and possibly the introspective task of reporting perceptual clarity is easier for participants than estimating accuracy, which may require both perceptual clarity as well as metacognitive knowledge of how this corresponds to different accuracy levels. At present, it is unclear whether introspection-based scales generated by participants also perform better in artificial grammar learning and attentional blink experiments.

Non-dichotomous introspective scales give more exhaustive results than dichotomous. They indicate less subliminal perception and they indicate a better correlation between accuracy and awareness. The scale steps on the 4-point PAS each correspond to accuracy levels that remain fairly stable across different conditions within the same experiment. Small amounts of unconscious processing seem to be found no matter which scale is used as the measure of awareness (although, in the early PAS experiments as well as in the blindsight experiment no significant above-chance performance was found).

In the review of recent findings using direct, introspective reports, it thus seems evident that the way in which participants are instructed to report has significant impact on results. This, we believe, is empirical support of the claim that introspective reports should be considered as distinct from other kinds of metacognitive reports, and that the distinctions shown in Table 4.1 is upheld. Further studies are necessary to conclude how strongly these distinctions should be interpreted.

In one interpretation, the distinction is represented at a neural level and captures a natural distinction of mental states. This version is cautiously suggested in Overgaard et al. [22] who report different neural activation patterns when asking participants to report introspectively about visual experiences in contrast to asking them to report in a non-introspective way (i.e. without attending directly to their experiences). However, a different interpretation would refrain from ontological commitments and limit the distinction to the methodological domain. According to this weaker version, we get different results with different instructions not because of distinctions in nature, but solely because of the differences in methodology. Although this conclusion immediately seems simpler to draw, it does demand a different ontology able to account for the empirical findings.

Regardless of the choice of interpretation, we believe that we have shown that some distinctions must be drawn between cognitive and conscious states, metacognition and introspection, and that these distinctions have important implications for how to think about and experiment on the mind.

Acknowledgments Both authors were supported by European Research Council.

References

1. Fleming SM, Weil RS, Nagy Z, Dolan RJ, Rees G (2010) Relating introspective accuracy to individual differences in brain structure. Science 329:1541–1543. doi:10.1126/science.1191883
2. Ramsøy TZ, Overgaard M (2004) Introspection and subliminal perception. Phenomenol Cogn Sci 3:1–23. doi:10.1023/B:PHEN.0000041900.30172.e8
3. Rounis E, Maniscalco B, Rothwell J, Passingham R, Lau H (2010) Theta-burst transcranial magnetic stimulation to the prefrontal cortex impairs metacognitive visual awareness. PCNS 1:165–175. doi:10.1080/17588921003632529
4. Neisser U (1967) Cognitive psychology. Meredith, New York
5. Shimamura AP (2000) Toward a cognitive neuroscience of metacognition. Conscious Cogn 9:313–323. doi:10.1006/ccog.2000.0450 (discussion 324–326)
6. Lyons W (1986) The disappearance of introspection. MIT Press, Cambridge

7. Marcel A (2003) Introspective report: trust, self knowledge and science. J Conscious Stud 10:167–186
8. Overgaard M (2006) Introspection in science. Conscious Cogn 15:629–633. doi:10.1016/j.concog.2006.10.004
9. Nisbett RE, Wilson TD (1977) Telling more than we can know: verbal reports on mental processes. Psychol Rev 84:231–259. doi:10.1037/0033-295X.84.3.231
10. Wilson TD (2003) Knowing when to ask: introspection and the adaptive unconscious. J Conscious Stud 10:9–10
11. Reingold EM, Merikle PM (1988) Using direct and indirect measures to study perception without awareness. Percept Psychophys 44:563–575
12. Sidis B (1898) The psychology of suggestion. Appleton, New York
13. Schwiedrzik CM, Singer W, Melloni L (2011) Subjective and objective learning effects dissociate in space and in time. Proc Natl Acad Sci 108:4506–4511. doi:10.1073/pnas.1009147108
14. Snodgrass M, Shevrin H (2006) Unconscious inhibition and facilitation at the objective detection threshold: replicable and qualitatively different unconscious perceptual effects. Cognition 101:43–79. doi:10.1016/j.cognition.2005.06.006
15. Dienes Z, Altmann GTM, Kwan L, Goode A (1995) Unconscious knowledge of artificial grammars is applied strategically. J Exp Psychol Learn Mem Cogn 21:1322–1338. doi:10.1037/0278-7393.21.5.1322
16. Persaud N, McLeod P, Cowey A (2007) Post-decision wagering objectively measures awareness. Nat Neurosci 10:257–261. doi:10.1038/nn1840
17. Sandberg K, Timmermans B, Overgaard M, Cleeremans A (2010) Measuring consciousness: is one measure better than the other? Conscious Cogn 19:1069–1078. doi:10.1016/j.concog.2009.12.013
18. Merikle PM (1982) Unconscious perception revisited. Percept Psychophys 31:298–301
19. Merikle PM, Joordens S (1997) Measuring unconscious influences. In: Cohen J, Schooler J (eds) Scientific approaches to consciousness. Erlbaum, Mahwah, pp 109–123
20. Reingold EM, Merikle PM (1990) On the Inter-relatedness of theory and measurement in the study of unconscious processes. Mind Lang 5:9–28. doi:10.1111/j.1468-0017.1990.tb00150.x
21. Snodgrass M (2002) Disambiguating conscious and unconscious influences: do exclusion paradigms demonstrate unconscious perception? Am J Psychol 115:545–579
22. Overgaard M, Rote J, Mouridsen K, Ramsøy TZ (2006) Is conscious perception gradual or dichotomous? A comparison of report methodologies during a visual task. Conscious Cogn 15:700–708. doi:10.1016/j.concog.2006.04.002
23. Sergent C, Dehaene S (2004) Is consciousness a gradual phenomenon? Evidence for an all-or-none bifurcation during the attentional blink. Psychol Sci 15:720–728. doi:10.1111/j.0956-7976.2004.00748.x
24. Sanders MD, Warrington EK, Marshall J, Weiskrantz L (1974) "Blindsight": vision in a field defect. Lancet 1:707–708
25. Stoerig P, Barth E (2001) Low-level phenomenal vision despite unilateral destruction of primary visual cortex. Conscious Cogn 10:574–587. doi:10.1006/ccog.2001.0526
26. Zeki S, Ffytche DH (1998) The Riddoch syndrome: insights into the neurobiology of conscious vision. Brain 121(Pt 1):25–45
27. Overgaard M, Fehl K, Mouridsen K, Bergholt B, Cleeremans A (2008) Seeing without Seeing? Degraded conscious vision in a blindsight patient. PLoS ONE 3:e3028. doi:10.1371/journal.pone.0003028
28. Tunney RJ, Shanks DR (2003) Does opposition logic provide evidence for conscious and unconscious processes in artificial grammar learning? Conscious Cogn 12:201–218
29. Tunney RJ (2005) Sources of confidence judgments in implicit cognition. Psychon Bull Rev 12:367–373
30. Dienes Z (2008) Subjective measures of unconscious knowledge. Prog Brain Res 168:49–64. doi:10.1016/S0079-6123(07)68005-4

31. Bernstein IH, Eriksen CW (1965) Effects of "subliminal" prompting on paired-associate learning 1:33–38
32. Cheesman J, Merikle PM (1986) Distinguishing conscious from unconscious perceptual processes. Can J Psychol Rev Can Psychol 40:343–367. doi:10.1037/h0080103
33. Scott RB, Dienes Z (2008) The conscious, the unconscious, and familiarity. J Exp Psychol Learn Mem Cogn 34:1264–1288. doi:10.1037/a0012943
34. Ruffman T, Garnham W, Import A, Connolly D (2001) Does eye gaze indicate implicit knowledge of false belief? Charting transitions in knowledge. J Exp Child Psychol 80:201–224. doi:10.1006/jecp.2001.2633
35. Shields WE, Smith JD, Guttmannova K, Washburn DA (2005) Confidence judgments by humans and rhesus monkeys. J Gen Psychol 132:165–186
36. Kornell N, Son LK, Terrace HS (2007) Transfer of metacognitive skills and hint seeking in monkeys. Psychol Sci 18:64–71. doi:10.1111/j.1467-9280.2007.01850.x
37. Kiani R, Shadlen MN (2009) Representation of confidence associated with a decision by neurons in the parietal cortex. Science 324:759–764. doi:10.1126/science.1169405
38. Seth AK (2008) Post-decision wagering measures metacognitive content, not sensory consciousness. Conscious Cogn 17:981–983. doi:10.1016/j.concog.2007.05.008
39. Clifford C, Arabzadeh E, Harris JA (2008) Getting technical about awareness. Trends Cogn Sci 12:54–58. doi:10.1016/j.tics.2007.11.009 (Regul. Ed.)
40. Dienes Z and Seth A (2009) Gambling on the unconscious: a comparison of wagering and confidence ratings as measures of awareness in an artificial grammar task. Conscious Cogn. doi:10.1016/j.concog.2009.09.009
41. Dienes Z and Seth AK (2010) Measuring any conscious content versus measuring the relevant conscious content: Comment on Sandberg et al. Conscious Cogn. doi:10.1016/j.concog.2010.03.009
42. Timmermans B, Sandberg K, Cleeremans A, Overgaard M (2010) Partial awareness distinguishes between measuring conscious perception and conscious content: reply to Dienes and Seth. Conscious Cogn 19:1081–1083. doi:10.1016/j.concog.2010.05.006
43. Song C, Kanai R, Fleming SM, Weil RS, Schwarzkopf DS, Rees G (2011) Relating inter-individual differences in metacognitive performance on different perceptual tasks. Conscious Cogn. doi:10.1016/j.concog.2010.12.011
44. Nieuwenhuis S, Kleijn R (2010) Consciousness of targets during the attentional blink: a gradual or all-or-none dimension? Atten Percept Psychophys 73:364–373. doi:10.3758/s13414-010-0026-1
45. Sandberg K, Bibby BM, Timmermans B, Cleeremans A, Overgaard M (2011) Measuring consciousness: task accuracy and awareness as sigmoid functions of stimulus duration. Conscious Cogn 20:1659–1675. doi:10.1016/j.concog.2011.09.002
46. Koch C, Preuschoff K (2007) Betting the house on consciousness. Nat Neurosci 10:140–141. doi:10.1038/nn0207-140

Chapter 5
The Highs and Lows of Theoretical Interpretation in Animal-Metacognition Research

J. David Smith, Justin J. Couchman and Michael J. Beran

Abstract Humans feel uncertain. They know when they do not know. These feelings and the responses to them ground the research literature on metacognition. It is a natural question whether animals share this cognitive capacity, and thus animal metacognition has become an influential research area within comparative psychology. Researchers have explored this question by testing many species using perception and memory paradigms. There is an emerging consensus that animals share functional parallels with humans' conscious metacognition. Of course, this research area poses difficult issues of scientific inference. How firmly should we hold the line in insisting that animals' performances are low-level and associative? How high should we set the bar for concluding that animals share metacognitive capacities with humans? This area offers a constructive case study for considering theoretical problems that often confront comparative psychologists. The authors present this case study and address diverse issues of scientific judgment and interpretation within comparative psychology.

Keywords Metacognition · Uncertainty monitoring · Metamemory · Comparative cognition · Decision-making

---

This chapter is adapted from: Smith JD, Couchman JJ, Beran MJ (2012) The highs and lows of theoretical interpretation in animal-metacognition research. Phil Trans R Soc B 367:1297–1309

J. D. Smith (✉)
Department of Psychology and Center for Cognitive Science, The University at Buffalo, State University of New York, Buffalo, NY 14260-4110, USA
e-mail: psysmith@buffalo.edu

J. J. Couchman
Department of Psychology, Albright College, Reading, PA 19604, USA

M. J. Beran
Language Research Center, Georgia State University, Atlanta, GA 30302, USA

S. M. Fleming and C. D. Frith (eds.), *The Cognitive Neuroscience of Metacognition*,
DOI: 10.1007/978-3-642-45190-4_5,

5.1 Introduction

Humans know when they do not know or remember. They respond well to uncertainty by deferring response and seeking information—for example, they Google. Humans' responses to uncertainty ground the literature on metacognition [1–6]. Metacognition is defined to be the monitoring and control of basic perceptual and cognitive processes. The theoretical assumption is that some minds can deploy a cognitive executive that oversees and optimizes thought and problem-solving. Researchers assess these metacognitive functions in humans by collecting judgments of confidence, feelings of knowing and tip-of-the-tongue experiences.

Humans' metacognitive capacity is linked to sophisticated aspects of mind. Metacognition can reveal a hierarchical structure to cognition, because often metacognitive processes regulate lower-level cognitive processes [7]. Metacognition reveals humans' awareness of their cognition [8, 9], because humans often reflect consciously on their cognitive states and declare them to others. Metacognition may also reveal humans' self awareness [10], because states like uncertainty are often imbued with self feelings (e.g., I don't know).

Metacognition's sophistication raises the question of whether it is uniquely human, and one of comparative psychology's current goals is to establish whether nonhuman animals (hereafter, animals) share this capacity [11–13]. If they do, it could bear on their consciousness and self-awareness [14], and it would affect many theoretical debates within comparative psychology. Given the question's importance, Smith and his colleagues inaugurated research on animal metacognition [15–19]. This area has been reviewed [11–13, 20], and active research continues in this area [21–39].

To explore animal metacognition, one cannot just adopt the usual human measures like feelings of knowing and tip-of-the-tongue states. Animals have no way to declare these states and feelings. For the same reason, the usual human measures are not suitable for young humans [40]. Instead, comparative researchers have built behavioral tasks that have two components. First, researchers make some trials difficult, to stir up something like an uncertainty state in animal minds. Second, researchers give animals a response apart from the task's primary discrimination responses that lets them decline to complete any trials they choose. This uncertainty response lets animals manage uncertainty and declare it behaviorally/observably. If animals are metacognitive and monitor internal uncertainty states or internal assessments of the probability of responding correctly, they should recognize difficult trials as doubtful or error-causing and decline those trials proactively and adaptively.

We will illustrate this approach to studying animal metacognition, introducing readers to the area and noting features of the experiments that raise issues of high- and low-level theoretical interpretation. Above all, we will have to deal with the question of whether animals employ metacognitive or nonmetacognitive strategies to avoid difficult trials. In the inaugural study [16], a dolphin (*Tursiops truncatus*) made a "high" response to a 2,100 Hz tone or a "low" response to any lower tone

(1,200–2,099 Hz). The frequency (Hz) of low trials was adjusted to constantly challenge the animal's psychophysical limit and to maximize difficulty and uncertainty within the task. The animal could respond "uncertain" to decline the trials he chose. Figure 5.1a shows that he assessed correctly when he was at risk for error, selectively declining the difficult trials near threshold. His uncertainty responses peaked near 2,086 Hz, 0.11 semitones from the standard high tone. Humans perform similarly in this task, and humans say their uncertainty responses reflect their high-level, conscious, metacognitive states of uncertainty.

The dolphin said nothing, but produced distinctive uncertainty behaviors. We carried out a factor analytic study of his ancillary behaviors on trials of different pitch. His hesitation–wavering behaviors peaked at his perceptual threshold, too (Fig. 5.1b). These hesitation behaviors could be additional behavioral symptoms of uncertainty.

Tolman [41] appreciated these hesitation behaviors— that he called "lookings and runnings back and forth"—because he thought they might operationalize animal consciousness for the behaviorist. That is a provocative way to frame this field's main question: do we accept Tolman's definition, and grant the dolphin high-level, metacognitive uncertainty in this perceptual task, or not?

This is a difficult problem of scientific inference. Analogous questions have attended the study of animals' counting, language, timing, self-awareness, theory of mind, and so forth. In fact, the animal-metacognition literature is a good case study in this inference problem. It raises issues that generalize constructively to other comparative research domains. We present this case study here.

5.2 Testing Low-Level Interpretations of Animal Metacognition

The dolphin met the criteria described by Hampton [29] for a metacognitive performance. There was an observable behavior (high and low responses) that could be scored as (in)correct. There was variation in the accuracy of primary responding across trial levels so that accuracy could be correlated to the use of the secondary, metacognitive response. There was a secondary, observable behavior (the uncertainty response) that might reflect monitoring processes overseeing the animal's primary responding. This secondary behavior was strongly (negatively) correlated across trial levels with the accuracy of the primary responses.

Yet, one is hesitant to immediately credit animals with high-level metacognitive capacities. Animal-metacognition research, like all comparative research, bears an interpretative burden given the tradition of explaining animals' behavior at the lowest psychological level [42]. Therefore, even given possibly metacognitive performances by some species, one must ask whether they might be explained using low-level, associative mechanisms. In fact, the possible low-level bases for uncertainty responses by animals—that is, the possibilities that these

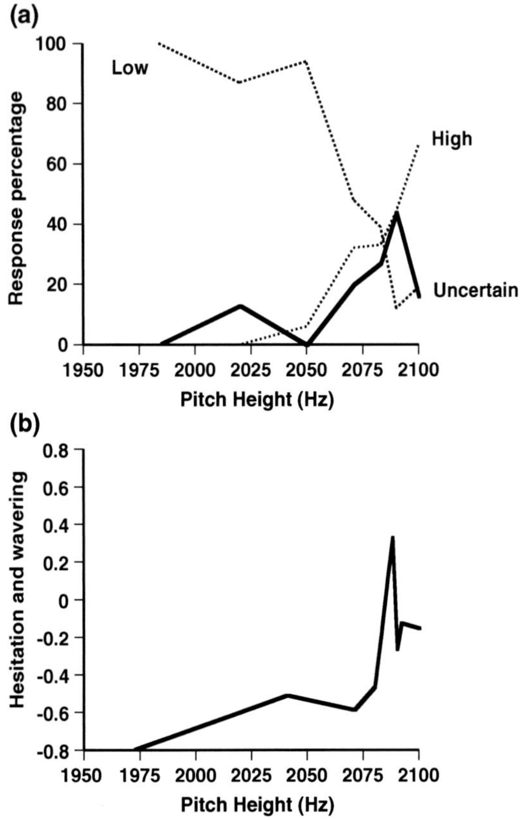

Fig. 5.1 **a** Performance by a dolphin in an auditory discrimination [16]. The dolphin swam to touch a *low* or *high* response. In addition, he could make an Uncertainty response to decline the trial. The pitch of the low tones was adjusted dynamically to titrate the dolphin's perceptual limit for distinguishing *low* and *high* tones and to examine his response pattern in detail within this region of maximum difficulty. The *horizontal axis* indicates the frequency (Hz) of the trial. The *low* and *high* response, respectively, was correct for frequencies of 1,200–2,099 Hz and 2,100 Hz. The latter trials are plotted as the *rightmost* data point for each *curve*. The *solid line* represents the percentage of trials receiving the uncertainty response at each pitch level. The percentage of trials ending with the low response (*dotted line*) or High response (*dashed line*) are also shown. **b** Four raters judged how much the dolphin slowed, wavered and hesitated for the trials within four video-taped sessions. Factor analysis was used to discern the simpler structure behind the four sets of ratings. The figure shows the dolphin's weighted overall Factor 1 behavior (*hesitancy, slowing, wavering*) for tones of different frequencies (Hz). Reprinted with permission from Smith et al. [16, p. 399, 402]. Copyright © 1995 by the American Psychological Association

responses were elicited by stimuli or entrained by reinforcement contingencies—was the principal theoretical issue through the first decade of animal-metacognition research [29, 34, 43]. To the extent that animals' uncertainty responses are triggered reactively by stimulus cues, by reinforcement histories, and the like, one would conclude that animals' uncertainty systems present a weaker analogue to human uncertainty, and that animals are not metacognitive. To the extent that animals' uncertainty responses turn out to be more highly cognitive—more executive, more controlled, more deliberate, perhaps even conscious—one would conclude that animals' uncertainty systems present a stronger analogue to human uncertainty, and that animals are metacognitive. Consequently, research has focused sharply on the low-level cues and processes that animals might use to achieve metacognitive performances, and on whether these can be disconfirmed as the behavioral cause of those performances.

In this section, we describe this decade of research, including the theoretical concerns raised and the empirical answers offered. Some aspects of the resulting low-level/high-level dialog represent comparative psychology at its best. The associative criticisms were disciplined and testable. They provoked new paradigms. They produced consensual answers and theoretical development in the field.

5.2.1 Reinforced Uncertainty Responses

One associative concern was that animals sometimes received food rewards or tokens for uncertainty responses [27, 28, 30, 36, 37, 44–46]. This approach could make the uncertainty response attractive solely for its reward properties, independent of any metacognitive role it plays in a task. This approach made it difficult to rule out low-level interpretations or to affirm metacognitive interpretations.

To address this concern, researchers removed the reward contingency for that response [19, 22, 35]. In one case [22], macaques (*Macaca mulatta*) judged whether arrays were less or more numerous than a session-specific numerosity. Numerosities nearer the boundary value were more difficult to classify. The uncertainty response only cleared the old trial and brought the next, randomly chosen trial. It offered no food reward, food token, trial hint, or easy next trial for its use. But monkeys still made uncertainty responses selectively for the trials near the boundary value on which they would most probably err. The associative concern about the immediate appetitive attractiveness of the uncertainty response cannot explain this result.

5.2.2 Reactions to Primary Stimulus Qualities

Another associative concern was that difficulty level within uncertainty-monitoring tasks was often perfectly correlated with the objective stimulus level—for example, tone height (Hz) in the dolphin's discrimination. Some stimuli would have caused animals frequent errors and ensured frequent penalty timeouts and lean rewards. These stimuli could have become aversive, and avoided through a default response that some mistook as a metacognitive response. This theoretical consideration recalls Hampton's [29] concern that environmental cue associations could underlie seemingly metacognitive performances. In general, our typology of possible low-level descriptions is similar to Hampton's typology.

To address this concern, researchers lifted uncertainty-monitoring tasks off the plane of concrete stimuli. In one case [15], macaques were allowed to make uncertainty responses in a same–different task. A same–different task—testing generalization over variable and novel stimulus contexts—requires some degree of abstraction beyond the absolute stimulus qualities that carry the relation. This abstractness explains why true same–different performances appear to be phylogenetically restricted [47] and why even nonhuman primates have distinctive weaknesses in same–different performance [48, 49].

Accordingly, macaques made same or different responses to pairs of rectangles that had the same or different pixel densities. To cause them difficulty, the size of the density difference on different trials was adjusted in a threshold paradigm to constantly challenge subjects' discrimination abilities. Moreover, same and different trials at several absolute pixel-density levels were intermixed to ensure a true relational performance. Yet, the macaques (Fig. 5.2) used the uncertainty response essentially identically to humans (a 0.97 cross-species correlation of the behavioral profiles), producing one of the closest correspondences between animals' and humans' performance. Shields et al. even reserved some regions of absolute density for use in immediate generalization tests to confirm the macaques' generalizable same–different performance. This illustrates the constructive use of transfer tests to show the representational generality of macaques' uncertainty processes, which might also indicate their similarity to humans' uncertainty processes. Uncertainty responses cannot have been triggered by low-level stimulus cues, because the performance survived immediate transfer tests and because in any case the relevant cue was abstract. Rather, uncertainty responses had to be prompted by the indeterminacy of the same–different relation instantiated by difficult and highly variable stimulus pairs.

Hampton [44] explored macaques' metamemory using a delayed matching-to-sample task. With longer delays between sample presentation and match-choice selection, matching performance decreased because monkeys remembered the sample less well. Monkeys selectively declined memory tests at long retention intervals when they had mostly forgotten the sample. In addition, one monkey performed better at each delay level on the trials he chose to complete than on the trials he was forced to complete. This undermines the explanation that the monkey was just reacting to long delays with an escape response. Instead, it suggests that he was monitoring some psychological signal of (not) remembering. Both monkeys also responded uncertain more, no matter the length of the retention interval, when blank trials occurred with no sample shown, guaranteeing that macaques could not know what to match.

These monkeys cannot have been conditioned to avoid particular concrete stimuli. The memory targets only applied for a single trial, so longer term avoidance learning was useless. Moreover, the uncertainty response was made with no visible sample stimulus to trigger an avoidance response. These macaques

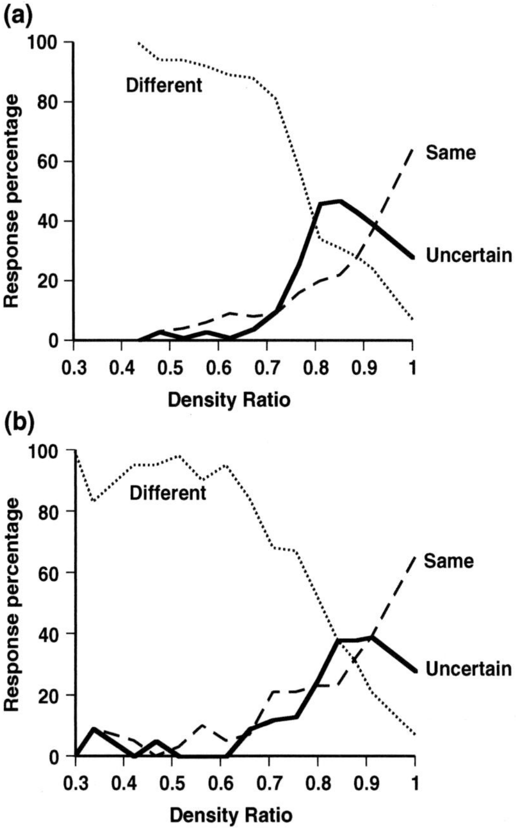

Fig. 5.2 **a** Performance by monkeys in a same–different discrimination [15]. The monkeys manipulated joysticks to make different or same responses, respectively, when two pixel boxes had the same or different internal density of lit pixels. In addition, they could make an uncertainty response to decline the trial. The pitch of the density difference on different trials was adjusted dynamically to titrate the monkeys' perceptual limit for distinguishing sameness from difference and to examine their response patterns in detail within this region of maximum difficulty. The *horizontal axis* gives the ratio between the densities of the two pixel boxes seen on each trial. The same response was correct for ratio 1—these trials are plotted as the rightmost data point for each curve. The different response was correct for all other trials. The *solid line* represents the percentage of trials receiving the uncertainty response at each density ratio. The percentages of trials ending with the different response (*dotted line*) or same response (*dashed line*) are also shown. **b** Performance by humans in the same–different discrimination, depicted in the same way. Reprinted with permission from Shields et al. [15, p. 158]. Copyright © 1997 by the American Psychological Association

showed a kind of metamemory. They appeared to monitor memory's contents to decline tests of weaker memories. This memory-strength signal—abstract, cognitive and nonassociative—is profoundly different from the stimulus signal available in traditional operant situations.

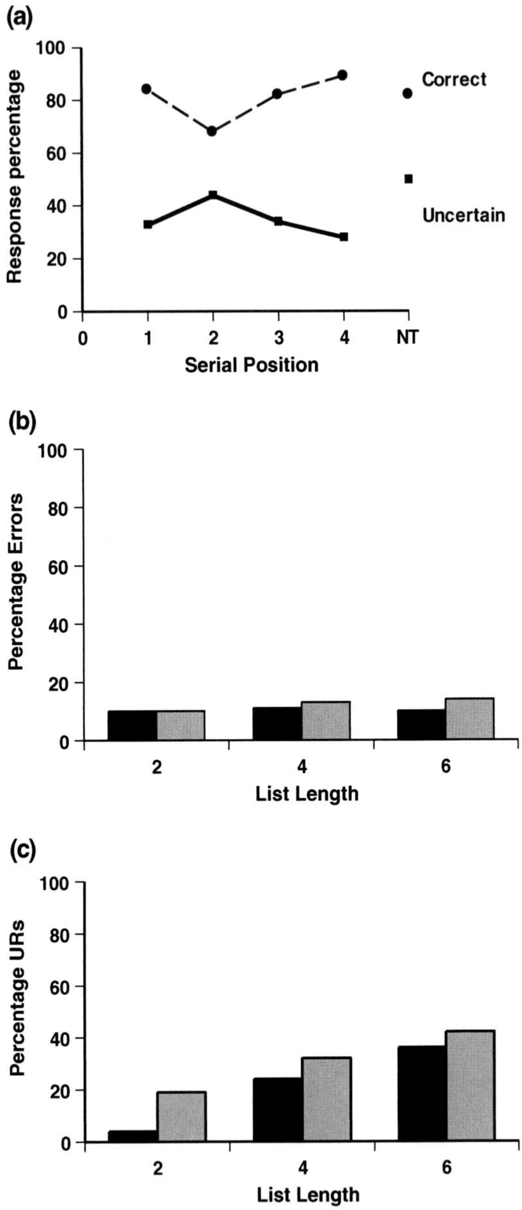

Fig. 5.3 **a** Performance by a macaque in a metamemory task [18]. NT denotes 'not there' trials in which the probe picture was not in the memory list of pictures. The serial position (*1–4*) of the probe picture in the list of pictures on 'there' trials is also given along the *x*-axis. The percentage of each type of trial that received the uncertainty response is shown (*solid line with squares*). The percentage correct (of trials on which the memory test was attempted) is also shown (*dashed line with circles*). Macaques responded "uncertain" most for the trials on which their memories were most indeterminate. **b** Percentage error rates by two monkeys (*black and gray bars*) when the difficulty of the memory test was increased by increasing the memory list from *2 to 6* pictures. **c** Percentage uncertainty responses (URs) by two monkeys when the difficulty of the memory test was increased in the same way. Reprinted with permission from Smith et al. [18, p. 236, 238]. Copyright © 1998 by the American Psychological Association

There are converging metamemory results [18, 30]. In one case [18], macaques proved able to adaptively decline memory tests of the most difficult serial positions in lists of to-be-remembered items (Fig. 5.3a).

In this experiment, they showed an additional kind of cognitive self-regulation. That is, as the metamemory task was made more difficult, macaques held their error rate near 10 % (Fig. 5.3b) by responding uncertain more in difficult task conditions (Fig. 5.3c). Thus, macaques accepted memory tests if they were 90 % certain of remembering. Finally, researchers have also assessed macaques' metamemory by using trans-cranial magnetic stimulation (TMS) to interfere with visual working memory [39]. TMS interfered with a macaque's matching-to-sample performance and also increased his uncertainty responding. It was not just some global TMS effect that caused this increase. The effect of TMS was hemisphere specific. That is, TMS caused an increase in uncertainty responding only when it occurred contralaterally to the presentation of the sample in the visual field (when it also probably maximally interfered with registering and remembering the to-be-remembered sample).

By dissociating difficulty and uncertainty from concrete stimuli, and thereby protecting against stimulus-based interpretations, all of these metamemory studies required animals to monitor uncertainty on a more abstract and cognitive level, producing results that the stimulus-based associative concern cannot explain.

5.2.3 Entrainment to Reinforcement Contingencies

A third concern centered on trial-by-trial feedback as always provided by the early paradigms. Every outcome could be associated with the stimulus–response pair that produced it. As also suggested by Hampton [29], animals might have been conditioned to make uncertainty responses when facing the trials that were associated with the worst stored reinforcement history.

To address this concern about entrained reinforcement gradients, researchers replaced trial-by-trial feedback with deferred feedback whereby animals worked for blocks of trials before receiving a performance evaluation [19, 26]. Moreover, in that evaluation, rewards and timeouts were bundled separately, so that the temporal sequence of trials completed and outcomes obtained bore no relation. This defeated the normal processes of association and conditioning. Animals could not know which trials had gone unreinforced through a clear and an immediate feedback signal, and so, based on the objective feedback of the task itself, they could not know which trials they had missed.

By this technique, researchers uncoupled objective performance from subjective difficulty. Animals had to set decision criteria and define response regions on their own—cognitively and decisionally. Yet, animals still made uncertainty responses proactively and adaptively under these circumstances. Emphasizing this uncoupling, the study by Smith et al. [19, p. 292, Fig. 8a, b] showed that there was no relationship between the proportion of uncertainty responses and the proportion of (in)correct responses across trial levels, as there would have to be if uncertainty responses were conditioned avoidance responses. Instead, there was a strong relationship between the proportion of uncertainty responses and the distance of the trial level from the animal's decisional breakpoint in the discrimination [19, p. 292, Fig. 8c, d], as there would be if the animal were monitoring difficulty or uncertainty.

These experiments dissociated for the first time strategies based in reinforcement history from strategies based in decisional difficulty. Animals can monitor uncertainty adaptively using the latter strategy. Therefore, the reinforcement-based concern about the uncertainty response is not fully justified. More generally, the deferred-feedback technique has broad potential applicability within comparative psychology. It forces animals to self-construe the task and to self-construct a task approach, and thus it provides a more cognitive read on their behavior.

5.2.4 Representational Specificity

Other research has tested the representational rigidity or narrowness of animals' uncertainty responses, expected if these responses are low-level behavioral reactions. For example, researchers [35] asked macaques to monitor uncertainty while multitasking. Four different difficult discriminations were randomly intermixed trial by trial. Despite this multitasking requirement, macaques were able to decline the difficult trials across domains. This shows that uncertainty responses are not reactive to just one well-trained trial type at a time. This kind of simultaneous transfer test suggests that uncertainty responses result from a general psychological signal that transcends a single task and perhaps is similar to the general psychological state of uncertainty that humans would bring to the same collection of discriminations.

Washburn et al. [38] tested another form of generalization in macaques' metacognitive performances, by asking whether they would respond uncertain adaptively on a novel task's first trial. The researchers adapted the learning-set paradigm [50], in which a new two-choice discrimination began every six trials. Macaques responded uncertain far more often on trial 1 of each problem than on trials 2–6, consistent with the fact that they could not know the answer on trial 1 but could know the answer on trials 2–6 (in this experiment, the uncertainty response revealed each discrimination's answer but gave no appetitive reward). This rapid, flexible application of the uncertainty response to new discrimination problems also strongly discourages associative interpretations and encourages metacognitive interpretations of the uncertainty response. It illustrates again the utility of using transfer to show the generality and flexibility with which some species use uncertainty-monitoring processes.

5.2.5 Phylogenetic Restrictions in Uncertainty Monitoring

Cross-species research also undermines lower-level, associative interpretations of the uncertainty response, sometimes through the failure of animals to respond adaptively. Capuchin monkeys (*Cebus apella*) represent another major primate

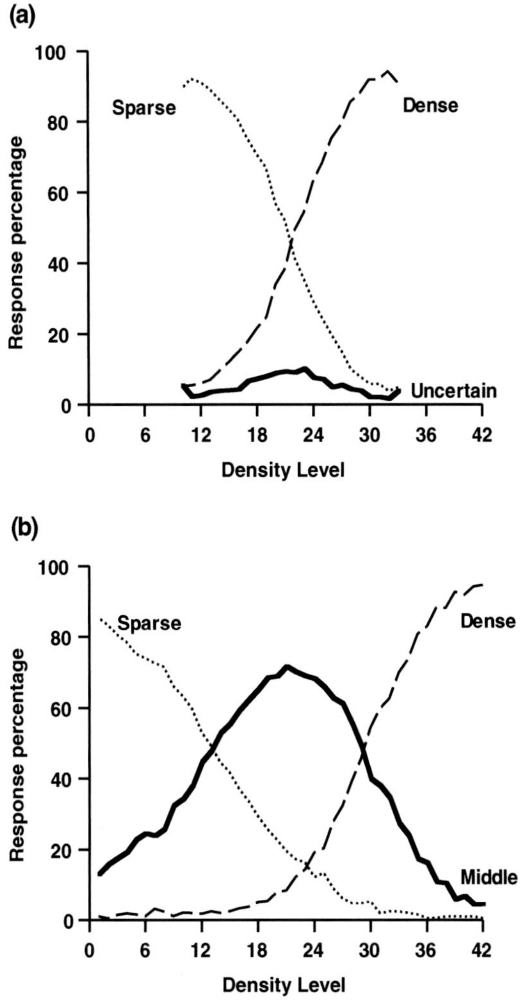

Fig. 5.4 **a** The performance of capuchin monkeys in a sparse–uncertain–dense task [23]. The *horizontal axis* indicates the density level of the box. The sparse and dense responses, respectively, were correct for boxes at density levels *1–21* and *22–42*. The *solid line* represents the percentage of trials receiving uncertainty responses at each trial level. The percentages of trials ending with the sparse response (*dotted line*) or dense response (*dashed line*) are also shown. **b** The performance of the same capuchin monkeys in the sparse–middle–dense task [23], depicted in a similar way. From Smith et al. [51, p. 48]

lineage (the New World primates). Researchers [23] tested capuchins' uncertainty monitoring along a sparse-to-dense perceptual continuum with the difficult and uncertain trials surrounding the discrimination's breakpoint. Strikingly, capuchins did not respond uncertain, though this sharply reduced their reward efficiency. A similar result was obtained when the error timeout was increased to 90 s, so that with each error capuchins potentially forfeited 30 trials and 30 food rewards (Fig. 5.4a).

In other sessions, capuchins performed a sparse–middle–dense task in which they could earn rewards or timeouts for (in)correctly making middle responses. Capuchins responded middle easily from the beginning of testing, in sharp contrast to their negligible uncertainty responding (Fig. 5.4b).

These two tasks were structured similarly—indeed, the same intermediate stimuli should have recruited middle and uncertain responses. Thus, the two tasks—strong mutual controls—produced a striking dissociation. The capuchins easily brought middle responses, but not uncertain responses—under the control of those intermediate stimuli.

Capuchins are such apt learners that they are known as the poor person's chimpanzee. If uncertainty responses were triggered by conflict, aversion, avoidance, fear, competing response strengths, reward maximization, hesitation-wavering behaviors, hesitation-wavering latencies, or any other first-order cue, capuchins would have used that cue to prompt adaptive uncertainty responses. Clearly, the mechanisms that underlie middle responding and uncertainty responding are different psychologically. And clearly, the uncertainty-monitoring capacities of capuchins and macaques are different as well, a conclusion that has been reached independently [21, 28, 31].

5.3 Interim Conclusion

The results considered in Sect. 5.2 have produced a growing consensus that some species have shown metacognition. 'Metamemory, the ability to report on memory strength, is clearly established in rhesus macaques (*M. mulatta*) by converging evidence from several paradigms' [37, p. 266]. 'Evidence for metacognition by nonhuman primates has been obtained in great apes and old world monkeys' [28, p. 575]. 'Substantial evidence from several laboratories converges on the conclusion that rhesus monkeys show metacognition in experiments that require behavioral responses to cues that act as feeling of knowing and memory confidence judgments' [32, p. 130]. This debate has shown some of comparative psychology's best practices, including interpretative conservatism, incisive criticism, testable low-level interpretations, disconfirmed low-level interpretations, and rapid empirical and theoretical progress toward a consensual conclusion.

5.4 Poor Interpretive Practices in Animal-Metacognition Research

However, not all the assertions of low-level processes have been disciplined and principled. There have been misunderstandings, shallow descriptive accounts, and misapplications of Morgan's canon. In this section, we discuss these poorer practices within this area of comparative psychology.

5.4.1 A Misconception About Formal Models

Researchers commonly use signal-detection models to describe animals' metacognitive performances [20, 34, 43, 52]. A misconception surrounds these models to which comparative psychologists should attend. The misconception is that if a formal model fits behavioral data, then one can and should interpret the data in a low-level, associative manner [53].

This supposition lacks a scientific basis. In signal detection models, the parameters, decision criteria and response regions are defined purely mathematically. These models do not specify cognitive representations, cognitive processes, levels of awareness or brain regions. The elements of the models are psychologically empty because they are purely mathematical. They cannot imply a low-level information-processing description: they imply no information-processing description.

It is a problem that one can be led by a model's simple mathematics to assume that it reflects simple psychological processes. There is no correlation of this kind. Signal-detection models would fit humans' metacognitive data perfectly well, even though humans often complete uncertainty-monitoring tasks using fully conscious cognition.

The broader implication is that it is not principled behaviorism to say that a model explains animals' behavior. Instead, we must reckon with the processes and representations that underlie the behavior, including their level in the animal's cognitive system and in its awareness.

5.4.2 A Misconception About Reinforcement's Benefits

Another misconception is that one can explain animals' uncertainty responses by saying that animals make them to reduce the time to the next reward. On this view, uncertainty responses are inherently low level and associative because they are about reward maximization. However, the reinforcement-maximization hypothesis is also psychologically empty. Though animals (and humans) may try to maximize rewards, the psychological question is how they do so. Reward-maximization processes are not necessarily low level. They could sometimes be linked to meta-level processes and representations [14]. Even a human's conscious, declarative metacognitive behaviors are compatible with reward maximization. Therefore, reward maximization cannot point to a low-level, information-processing description: it points to no information-processing description.

It is also a problem that one can be led by the simple premise of reward maximization to mistakenly assume that it reflects low-level processes. There is no correlation of this kind. In fact, we saw in Sect. 5.2 that low-level reward maximization—using traditional stimulus and reinforcement cues—does not fit the data.

High-level reward maximization might explain the data, if animals choose to complete easy trials that bring immediate reward and decline difficult trials they believe could produce error and reinforcement delay. Only in this way will they avoid difficult trials (speeding reinforcement) without avoiding easy trials (slowing reinforcement). But difficulty monitoring is a higher level, metacognitive process. Thus, reward maximization using the uncertainty response is not evidence against metacognition. To the contrary, it shows the animal using its metacognitive understanding productively.

The broader implication is that it is not principled behaviorism to say that animals make responses because they have a benefit. Instead, one must describe how the animal gets to the benefit psychologically—how its mind produces the benefit.

5.4.3 A Misconception About Present Stimuli

A third misconception is that uncertainty tasks can only reflect metacognition when humans or animals respond with the relevant stimuli absent. The idea is that stimulus absence prevents organisms from responding directly to stimulus properties, and ensures that they must (if they can) represent their mental states in some way and make a judgment based on those.

The animal-metacognition literature transcends this idea factually. Macaques have performed both stimulus-present and stimulus-absent metamemory tasks [18, 44]. The results converged strongly, and modeling showed that macaques in those studies had the same memory-strength criterion for choosing to complete memory trials [20]. Metacognitive judgments unfold the same whether the stimuli are present or absent.

On reflection, this convergence will be intuitive. As a student considers a multiple-choice test question, the question and response alternatives are fully visible. But he or she will still make metacognitive assessments (Where in the book was that material? Is [b] a lure? Do I know this or should I skip on to use time better? Should I change majors?). Present stimuli do not dampen metacognition [54]. Indeed, the mind could be freed toward more efficient metacognition when it is not occupied with stimulus maintenance.

It is another problem that one can be led by present stimuli to assume that the presence of those stimuli implies low-level psychological processes. There is no correlation of this kind. The broader implication is that it is not principled behaviorism to suppose that a stimulus-present task is low-level and associative. By doing so, one decides the issue using an associative bias that lacks a scientific basis. Instead, one must find out how the animal thinks about the present stimuli; and the level that thinking occupies in its cognitive system. One's preferences for associative explanations cannot provide those answers, but science may provide them.

Low-level/high-level disputes especially occur, in the animal-metacognition literature and elsewhere, when the facts are mixed and there is a temptation to tie-break using a theoretical preference. However, mixed data patterns especially deserve no strong interpretation. In these situations, there is a constructive place for agnostic silence while the facts accumulate and the empirical reality asserts itself. We can wait and see, while we document animals' capacities: here is what they do and fail to do.

Reading the animal-metacognition literature in this way is illuminating. Adaptive uncertainty responses can be independent of stimuli and reinforcement. They are used flexibly, during metacognitive multitasking, and even selectively on the first trial of novel tasks while animals discover what to do. They are cognitive and decisional, available alike for difficult same–different judgments and memory reports. It is an extraordinary set of performances, no matter the interpretative lens one views it through.

Even Morgan [42, p. 59] may have expressed agnosticism. He said: "In no case is an animal activity to be interpreted as the outcome of the exercise of a higher psychical faculty, if it can be fairly interpreted as the outcome of the exercise of one which stands lower in the psychological scale." He also said: "it is clear that any animal may be at a stage where certain higher faculties have not yet been evolved from their lower precursors; and hence we are logically bound not to assume the existence of these higher faculties until good reasons shall have been shown for such existence." He does urge caution in making the high-level attribution. But he neither strongly denies the high-level capacity nor strongly asserts the low-level explanation. What he says is consistent with applying a razor of silence—wait and see—rather than a razor of denial [55].

5.4.4 Selective Criticism

An additional problem is that some pursue low-level interpretations of metacognitive performances by being selective—they only discuss the paradigms that allow the criticism. This practice disregards a growing empirical literature.

Disciplined theoretical interpretations must treat the totality of the research findings. Research findings bootstrap off of one another. Some early findings may have had a potential low-level interpretation, as we have discussed. But once later findings address the issue, the original study can regain some lustre, because now the parsimonious explanation might be that the older finding also deserved a high-level interpretation. If the species is metacognitive in one task, it is not parsimonious to suppose that it became qualitatively nonmetacognitive in a closely related task.

In effect, the later task fulfils Morgan's addendum to the canon, which is extremely important though often overlooked. He said [42, p. 59]: "To this, however, it should be added, lest the range of the principle be misunderstood, that the canon by no means excludes the interpretation of a particular activity in terms of the higher processes, if we already have independent evidence of the occurrence of these higher processes in the animal under observation." The later study provides the independent evidence, and it should affect one's theoretical interpretation of the earlier study. A philosophical analysis of Morgan's canon [56] reached a similar conclusion.

5.4.5 Shopping Associative Mechanisms

An additional concern is that there has been a kind of associative musical chairs in the field of animal metacognition. By turns, stimulus aversion/avoidance, reinforcement history, reward maximization and other associative explanations have been asserted. It betrays a bias to give associative theory many bites of the apple like this, because it reveals an insistence to find a low-level interpretation. This problem becomes worse, as we have discussed, when the associative hypotheses become nonpsychological and less principled. It also becomes worse when different associative mechanisms are used to explain different individual findings (i.e., task *A* and *B*, respectively, need associative interpretations *X* and *Y*). It is not parsimonious to depend on multiple low-level descriptions of animals' performance across tasks, when a basic metacognitive process explains everything simply and naturally, using an adaptive capacity with which cognitive evolution would likely have endowed some species.

5.4.6 Implausible Dualism

Finally, it is an implausible scientific dualism that humans and animals are qualitatively metacognitive and associative, respectively [57]. In no other instance, be it younger versus older human children, or younger versus older human adults, in which the groups' performance profiles correlated at 0.97 (Fig. 5.2) [15], would one think to offer qualitatively different low and high-level interpretations of the behavioral data. Instead, one would naturally interpret similar performances similarly. Thus, our literature provides a case wherein the weight of parsimony has shifted toward the metacognitive interpretation, and the burden of proof has shifted toward associative theorists to demonstrate the necessity for, and the sufficiency of, low-level cues in supporting animals' uncertainty performances. That burden has not been met.

Remember, too, that macaques and humans share evolutionary histories, homologous brain structures and so forth. This also makes it implausible that humans would produce their highly similar graph in a qualitatively different way. As De Waal [58, p. 316] said: "the most parsimonious assumption concerning nonhuman primates is that if their behavior resembles human behavior the psychological and mental processes are similar." Though surely there are limits on this application of evolutionary parsimony, it is at least likely that, if humans perform metacognitively, this provides some evidence that nonhuman primates use similar information-processing mechanisms [59].

One could argue that metacognition had no evolutionary depth, no phases in its development, and no antecedents. To the contrary, we suppose that metacognition had some evolutionary course of development. This predicts psychological continuities between macaques and humans in this capacity, just as we know there are biological continuities. However, this does not mean that macaques must have every conscious and self-aware facet of humans' metacognition. Those facets could have been the add-ons of human evolution as the metacognitive capacity matured and flowered. These are important remaining theoretical questions for the field to explore.

5.5 Detecting a Metacognitive Signal Within Animal Minds

Figure 5.5 summarizes the theoretical situation in the animal-metacognition literature from the perspective of signal-detection theory. The signal-detection framework is apt because we are evaluating whether research has yet detected the signal from animal minds of a higher-level cognitive capacity called metacognition. Animals' true metacognition (Fig. 5.5a)—which we cannot see into their minds to directly confirm—will produce a distribution of cognitive performances across paradigms that appear sometimes more or less cognitively sophisticated. So will their true associative capacity (Fig. 5.5b). But the appearances presented by these capacities may overlap, creating a difficult interpretative problem. Therefore, through this decision space, behavioral analysts place a decision criterion or theoretical dividing line. Performances above the line meet the theoretical grade and are interpreted as metacognitive. Performances below the line do not and are deemed to reflect lower-level, associative processes.

Within this decision space, there are four possible interpretative outcomes. Hits occur to the right in Fig. 5.5a when the scientist correctly concludes for metacognition. Correct rejections occur to the left in Fig. 5.5b when the scientist correctly concludes for an associative mechanism. Hits and correct rejections are salutary scientific events.

Misses occur to the left in Fig. 5.5a when the scientist wrongly concludes against metacognition. In the illustration, about 75 % of true metacognitive performances by animals would be interpreted away. False alarms occur to the right in Fig. 5.5b when the scientist wrongly concludes for metacognition. In the illustration, about 3 % of all associative performances by animals would falsely be called metacognitive. Misses and false alarms are infelicitous scientific events.

The decisional stance within the animal-metacognition literature, as with all domains of comparative psychology, followed Morgan's interpretative lead. His canon was designed to counter the anecdotal reports of animal intelligence and the introspective methods of mental attribution that had produced an anthropomorphic bias. This interpretative lead yielded a distinctive and familiar theoretical culture. Our literature set the high decision criterion shown in Fig. 5.5 for accepting the presence of metacognition, so that few animal performances exceeded it. There was skepticism about animal metacognition. There was a lack of agnosticism. The preference for associative interpretations was used to break interpretative ties.

One sees this culture operating when mathematical parameters are deemed to reflect low-level processes, and when casual reinforcement-maximization

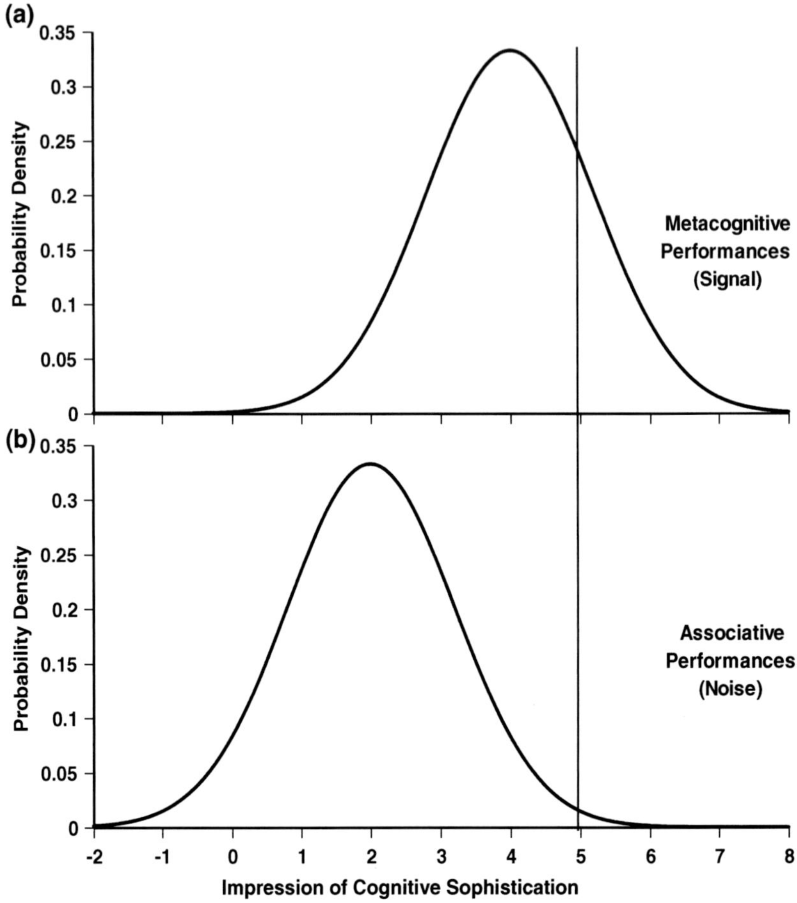

Fig. 5.5 A signal-detection portrayal of theoretical inference within the animal-metacognition literature. Across paradigms, animals' **a** metacognitive performances and **b** associative performances create distributed impressions of cognitive sophistication along the *x*-axis. Current standards of scientific inference engender a criterion point, above which performances are deemed to be metacognitive. From this criterion arise the four possible scientific outcomes: hits (metacognitive performances correctly called metacognitive), correct rejections (associative performances correctly called nonmetacognitive), misses (metacognitive performances incorrectly labelled associative), and false alarms (associative performances incorrectly labelled metacognitive)

hypotheses—not rooted in a concrete information-processing description—are mistaken for low-level processes. One sees it operating when stimulus-based tasks are automatically assumed to elicit reactive processes, and when behavioral analysts persistently try out different low-level mechanisms, taking multiple bites from the associative apple.

In Sects. 5.2 and 5.3 of this article, we praised some elements of this decisional stance within our field—it produced strong theoretical progress. In Sect. 5.4, we pointed out some flawed aspects of this scientific culture. Now we consider some structural weaknesses within this scientific culture that produce a complementary interpretative bias to the old, anthropomorphic bias.

First, the high-threshold stance of the animal metacognition literature guarantees that we will detect less accurately the true psychological signals issuing from animal minds. The highest level of overall correct responding in a detection experiment comes from a criterion level that is midway between the extremes, not at an extreme position. The criterion chosen in the animal-metacognition literature could possibly be expected to double the incorrect scientific conclusions reached in our area. This is a serious problem that has rarely been noted in discussions of scientific inference within the animal metacognition literature or within comparative psychology more broadly.

Second, the high-criterion stance copes poorly with the fact that the two types of interpretative error inexorably trade off with one another. The more avoidant we are of false alarms, creating a high bar for a metacognitive theoretical interpretation, the more misses we will experience because many actual metacognitive performances will not clear the high-bar inferential hurdle.

One can roughly quantify this tradeoff using Fig. 5.5. In Fig. 5.5b, as the decisional threshold moves rightward near its position in the figure, one continues to avoid a tiny set of additional false alarms, as one correctly puts more of the vanishing tail of the associative distribution to the left of the criterion. But, in exchange, in Fig. 5.5a, one moves the line through the probability-dense center of that Gaussian distribution. As a result, one displaces large numbers of true metacognitive events to the left of the decision criterion, and thus we would interpret away these metacognitive performances and increase the number of misses. There would be a many-to-one exchange of misses incurred to false alarms avoided. This is a lousy tradeoff. This would only be acceptable scientific practice if we were for some reason many times more accepting of misses than of false alarms. Of course, comparative psychologists have historically been asymmetrically avoidant of false alarms.

The third problem is that the decisional stance in our literature has caused misses to receive less attention. Indeed, rarely does anyone warn of the dangers of misses in the scientific interpretation of animals' uncertainty performances (or their performances in other cognitive domains). We issue that warning here. It is easy to specify in detail the dangers of misses, and we believe these dangers are serious.

Misses in animal-metacognition research are as wrong as false alarm interpretations. Misses create artificial discontinuities between human and animal minds. Misses may cause us to underestimate the experience of pain and suffering by animals and threaten the ethical conduct of animal studies. These artificial discontinuities can blind us to the origins of human capacities and to their emergence during phylogeny. Misses make it seem that animal models have no place in studying human capacities, because animal minds are qualitatively different and low level. As a result, misses downgrade the relevance of animal research. They downgrade its fundability, too. The more that animals are qualitatively different, the less that animal studies have to contribute to issues of humans' mental health and psychological functioning. Misses also make animal research less accessible and interesting to the wider academic community. They isolate comparative science and reduce its societal impact and footprint. These isolating effects have become increasingly clear and problematic over the last 20 years. Indeed, one can see that some elegant fields of comparative study have retreated—like beautiful glaciers—to the higher elevations of behavioral analysis where few seek access.

5.6 A Middle Ground for Interpretation in Animal-Metacognition Research

In contrast, the animal-metacognition literature is struggling to achieve a middle ground of theoretical interpretation. We will end by describing this middle ground as it presents itself in our field.

First, research encourages the conclusion that animals' uncertainty responses are qualitatively different psychologically from their primary perceptual responses (e.g., the dolphin's low/high responses). Acknowledging this difference seems necessary to explain the dissociation between uncertain and middle responses [23] and the differences in uncertainty responding between macaques and capuchins [24]. Notice, though: qualitatively different does not mean qualitatively conscious, self-aware, and so forth.

Second, research encourages the conclusion that animals' uncertainty responses are not appropriately considered associative responses in the traditional sense. They can be independent from stimuli, reinforcement, and so forth. Uncertainty responses should probably be elevated interpretatively to the level of cognitive-decisional processes in animals' minds.

Third, the research also shows that uncertainty responses serve animals on the first trial of novel tasks, in abstract tasks, in metamemory tasks and even during multitasking. They are sometimes used with so much flexibility and agility that they appear to have some continuities with humans' explicit and declarative cognitive processes—that is, those processes that let humans' reflective minds turn on a dime. Notice: the presence of some continuities does not imply the presence of all continuities.

In conceptualizing these continuities psychologically, some have followed Shiffrin and Schneider [60] by noting that uncertainty-monitoring tasks are inconsistently mapped. That is, they feature indeterminate mental representations that map unreliably onto responses so that those representations become inadequate guides to behavior. The animal must engage higher levels of controlled cognitive processes to adjudicate the indeterminacy and choose adaptive behaviors.

One might say that uncertainty responses represent controlled decisions, at the limit of perception or memory, to decline difficult trials. This is a careful statement of our field's middle ground. It grants animals' uncertainty responses some deserved cognitive sophistication, without attributing to them all the sophisticated features that human metacognition can show.

Thus, the animal-metacognition literature is seeking a moderate decisional stance by which it avoids extreme interpretations depending on either low-level associations or florid, conscious metacognition. Instead, it has characterized in more specific and more accurate information-processing terms the mental representations and cognitive processes that underlie animals' metacognitive performances.

This stance confers benefits on the research area, by increasing the theoretical and empirical scope granted to researchers. For one example, researchers are now freed to try to pinpoint the nature and level of the controlled uncertainty processes that animals show. To this end, our laboratory is asking whether uncertainty responses reflect an executive cognitive utility that especially requires attentional resources or working-memory capacity. We are also asking whether macaques can experience sudden realizations of knowing. Through empirical approaches of this kind, one can continue to cautiously elevate one's interpretation of the uncertainty response, showing that it is higher-level, attentional, executive and even perhaps conscious.

These approaches were held in abeyance in the early years of animal-metacognition research, when the focus was on the associative content of uncertainty responses. A high decisional threshold is not just a quantitative threshold by which a field expresses its general conservatism. To the contrary, the decisional threshold qualitatively changes the nature of the theoretical discussion and affects the kind of research questions that are naturally asked. A field has more scope to ask diverse empirical questions given a more temperate theoretical climate.

Likewise, our field has gained more scope to consider human and animal metacognitive capacities in relation to one another. What are the benefits and affordances of language-based metacognition that is propositionally encoded and in which animals cannot share? What is the essential nature of metacognition that can occur without language and propositions? Do humans feel like uncertain selves in metacognition tasks in ways that monkeys do not? Why, when, and how did conscious cognitive regulation come to play a substantial role within humans' cognitive system? These questions open up as one honours the homologies in the uncertainty-monitoring performances of humans and animals. In these homologies, one also clearly sees the value of animal models for human metacognition, and the possibility of searching for biochemical blocks that might be removed and biochemical enhancers that might be applied to improve metacognitive regulation.

The animal-metacognition literature has also begun to instantiate the distinctive theoretical premise that metacognition is not all-or-none. A great deal of energy has been spent debating the qualitative choice: do animals have metacognition or are they being associative? However, there is a constructive theoretical middle ground wherein one grants organisms a basic uncertainty-monitoring capacity without overinterpreting what they do. In this middle ground may lie the phylogenetic emergence of human metacognition and the ontogenetic emergence of metacognition in human development.

This perspective grants animal-metacognition research strong links to and implications for metacognition research in human development. The behavioral animal paradigms expand the range of metacognition paradigms available for testing young human children. Using them, researchers may uncover the earliest developmental roots of human metacognition [40]. The animal paradigms can also be used to explore the metacognitive capacities of language-delayed and autistic children, or children with mental retardation. It is an important possibility that there might be more basic forms of cognitive regulation (more implicit; less language-based) that could be preserved or fostered in children who are challenged in the highest-level aspects of metacognition.

Thus, one sees that a middle ground of theoretical interpretation is not the compromise of weakness. It balances our field better between the two types of inferential errors. It grants the field more theoretical scope. It opens new lines of research, concerning psychological content, consciousness, human origins, language affordances, and so forth. It broadens our field, granting it outreach to issues of human development and psychological well-being. It makes the research in animal metacognition accessible to a wider range of interested but nonexpert consumers of science in the public domain. And, remarkably, the empirical picture in our field makes plain that, in addition, we are now reading more accurately than ever the uncertain signals emanating from animal minds.

Acknowledgments The preparation of this article was supported by Grant 1R01HD061455 from NICHD and Grant BCS-0956993 from NSF.

References:

1. Dunlosky J, Bjork RA (eds) (2008) Handbook of memory and metamemory. Psychology Press, New York
2. Flavell JH (1979) Metacognition and cognitive monitoring: a new area of cognitive-developmental inquiry. Am Psychol 34:906–911. doi:10.1037/0003-066X.34.10.906
3. Koriat A (1993) How do we know that we know? The accessibility model of the feeling of knowing. Psychol Rev 100:609–639. doi:10.1037/0033-295X.100.4.609
4. Metcalfe J, Shimamura A (1994) Metacognition: knowing about knowing. Bradford Books, Cambridge
5. Nelson TO (ed) (1992) Metacognition: core readings. Allyn and Bacon, Toronto
6. Schwartz BL (1994) Sources of information in metamemory: judgments of learning and feelings of knowing. Psychon Bull Rev 1:357–375. doi:10.3758/BF03213977
7. Nelson TO, Narens L (1990) Metamemory: a theoretical framework and new findings. In: Bower GH (ed) The psychology of learning and motivation, vol 26. Academic Press, New York, pp 125–173
8. Koriat A (2007) Metacognition and consciousness. In: Zelazo PD, Moscovitch M, Thompson E (eds) The Cambridge handbook of consciousness. Cambridge University Press, Cambridge, pp 289–325

9. Nelson TO (1996) Consciousness and metacognition. Am Psychol 51:102–116. doi:10.1037/0003-066X.51.2.102
10. Gallup GG (1982) Self-awareness and the emergence of mind in primates. Am J Primatol 2:237–248. doi:10.1002/ajp.1350020302
11. Kornell N (2009) Metacognition in humans and animals. Curr Dir Cogn Sci 18:11–15. doi:10.1111/j.1467-8721.2009.01597.x
12. Smith JD (2009) The study of animal metacognition. Trends Cogn Sci 13:389–396. doi:10.1016/j.tics.2009.06.009
13. Smith JD, Beran MJ, Couchman JJ (2012) Animal metacognition. In: Zentall T, Wasserman E (eds) Comparative cognition: experimental explorations of animal intelligence. Oxford University Press, Oxford
14. Shea N, Heyes C (2010) Metamemory as evidence of human consciousness: the type that does the trick. Biol Phil 25:95–110. doi:10.1007/s10539-009-9171-0
15. Shields WE, Smith JD, Washburn DA (1997) Uncertain responses by humans and rhesus monkeys (*Macaca mulatta*) in a psychophysical same–different task. J Exp Psychol Gen 126:147–164. doi:10.1037/0096-3445.126.2.147
16. Smith JD, Schull J, Strote J, McGee K, Egnor R, Erb L (1995) The uncertain response in the bottlenosed dolphin (*Tursiops truncatus*). J Exp Psychol Gen 124:391–408. doi:10.1037/0096-3445.124.4.391
17. Smith JD, Shields WE, Schull J, Washburn DA (1997) The uncertain response in humans and animals. Cognition 62:75–97. doi:10.1016/S0010-0277(96)00726-3
18. Smith JD, Shields WE, Allendoerfer KR, Washburn WA (1998) Memory monitoring by animals and humans. J Exp Psychol Gen 127:227–250. doi:10.1037/0096-3445.127.3.227
19. Smith JD, Beran MJ, Redford JS, Washburn DA (2006) Dissociating uncertainty states and reinforcement signals in the comparative study of metacognition. J Exp Psychol Gen 135:282–297. doi:10.1037/0096-3445.135.2.282
20. Smith JD, Shields WE, Washburn DA (2003) The comparative psychology of uncertainty monitoring and metacognition. Behav Brain Sci 26:317–339. doi:10.1017/S0140525X03000086
21. Basile BM, Hampton RR, Suomi SJ, Murray EA (2009) An assessment of memory awareness in tufted capuchin monkeys (*Cebus apella*). Anim Cogn 12:169–180. doi:10.1007/s10071-008-0180-1
22. Beran MJ, Smith JD, Redford JS, Washburn DA (2006) Rhesus macaques (*Macaca mulatta*) monitor uncertainty during numerosity judgments. J Exp Psychol Anim Behav Process 32:111–119. doi:10.1037/0097-7403.32.2.111
23. Beran MJ, Smith JD, Coutinho MVC, Couchman JJ, Boomer J (2009) The psychological organization of 'uncertainty' responses and 'middle' responses: a dissociation in capuchin monkeys (*Cebus apella*). J Exp Psychol Anim Behav Process 35:371–381. doi:10.1037/a0014626
24. Beran MJ, Smith JD (2011) Information seeking by rhesus monkeys (*Macaca mulatta*) and capuchin monkeys (*Cebus apella*). Cognition 120:90–105. doi:10.1016/j.cognition.2011.02.016
25. Call J (2010) Do apes know that they can be wrong? Anim Cogn 13:689–700. doi:10.1007/s10071-010-0317-x
26. Couchman JJ, Coutinho MVC, Beran MJ, Smith JD (2010) Beyond stimulus cues and reinforcement signals: a new approach to animal metacognition. J Comp Psychol 124:356–368. doi:10.1037/a0020129
27. Foote A, Crystal J (2007) Metacognition in the rat. Curr Biol 17:551–555. doi:10.1016/j.cub.2007.01.061
28. Fujita K (2009) Metamemory in tufted capuchin monkeys (*Cebus apella*). Anim Cogn 12:575–585. doi:10.1007/s10071-009-0217-0
29. Hampton RR (2009) Multiple demonstrations of metacognition in nonhumans: converging evidence or multiple mechanisms? Comp Cogn Behav Rev 4:17–28

30. Kornell N, Son L, Terrace H (2007) Transfer of metacognitive skills and hint seeking in monkeys. Psychol Sci 18:64–71. doi:10.1111/j.1467-9280.2007.01850.x
31. Paukner A, Anderson JR, Fujita K (2006) Redundant food searches by capuchin monkeys (*Cebus apella*): a failure of metacognition? Anim Cogn 9:110–117. doi:10.1007/s10071-005-0007-2
32. Roberts WA, Feeney MC, McMillan N, MacPherson K, Musolino E (2009) Do pigeons (*Columba livia*) study for a test? J Exp Psychol Anim Behav Process 35:129–142. doi:10.1037/a0013722
33. Shields WE, Smith JD, Guttmannova K, Washburn DA (2005) Confidence judgments by humans and rhesus monkeys. J Gen Psychol 132:165–186
34. Smith JD, Beran MJ, Coutinho MVC, Couchman JJ (2008) The comparative study of metacognition: sharper paradigms, safer inferences. Psychon Bull Rev 15:679–691. doi:10.3758/PBR.15.4.679
35. Smith JD, Redford JS, Beran MJ, Washburn DA (2010) Monkeys adaptively monitor uncertainty while multi-tasking. Anim Cogn 13:93–101. doi:10.1007/s10071-009-0249-5
36. Suda-King C (2008) Do orangutans (*Pongo pygmaeus*) know when they do not remember? Anim Cogn 7:239–246
37. Sutton JE, Shettleworth SJ (2008) Memory without awareness: pigeons do not show metamemory in delayed matching-to-sample. J Exp Psychol Anim Behav Process 34:266–282. doi:10.1037/0097-7403.34.2.266
38. Washburn DA, Smith JD, Shields WE (2006) Rhesus monkeys (*Macaca mulatta*) immediately generalize the uncertain response. J Exp Psychol Anim Behav Process 32:185–189
39. Washburn DA, Gulledge JP, Beran MJ, Smith JD (2010) With his memory erased, a monkey knows he is uncertain. Biol Lett 6:160–162. doi:10.1098/rsbl 2009.0737
40. Balcomb FK, Gerken L (2008) Three-year-old children can access their own memory to guide responses on a visual matching task. Dev Sci 11:750–760. doi:10.1111/j.1467-7687.2008.00725.x
41. Tolman EC (1927) A behaviorist's definition of consciousness. Psychol Rev 34:433–439. doi:10.1037/h0072254
42. Morgan CL (1906) An introduction to comparative psychology. Walter Scott, London
43. Jozefowiez J, Staddon JER, Cerutti D (2009) Metacognition in animals: how do we know that they know? Comp Cogn Behav Rev 4:29–39. doi:10.3819/ccbr.2009.40003
44. Hampton RR (2001) Rhesus monkeys know when they remember. Proc Natl Acad Sci USA 98:5359–5362. doi:10.1073/pnas.071600998
45. Inman A, Shettleworth SJ (1999) Detecting metamemory in nonverbal subjects: a test with pigeons. J Exp Psychol Anim Behav Process 25:389–395. doi:10.1037/0097-7403.25.3.389
46. Kiani R, Shadlen MN (2009) Representation of confidence associated with a decision by neurons in the parietal cortex. Science 324:759–764. doi:10.1126/science.1169405
47. Premack D (1978) On the abstractness of human concepts: why it would be difficult to talk to a pigeon. In: Hulse SH, Fowler H, Honig WK (eds) Cognitive processes in animal behavior. Erlbaum, Hillsdale, pp 423–451
48. Katz JS, Wright AA, Bachevalier J (2002) Mechanisms of same/different abstract-concept learning by rhesus monkeys (*Macaca mulatta*). J Exp Psychol Anim Behav Process 28:358–368. doi:10.1037/0097-7403.28.4.358
49. Fleming TM, Beran MJ, Washburn DA (2006) Disconnect in concept learning by rhesus monkeys: judgment of relations and relations-between-relations. J Exp Psychol Anim Behav Process 33:55–63. doi:10.1037/0097-7403.33.1.55
50. Harlow HF (1949) The formation of learning sets. Psychol Rev 56:51–65. doi:10.1037/h0062474
51. Smith JD, Beran MJ, Couchman JJ, Coutinho MVC, Boomer J (2009) The curious incident of the capuchins. Comp Cogn Behav Rev 4:61–64. doi:10.3819/ccbr.2009.40008
52. Crystal JD, Foote AL (2009) Metacognition in animals. Comp Cogn Behav Rev 4:1–16. doi:10.3819/ccbr.2009.40001

53. Kepecs A, Mainen ZF (2012) A computational framework for the study of confidence in humans and animals. Philos Trans R Soc B 367:1322–1337. doi:10.1098/rstb 2012.0037
54. Higham PA (2007) No special K! A signal detection framework for the strategic regulation of memory accuracy. J Exp Psychol Gen 136:1–22. doi:10.1037/0096-3445.136.1.1
55. Sober E (2009) Parsimony arguments in science: a test case for naturalism. Proc Address Am Philos Assoc 83:117–155
56. Sober E (1998) Morgan's canon. In: Cummins D, Allen C (eds) The evolution of mind. Oxford University Press, New York, pp 224–242
57. Smith JD (2006) Species of parsimony in comparative studies of cognition. In: Washburn D (ed) Primate perspectives on behavior and cognition. American Psychological Association, Washington, DC, pp 63–80
58. De Waal FBM (1991) Complementary methods and convergent evidence in the study of primate social cognition. Behavior 118:297–320. doi:10.1163/156853991X00337
59. Sober E (2012) Anthropomorphism, parsimony, and common ancestry. Mind Lang 27(3): 229–238
60. Shiffrin RM, Schneider W (1977) Controlled and automatic human information processing: II. Perceptual learning, automatic attending and a general theory. Psychol Rev 84:127–190. doi:10.1037/0033-295X.84.2.127

Part II
Computational Approaches to Metacognition

Chapter 6
A Computational Framework for the Study of Confidence Across Species

Adam Kepecs and Zachary F. Mainen

Abstract Confidence judgments, the self-assessment of the quality of a subject's knowledge, are considered a central example of metacognition. Prima facie, introspection, and self-report appear the only way to access the subjective sense of confidence or uncertainty. Could confidence be also studied in nonhuman animals so one could probe its neural substrates? Indeed, behavioral paradigms that incentivize animals to evaluate and act upon their own confidence can yield implicit reports of confidence. Here, we suggest that a computational approach can clarify the issues involved in interpreting these tasks and provide a much-needed springboard for advancing the scientific understanding of confidence. We first review relevant theories of probabilistic inference and decision making. We then critically discuss behavioral tasks employed to measure confidence in animals and show how quantitative models can help to constrain the computational strategies underlying confidence-reporting behaviors. In our view, post-decision wagering tasks with continuous measures of confidence appear to offer the best available metrics of confidence. Since behavioral reports alone provide a limited window into mechanism, we argue that progress calls for measuring the neural representations and identifying the computations underlying confidence reports. We present a case study using such a computational approach to study the neural correlates of decision confidence in rats. This work shows that confidence assessments may be considered higher order, but can be generated using

This chapter is adapted from: Kepecs A, Mainen ZF (2012) A computational framework for the study of confidence in humans and animals. Phil Trans R Soc B 367:1322–1337

A. Kepecs (✉)
Cold Spring Harbor Laboratory, 1 Bungtown Road, Cold Spring Harbor, New York 11724, USA
e-mail: kepecs@cshl.edu

Z. F. Mainen
Champalimaud Neuroscience Programme, Champalimaud Centre for the Unknown, Av. Brasília s/n 1400-038 Lisbon, Portugal

S. M. Fleming and C. D. Frith (eds.), *The Cognitive Neuroscience of Metacognition*, 115
DOI: 10.1007/978-3-642-45190-4_6, © Springer-Verlag Berlin Heidelberg 2014

elementary neural computations that are available to a wide range of species. Finally, we discuss the relationship of confidence judgments to the broader behavioral uses of confidence and uncertainty.

6.1 Introduction

In face of the pervasive uncertainty and variability in the world, the evaluation of confidence in one's beliefs is a critical component of cognition. As humans we intuitively assess our confidence in our percepts, memories, and decisions all the time, seemingly automatically. Nevertheless, confidence judgments also seem to be part of a reflective process that is deeply personal and subjective. Therefore, a natural question arises: Does assessing confidence—knowledge about subjective beliefs—constitute an example of the human brain's capacity for self-awareness? Or is there a simpler explanation that might suggest a more fundamental role for confidence in brain function across species?

Prima facie, introspection, and self-report appear the only way to access the subjective sense of confidence or uncertainty. Indeed, confidence judgments have been long studied as a central example of metacognition [1–3]. In this light confidence judgments are viewed as a monitoring process reporting on the quality of internal representations of perception, memory, or decisions that reflect a uniquely human cognitive capacity [1–3], requiring advanced neural architecture available only in the brains of higher primates [1, 4, 5]. Moreover, as subjective reports about our beliefs, confidence judgments have even been used as indices of conscious awareness.

Against this backdrop of studies emphasizing the apparent association of confidence with the highest levels of cognition, a recent line of research has attempted to show that nonhuman animals are also capable of confidence judgments, with mixed and sometimes contentious results [5]. How could animals possibly think about their thoughts and report their confidence? And even if they did, how could one even attempt to establish this without an explicit self-report?

While such philosophically charged debates persist, an alternative approach, rooted in computational theory, is taking hold. According to this view, assessment of the certainty of beliefs can be considered to be at the heart of statistical inference. Formulated in this way, assigning a confidence value to a belief can often be accomplished using relatively simple algorithms that summarize the consistency and reliability of the supporting evidence [6]. Therefore, it should come as no surprise that in the course of building neurocomputational theories to account for psychophysical phenomena, many researchers came to the view that probabilistic reasoning is something that nervous systems do as a matter of their construction [7–11].

In this light, confidence reports might reflect a readout of neural dynamics that is available to practically any organism with a nervous system. Hence, representations of confidence might not be explicit or anatomically segregated [12–14]. Although statistical notions can account for the behavioral observations used to index metacognition, it remains to be seen whether there are aspects of metacognition that will require expansion of this framework. For instance, it may be that while choice and confidence are computed together, confidence is then relayed to a brain region serving as a clearinghouse for confidence information from different sources. Confidence representations in such region can be viewed as metacognitive but nevertheless may still require only simple computations to generate.

Here, we argue that progress in understanding confidence judgments requires us to place the study of confidence on a solid computational foundation. A computational approach can clarify the issues involved in interpreting behavioral data and provide a much-needed springboard for advancing the scientific understanding of confidence. We first review relevant theories of probabilistic inference and decision making that provide an emerging computational framework for confidence. We then critically discuss behavioral tasks employed to measure confidence in animals and show how quantitative models can help to constrain the computational strategies underlying confidence-reporting behaviors. Our review also comes from the vantage point of two neuroscientists: behavioral reports alone provide a limited window into mechanism, thus we advocate opening up the brain's "black box" to search for neural circuits mediating metacognition. We present a case study using such an approach to study the neurobiological basis of decision confidence in rats. We conclude by discussing the relationship of confidence judgments to the wider behavioral uses of confidence and uncertainty.

6.2 Behavioral Reports of Confidence in Humans and Other Animals

The topic of behavioral studies of confidence is broad and therefore we will largely limit our discussion to confidence about simple psychophysical decisions and focus mostly on animal behavior. Behavioral reports of confidence in humans can be explicit, usually verbal or numerical self-reports, and these are usually taken at face value. In contrast, in nonhuman animals, only implicit behavioral reports are available. This has led to interpretational difficulties, a topic we address next.

6.2.1 Explicit Reports of Confidence in Humans

The most straightforward behavioral paradigm for testing confidence is to ask subjects to assign a numerical rating of how sure they are in their answer [13–20]. Indeed, humans performing psychophysical discrimination tasks can readily assign appropriate confidence ratings to their answers [21]. By appropriate rating we mean that performance accuracy in humans is well correlated with the self-reported confidence measures. Self-reported confidence also correlates with choice reaction times [15, 16]. It should be noted that confidence reports are not always perfectly calibrated; there are systematic deviations found such as overconfidence when decisions are difficult and underconfidence when they are easy [21–23]. Clearly, it is not possible to ask animals to provide explicit confidence reports, therefore animal studies need to employ more sophisticated tasks designed to elicit implicit reports of confidence.

6.2.2 "Uncertain Option" Task

A widely used class of tasks extends the two-choice categorization paradigm by adding a third choice, the "uncertain option," to the available responses. Interestingly, the scientific study of this paradigm originated in experimental psychology along with quantitative studies of perception [24–27]; however, these early attempts were found to be too subjective [25, 26, 28] and were soon abandoned in favor of "forced choice" tasks to quantify percepts based on binary choices. After nearly a century of neglect, a series of studies by Smith et al. [5] reinvigorated the field of confidence judgments using these paradigms in both human and nonhuman subjects with the goal of placing the notion of subjective confidence on scientific footing.

In the first of the modern studies using this paradigm, Smith and colleagues used a perceptual categorization paradigm [29] (Fig, 6.1a; see also Smith et al., this volume). A sensory stimulus is presented along a continuum (e.g., frequency of an auditory tone) that needs to be categorized into two classes based on an arbitrarily chosen boundary (above or below 2.1 kHz). Subjects are then given three response options: left category, right category, or uncertain response. As might be expected, subjects tend to choose the uncertain option most frequently near the category boundary. Post-experimental questionnaires indicate that humans choose the uncertain option when they report having low confidence in the answer.

Importantly, the design of the task allowed Smith and colleagues to ask the same questions to nonhuman animals and hence to reignite an age-old enquiry about the cognitive sophistication of different animal species. In order to test animals in this task, the different options were linked with different reward contingencies: correct choices are rewarded with one unit of food reward, uncertain responses are rewarded with a smaller amount of food while incorrect choices lead to omission of reward and a timeout punishment. Smith and colleagues [29–34] studied monkeys, dolphins, and rats and compared their performance to humans. Under these conditions, monkeys and dolphins showed a qualitative similarity in response strategies as well as a quantitative agreement in the response distributions of animals and humans. The striking similarities of dolphin and monkey behavior to that of humans suggested that these animals possess more sophisticated cognitive architectures than previously appreciated. Interestingly, their studies also failed to show that evolutionarily "simpler" animals such as rats could perform confidence judgments [30]. The authors concluded that this might reflect a failure to find a suitable task for accessing the appropriate abilities in these species.

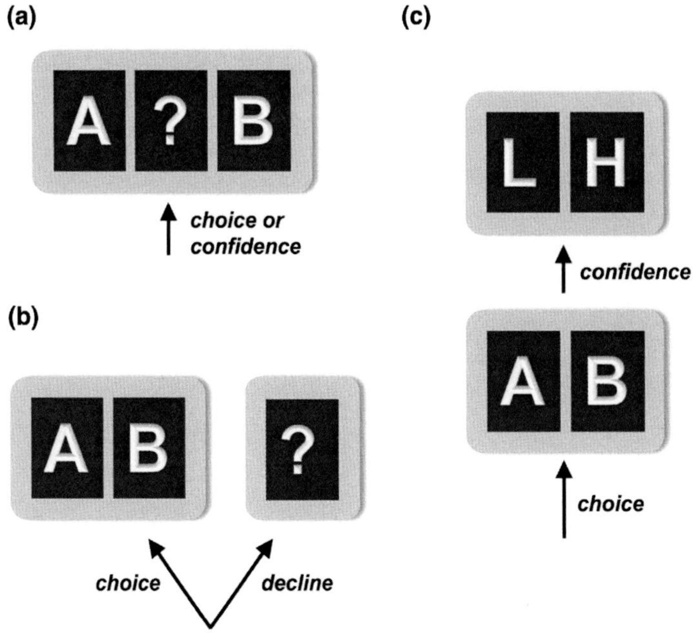

Fig. 6.1 Behavioral tasks for studying confidence in animals. **a** In uncertain option tasks there are three choices, the two categories, A and B, and the uncertain option. **b** In decline option or opt-out tasks, there is first a choice between taking the test or declining it, then taking the test and answering A or B. In a fraction of trials the option to decline is omitted. **c** In post-decision wagering for every trial there is first a two-category discrimination, A or B, and then a confidence report, such as low or high options

Indeed, pigeons can respond similarly to primates in such tasks [35], although they fail to exhibit other uncertainty monitoring behaviors (see below).

A similar opt-out task was used by Komura et al. [36] to study the neural representation of confidence in the pulvinar region. The authors started with a motion categorization task in which monkeys were presented with a cloud of dots, each either red or green, which were moving up or down. Monkeys were trained to report the direction of the moving cloud of the target color that was cued. When all the dots were moving in one direction, the monkey had an easy choice, whereas when the mixture of upward and downward moving dots was nearly balanced the choice was more difficult. Correct choices were rewarded with a drop of juice. To make use of confidence information, the monkeys were also given a third choice: to opt out of the categorization task and receive a smaller but certain reward. As expected, monkeys used the safe, opt-out choice in proportion to the stimulus difficulty.

A weakness in the design of such opt-out tasks for confidence reporting is the possibility that uncertain responses could be simply associated with those stimuli intermediate to the extreme category exemplars. In other words, the two-alternative plus uncertain option task can be alternatively viewed as a three-choice decision task (left/right/uncertain, Fig. 6.1a), which can be solved simply by learning appropriate stimulus–response categories without necessitating confidence estimates. This criticism was first addressed with task variants that require same–different discrimination [31]. In this version of the task there are no external stimuli that can be associated with the uncertain option. Nevertheless, from a computational perspective the difference between the two stimuli is represented in the brain, then again the uncertain response can simply be associated with a class of such difference representations. To argue against such associative mechanisms, Smith et al. [37] also reported on a task version in which monkeys were only rewarded at the end of a block of trials, so that individual responses were not reinforced. However, while this manipulation does rule out the simplest forms of reward learning, it is still compatible with more sophisticated forms [38].

6.2.3 "Decline Option" Tasks

Hampton introduced a memory test that included a "decline" option rather than three choices [39] (Fig. 6.1b). In this test, monkeys performed a delayed-matching-to-sample task using visual stimuli. At the end of the delay, subjects were presented with the option of accepting or declining the discrimination test. Once a subject accepted the test, it received either a food reward for correct choices or a timeout punishment for error choices. After declining the test, the subject received a nonpreferred food reward without a timeout. Therefore, the optimal strategy was to decline when less certain, and indeed, monkeys tended to decline the discrimination more often for longer delays. However, since the difficulty was determined solely by the delay, the decline option could be learned simply by associating it with longer delays for which performance was poorer. To circumvent this problem, Hampton introduced forced-choice trials in which the subject had no choice but to perform the discrimination test. If longer delays were simply associated with the choice to decline the test, then decline responses would be equally likely regardless of performance. However, he found a systematic increase in performance in freely chosen trials compared to forced-choice trials, consistent with the idea that monkeys are monitoring the likelihood of being correct, rather than associating delays with decline responses.

A similar decline option test was used by Kiani and Shadlen [40] in macaque monkeys, who made binary decisions about the direction of visual motion. On some trials, after stimulus presentation, a third "opt out" choice was presented for which monkeys received a smaller but guaranteed reward. They found that frequency of choosing the "uncertain option" increased with increasing stimulus difficulty and with shorter stimulus sampling. Moreover, as observed by Hampton, monkeys' performance on trials in which they declined to opt out was better than when they were forced to perform the discrimination.

Inman and Shettleworth [41], and Teller [42] tested pigeons using similarly designed "decline" tests with a delayed-matching-to-sample task. They observed that the rates of choosing the decline option slightly increased with delay duration. Because performance decreased with delay, decline choices also increased as performance decreased. However, there was no difference in the performance on forced-choice versus free-choice (i.e., nondeclined) trials. These results were interpreted as arguing against the metacognitive abilities of pigeons [43].

After these negative results with pigeons and rats in uncertain and decline option tasks, it came as a surprise that Foote and Crystal reported that rats have metacognitive capacity [44]. Similar to the design of Hampton, they used a task with freely chosen choice trials with a decline option interleaved with forced-choice trials. But unlike Hampton, Foote and Crystal used an auditory discrimination task without an explicit memory component. They found that decline option choices increased in frequency with decision difficulty and that free-choice performance was better than forced-choice performance on the most difficult stimuli, arguing against associative learning of decline choices.

The argument that this class of task tests confidence-reporting abilities chiefly rests on the decrease in performance on forced-choice trials compared to freely chosen choice trials. The results for rats showed a small change in performance (<10 %) and only for the most difficult discrimination type. An alternative explanation is that attention or motivation waxes and wanes and animals' choices and performance are impacted by their general "vigilance" state [45]. When their vigilance is high animals would be expected to choose to take the test and perform well compared to a low vigilance state when animals would tend to decline and accept the safer, low value option. Although being aware of one's vigilance state may be considered as a form of metacognition, it is distinct from mechanisms of confidence judgments.

6.2.4 Post-decision Wagering

The "uncertain option" and "decline option" tasks have the weakness that either a choice report or a confidence report is collected in each trial but not both. The "post-decision wager" paradigms improve on this by obtaining both choice and confidence on every trial [46, 47]. The central feature of this class of tasks is that after the choice is made, confidence is assessed by asking a subject to place a bet on her reward (Fig. 6.1c). The probability of betting high (or the amount wagered) on a particular decision serves as the index for confidence. Persaud and colleagues used this paradigm to test a subject with blindsight, who had lost nearly his entire left visual cortex yet could make visual discriminations in his blind field despite having no awareness. Using a post-decision wagering paradigm they found that wagers were better correlated with the subjects' explicit self-reported visibility of the stimulus than with actual task performance [46]. Hence the authors argued that post-decision wagers not only provided an index of confidence, but also served as an objective assessment of "awareness," independent of perceptual performance.

Leaving aside the thorny issue of whether post-decision wagers can be used to study awareness [48, 49], for the purposes of studying confidence judgments, the wagering paradigm has many attractive features [46]. Wagers provide a means to make confidence reports valuable and hence by providing appropriate reward incentives animals can be trained to perform post-decision wagering [33, 50–53]. For instance, Middlebrooks and Sommer [53] tested rhesus monkeys on a simple visual discrimination task that was followed by a betting stage. Choosing the high wager option either earned a large reward for correct choices or a timeout punishment after incorrect choices. Choosing the low wager option earned a small reward independent choice correctness. The authors showed that the proportion of high wagers decreased with perceptual difficulty and was correlated with choosing the correct option.

One caveat with post-decision wagering paradigms is that because the payoff matrix interacts with the level of confidence to determine the final payoff, care must be taken with the design of the matrix. It has been observed that in the study of Persaud et al. [46], the optimal strategy for the payoff matrix was to always bet high regardless of the degree of confidence [54, 55]. Subsequently, Fleming and colleagues established that human subjects casting post-decision wagers in this task display loss aversion [55]. Although subjects were in fact found to vary their wager with uncertainty, it would be difficult to disambiguate a suboptimal wagering strategy from the lack of appropriate confidence estimates [54, 56]. Therefore, the design of the payoff matrix and an independent evaluation of the wagering strategy are important considerations to determine whether wagering reflects confidence. Using a continuum of wagers instead of a binary bet (certain/uncertain) somewhat mitigates the difficulty of finding an optimal payoff matrix (see below). A second concern about the post-decision wager task is that bets might be placed by associating the optimal wager with each stimulus using reinforcement learning. In particular, each two-choice plus wager test could be transformed into a four-choice test where distinct stimuli could be associated with distinct responses. For instance, in a motion direction categorization, weak versus strong motion to one direction could be associated with low versus high wagers. Importantly, however, this concern can be alleviated by appropriate analysis of behavioral data, as we will discuss below. Moreover, regardless of concerns over the optimality of wagering, animals can be trained to perform this kind of task [33, 50–53], providing a rich class of confidence-reporting behaviors.

6.2.5 Decision Restart and Leaving Decision Tasks

Recently, Kepecs et al. [57] introduced a behavioral task similar to post-decision wagering that can be used in animals. They trained rats to perform a two-alternative forced-choice olfactory discrimination task. In this task, subjects initiated a trial by entering a central odor port. This triggered the delivery of a stimulus comprising a binary odor mixture. The subject was rewarded for responding to the left or to the right depending on the dominant component. In order to vary the decision difficultly, the ratio of the two odors was systematically varied. Rats performed at near chance level for 50/50 odor mixtures and nearly perfectly for pure odors. To assess confidence, reward was delayed by several seconds while subjects were given the option to "restart" trials by leaving the reward port and reentering the odor sampling port. In other words, after the original decision about the stimulus, rats were given a new decision whether to stay and risk no reward with timeout punishment, or leave and start a new, potentially easier trial. An important feature of this task, similar to the post-decision wagering and confidence rating tasks, is that the choice and confidence reports are collected in the same trial, an issue we will further discuss below.

It was observed that the probability of aborting and restarting trials increased with stimulus difficulty and that discrimination performance was better on trials in which the subjects waited for the reward compared to trials in which they reinitiated. Moreover, as we will describe in more detail below, the probability to reinitiate choices matched the pattern expected for confidence judgments. In particular, even for trials of the same stimulus, Kepecs et al. [57] observed that performance was systematically higher for correct than for error trials. This correct/error analysis permitted the authors to circumvent the criticism that reinforcement learning on stimuli could explain the pattern of behavior (see below).

One difficulty with this task is that its parameters (e.g., reward delay) need to be carefully tuned in order to have a reasonable balance between restarted and non-restarted trials. This difficulty, in part, can be traced to the issue of choosing an appropriate payoff matrix in post-decision wagering paradigms discussed above. The cost of waiting and the value of restarts must be chosen to be just right and are difficult to infer a priori. For instance, in some parameter regimes rats never restarted trials, while in others they always restarted until an easier stimulus was provided (unpublished observations). A related issue with binary wagers is that in each trial only a single bit of information is gained about decision confidence.

Both of these issues can be mitigated using task versions that provide graded reports of decision confidence (Fig. 6.2a). Rats were trained in a task variant we call the "leaving decision" task. In this version we delay reward delivery using an exponential distribution of delays (to keep reward expectancy relatively constant) and measure the time an animal is willing to wait at the choice ports (Fig. 6.2a). Incorrect choices are not explicitly signaled and hence rats eventually leave the choice ports to initiate a new trial. In order to measure confidence for correct choices we introduce a small fraction ($\sim$10–15 %) of catch trials for which rewards are omitted. The waiting time at the reward port before the leaving decision (obtained for all incorrect and a fraction of correct trials) provides a graded measure reflecting decision confidence (Fig. 6.2c). Waiting time, naturally, also depends on when the animal is expecting the reward delivery. However, we found that the relative patterns of waiting times were systematically related to decision parameters (Fig. 6.2c, d) for a range of reward delay distributions providing a behaviorally robust proxy for decision confidence. Indeed, an accurate estimate of decision confidence modulating the waiting time will help to maximize

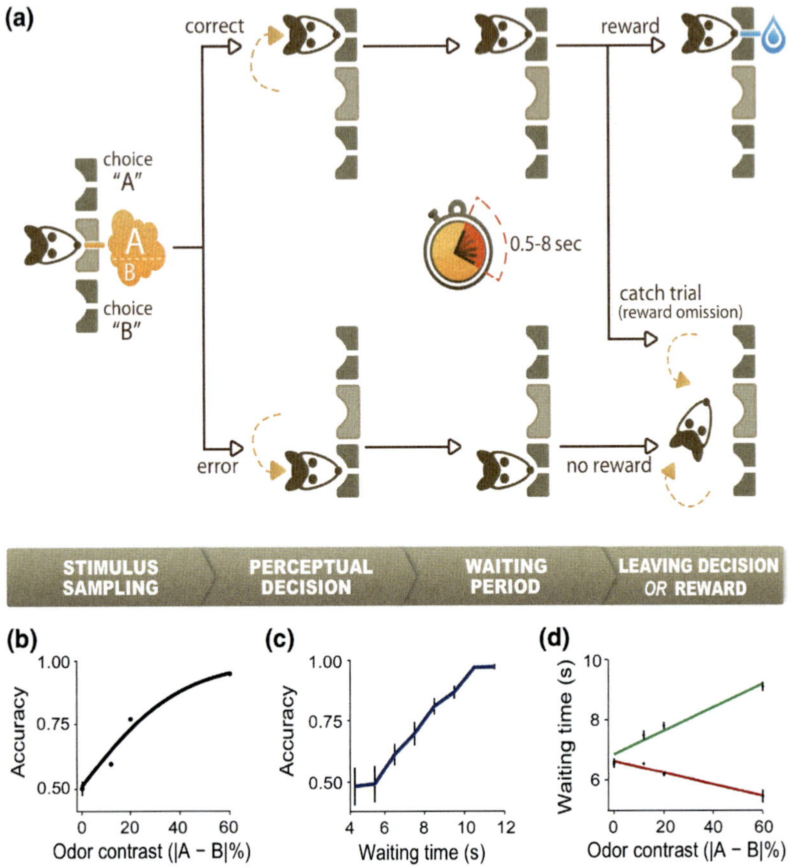

Fig. 6.2 Leaving decision tasks to studying confidence in animals. **a** Schematic of the behavioral paradigm. To start a trial, the rat enters the central odor port and after a pseudorandom delay of 0.2–0.5 s an unequal mixture of two odors is delivered. Rats respond by moving to the A or B choice port, where a drop of water is delivered after a 0.5–8 s (exponentially distributed) waiting period for correct decisions. In catch trials ($\sim 10\ \%$ of correct choices) the rat is not rewarded and no feedback is provided. Therefore, the waiting time can be measured (from entry into choice ports until withdrawal) for all error and a subset of correct choices. **b** Psychometric function for an example rat. **c** Choice accuracy as a function of waiting time. For this plot we assumed that the distribution of waiting times for correct catch trials is a representative sample for the entire correct waiting time distribution. **d** Mean waiting time as a function of odor mixture contrast and trial outcome (correct/error) for an example rat

reward rate, while also minimizing effort and opportunity costs incurrent by waiting. Because the animals' cost functions are difficult to infer it is challenging to make quantitative predictions about the optimal waiting time. Nevertheless, based on reasonable assumptions we expect accuracy to be monotonically related to waiting time, in agreement with our observations (Fig. 6.2c). Note that although waiting time can also be measured in the "decision restart" task variant, for all restarted trials, we only observed a weak relationship between waiting time and confidence, presumably because many trials were restarted early after the mandatory wait time.

6.2.6 Looking Tests

In addition to tests based on psychophysical methodologies, the metacognitive abilities of animals have also been addressed using more ethologically minded behaviors. So-called "looking" paradigms take advantage of the fact that during foraging animals may naturally seek information about where foods might be located [58, 59]. Such information seeking can be considered an assessment of the animal's state of knowledge: the less certain they are about their given state of knowledge the more likely they will seek new information [60]. Indeed, chimpanzees, orangutans, macaques, capuchins, and human children all show a tendency to seek new information when specifically faced with uncertainty [59, 61–63]. By "looking" more frequently in the appropriate situations, these species can demonstrate knowledge about their own belief states. In contrast, dogs have so far failed to show such information-seeking behavior [64, 65]. However, in the case of such a failure, it remains possible that the setup wasn't ecologically relevant for the species in question.

It may be useful to consider "looking" tests as an instance of a more general class of behaviors in which confidence assessment can be useful to direct information seeking or exploration. Given the limitations of experimental control possible in ethological settings, it would be profitable to transform the looking tests into psychophysical paradigms where the confidence can be read out by choices to seek out more information [66].

6.2.7 Criticisms of Confidence-Reporting Behaviors

Many of these studies discussed above triggered controversies; some of the criticisms have been highlighted above. To summarize, critics have systematically attempted to come up with alternative explanations for the performance of nonhumans animals that do not require uncertainty monitoring. The primary thrust of these critiques has been that some confidence-reporting tasks can be solved by learning appropriate stimulus–response categories without the need for true uncertainty monitoring [67]. A second important criticism is that a behavioral report can in some cases arise from reporting the level of a cognitive variable such as "motivation," "attention," or "vigilance" that impacts performance, rather than confidence per se [39]. Thus, a simpler mechanism might be sufficient to account for the observed behavior without invoking confidence or metacognition.

However, we have seen that these alternative classes of explanation, while very important to address, are being tackled through increasingly sophisticated task designs [57, 61, 68].

A third, somewhat different, line of criticism has questioned the similarity of various confidence tests to confidence-reporting tests that can be performed by humans [4, 69, 70]. We find this line of criticism much less compelling. For instance, it has been suggested that if long-term memory is not required then a task cannot be considered metacognitive. Yet, from a neuroscientific (mechanistic) perspective, it is hard to see the relevant difference between a memory representation and a perceptual representation. A second argument has been that generalization across tasks is an important requirement [51, 61, 71]. This argument is akin to the claim that a speaker of a single language, such as English, does not demonstrate linguistic competence until she is also shown to generalize to another languages, such as Hungarian. Clearly, cognitive flexibility (and knowledge of Hungarian) are advantageous skills but not necessary to demonstrate linguistic competence. Likewise, it may be expected that confidence reporting, like other sophisticated cognitive capabilities may not be solved in a fully general form by most animals or even humans. Finally, some studies have been criticized on the basis of the number of subjects (e.g., "2 monkeys alive have metacognition") [72]. We find this criticism somewhat out of line, especially considering that it has long been routine in monkey psychophysical and neurophysiological studies to use only two subjects. The legitimacy of extrapolating from few subjects is based in part on the argument that individuals of a species share a common neural, and hence cognitive, architecture.

We conclude this section by noting with some puzzlement that it has rarely, if ever, been suggested that human behavioral reports of confidence themselves might be suspect. Why should it be taken for granted that self-reported confidence judgments in humans require an instance of metacognition and uncertainty monitoring processes? Ultimately, whether applied to human or to other animals, we are stuck with observable behavior. That human behavioral reports can have a linguistic component while animal reports cannot does not justify two entirely distinct sets of rules for human versus animal experiments. Regardless of the species of the subject, we ought to determine whether a particular behavioral report can be implemented through a simpler mechanism, such as associative learning. In order to best make this case it is critical to be very careful about how confidence behavior is defined. To do so, we will argue that semantic definitions need to be dropped in favor of formal (mathematical) ones. It is to this topic that we turn next.

6.3 Computational Perspective on Confidence Judgments

The study of decision making provides important insights and useful departure points for a computational approach to uncertainty monitoring and confidence judgments. Smith and colleagues in their groundbreaking review advocated and initiated a formal approach to study confidence judgments [5]. We argue that this approach can be taken further to provide a mathematically formal and quantitative foundation. That is because formal definitions can yield concrete, testable predictions without resorting to semantic arguments about abstract terms [73, 74]. To seek a formal basis for confidence judgments, we will first consider computational models of simpler forms of decision making [75].

From a statistical perspective a two-choice decision process can be viewed as an hypothesis test. In statistics each hypothesis test can be paired with an interval estimation problem to compute the degree of confidence in the hypothesis [6]. Perhaps the most familiar quantitative measure of confidence is the P value that can be computed for a hypothesis test. Indeed, the notion of confidence is truly at the heart of statistics, and similarly it should be at the center of attention for decision making as well. Moreover, statistical analysis provides a solid departure point for any attempt to seek psychophysical or neural evidence for confidence.

We begin with the core idea that confidence in a decision can be mechanistically computed and formalized in appropriate extensions of decision models. First, we will define confidence and then discuss how to derive (compute) it.

6.3.1 Defining Confidence

Confidence can be generally defined as the degree of belief in the truth of a proposition or the reliability of a piece of information (memory, observation, prediction). Confidence is also a form uncertainty, and, previously, several classifications of uncertainty have been discussed. In psychology, external and internal uncertainties have been referred to as "Thurstonian" and "Brunswikian" uncertainty, respectively [76, 77]. In economics, there are somewhat parallel notions of "risk" and "ambiguity" [78–82]. Risk refers to probabilistic outcomes with fully known probabilities, while in the case of ambiguity, the probabilities are not known.

Here, we focus on decision confidence, an important instance of uncertainty, which summarizes the confidence associated with a decision. Decision confidence can be defined, from a theoretical perspective, as an estimate by the decision maker of the probability that a decision taken is correct. Note that we also use "decision uncertainty" interchangeably, after a sign change, with "decision confidence."

6.3.2 Bayesian and Signal Detection Models for Decision Confidence

Signal detection theory (SDT) and Bayesian decision theory [6] provide quantitative tools to compare the quality of stimulus representation in the neurons and with variability in behavioral performance [83]. These quantitative approaches have provided a strong basis for probing the neural mechanisms that underlie perception [84].

We begin with a model for a two-alternative decision, such as the olfactory mixture discrimination paradigm discussed above. For the purposes of our argument we will consider a simplified case of determining whether stimulus, s is above or below a given boundary, b. Let us assume that b is fixed and stimulus s is corrupted by Gaussian noise (Fig. 6.3a). SDT provides a language for analyzing such decisions under uncertainty and has been used previously to derive predictions for decision confidence [57, 85]. In each trial, the observer draws a sample s_i from the stimulus distribution (Fig. 6.3a). The choice can be made simply by determining whether $s_i < b$ or $b > s_i$. Confidence, d, is a function of the distance between these variables, $d_i = |s_i - b|$.

This model yields specific predictions about how the representation of confidence relates to other variables. First, by definition confidence predicts the probability of a choice being correct, in keeping with the intuitive notion of "confidence" (Fig. 6.3c). Second, when computed as a function of the stimulus difficulty and choice outcome (the observables in a 2AFC task), the model predicts a distinctive and somewhat counterintuitive "X" pattern (Fig. 6.3b), in which confidence increases with signal to noise for correct choices while decreasing for error choices. Examining the stimulus and boundary configurations that could lead to a given choice offers an intuition behind this (Fig. 6.3a). For correct choices, distance between stimulus distribution and the category boundary increases as the stimulus becomes easier. For error choices, however (which happen when a stimulus is perceived to be on the wrong side of the boundary), the distance between sampled stimulus and categorization boundary tends to be smaller for easy stimuli because the overlapping area of the two distributions becomes smaller. In other words, for easier stimuli errors are rare, but in those cases where they do occur the decision maker cannot have been very confident.

Note that these patterns are not only robust to different stimulus distributions, but can also be derived from other classes of models. For instance it can be simply generalized to Bayesian decision theory, where the absolute value of the log posterior ratio can be used as a measure of the "decision distance," d, above, and provide an estimate of confidence [86]. Similarly, models based on integration of evidence [57, 87–89], attractor networks [90, 91], and even support vector machine classifiers [92] make similar predictions about confidence, as we discuss next.

6.3.3 *Other Models of Decision Confidence: Integrators, Attractors, and Classifiers*

We have seen how natural it is to introduce a notion of confidence in SDT. Integrator or drift–diffusion models of decision making can be seen as adding a temporal dimension to SDT [93, 94], and we can extend these models in similar ways. This is most natural to examine for the "race" variant of the integrator

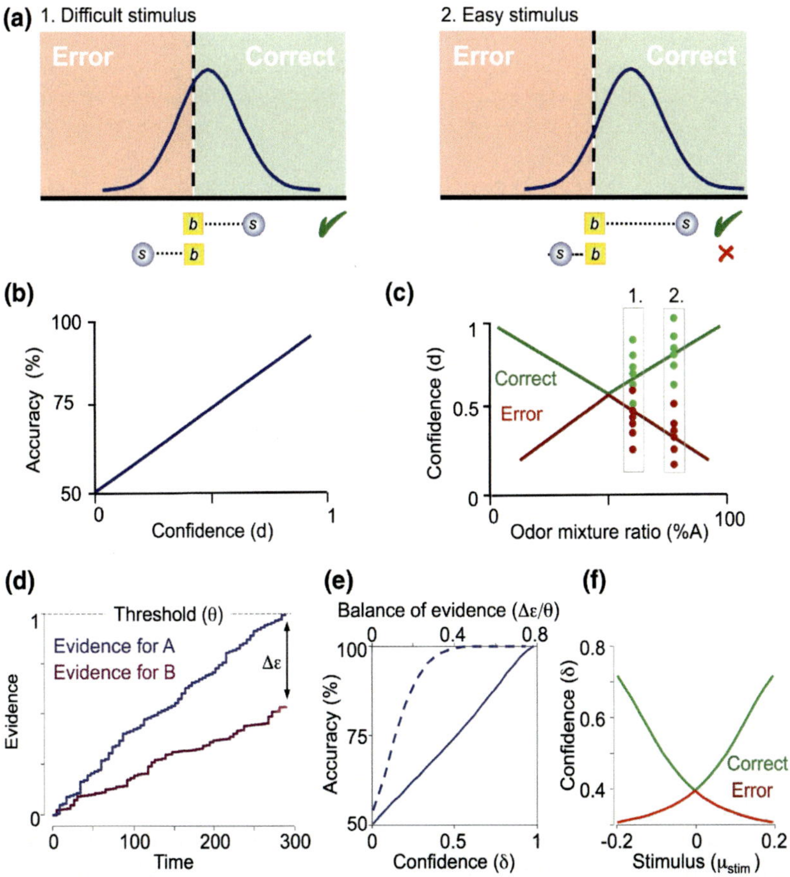

Fig. 6.3 Computational models for choice and confidence. (**a**–**c**) Signal detection theory model for choice and confidence. **a** On a given trial a decision maker has to determine whether stimulus s is above or below the boundary (*dashed line*). *Left panel* shows a stimulus distribution close the boundary and hence difficult decisions on average, the right panel shows stimulus distribution with smaller overlap. Decision confidence can be estimated by computing the distance of the stimulus s (*yellow*) and boundary, b, (*blue*), $d = |s - b|$. Error choices occur when s is to the *left* of b (where the stimulus distribution extends into the *red shaded* region). When the stimulus distribution is easier (*right*), the *red* region under the curve shrinks and the *green* region expands. Thus, the maximum distance between s and b for error choices is lower and the maximum for correct choices higher, so confidence estimates average lower for the rare easy-stimulus error than for difficult stimulus errors. **b** The resulting confidence estimate, d, can be used to predict choice accuracy. **c** Confidence estimates (average d) show a characteristic "X" pattern plotted as a function of stimulus difficulty and choice correctness. **d**–**e** Race model of decision making. **d** Schematic showing the accumulation of evidence in two integrators up to a threshold. **e** Calibration of "balance of evidence" yields veridical confidence estimate. **f** Prediction for confidence estimate as a function of observables

model where evidence for and against a proposition accumulates in separate decision variables (Fig. 6.3d). Vickers proposed that confidence in a decision may be computed as the "balance of evidence," the difference between the two decision variables at decision time [95]. This distance can be transformed into a veridical estimate of confidence with qualitatively the same properties as the estimate from SDT models (Fig. 6.3e, f).

Another class of models where similar notions of confidence can be applied is classifier models from machine learning theory. For instance, support vector machines (SVM) are a class of algorithms that learn by example to assign labels to objects. In a probabilistic interpretation of SVM classifiers [92], the size of the margin for a sample (distance of the separating hyperplane to the sample) is proportional to the likelihood of that point belonging to a class given the classifier (separating hyperplane). This yields what is known technically as a measure of the "posterior variance of the belief state given the current model," and, in other words, an estimate of confidence about the category [92, 96]. Beyond providing a good prediction of classification accuracy, this confidence measure also yields the same qualitative "X" patterns when plotted as function of stimulus and outcome.

These normative models can account for how confidence may be computed algorithmically but not for how confidence can be used to make choices, such as the confidence-guided restart decisions discussed above. Insabato et al. [91, 97] introduced a two-layer attractor network based on integrate-and-fire neurons that can accomplish this. The first network is a competitive decision-making module whose dynamics compute a categorical choice based on a noisy (uncertain) stimulus. This decision network then feeds into a second attractor network in which a "low confidence" and a "high confidence" neuron pool compete with the winning population representing the decision to stay or restart, respectively. The attractor networks are not handcrafted and tuned for this purpose but rather based on generic decision networks [97, 98] that have been used to account for other decision processes. Interestingly the design of this model suggests a generic architecture in which one network monitors the confidence of another, similar to cognitive ideas about uncertainty monitoring [99].

Some of the specific models presented above can be interpreted as normative models, prescribing how the computation of confidence ought to be done based on some assumptions. In this sense they are useful for describing what a representation of confidence or its behavioral report should look like. Some of these models are also generative and can be taken literally as an algorithm that neural circuits might use to compute confidence. These models may be also useful in considering criticisms, such as the argument that some metacognitive judgments can be performed simply by stimulus–response learning based on internal stimulus representations. At least for certain behavioral tasks, as we have shown, this is not the case since mapping confidence requires a specific, highly nonlinear form of the subjective stimulus beyond simple forms of associative learning. Although more sophisticated algorithms might compute confidence in different ways, ultimately what characterizes confidence is that it concerns a judgment about the quality of a subjective internal variable, notwithstanding how that response is learned. In this sense the models presented here can serve as normative guideposts.

Taken together these computational modeling results establish, first, that computing confidence is not difficult and can be done using simple operations and, second, that the results are nearly independent of the specific model used to derive it, and finally that it requires computations that are distinct from the computation of other decision variables such as value or evidence.

6.3.4 Veridical Confidence Estimates and Calibration by Reinforcement Learning

Thus far, we discussed how simple models can be used to compute decision confidence on a trial-by-trial basis. However, we dodged the question of how to find the appropriate transform function that will result in a veridical confidence estimate. First, what do we mean by an estimate being veridical? A veridical estimate is the one that correctly predicts the probability of its object. For instance, in Bayesian statistics the posterior probability of an event is a veridical estimate of confidence. In our examples above, we call estimates veridical if they are linearly related to accuracy. Of course, most confidence judgments are not entirely veridical; in fact people tend to systematically overestimate or underestimate their confidence. While such systematic deviations have been extensively studied, they are likely to involve a host of emotional and social factors that we wish to leave aside for now. Rather we focus on the basic computational question of how can naïve confidence estimates be tuned at all so they roughly correspond to reality [100].

To obtain veridical confidence estimates, it is necessary to calibrate the transfer function (e.g., Fig. 6.3e). We can assume that the calibration transform changes on a much slower time scale compared to variations in confidence, and hence this computation boils down to a function learning problem. Therefore, a subject can use reinforcement learning, based on the difference between the received and predicted outcome (derived from confidence), to learn the appropriate calibration function. Interestingly, consistent with this proposal, experiments show that confidence ratings in humans become more veridical with appropriate feedback [17]. Note that this use of reinforcement learning to calibrate confidence still relies on a trial-to-trial computation of a confidence estimate.

6.3.5 Applying Predictions of Confidence Models to Confidence-Reporting Tasks

These computational foundations for confidence emphasize the separation between *how* a particular representation is computed and *what* function it ultimately serves. But in order to study confidence nearly all behavioral tasks exploit the fact that animals try to maximize their reward and therefore incentives are set up so that maximizing reward requires the use of confidence information. There is a multitude of possible approaches, for instance using the idea that confidence can also be used to drive information-seeking behavior [66, 101, 102], which is exploited in the more ethologically configured "looking" paradigms discussed above. Clearly, confidence signals can have many functions, and correspondingly many psychological labels. Therefore, our first goal is not to study how confidence is functionally used ("reward maximization" or "information seeking") but rather its algorithmic origin, how it was computed. To accomplish this we can use the computational models introduced above to formally link the unobservable internal variable, confidence, to observable variables, such as stimuli and outcomes. This general strategy is beginning to be used to infer various decision variables such as subjective value representations [101, 103–107].

In order to argue that rats reported their confidence by restarting, we showed that the probability of restarting was not only dependent on stimulus difficulty but also the correctness of the choice. Figure 6.2d shows the observed (folded) "X" pattern for the "leaving decision" task variant. This pattern of data is critical in that it can rule out the two main criticisms discussed above with reference to confidence tests. First, this pattern cannot be explained by assuming that reinforcement learning assigned a particular degree of confidence to each stimulus. That is because correct and error choices for the same stimulus are associated with different confidence measures. Note we do not exclude the possibility that reinforcement learning processes may be used to calibrate confidence on a slower timescale as discussed above. In this respect, fitting confidence reports to reinforcement learning models is useful to rule out the contribution of such process to correct/error differences in confidence reports. Second, since this pattern is "non-factorizable," it cannot be reproduced by independently combining stimulus difficulty effects, manipulated by the experimenter, with a waxing and waning internal factor, such as vigilance or attention. This rules out the alternative explanation used for the "decline option" tests [39, 40] according to which decline choices follow a stimulus difficulty factor times an attention or memory-dependent factor. The leaving decision version of this task enables even stronger inferences, because waiting time is a graded variable. Indeed, as expected for a proxy for confidence, waiting time predicts decision accuracy (Fig. 6.2c). Moreover, these trial-to-trial confidence reports can be directly fit to alternative models, such as those based on reinforcement learning in an attempt to exclude them [57].

It is interesting to note that the same method of separating correct and error choices could be applied to the "post-decision wagering" test [46]. While appropriate data from these tasks, i.e., sorted by both correct and error as well as by difficulty, may already be available, to our knowledge they have not been reported in this way.

Also note that some other tasks, such as the "decline option" test or "uncertain option" task, do not admit this possibility because one obtains either an answer or a confidence judgment, but not both, in any given trial. If the animal declines to take the test, you get no answer, so you have no error trials to look at. As a result only weak inferences are possible leaving us with a plethora of alternative explanations for the observed data [3, 4, 47, 69, 70].

To summarize the past two sections, the lesson we take from these studies and the related debates is three-fold. First, confidence-reporting tasks should collect data about the choice and the confidence associated with it *in the same trial* and for as large a fraction of trials as possible. Lacking this, it is difficult if not impossible to rule out alternative mechanisms. Second, the confidence readout should ideally be a *graded* variable. Finally, we believe that to call a particular behavior a "confidence report" we need to drop semantic definitions and focus on formal accounts of confidence by fitting appropriate models to the behavioral data.

6.4 A Case Study of Decision Confidence in Rat Orbitofrontal Cortex

Although computational models can be used to rule out and to some degree infer certain computational strategies, behavioral reports alone provide fundamentally limited evidence about the mechanisms generating that report. Therefore, ultimately we need to look into the brain and attempt to identify the necessary neural representations and processes underlying the assumed computations. Recent studies have identified single neuron correlates of confidence in a handful of brain regions, the orbitofrontal cortex of rats, and the dorsal pulvinar, parietal cortex, and supplementary eye field regions of rhesus monkeys. In the following we will focus on a study conducted by us, together with several colleagues, that illustrates the application of the computational approach described above.

6.4.1 Representation of Decision Confidence in Orbitofrontal Cortex

We wondered if orbitofrontal cortex, an area involved in representing and predicting decision outcomes, carries neural signals related to confidence [108–110]. Our principal strategy was to look for neural correlates of confidence and try to understand their origin in a mechanistic framework [57]. We recorded neural firing in OFC while rats performed the olfactory mixture categorization task described above and focused our analysis on the reward anticipation period. This is the period of waiting for a rat at the reward port, after a choice was made but before any feedback was received about choice correctness. Figure 6.4 shows an example neuron whose firing rate during this anticipation period signals decision confidence.

How did we establish that this is confidence and not some other variable? Similar to our approach to analyzing the confidence-reporting behavior, we first

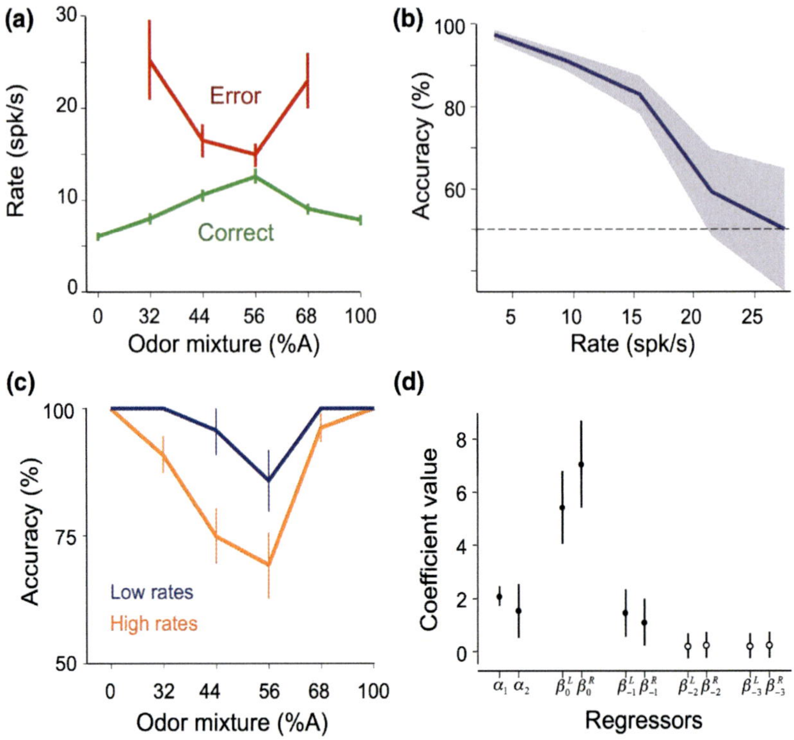

Fig. 6.4 Neural correlates of decision confidence in rat orbitofrontal cortex. Firing rate analyses for a single OFC neuron. The rate was computed during the reward anticipation period, after the animal made its choice and before it received feedback. Note that this neuron increases its rate with decreasing confidence, hence it signals "decision uncertainty." **a** Tuning curve as a function of stimulus and outcome. **b** Firing rate predicts accuracy and even the highest rates predict above chance performance. **c** Psychometric function conditioned on firing rate. Trials were divided into high or low rate trials based on a median split. **d** Regression analysis of firing rate based on reward history. Coefficient α_1 is an offset term, α_2 is stimulus difficulty, and β coefficients represent outcomes (correct/error) divided by left/right choice side and a function of recent trial history (current trial = 0). Note that the largest coefficients are for the current trial and beyond the past trial the coefficients are not significantly different from zero (*empty circles*)

plotted firing rates as a function of the stimulus and outcome (two observables). We noticed the same non-factorizable "X" pattern as a function of stimulus and outcome that are a key prediction of confidence models (Fig. 6.4a). This is also the pattern of behavioral responses we observed during leaving decisions. Second, the firing rates predict accuracy, being highest on average for trials associated with chance performance and lowest on average for trials with near perfect performance (Fig. 6.4b). This is essentially a definition of confidence (or rather its inverse in this case, decision uncertainty). Third, to show that these neurons predict accuracy beyond what is knowable from the stimulus, we plotted behavioral accuracy as a function of stimuli conditioned on low and high firing rates (Fig. 6.4c). This shows that the correlation between firing rates and performance is not solely due to stimulus-information, because knowing even a single bit about firing rates (high or low) can significantly improve behavioral predictions. Fourth, the firing rates could not be explained by recent reward history as determined by a regression analysis, showing that reinforcement learning based on past experiences with outcomes did not produce this pattern. Note, that some forms of performance fluctuations coupled to reinforcement learning could result in different firing rates for correct and error choices. Therefore, while the "X" pattern is suggestive, it is crucial to explicitly rule out these history-based mechanisms. Although in principle this analysis could be applied to binary choices as well, it is more powerful for continuous variables like firing rate. This result implies that firing rates were produced by a process that uses information mostly from the current trial. This is consistent with confidence estimation but not with reinforcement-based predictions. Taken together, the most parsimonious explanation of these data is that neurons have access to a measure of decision confidence.

We found over 20 % of neurons in OFC had correlates like the neuron described above that can be called "decision uncertainty," while about 10 % carried a signal of the opposite sign, "decision confidence." This confidence signal is a scalar quantity, which is why it is surprising that so many OFC neurons encoded this single variable. OFC supports a broad range of functions and we expect that it will encode other variables as well. The current understanding of OFC suggests that it is mainly involved in outcome prediction [108, 109, 111]. Predicting outcomes is difficult and different situations call for different computational mechanisms. In a psychophysical decision task like the one we used, the only source of stochasticity is the decision of the animal. Therefore, an estimate of confidence provides the appropriate prediction of trial outcome. At the same time, we expect that OFC neurons will incorporate different signals as needed to make outcome predictions, as we discuss elsewhere [112].

At this point it is important to return to semantics. First, we use the word confidence to refer to the formal notion of decision confidence, which happens to overlap to a large degree with our intuitive notion of confidence. Second, we wish to emphasize strongly that we are not labeling these neurons as representing "confidence" or anything else; the claim being made is that they must have had *access* to mechanisms that computed an estimate of confidence. In this sense what we established is not a neural correlate of a behavior but rather a computational basis for a neural signal. This is both a strength and a shortcoming of our previous study. We did not show that the neurons' firing correlates with the confidence behavior on a trial-by-trial basis, and in this sense we did not establish a neural correlate of an observed behavior. On the other hand, there is a long and troubled history of labeling neurophysiological signals with psychological concepts based on a behavioral correlate alone [113]. Indeed, our neurons may also be correlated with "anxiety," "arousal," or "exploration". And in fact these concepts can be related to different uses of uncertainty, and may all turn out to be neural correlates in some behavioral conditions. Rather our interpretation hinges on the only class of computational mechanisms that we found could successfully explain the observed firing patterns. This also implies that our claims can be disproved, either by showing that alternative models, without computing confidence, can also account for our data, or make predictions based on these models that are inconsistent with our data.

Although we use decision confidence in the formally defined sense, as the likelihood that a choice was correct based on the available evidence, this definition overlaps to some extent with our intuitive notion about confidence. Nevertheless, it will be important to directly assess how formal definitions of confidence and implicit confidence reports by animals correspond to the human notion of subjective confidence.

6.5 Behavioral Uses of Confidence Estimates

Until now we discussed confidence in a limited context focusing on explicit or implicit reports, and argued that formalized notions based on statistics provide a useful way forward. Placing this topic in a broader context, there is a vast literature in neuroscience, psychology, economics, and related fields showing that uncertainty and confidence are critical factors in understanding behavior [80, 82, 114–118]. Most of these fields use formal notions of confidence, and while impacting behavior in lawful ways this need not always correlate with a conscious sense of confidence. To highlight this dissociation we briefly point to a set of examples where confidence signals are used to guide behavior in the apparent absence of awareness. Interestingly, in these cases humans and nonhuman animals seem to be on par in the uses of uncertainty. Note, however, that in several of the examples although the requisite monitoring processes might use uncertainty, an explicit report may not be available. It will be valuable to examine how the uses of uncertainty and confidence in these behavioral situations are related to the metacognitive notion of confidence [119].

6.5.1 Foraging and Leaving Decisions

As animals search for food they must continually assess the quality of their current location and the uncertainty about future possibilities to find new food items [120]. In other words, they must continually decide whether to stay or to go, depending on their level of confidence in the current location or "patch." Behavioral observations suggest that the time an animal spends at a particular patch depends not only on the mean amount of food but also the variability [121, 122]. Optimal foraging theory can account for these features by incorporating information about variability and other costs [120, 123]. The patch allocation, i.e., the time animals spend at a particular location, should depend on the uncertainty of the estimate of its value [120]. Similarly, our "decision restart" and "leaving decision" tasks provide a psychophysical instantiation of a foraging decision: whether to stay at the reward port or leave and to start a new trial. The optimal solution here also depends on uncertainty concerning the immediately preceding perceptual decision, which determines the outcome. Therefore, foraging decisions are an example where uncertainty estimates are directly turned into actions, an on-line use of uncertainty as a decision variable.

6.5.2 Active Learning and Driving Knowledge Acquisition

When you are not confident about something, it's a good time to learn. A subfield of metacognition refers to this as "judgments of learning" [124, 125]. The notion is that in order to figure out how much and what to learn, one needs to have meta-representations [126, 127]. Interestingly, the field of machine learning in computer science uses a very similar but quantitative version of this insight. Statistical learning theory proposes that "active learners" use not only reinforcements but also their current estimates of uncertainty to set the size of updates, i.e., learn more when uncertain and less when certain [128]. For instance, the Kalman-filter captures the insight that learning rate ought to vary with uncertainty [129]. Simplified versions of the Kalman-filter have been used to account for a range of findings in animal learning theory about how stimulus salience enhances learning [130–132].

One of the key uses of estimating uncertainty or knowing your confidence is to drive information-seeking behavior so as to reduce the level of uncertainty. This is related both to foraging decisions as well as the active learning examples given immediately above. The basic idea here is that the value of information is related to the uncertainty of the agent [102]. When the agent is very confident about the state of the world then the value of information is low [133, 134]. When the agent is less confident, then the value of information is high. Thus, when faced with a decision of how much time to allocate to information gathering or a decision between exploiting current information versus acquiring a new piece of information, we would expect that a representation of uncertainty might be particularly useful [135]. In other words, we expect the value of exploration to decrease proportionately with the current confidence in that piece of information.

6.5.3 Statistical Inference and Multisensory Integration

Perhaps the most ubiquitous and important use of uncertainty is in the process of statistical inference: using pieces of partly unreliable evidence to infer things about the world [6]. In principle, probability theory, or more specifically Bayesian inference tells us how one ought to reason in the face of uncertainty [8]. What this theory says in a nutshell is that evidence must be weighted according to its confidence (or inversely according to its uncertainty). There are a growing number of examples of statistically optimal behavioral performance (in the sense of correctly combined evidence), mainly in humans [136–138]. Since these examples involve primarily "low-level" sensorimotor tasks to which humans often have no explicit access, it is not clear whether explicit ("metacognitive") access to confidence estimates are relevant. For example, a tennis player is unlikely to be able to report his relative confidence in the prior of where he expected the ball to fall as compared to his confidence in the evidence provided by the ball's present trajectory. Nevertheless, his swing is accurate. It has been suggested that such problems reflect a probabilistic style of neural computation, one that would implicitly rather than explicitly represent confidence [9–11, 139, 140].

6.6 Summary

Confidence judgments appear to us as personal, subjective reports about beliefs, generated by a process of apparent self-reflection. If so, then could animals also experience a similar sense of certainty? And is it even possible to ask this question as a pragmatic neuroscientist with the hope of finding an answer?

Undeniably, the concepts of "confidence" and "uncertainty" have established meanings in the context of human subjective experience. The importance of these concepts greatly motivates our research and therefore it is important to assess the relationship between formally defined measures and the subjective entity in humans. Of course, it is impossible for us to know whether animals, in any of the tasks discussed above, ever feel the same sense of uncertainty that we humans do. In fact, some philosophers argue that it is impossible for us to know whether anyone else experiences the same sense of uncertainty—the problem of other minds. But in practice it appears that verbal sharing of confidence information in humans can achieve the same goal of metacognitive alignment [141].

Before we can approach these vexing questions from a scientific perspective, it is important to establish that there is no justification in having distinct rules for interpreting human and animal experiments. Behavioral experiments, both in humans and in animals, need to be interpreted based on observables and not subjective experiences only accessible via introspection. This demands a behaviorist perspective but also new tools to go beyond old-fashioned, denialist behaviorism so that we are able to study a variable that is not directly accessible to measurement. This is possible using model-based approaches that enable one to link hidden, internal variables driving behavior to external, observable variables in a quantitative manner. Such formal approaches not only enable us to drop semantic definitions, but also to go beyond fruitless debates. Models are concrete: they can be tested, disproved, and iteratively improved, moving the scientific debate forward.

Here we outlined an approach to studying confidence predicated on two pillars: an appropriately designed behavioral task to elicit implicit reports of confidence, and a computational framework to interpret the behavioral and neuronal data. Establishing a confidence-reporting behavior requires us to incentivize animals to use confidence, for instance by enabling animals to collect more reward or seek out valuable information based on confidence. We saw that in order to interpret behavior and rule out alternative explanations, it is crucial to use tasks where data about the choice and the confidence associated with it are collected simultaneously in the same trial. Moreover, it is advantageous, although not required, that the confidence report is a graded variable and the task provides a large number of trials for quantitative analysis. To begin to infer behavioral algorithms for how confidence may have been computed (or whether it was), we presented a normative theoretical framework and several computational models.

In so doing, we have tried to lift the veil of this murky, semantically thorny subject. By showing that confidence judgments need not involve mysterious acts of self-awareness but something more humble like computing the distance between two neural representations, we hope to have taken a step toward reducing the act of measuring the quality of knowledge to something amenable to neuroscience, just as the notion of *subjective value* and its ilk have been [107, 142–144]. Indeed, recent studies on the neural basis of confidence have brought a neurobiological dawn to this old subject [18, 19, 36, 40, 53, 57, 97]. We also believe that as a consequence of this demystification, animals may be put on a more even footing with humans, at least with respect to the confidence-reporting variety of metacognition. Yet this may reflect as much a humbling of our human abilities as a glorification of the animal kingdom.

Acknowledgments We are grateful to our collaborators and members of our groups for discussions. Preparation of this article was supported by the Klingenstein, Sloan, Swartz, and Whitehall Foundations to A.K.

References

1. Flavell JH (1979) Metacognition and cognitive monitoring: a new area of cognitive-developmental inquiry. Am Psychol 34:906–911
2. Metcalfe J, Shimamura AP (1994) Metacognition: knowing about knowing. MIT Press, Cambridge
3. Bjork RA (1994) Memory and metamemory considerations in the training of human beings. In: Metacognition: knowing about knowing. MIT Press, Cambridge
4. Metcalfe J (2008) Evolution of metacognition. In: Handbook of metamemory and memory, pp 185–205, 29–46
5. Smith JD, Shields WE, Washburn DA (2003) The comparative psychology of uncertainty monitoring and metacognition. Behav Brain Sci 26:317–339 (discussion 340–373)
6. Cox DR (2006) Principles of statistical inference. Cambridge University Press, Cambridge
7. Zemel RS, Dayan P, Pouget A (1998) Probabilistic interpretation of population codes. Neural Comput 10:403–430

8. Rao RPN, Olshausen BA, Lewicki MS (2002) Probabilistic models of the brain: perception and neural function. The MIT Press, Cambridge
9. Knill DC, Pouget A (2004) The Bayesian brain: the role of uncertainty in neural coding and computation. Trends Neurosci 27:712–719
10. Ma WJ, Beck JM, Latham PE, Pouget A (2006) Bayesian inference with probabilistic population codes. Nat Neurosci 9:1432–1438
11. Rao RP (2004) Bayesian computation in recurrent neural circuits. Neural Comput 16:1–38
12. Higham PA (2007) No special K! A signal detection framework for the strategic regulation of memory accuracy. J Exp Psychol Gen 136:1–22
13. Charles L, Van Opstal F, Marti S, Dehaene S (2013) Distinct brain mechanisms for conscious versus subliminal error detection. Neuroimage 73:80–94
14. McCurdy LY, Maniscalco B, Metcalfe J, Liu KY, de Lange FP, Lau H (2013) Anatomical coupling between distinct metacognitive systems for memory and visual perception. J Neurosci 33:1897–1906
15. Johnson DM (1939) Confidence and speed in the two-category judgment. Columbia University, Ney York
16. Festinger L (1943) Studies in decision: I. Decision-time, relative frequency of judgment and subjective confidence as related to physical stimulus difference. J Exp Psychol 32:291–306
17. Baranski JV, Petrusic WM (1994) The calibration and resolution of confidence in perceptual judgments. Percept Psychophys 55:412–428
18. Fleming SM, Weil RS, Nagy Z, Dolan RJ, Rees G (2010) Relating introspective accuracy to individual differences in brain structure. Science 329:1541–1543
19. Yokoyama O, Miura N, Watanabe J, Takemoto A, Uchida S, Sugiura M, Horie K, Sato S, Kawashima R, Nakamura K (2010) Right frontopolar cortex activity correlates with reliability of retrospective rating of confidence in short-term recognition memory performance. Neurosci Res 68:199–206
20. Fleming SM, Huijgen J, Dolan RJ (2012) Prefrontal contributions to metacognition in perceptual decision making. J Neurosci 32:6117–6125
21. Gigerenzer G, Hoffrage U, Kleinbolting H (1991) Probabilistic mental models: a Brunswikian theory of confidence. Psychol Rev 98:506–528
22. Klayman J, Soll JB, González-Vallejo C, Barlas S (1999) Overconfidence: it depends on how, what, and whom you ask. Organ Behav Hum Decis Process 79:216–247
23. Finn B (2008) Framing effects on metacognitive monitoring and control. Mem Cognit 36:813–821
24. Angell F (1907) On judgments of "like" in discrimination experiments. Am J Psychol 253–260
25. Watson CS, Kellogg SC, Kawanishi DT, Lucas PA (1973) The uncertain response in detection-oriented psychophysics. J Exp Psychol 99:180–185
26. Woodworth RS (1938) Experimental psychology. Henry Holt and Company Inc, New York
27. Peirce CS, Jastrow J (1885) On small differences of sensation. Mem Natl Acad Sci 3:73–83
28. George SS (1917) Attitude in relation to the psychophysical judgment. Am J Psychol 28:1–37
29. Smith JD, Schull J, Strote J, McGee K, Egnor R, Erb L (1995) The uncertain response in the bottlenosed dolphin (Tursiops truncatus). J Exp Psychol Gen 124:391–408
30. Smith JD, Schull J (1989) A failure of uncertainty monitoring in the rat. (unpublished data)
31. Shields WE, Smith JD, Washburn DA (1997) Uncertain responses by humans and rhesus monkeys (Macaca mulatta) in a psychophysical same-different task. J Exp Psychol Gen 126:147–164
32. Smith JD, Shields WE, Schull J, Washburn DA (1997) The uncertain response in humans and animals. Cognition 62:75–97
33. Shields WE, Smith JD, Guttmannova K, Washburn DA (2005) Confidence judgments by humans and rhesus monkeys. J Gen Psychol 132:165–186

34. Beran MJ, Smith JD, Redford JS, Washburn DA (2006) Rhesus macaques (Macaca mulatta) monitor uncertainty during numerosity judgments. J Exp Psychol Anim Behav Process 32:111–119
35. Sole LM, Shettleworth SJ, Bennett PJ (2003) Uncertainty in pigeons. Psychon Bull Rev 10:738–745
36. Komura Y, Nikkuni A, Hirashima N, Uetake T, Miyamoto A (2013) Responses of pulvinar neurons reflect a subject's confidence in visual categorization. Nat Neurosci 16:749–755
37. Smith JD, Beran MJ, Redford JS, Washburn DA (2006) Dissociating uncertainty responses and reinforcement signals in the comparative study of uncertainty monitoring. J Exp Psychol Gen 135:282–297
38. Sutton RS, Barto AG (1998) Reinforcement learning: an introduction. MIT Press, Cambridge
39. Hampton RR (2001) Rhesus monkeys know when they remember. Proc Natl Acad Sci USA 98:5359–5362
40. Kiani R, Shadlen MN (2009) Representation of confidence associated with a decision by neurons in the parietal cortex. Science 324:759–764
41. Inman A, Shettleworth SJ (1999) Detecting metameory in nonverbal subjects: a test with pigeon. J Exp Psychol Anim Behav Process 25:389–395
42. Teller SA (1989) Metamemory in the pigeon: prediction of performance on a delayed matching to sample task. Reed College
43. Sutton JE, Shettleworth SJ (2008) Memory without awareness: pigeons do not show metamemory in delayed matching to sample. J Exp Psychol Anim Behav Process 34:266
44. Foote AL, Crystal JD (2007) Metacognition in the rat. Curr Biol 17:551–555
45. Kepecs A (2013) The uncertainty of it all. Nat Neurosci 16:660–662
46. Persaud N, McLeod P, Cowey A (2007) Post-decision wagering objectively measures awareness. Nat Neurosci 10:257–261
47. Persaud N, McLeod P (2008) Wagering demonstrates subconscious processing in a binary exclusion task. Conscious Cogn 17:565–575
48. Sahraie A, Weiskrantz L, Barbur JL (1998) Awareness and confidence ratings in motion perception without geniculo-striate projection. Behav Brain Res 96:71–77
49. Rajaram S, Hamilton M, Bolton A (2002) Distinguishing states of awareness from confidence during retrieval: evidence from amnesia. Cognitive, Affect Behav Neurosci 2:227–235
50. Son LK, Kornell N (2005) Meta-confidence judgments in rhesus macaques: explicit versus implicit mechanisms. In: Terrace HS, Metcalfe J (eds) The missing link in cognition: origins of self-reflective consciousness. Oxford University Press, Oxford, pp 296–320
51. Kornell N, Son LK, Terrace HS (2007) Transfer of metacognitive skills and hint seeking in monkeys. Psychol Sci 18:64–71
52. Middlebrooks PG, Sommer MA (2010) Metacognition in monkeys during an oculomotor task. J Exp Psychol Learn Mem Cogn 37:325–337
53. Middlebrooks PG, Sommer MA (2012) Neuronal correlates of metacognition in primate frontal cortex. Neuron 75:517–530
54. Clifford CW, Arabzadeh E, Harris JA (2008) Getting technical about awareness. Trends Cogn Sci 12:54–58
55. Fleming SM, Dolan RJ (2010) Effects of loss aversion on post-decision wagering: implications for measures of awareness. Conscious Cogn 19:352–363
56. Schurger A, Sher S (2008) Awareness, loss aversion, and post-decision wagering. Trends Cogn Sci 12:209–210 (author reply 210)
57. Kepecs A, Uchida N, Zariwala HA, Mainen ZF (2008) Neural correlates, computation and behavioural impact of decision confidence. Nature 455:227–231
58. Call J, Carpenter M (2001) Do apes and children know what they have seen? Animal Cognition 3:207–220

59. Hampton RR, Zivin A, Murray EA (2004) Rhesus monkeys (Macaca mulatta) discriminate between knowing and not knowing and collect information as needed before acting. Anim Cogn 7:239–246
60. Radecki CM, Jaccard J (1995) Perceptions of knowledge, actual knowledge, and information search behavior. J Exp Soc Psychol 31:107–138
61. Basile BM, Hampton RR, Suomi SJ, Murray EA (2009) An assessment of memory awareness in tufted capuchin monkeys (Cebus apella). Anim Cogn 12:169–180
62. Brauer J, Call J, Tomasello M (2007) Chimpanzees really know what others can see in a competitive situation. Anim Cogn 10:439–448
63. Suda-King C (2008) Do orangutans (Pongo pygmaeus) know when they do not remember? Anim Cogn 11:21–42
64. Brauer J, Call J, Tomasello M (2004) Visual perspective taking in dogs (*Canis familiaris*) in the presence of barriers. Appl Anim Behav Sci 88:299–317
65. Brauer J, Kaminski J, Riedel J, Call J, Tomasello M (2006) Making inferences about the location of hidden food: social dog, causal ape. J Comp Psychol 120:38–47
66. Bromberg-Martin ES, Hikosaka O (2009) Midbrain dopamine neurons signal preference for advance information about upcoming rewards. Neuron 63:119–126
67. Shettleworth SJ, Sutton JE (2003) Metacognition in animals: it's all in the methods. Behav Brain Sci 23:353–354
68. Smith JD, Beran MJ, Couchman JJ, Coutinho MV (2008) The comparative study of metacognition: sharper paradigms, safer inferences. Psychon Bull Rev 15:679–691
69. Schwartz BL, Metcalfe J (1994) Methodological problems and pitfalls in the study of human metacognition. In: Metacognition: knowing about knowing 93–113
70. Kornell N (2009) Metacognition in humans and animals. Curr Dir Psychol Sci 18:11–15
71. Washburn DA, Smith JD, Shields WE (2006) Rhesus monkeys (Macaca mulatta) immediately generalize the uncertain response. J Exp Psychol Anim Behav Process 32:185–189
72. Metcalfe J (2004) Drawing the line on metacognition. Behav Brain Sci 26:350–351
73. Glimcher PW (2008) Understanding risk: a guide for the perplexed. Cogn Affect Behav Neurosci 8:348–354
74. Gold JI, Shadlen MN (2001) Neural computations that underlie decisions about sensory stimuli. Trends Cogn Sci 5:10–16
75. Dayan P, Daw ND (2008) Decision theory, reinforcement learning, and the brain. Cogn Affect Behav Neurosci 8:429–453
76. Juslin P, Olsson H (1997) Thurstonian and Brunswikian origins of uncertainty in judgment: a sampling model of confidence in sensory discrimination. Psychol Rev 104:344–366
77. Vickers D, Pietsch A (2001) Decision making and memory: a critique of Juslin and Olsson's (1997) sampling model of sensory discrimination. Psychol Rev 108:789–804
78. Knight F (1921) Risk, ambiguity, and profit. Houghton Mifflin, Boston
79. Hsu M, Bhatt M, Adolphs R, Tranel D, Camerer CF (2005) Neural systems responding to degrees of uncertainty in human decision-making. Am Assoc Adv Sci 310:1680–1683
80. Glimcher PW (2003) Decisions, uncertainty, and the brain: the science of neuroeconomics. MIT Press, Cambridge
81. McCoy AN, Platt ML (2005) Risk-sensitive neurons in macaque posterior cingulate cortex. Nat Neurosci 8:1220–1227
82. Platt ML, Huettel SA (2008) Risky business: the neuroeconomics of decision making under uncertainty. Nat Neurosci 11:398–403
83. Green DM, Swets JA (1966) Signal detection theory and psychophysics. Wiley, London
84. Parker AJ, Newsome WT (1998) Sense and the single neuron: probing the physiology of perception. Annu Rev Neurosci 21:227–277
85. Maniscalco B, Lau H (2012) A signal detection theoretic approach for estimating metacognitive sensitivity from confidence ratings. Conscious Cogn 21:422–430
86. Kepecs A, Mainen ZF (2012) A computational framework for the study of confidence in humans and animals. Philos Trans R Soc Lond B Biol Sci 367:1322–1337

87. Vickers D (1970) Evidence for an accumulator model of psychophysical discrimination. Ergonomics 13:37–58
88. Moreno-Bote R (2010) Decision confidence and uncertainty in diffusion models with partially correlated neuronal integrators. Neural Comput 22:1786–1811
89. Zylberberg A, Barttfeld P, Sigman M (2012) The construction of confidence in a perceptual decision. Front Integr Neurosci 6:79
90. Rolls ET, Grabenhorst F, Deco G (2010) Decision-making, errors, and confidence in the brain. J Neurophysiol 104:2359–2374
91. Insabato A, Pannunzi M, Rolls ET, Deco G (2010) Confidence-related decision making. J Neurophysiol 104:539–547
92. Sollich P (2002) Bayesian methods for support vector machines: evidence and predictive class probabilities. Mach Learn 46:21–52
93. Mazurek ME, Roitman JD, Ditterich J, Shadlen MN (2003) A role for neural integrators in perceptual decision making. Cereb Cortex 13:1257–1269
94. Bogacz R, Brown E, Moehlis J, Holmes P, Cohen JD (2006) The physics of optimal decision making: a formal analysis of models of performance in two-alternative forced-choice tasks. Psychol Rev 113:700–765
95. Vickers D, Packer J (1982) Effects of alternating set for speed or accuracy on response time, accuracy and confidence in a unidimensional discrimination task. Acta Psychol (Amst) 50:179–197
96. Tong S, Koller D (2002) Support vector machine active learning with applications to text classification. J Mach Learn Res 2:45–66
97. Rolls ET, Grabenhorst F, Deco G (2010) Choice, difficulty, and confidence in the brain. Neuroimage 53:694–706
98. Wang XJ (2002) Probabilistic decision making by slow reverberation in cortical circuits. Neuron 36:955–968
99. Timmermans B, Schilbach L, Pasquali A, Cleeremans A (2012) Higher order thoughts in action: consciousness as an unconscious re-description process. Philos Trans R Soc Lond B Biol Sci 367:1412–1423
100. Juslin P, Olsson H (1999) Computational models of subjective probability calibration. In Juslin P, Montgomery H (eds) Judgment and decision-making: Neo-Brunswickian and process-tracing approaches. Lawrence Erlbaum Associates Inc., Mahwah NJ
101. Daw ND, O'Doherty JP, Dayan P, Seymour B, Dolan RJ (2006) Cortical substrates for exploratory decisions in humans. Nature 441:876–879
102. Cohen JD, McClure SM, Yu AJ (2007) Should I stay or should I go? How the human brain manages the trade-off between exploitation and exploration. Philos Trans R Soc B: Biol Sci 362:933
103. Pessiglione M, Seymour B, Flandin G, Dolan RJ, Frith CD (2006) Dopamine-dependent prediction errors underpin reward-seeking behaviour in humans. Nature 442:1042–1045
104. Corrado G, Doya K (2007) Understanding neural coding through the model-based analysis of decision making. J Neurosci 27:8178
105. Sugrue LP, Corrado GS, Newsome WT (2004) Matching behavior and the representation of value in the parietal cortex. Science 304:1782–1787
106. Barraclough DJ, Conroy ML, Lee D (2004) Prefrontal cortex and decision making in a mixed-strategy game. Nat Neurosci 7:404–410
107. Lau B, Glimcher PW (2005) Dynamic response-by-response models of matching behavior in rhesus monkeys. J Exp Anal Behav 84:555–579
108. Rolls ET, Grabenhorst F (2008) The orbitofrontal cortex and beyond: from affect to decision-making. Prog Neurobiol 86:216–244
109. Schoenbaum G, Roesch M (2005) Orbitofrontal cortex, associative learning, and expectancies. Neuron 47:633–636
110. Elliott R, Friston KJ, Dolan RJ (2000) Dissociable neural responses in human reward systems. J Neurosci 20:6159–6165

111. Wallis JD (2007) Orbitofrontal cortex and its contribution to decision-making. Annu Rev Neurosci 30:31–56
112. Mainen ZF, Kepecs A (2009) Neural representation of behavioral outcomes in the orbitofrontal cortex. Curr Opin Neurobiol 19:84–91
113. Maunsell JH (2004) Neuronal representations of cognitive state: reward or attention? Trends Cogn Sci 8:261–265
114. Kahneman D, Slovic P, Tversky A (1982) Judgment under uncertainty: heuristics and biases. Cambridge University Press, Cambridge; New York
115. Kaelbling LP, Littman ML, Cassandra AR (1998) Planning and acting in partially observable stochastic domains. Artif Intell 101:99–134
116. Volz KG, Schubotz RI, von Cramon DY (2005) Variants of uncertainty in decision-making and their neural correlates. Brain Res Bull 67:403–412
117. Yu AJ, Dayan P (2005) Uncertainty, neuromodulation, and attention. Neuron 46:681–692
118. Daw ND, Niv Y, Dayan P (2005) Uncertainty-based competition between prefrontal and dorsolateral striatal systems for behavioral control. Nat Neurosci 8:1704–1711
119. Fleming SM, Dolan RJ, Frith CD (2012) Metacognition: computation, biology and function. Philos Trans R Soc Lond B Biol Sci 367:1280–1286
120. Stephens DW, Krebs JR (1986) Foraging theory. Princeton University Press, Princeton
121. Bernstein C, Kacelnik A, Krebs JR (1988) Individual decisions and the distribution of predators in a patchy environment. J Anim Ecol 57:1007–1026
122. Hayden BY, Pearson JM, Platt ML (2011) Neuronal basis of sequential foraging decisions in a patchy environment. Nat Neurosci 14:933–939
123. Kamil AC, Misthal RL, Stephens DW (1993) Failure of simple optimal foraging models to predict residence time when patch quality is uncertain. Behav Ecol 4:350–363
124. Nelson TO, Narens L (1990) Metamemory: a theoretical framework and new findings. Psychol Learn Motiv: Adv Res Theory 26:125–173
125. Son LK, Metcalfe J (2000) Metacognitive and control strategies in study-time allocation. J Exp Psychol Learn Mem Cogn 26:204–221
126. Son LK (2004) Spacing one's study: evidence for a metacognitive control strategy. J Exp Psychol Learn Mem Cogn 30:601–604
127. Son LK, Sethi R (2006) Metacognitive control and optimal learning. Cogn Sci: Multi J 30:759–774
128. Schohn G, Cohn D (2000) Less is more: active learning with support vector machines. In: Machine learning-international workshop then conference, pp 839–846
129. Sutton RS (1992) Gain adaptation beats least squares. In: Proceedings of the 7th Yale workshop on adaptive and learning systems, pp 161–166
130. Dayan P, Kakade S (2001) Explaining away in weight space. Advances in neural information processing systems, pp 451–457
131. Pearce JM, Hall G (1980) A model for Pavlovian learning: variations in the effectiveness of conditioned but not of unconditioned stimuli. Psychol Rev 87:532–552
132. Courville AC, Daw ND, Touretzky DS (2006) Bayesian theories of conditioning in a changing world. Trends Cogn Sci 10:294–300
133. Dearden R, Friedman N, Russell S (1998) Bayesian Q-learning. In: Proceedings of the national conference on artificial intelligence. Wiley, pp 761–768
134. Strens M (2000) A Bayesian framework for reinforcement learning. In: Machine learning-international workshop then conference, pp 943–950
135. Knight FH (1921) Risk, uncertainty and profit. Boston and New York
136. Ernst MO, Bülthoff HH (2004) Merging the senses into a robust percept. Trends Cogn Sci 8:162–169
137. Körding KP, Wolpert DM (2004) Bayesian integration in sensorimotor learning. Nature 427:244–247
138. Graf EW, Warren PA, Maloney LT (2005) Explicit estimation of visual uncertainty in human motion processing. Vision Res 45:3050–3059

139. Körding KP, Wolpert DM (2006) Bayesian decision theory in sensorimotor control. Trends Cogn Sci 10:319–326
140. Deneve S (2008) Bayesian spiking neurons I: inference. Neural Comput 20:91–117
141. Bang D, Mahmoodi A, Olsen K, Roepstorff A, Rees G, Frith C, Bahrami B (2014) What failure in collective decision-making tells us about metacognition. In: Fleming SM, Frith C (eds) The cognitive neuroscience of metacognition. Springer, Berlin
142. Montague PR, Berns GS (2002) Neural economics and the biological substrates of valuation. Neuron 36:265–284
143. Sugrue LP, Corrado GS, Newsome WT (2005) Choosing the greater of two goods: neural currencies for valuation and decision making. Nat Rev Neurosci 6:363–375
144. Padoa-Schioppa C, Assad JA (2006) Neurons in the orbitofrontal cortex encode economic value. Nature 441:223–226

Chapter 7
Shared Mechanisms for Confidence Judgements and Error Detection in Human Decision Making

Nick Yeung and Christopher Summerfield

Abstract People give accurate evaluations of their own choices and decisions: they are often aware of their mistakes without needing feedback, and report levels of confidence in their choices that correlate with objective performance. These metacognitive judgements guide current and future behaviour, helping people to avoid making the same mistake twice and to evaluate whether they have enough information on which to base a reliable choice. Here we review progress in characterising the neural and mechanistic basis of these related aspects of metacognition—confidence judgements and error monitoring—and identify several points of convergence between methods and theories in the two fields. This convergence promises to resolve key debates in the separate literatures, to identify productive new lines of enquiry, but also to highlight shared limitations in the two research fields. In particular, future theories of choice and metacognitive evaluation may need to look beyond simple, discrete decisions to model the structure and fluidity of real-world decisions and actions that are embedded in the broader context of evolving behavioural goals.

Keywords Decision making · Confidence · Metacognition · Error monitoring

---

This chapter is adapted from: Yeung N, Summerfield C (2012) Metacognition in human decision-making: confidence and error monitoring. Phil Trans R Soc B 367:1310–1321

N. Yeung (✉) · C. Summerfield
Department of Experimental Psychology, University of Oxford, South Parks Road, Oxford OX1 3UD, UK
e-mail: nicholas.yeung@psy.ox.ac.uk

C. Summerfield
e-mail: christopher.summerfield@psy.ox.ac.uk

S. M. Fleming and C. D. Frith (eds.), *The Cognitive Neuroscience of Metacognition*, DOI: 10.1007/978-3-642-45190-4_7, © Springer-Verlag Berlin Heidelberg 2014

7.1 Metacognition in Human Decision Making

Consider a jury member in court deciding whether the defendant standing before them is innocent or guilty, a doctor deciding whether to prescribe a patient a certain course of treatment, or a trust manager deciding whether to invest in a particular stock. Common to all of these situations is that the protagonist is faced with a choice based on variable and potentially unreliable evidence, and must weigh this evidence together to reach a decision.

Psychologists and neuroscientists have become increasingly interested in the mechanisms underlying these kinds of decisions under uncertainty. Important progress has been made based on the assumption that the sorts of complex, nuanced choices mentioned above—whether to convict, prescribe, or invest—share something crucially in common with simple perceptual decisions that have been rigorously investigated in the fields of visual psychophysics and sensory neuroscience in the past few decades [21]. This research programme has been informed by computational models that offer a formal account of how categorical decisions are made in the face of ambiguous or unreliable perceptual evidence [7, 54, 62, 69], and has been given substantial impetus by findings that these models predict intricate details of the observed behaviour and neural activity in humans and other animals performing simple decision tasks [21, 54].

In this chapter, we review attempts to apply the same formal models to investigate metacognitive processes in decision making, focusing in particular on confidence judgements and error monitoring. Consider, for example, the case of the doctor making a decision about diagnosis and treatment. Accompanying this decision will be a subjective sense of confidence: the doctor may feel sure in her diagnosis and correspondingly confident about which drug to prescribe; or she may feel unsure and opt to run additional tests before settling on a line of treatment. With this new information, she may decide that her original line of thinking was wrong, and change her diagnosis and prescription accordingly.

Curiously, despite universal agreement that an accompanying sense of confidence is a subjectively salient property of almost all our decisions, there is currently little consensus about how we might incorporate decision confidence into formal models of choice behaviour or explore its biological substrates. Fundamental questions remain unanswered. For example, is the information that gives rise to the 'second-order' estimate of confidence identical to that determining the 'first-order' choice itself? Why are we generally more sure that we are correct than that we have made an error, even for difficult choices? Why do we sometimes appear to change our mind? Are these changes of mind necessarily accompanied by awareness that the initial choice was incorrect? In what follows, we discuss theories of confidence and error monitoring in the context of formal models of decision making, and point towards potentially fruitful avenues for research that draw upon common themes in these two literatures.

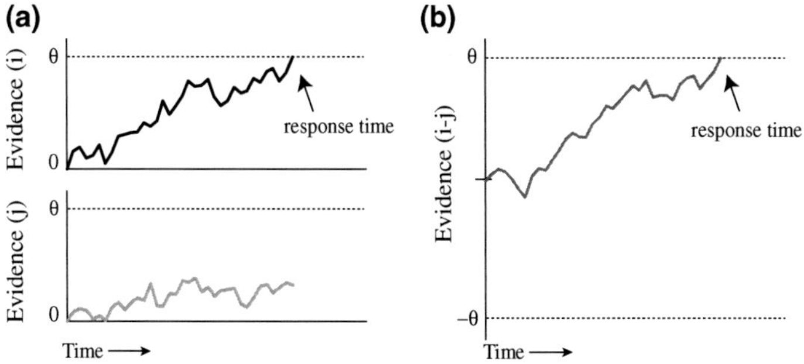

Fig. 7.1 Formal models of decision making. **a** The race model, in which separate counters accumulate evidence (the DV, *y*-axis) over time (*x*-axis) until one counter reaches threshold, θ. **b** The drift–diffusion model, in which a single DV symmetrically accumulates evidence for and against two choices. A decision is triggered when evidence reaches threshold, θ or $-\theta$, for one of those choices

7.2 Decision Confidence

7.2.1 Formal Models of Perceptual Choice

Even under good viewing conditions, visual information is corrupted by multiple sources of noise and uncertainty, arising both in the external world and in stochastic neural processes. A sensible way to increase the signal-to-noise ratio is to sample the external world repeatedly and integrate over time, making a decision only when the information is considered to be of sufficient quality [69]. This idea forms the basis of a class of model in which binary choices are described via an accumulation-to-bound mechanism, with successive samples of information totted up until they reach a criterion level, or *bound*, upon which a commitment to action is made (Fig. 7.1).

Variants of this model make differing assumptions about exactly what quantity is integrated *en route* to a decision—i.e., about precisely how the 'decision variable' (DV) is composed. In one canonical model (Fig. 7.1a), evidence is accumulated separately for each possible decision in a race to the terminating bound, with the DV (v) for each choice i updated at each sample t with an increment composed of two quantities: δ, a term encoding the strength of evidence for that choice, and η, a Gaussian noise term:

$$v_{i,t} = v_{i,t-1} + \delta_i + \eta_{i,t} \tag{7.1}$$

Decisions are made when the DV exceeds a fixed deviation from zero, θ, such that during evidence accumulation:

$$v_i < \theta \tag{7.2}$$

A prominent alternative to the simple race model is the drift diffusion model (DDM), in which a single DV encodes the difference in accumulated evidence for the two choices, i and j, thus capturing the intuition that evidence favouring one choice should simultaneously be taken as evidence against the alternative [54] (Fig. 7.1b).

$$v_t = v_{t-1} + \delta_{i-j} + \eta_t \tag{7.3}$$

Here, δ is a linear drift term whose magnitude and valence varies with the relative strength of evidence in favour of the two options. Evidence is accumulated towards symmetric positive and negative bounds θ and $-\theta$, for options i and j, respectively, such that during evidence accumulation:

$$-\theta < v < \theta \tag{7.4}$$

Framed as such, the race model and DDM form two opposing ends of a continuum of models that differ only in the degree of cross-talk between evidence accumulators, with complete independence in the race model and complete interdependence in the DDM [7, 43, 65, 72]. Substantial research effort has been devoted to determining which point on the continuum best characterises human decision making, with some consensus—but far from universal agreement—that the DDM has much to recommend it as both a normative and descriptive account of categorical choice. Thus, the DDM has been successfully applied to decision making in a range of cognitive domains—from low-level perceptual decisions to retrieval of facts from semantic memory, to economic decision making [11, 52]—while implementing a statistically optimal mechanism for decision making [7]. Nevertheless, both the race model and DDM have featured prominently in attempts to extend this formal approach to understand the subjective evaluation of confidence that accompanies each decision.

7.2.2 Models of Decision Confidence

Human subjects readily report their subjectively experienced level of confidence in a perceptual decision—for example, when asked to do so using a numerical scale after each choice—and these judgements are tightly, albeit imperfectly, linked to their objective performance [3, 24, 67]. The very earliest investigations of decision confidence revealed that, perhaps unsurprisingly, people are more certain about their perceptual choices when sensory inputs are stronger [17], and when they are given longer to sample sensory information [66]. It thus follows that confidence should reflect both the quality of the evidence (represented by the drift rate, δ) and the quantity of evidence at the choice point (represented by the absolute value of the decision bound, $|\theta|$).

How, though, can we incorporate these basic intuitions, as well as more detailed empirical evidence collected in recent years, into the formal frameworks offered by the DDM and other quantitative models of perceptual choice? Much of the research on this topic has hinged on a simple question [2, 4, 72]: Can confidence be read out directly from some feature of the decision process at the time of choice (decisional locus models), or do confidence judgements depend on further processing that might be sensitive to new information arriving beyond the decision point (post-decisional locus models)?

7.2.2.1 Decisional Locus Theories

An attractively simple hypothesis is that factors governing confidence in a choice are inherent in the processes that led to the choice in the first place. This hypothesis is plausible given that choice and confidence both relate systematically to the quality and quantity of perceptual evidence available during decision making. However, we can immediately rule out several candidate decision parameters as underlying confidence judgements. First, confidence cannot simply reflect the strength of evidence accumulated in favour of the chosen option: In the diffusion-to-bound models described above, this value is fixed at $|\theta|$ and therefore all choices should be made with precisely the same confidence (corresponding to the evidence level required to reach the bound). Meanwhile, any model proposing that confidence directly reflects evidence quality (δ) implicitly assumes that observers have direct access to this quantity—which, if they had, would obviate the need for a sequential sampling approach in the first place (see [46] for an excellent recent review).

One parameter that might transparently encode the quality and quantity of evidence is the time taken to reach a decision, and indeed it is often hypothesized that decision time could provide a frugal cue to confidence—the faster a decision was made, the more confident one should be that it is correct [2, 24, 52, 68, 72]. Within evidence accumulation models, decision time could be measured directly [43, 52] or could be inferred from features of the accumulation process such as the number of vacillations between choices in the DDM [2] or the amount of accumulated evidence in a race model [72]. However, several lines of evidence argue against the use of decision time as a cue to confidence. For example, subjects are consistently less confident when they err than when they respond correctly, even when decision times are equated, and they tend to be (rightly) more confident when they are given time to respond than when pressed for speed [67].

A decisional locus hypothesis that can account for both these findings is the *balance of evidence* view, which proposes that confidence reflects the relative strength of evidence for the chosen and unchosen options at the time of the decision [67]. This idea can be formulated simply within a race model with confidence, C, calculated as:

$$C = v_{i,T} - v_{j,T} = \theta - v_{j,T} \tag{7.5}$$

where i is the chosen option, j is the alternative, and T is the time of decision. This formulation is a close analogue of the most common interpretation of confidence within static signal detection models of perceptual choice [18, 40]. In signal detection models, decisions are made by comparing an internal evidence signal, s, to a pre-defined criterion category boundary, b, with confidence given as a monotonic function of the distance between the two ($|b - s|$). This distance metric turns out to behave much like balance of evidence in the race model [31] (see also Kepecs and Mainen, this volume): The formulations differ largely in terms of whether initial choice depends on an absolute threshold of evidence for one option (θ in the race model) or a difference threshold on evidence for both options (b in signal detection).

Two recent single-cell recording studies claim to have identified neurons encoding subjective decision confidence as envisioned by the balance of evidence hypothesis [31, 34]. The experimental set-ups were quite different: Kepecs et al. [31] measured sustained post-decision activity in the orbitofrontal cortex (OFC) of rats performing an odour discrimination task; Komura et al. [34] recorded dynamic decision-related activity in the pulvinar nucleus of the thalamus in monkeys performing a motion discrimination task. Yet a subset of cells recorded in the two studies exhibited a strikingly similar firing pattern, one that conformed to a distinctive prediction of the balance of evidence hypothesis. Specifically, this hypothesis predicts that confidence should not only be greater following correct choices than errors, but also that as decisions become easier (e.g., as the prevailing direction in a motion discrimination task becomes stronger) confidence in correct choices should increase but confidence in errors should decrease. Komura et al.'s pulvinar neurons tracked precisely this pattern, as if encoding decision confidence. Kepecs et al.'s OFC neurons showed the inverse pattern, as if encoding uncertainty (the inverse of confidence). Together these studies provide suggestive evidence that confidence is read out directly from the decision process as the balance of evidence for the competing options (for converging evidence from human neuroimaging, see [14]).

Although predominant as a model of the decision process, the DDM framework offers no balance of evidence readout of confidence because in this model there is no separate accumulation of evidence for chosen and unchosen options [43, 72]. However, in another influential single-cell recording study, Kiani and Shadlen [32] proposed an alternative decisional locus model of confidence and its neural basis that is consistent with the DDM. They recorded from neurons in the lateral intraparietal cortex (LIP) of macaques indicating a motion discrimination response with a saccade to one of two targets. In this task, LIP neurons whose receptive field overlaps with the chosen target display a characteristic acceleration of spiking activity whilst the monkey views the motion stimulus [61]. This build-up scales with signal strength, and terminates when a saccadic response is initiated, prompting the view that LIP firing rates encode a neural representation of the DV proposed by the DDM [21].

The novelty of Kiani and Shadlen's study was that the monkey was offered a 'sure bet' option on a fraction of trials, such that a certain but lower-valued reward could be obtained via a saccade to a third 'opt-out' response. The data showed that not only did the monkeys use this option judiciously—tending to opt-out when the stimulus was weak or ambiguous—but also that LIP firing rates on low-confidence trials were substantially attenuated (an effect recently replicated in Komura et al.'s pulvinar neurons, as well as in supplementary eye field neurons studied by [41]). Importantly, Kiani and Shadlen model their data using the DDM framework, arguing that confidence reflects the quantity of evidence discounted by the time needed to reach that level of evidence (a simple heuristic estimate of evidence quality):

$$C = v_t/f(t) \tag{7.6}$$

where *f*(*t*) is a monotonically increasing function that reflects the posterior odds of giving a correct response, a function that is presumed to be learnt through the experienced history of reward and punishment. This model provides a straightforward account of the first-order finding that confidence scales with evidence strength. However, by coupling confidence so tightly to evidence strength, the theory may struggle to explain other benchmark findings discussed above, such as the difference in confidence between correct responses and errors (which were not studied by Kiani and Shadlen, since 'correct' and 'error' are undefined for opt-out trials; cf. [41]).

Taken together, these single-cell recording studies present diverging views on both the algorithm and neural implementation of confidence judgements. However, they agree on a crucial point with each other and with most commonly articulated theories of confidence, that confidence is intrinsic to the decision process that led to the choice.

7.2.2.2 Post-Decisional Locus Theories and Changes of Mind

Although decisional locus theories dominate current thinking about confidence judgements, a scattering of contrary views have been presented. One proposal is that stimulus information is sampled twice in parallel, with one sample governing choice and the other confidence [28], perhaps within separate automatic and controlled pathways for action selection [12]. However, of particular relevance to the present discussion is the suggestion that confidence judgements depend crucially on post-decisional processing within the same system that gave rise to the initial choice. This hypothesis is not new (e.g., [2, 4], but has only recently been articulated in quantitative detail [46]).

A key assumption of this hypothesis is that evidence accumulation continues beyond the point at which an initial choice is made. Resulaj et al. [55] report convincing behavioural evidence for this idea. In their study, human subjects indicated the direction of a random dot motion stimulus by moving a handle to a leftwards or rightwards target some 20 cm away. This design allowed the researchers to isolate trials on which subjects began to move towards one target but then changed their mind and veered off towards the other. Careful behavioural analyses indicated that these change-of-mind trials tended to occur when, due to stochasticity in the stimulus display, motion energy initially favoured one choice but subsequently came to favour the switched-to alternative. Because the motion stimulus offset once movement began, subjects must have capitalised on the balance of information in the immediate pre-decision period. Notably, although this motion information was available prior to the decision, the switch occurred only once movement initiation began, suggesting that evidence accumulation continued beyond the point at which the initial choice was made.

To account for these and other phenomena, several researchers have proposed models in which, in contrast to the classical decision models, evidence accumulation continues even beyond the choice point [2, 4, 46]. Resulaj et al. propose that their data can be explained by just this type of model, with changes of mind occurring when latent information in the processing pipeline drives the DV across a second, 'change-of-mind' bound (for formal implementation, see [1]). A related account, the two-stage dynamic signal detection (2DSD) model [46], likewise proposes that the diffusion process continues beyond initial choice, with decision confidence reflecting the absolute value of the DV at the post-decision point at which a second-order decision is required.

Post-decisional processing models can account for a broad range of findings concerning decision confidence: First, they correctly predict that observers will change their mind more often from incorrect to correct responses than vice versa, because beyond the bound the DV on error trials will tend to regress towards the mean, whereas after correct responses it will continue to grow, driven by the true underlying drift rate. This observation also naturally explains another conundrum associated with decision confidence: that second-order confidence is generally higher for correct trials than incorrect trials. In addition, post-decisional models can accommodate evidence that people exhibit systematic variation in the time they take to give confidence judgements, being faster when they are sure than when they feel they are guessing, a finding that would be puzzling if confidence can be read off directly from the decision process itself [4].

7.3 Error Monitoring

People are often aware of their own mistakes, for example in choice RT tasks when time pressure is applied to induce errors in simple judgements. *Error monitoring* is the metacognitive process by which we are able to detect and signal our errors as soon as a response has been made. This process allows our actions to be shaped by their outcomes both in the short term, for example by responding more cautiously to avoid further errors, and in the longer term, through gradual learning of appropriate stimulus–response contingencies. Contrasting with theories of confidence, models of error monitoring argue almost exclusively for a post-decisional locus for metacognitive evaluation, following Rabbitt and colleagues pioneering work beginning in the 1960s.

7.3.1 Post-Decision Processing in Error Monitoring

Rabbitt's studies established that people can very reliably detect and correct their own errors without explicit feedback [47], but that this ability is impaired when stimulus duration is reduced [51], suggesting its dependence on continued processing of the stimulus after an initial error (which is curtailed when stimuli are very briefly presented). Error monitoring is also impaired when subsequent stimuli appear soon after the initial response [48], and responses to those later stimuli are themselves postponed following errors [29], consistent with the notion that this monitoring involves the same mechanisms as the initial decision.

Summarizing these findings, Rabbitt likened evidence accumulation in decision making to votes in a committee, in which incorrect decisions are sometimes made on the basis of incomplete information, but "as subsequent votes come in, a more accurate consensus will accumulate and the earlier mistake will become apparent" [51]. Thus, errors are characterised by biphasic evidence accumulation, with initial accumulation in favour of the incorrect response followed by later drift towards the correct decision (as the trial-average drift rate regresses to the true mean). By contrast, continued evaluation following correct responses tends simply to reinforce the original decision. This model is an obvious precursor to more recent post-decisional processing accounts of confidence [46] and changes of mind [55].

All subsequent models of error detection have adopted Rabbitt's broad framework, with debate focusing instead on the precise mechanism by which post-decision processing leads to error detection. Figure 7.2 illustrates some key model variants. Within a standard DDM framework, errors could be detected as successive crossings of decision boundaries for the two competing responses [16, 63] or as "double crossings" of a single decision bound [30]—both close relatives of Resulaj et al.'s notion of a change-of-mind bound. Errors can also be detected in terms of the occurrence of uncertainty—or *conflict*—in the decision process after an initial response [70], or as inconsistency between the outcomes of parallel decision processes at different processing stages (e.g., perceptual categorization and response selection) [12, 26].

These theories suggest that error detection depends crucially on error correction, but it turns out that detection and correction of errors are at least partly dissociable. Thus, whereas corrections can be extremely fast, occurring within 10–20 ms of the initial error [49], and may be produced even when subjects are instructed to avoid doing so [50], explicit detection and signalling of errors is much slower, voluntary and more prone to interference by distracting tasks [48]. Indeed, people sometimes remain unaware of errors that they nevertheless correct [45]. These differences suggest that explicit error detection is not an immediate

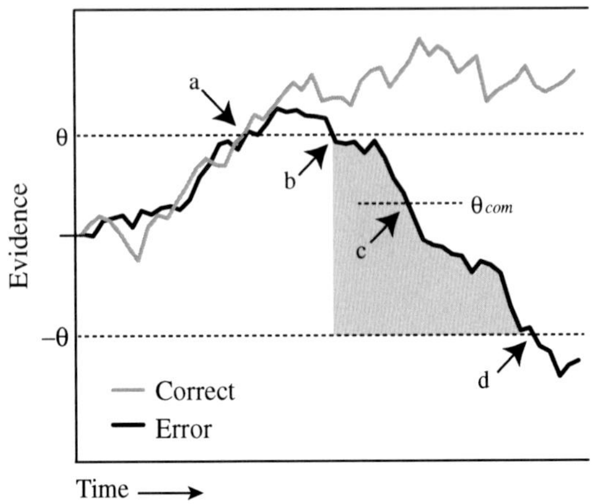

Fig. 7.2 Theories of error detection within the DDM framework. The drift diffusion process is illustrated schematically for two trials, one in which decision θ is the correct response and one trial in which this decision is incorrect. Both decisions occur at the same time point (*a*). Following the correct response, post-decision processing continues to accumulate in favour of the decision just made. Following errors, the drift rate regresses to its true mean, causing the DV to re-cross the decision bound (*b*) subsequently cross a change-of-mind bound (*c*) and finally cross the originally correct decision bound, $-\theta$ (*d*). The *grey-shaded* area indicates a period of uncertainty, or conflict, between the re-crossing of the θ bound (*b*) and later crossing of the $-\theta$ bound (*d*)

and necessary consequence of post-decision correction: Further processing or evaluation must intervene between initial correction and later explicit awareness that an error has occurred. Consistent with this analysis, recent investigations have identified dissociable neural correlates of error correction and detection.

7.3.2 Neural Substrates of Error Monitoring

Research interest in error monitoring increased substantially following the discovery of scalp EEG potentials that reliably occur time-locked to decision errors. Most studies have used a flanker task in which subjects perform a speeded categorization of a central target (e.g., *H* or *S*) while ignoring flanking distractors that are sometimes compatible (e.g., *HHH*) and sometimes incompatible (e.g., *SHS*) with that target. With modest speed pressure, error rates on incompatible trials can exceed 20 %. Following such errors, a negative event-related potential over frontocentral sites is observed within 100 ms of the incorrect response, followed by a later positive wave peaking 200–400 ms later over parietal cortex [16]. Labelled the error-related negativity (ERN/Ne) and error positivity (Pe), respectively, these EEG components have been widely studied to provide insight into error monitoring in healthy and clinical populations.

Converging evidence identifies anterior cingulate cortex (ACC) as the source of the ERN. For example, in simultaneous EEG-fMRI recordings, single-trial ERN amplitude correlates reliably only with activity in a focused ACC source [15]. The source of the Pe is less well characterised, but evidence that it is a variant of the well-studied P3 component [56] would imply widely distributed neural generators in parietal and prefrontal cortex: The P3 has been suggested to reflect the distributed action of norepinephrine released by the locus coeruleus brainstem nucleus in response to motivationally salient events during decision making [44].

Debate continues into the precise functional significance of the ERN and Pe, but it is now clear that the two components dissociably map onto processes related to error correction and detection, respectively. Thus, whereas ERN amplitude varies with both the speed [57] and probability [20] of error correction, Pe amplitude is insensitive to the strength of correcting activity when error detection rates are controlled [27]. Conversely, although both ERN and Pe covary with subjective ratings of response accuracy [20, 59], correlations involving the ERN disappear when the two components are carefully dissociated (e.g., [45, 64]). Collectively, these findings suggest that whereas the ERN directly indexes automatic post-decision processes leading to rapid error correction, the later Pe is selectively associated with explicit detection and signalling of errors. These results thus provide converging evidence for the view that error correction and detection reflect distinct processes.

7.3.3 Impact of Error Monitoring on Behaviour

A parallel line of research has considered the impact of error monitoring on future behaviour. Much of this work has focused on the finding that subjects usually respond more slowly on trials immediately following errors [47], a strategic adaptation to prevent further errors [8]. EEG studies have subsequently shown that the degree of observed slowing scales with the magnitude of error-related ERN/Pe activity [20]. Computational models implementing error-related control over distance-to-bound, $|\theta|$, account for detailed properties of empirically observed post-error slowing: In one class of model, detection of response uncertainty (conflict) immediately following error commission leads to an increase in the bound—and, hence, more cautious responding—on subsequent trials [10]. Recent extensions of this idea suggest that conflict detection may also be used to adjust the bound dynamically even as a decision is being made [8, 53].

Error signals not only support subtle adjustments that optimise online decision making; they also play a key role in longer-term adjustments during learning. Holroyd and Coles [26], for example, suggest that the ERN reflects reinforcement learning of action values. They showed that the ERN migrates in time as new stimulus–response mappings are learned, from initially being triggered by environmental feedback to later being driven by internal representations of the learned mappings, a pattern that mimics the migration of dopaminergic reward prediction error signals from unconditioned rewards to predictive stimuli during conditioning [60]. Meanwhile, fMRI activity in ACC and neighbouring cortex at the time of an incorrect response has been shown to predict response accuracy on later presentations of the relevant stimulus [25].

Most studies of post-error adjustments have focused on the ERN and ACC, reflecting wide interest in the role of ACC in reinforcement learning [26, 58], rather than on the later Pe component. However, the ERN and Pe typically covary across conditions and, when the two components are dissociated, post-error adjustments are only observed following detected errors on which a Pe component is present [45], suggesting that the latter may be a more direct correlate of the learning mechanisms by which future behaviour is adapted following an error.

7.4 Integrative Models of Decision Confidence and Error Monitoring

The preceding review has dealt separately with confidence and errors, reflecting the surprising lack of cross-fertilisation between the respective literatures to date. Yet the degree of methodological and conceptual overlap is obvious. Confidence judgements and error monitoring both entail a metacognitive evaluation of a decision just made, with only the polarity reversed: Whereas studies of confidence ask subjects whether they made a *correct* choice, studies examining error monitoring have tended to ask subjects the converse question—i.e., to report the likelihood that they made an *error*. Conceptually, we have already noted the similarity between post-decisional locus theories in the two domains.

Nevertheless, the two literatures do diverge in certain respects. In particular, decision confidence has mostly been studied using tasks that are challenging even without time pressure. Under these circumstances, subjects are sometimes sure they responded correctly and sometimes unsure, but rarely certain they made a mistake [46]. In contrast, error monitoring has mostly been studied using simple but time-pressured tasks in which subjects tend to be aware of their errors. Framed in terms of formal models of the decision process, the distinction concerns whether errors and sensitivity to processing noise arise because of low drift rate, δ, in the case of perceptual ambiguity, or adoption of a low threshold, θ, to engender fast responding. However, the distinction is one of degree rather than kind. A handful of studies have blurred the distinction entirely, asking subjects to rate their confidence ranging from "certainly correct" to "certainly wrong" [3] or to evaluate their accuracy with varying levels of confidence [59], without apparent disquiet among experimental subjects or journal reviewers. Indeed, such methods may be preferable to commonly used confidence scales that leave ambiguous whether subjects should use low ratings to indicate that they were wrong or merely that they were uncertain (cf. [4]).

Overall, then, the separation between the two literatures appears to be a historical curiosity rather than a principled distinction. We suggest that confidence judgements and error detection are at least overlapping, and perhaps even indistinguishable processes, thus forming a smooth continuum from certainty of correctness at one end to certainty of error at the other. This claim, though seemingly straightforward, has important implications for research in both fields.

7.4.1 Converging Implications

A first, striking implication is that if metacognitive evaluations of confidence and error form a continuum, we can immediately rule out the most popular class of theories of confidence—decisional locus models—because there is no coherent way for a decision process to yield a particular choice concurrent with a judgement that this choice is wrong: committing to a choice implies a belief that this choice is the right one. Thus, only post-decisional locus theories can explain error judgements. However, even post-decisional theories of confidence will need modification to accommodate evidence from study of error monitoring that metacognitive awareness (cf. error detection and the Pe) cannot be reduced simply to post-decision processing (cf. error correction and the ERN)—the two are at least partly dissociable.

Convergence with work on confidence judgements has similarly stark implications for current theories of error monitoring. Specifically, whereas confidence is near-universally characterised as varying along a continuum, and formalised as such in accounts such as the balance of evidence [67] and two-stage dynamic signal detection (2DSD) models [46], error detection is often characterised as an all-or-none process [26, 30, 63]. Thus, according to many current theories of error monitoring, binary yes/no error judgements are an intrinsic feature of the monitoring system rather than a reflection of the arbitrary metacognitive decision that subjects are asked to make. As such, these theories cannot explain how subjects are able to express graded confidence in their accuracy judgements [59, 64], and can therefore be discarded if the present line of reasoning is correct.

As well as helping to adjudicate major theoretical disputes in the two research literatures, integration of work on confidence and errors suggests fruitful avenues for future research. One such avenue concerns the impact of confidence judgements on future actions: Whereas research on confidence has largely focused on how confidence evaluations are derived, a major focus of error monitoring research has been on the impact of metacognitive evaluation on future behaviour both in the short [10, 47] and long-term [25, 26]. Borrowing these insights, we might predict that graded estimates of confidence could support finer-grained control of behaviour than can be achieved through binary categorization of responses as correct or incorrect. For example, confidence judgements might be used for parametric control of response threshold, with graded increases in caution adopted as confidence drops (cf. post-error slowing; [47]), or of learning rate, so that greater attention is paid to feedback in uncertain environments (cf. error-driven reinforcement learning; [26]).

Another avenue concerns the nature of the evaluative process: Whereas error monitoring is often conceptualised as a crude binary choice, studies of decision confidence indicate that metacognitive evaluations are at least as complex, nuanced, and sensitive to bias and expectation as the first-order decisions on which they are based [18, 40]. There is important work to be done in understanding how complex evaluations of this kind in error monitoring can be translated into categorical decisions about current and future behaviour (Was I right or wrong? How can I atone or compensate for my mistake?).

7.4.2 Shared Limitations

Studies of confidence and error monitoring promise to be mutually very informative, but conceptual and methodological overlap also means that the two research areas share common weaknesses. In particular, like the models of decision making on which they are based, current theories in both domains have focused almost exclusively on decisions that are discrete and punctate in time: a decision is made when the bound is reached [54]; errors detected when a change-of-mind bound is crossed [55]; and confidence estimated at the time of a metacognitive probe [46]. Characterising behaviour as a series of discrete decisions is a useful convenience. However, it is not clear whether the findings will scale up to explain real-world decisions and actions that are fluid, temporally extended and embedded in the broader context of evolving behavioural goals. In this regard, we see two particular limitations to current approaches.

First, many decisions that initially appear discrete and categorical turn out, on closer inspection, to be graded and transitory. For example, overt responses occur tightly time-locked to threshold-crossings in cortical motor activity (e.g., [21, 22]), suggestive of a fixed decision point after which an action is initiated. However, finer-grained analyses reveal graded and continuous information flow at every stage. Thus, during the course of a single decision, motor cortex activity may first favour one response then another; small EMG twitches in one finger may be followed by full movements of another; and overt actions themselves may vary in force in a graded manner, for example with errors executed less forcefully than correct actions [19, 22, 23, 55]. Meanwhile, categorical or economic judgements about visual information are often preceded by exploratory eye movements, which may themselves constitute interim decisions *en route* to the eventual choice [35]. In such systems, there is no single, final decision point that could mark the beginning of metacognitive evaluation.

Second, human decision making has a continuous quality when viewed over longer timescales, with individual decisions chained into sequences that serve longer-term behavioural goals. Thus, actions that reflect definitive choices at a lower goal level (e.g., saccadic fixations, manual responses) may constitute reversible, interim choices at a higher goal level (e.g., selection of an initial solution path that is later abandoned) [37, 42]. This form of hierarchical structure is built into many recent theories of the computational and neural basis of action selection [9, 33]. Recent findings indicate that metacognitive processes are similarly sensitive to this hierarchical structure [38]. For example, errors that are equally discrepant in terms of low-level actions are treated very differently according to their impact on global task performance [36]. Little is currently known about the mechanisms by which metacognitive judgements might be embedded in ongoing higher-level behaviour in this way.

Thus, a crucial limitation of current metacognitive theories is that they do not reflect the way that confidence in our actions contributes to temporally extended, structured behaviour. Rather, they consider that decisions are evaluated in an all-or-nothing fashion, for example when a post-decision process crosses a metacognitive decision bound. In what follows, we consider another way of thinking about decision confidence that is not subject to these limitations.

7.5 Confidence and Precision

7.5.1 Evidence Reliability

Formal accounts of categorical decisions, such as the DDM, are often illustrated by analogy to court of law, in which the jury weighs up evidence favouring the innocence or guilt of the defendant [21]. However, this analogy also highlights an inconsistency between current decision models and choices made in the real world: that in the latter, we usually consider the extent to which we *trust* the evidence relevant for a decision. For example, in a law court, evidence from a trustworthy source (e.g., official telephone records) might weigh more heavily in a jury's deliberations than evidence from an unreliable source (e.g., a witness with a vested interest). Yet most currently popular models of perceptual decisions offer no way of expressing the trust or distrust associated with the evidence accumulated: All sources of evidence are combined in the common currency of the DV, which then gives the strength of evidence as a simple scalar value (e.g., as the vertical location of the diffusing particle in the DDM).

Mathematically, the reliability of evidence is orthogonal to its strength, because the mean and variance represent different moments of a probability distribution. Bayesian models that exploit this point—representing evidence as a probability distribution with a given mean (evidence strength) and variance (evidence reliability)—have been used to calculate ideal estimates of expected value in economic choice tasks [6]. Applied to perceptual decision tasks, the notion would be that the diffusing particle of the DDM is more accurately conceived of as a probability distribution that evolves across samples (Fig. 7.3), with the central tendency of that distribution analogous to the vertical location of the particle. Crucially, the variance of this distribution provides additional information—specifically, a representation of evidence reliability in terms of the *precision* (i.e., the inverse of the standard

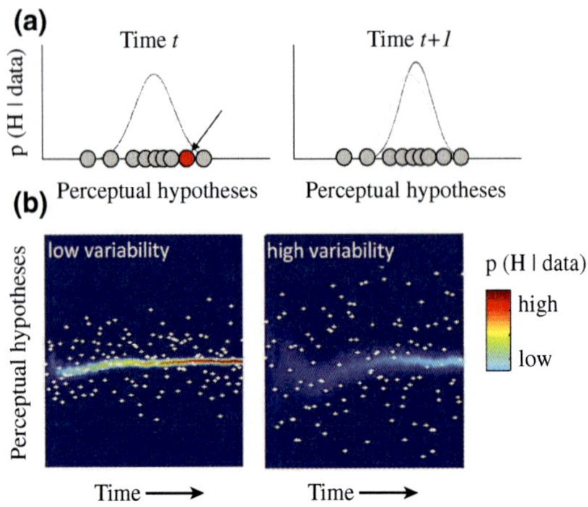

Fig. 7.3 Schematic representation of a model in which both the mean and variance of information in an array are estimated in a serial sampling framework. **a** the left panel shows the posterior probability distribution p (H | data) over a continuous space of possible perceptual hypotheses (e.g., these dots are moving to the right with 30 % coherence; this signal is 40 % visible; etc.) at a given time, t. This distribution reflects the evidence sampled from the stimulus thus far, i.e., between onset and time t (*light grey dots*). The new sample received at time t is marked with an arrow. Right panel: at time $t + 1$, this distribution (*light grey*) is updated in the light of the newly sampled information, giving rise to a new probability distribution. In this model, confidence is reflected in the precision of the posterior distribution, i.e., the reciprocal of its standard deviation. **b** The evolving posterior probability distribution over perceptual hypotheses (y-axis) for each successive time point (x-axis). The posterior distribution is updated following the arrival of successive samples with low variance (*left panel*) or high variance (*right panel*). Precision of the probabilistic representation of evidence strength increases more rapidly for the low-variability samples

deviation) of the mean. This information about precision is lost in the standard DDM because the decision variable (i.e., v_t in Eq. 7.3) retains no history of its fluctuations across trials, only its momentary value.

Recently, it has been shown that neural network models in which LIP neurons encode the full posterior probability distribution associated with a stimulus can capture behaviour and neural dynamics occurring during discrimination tasks in primates [5]. This result suggests that evidence reliability may be encoded in the variance of firing rates across a neural population, in much the same way that evidence strength is encoded in the mean firing rate [39]. Meanwhile, psychophysical studies have shown that human observers are highly sensitive to evidence variability, for example when asked to judge the average feature (e.g., colour) of an array of multiple elements [13]: Observers are slower to discriminate more variable arrays, a result that is predicted by the precision account but not by standard accumulation models such as the DDM, and those observers tend to downweight outlying evidence much as statisticians exclude outliers from a sample of data.

7.5.2 Reliability as a Cue to Confidence

Integrated coding of evidence strength and reliability provides an intriguing new implementation of decision confidence, as the precision of evidence accumulated during a single trial. In common with other proposed bases of confidence judgements, precision estimates are intrinsic to the decision process and are systematically related to evidence quality and quantity. However, in contrast to most other proposed mechanisms, precision is available continuously and instantaneously. As such, it provides an attractive basis for metacognitive evaluation in the kinds of temporally extended tasks discussed above, in which no discrete decision point divides cognitive decisions from metacognitive evaluation.

This hypothesis, though speculative, has several attractive features. First, it is consistent with the evidence described above on post-decision processing because precision estimates would continue to evolve beyond the decision point with continued processing, with high levels of variability indicative of a change in the estimated underlying drift rate—that is, with detection of an error. Second, this model is able to describe situations in which evidence quality varies even within a single trial [5], something that standard models cannot achieve. In fact, by keeping track of the likely variability of information in the external world, Bayesian accounts can optimally distinguish true state-changes in the generative information giving rise to the senses from noise [71]. Precision estimates are thus particularly useful in situations where the causes of perceptual evidence may change unpredictably over time, and as such may provide a better account of the sort of fluid, ongoing sensorimotor integration that characterises everyday activities.

Finally, this conception of decision confidence makes direct contact with broader theories of the role of metacognitive evaluation in behavioural control. In particular, because precision is closely related to the concept of decision conflict [10], the theory can inherit ideas from research on conflict about how precision estimates might be used to guide both current performance (e.g., through dynamic modulation of decision bounds; [8, 53]) and future behaviour (e.g., through modulation of learning rate in relation to environmental signals of success or failure; [6]). As such, the precision model not only provides a formally specified account of decision confidence, but also leads to immediate suggestions about the use of confidence judgements in the optimization of behaviour.

7.6 Conclusion

Formal models of decision making have proven extremely valuable in understanding human and animal decision making, by situating experimental observations of behaviour and neural activity within a precisely specified and normatively motivated framework. Direct extensions of these models have proven similarly useful in probing the metacognitive processes by which we evaluate and express our degree of confidence in our decisions. In particular, significant convergence in methods and theories of decision confidence and error monitoring suggest that common principles may govern different types of metacognitive judgements. This convergence helps to resolve several substantive debates that have flourished in each separate field, in particular by favouring theories in which graded metacognitive evaluation emerges from continued processing beyond the initial choice.

There is nonetheless important scope for current models to consider decision making and metacognitive evaluation in situations that encompass not only simple, punctate choices but also the kinds of extended, goal-directed decisions and actions that typify human behaviour outside the experimental lab. We have proposed one such extension: the hypothesis that people are sensitive not only to the strength of evidence they encounter as they make a decision, but also to the reliability of that evidence. Future developments in theories of human decision making promise to have similarly significant implications for our understanding of the way in which people evaluate their decisions in the service of adapting and optimising those decisions in the face of an uncertain, complex and ever-changing environment.

Acknowledgments CS was supported by a grant from the Wellcome Trust (WT092646AIA). We thank Vincent de Gardelle and Annika Boldt for valuable discussion that guided our thinking on this topic.

References

1. Albantakis L, Deco G (2011) Changes of mind in an attractor network of decision-making. PLoS Comput Biol 7(6):e1002086. doi:10.1371/journal.pcbi.1002086
2. Audley RJ (1960) A stochastic model for individual choice behavior. Psychol Rev 67:1–15
3. Baranski JV, Petrusic WM (1994) The calibration and resolution of confidence in perceptual judgments. Percept psychophysics 55(4):412–428
4. Baranski JV, Petrusic WM (1998) Probing the locus of confidence judgments: experiments on the time to determine confidence. J Exp Psychol Hum Percept Perform 24(3):929–945
5. Beck JM, Ma WJ, Kiani R, Hanks T, Churchland AK, Roitman J, Shadlen MN, Latham PE, Pouget A (2008) Probabilistic population codes for bayesian decision making. Neuron 60(6):1142–1152. doi:10.1016/j.neuron.2008.09.021 S0896 6273(08)00803-9[pii]
6. Behrens TE, Woolrich MW, Walton ME, Rushworth MF (2007) Learning the value of information in an uncertain world. Nat Neurosci 10(9):1214–1221
7. Bogacz R, Brown E, Moehlis J, Holmes P, Cohen JD (2006) The physics of optimal decision making: a formal analysis of models of performance in two-alternative forced-choice tasks. Psychol Rev 113(4):700–765. doi:10.1037/0033-295X.113.4.700
8. Bogacz R, Wagenmakers EJ, Forstmann BU, Nieuwenhuis S (2010) The neural basis of the speed-accuracy tradeoff. Trends Neurosci 33(1):10–16. doi:10.1016/j.tins.2009.09.002
9. Botvinick MM (2008) Hierarchical models of behavior and prefrontal function. Trends Cogn Sci 12(5):201–208. doi:10.1016/j.tics.2008.02.009 S1364-6613(08)00088-0 [pii]
10. Botvinick MM, Braver TS, Carter CS, Barch DM, Cohen JD (2001) Conflict monitoring and cognitive control. Psychol Rev 108:624–652
11. Busemeyer JR, Johnson JG (2004) Computational models of decision making. In: Koehler D, Harvey N (eds) Handbook of judgment and decision making. Blackwell, Oxford, pp 133–154

12. Charles L, Van Opstal F, Marti S, Dehaene S (2013) Distinct brain mechanisms for conscious versus subliminal error detection. Neuroimage 73:80–94. doi:10.1016/j.neuroimage.2013.01.054
13. de Gardelle V, Summerfield C (2011) Robust averaging during perceptual judgment. Proc Natl Acad Sci USA 108(32):13341–13346. doi:10.1073/pnas.1104517108
14. De Martino B, Fleming SM, Garrett N, Dolan RJ (2013) Confidence in value-based choice. Nat Neurosci 16(1):105–110. doi:10.1038/nn.3279
15. Debener S, Ullsperger M, Siegel M, Fiehler K, von Cramon DY, Engel AK (2005) Trial-by-trial coupling of concurrent electroencephalogram and functional magnetic resonance imaging identifies the dynamics of performance monitoring. J Neurosci 25(50):11730–11737
16. Falkenstein M, Hohnsbein J, Hoorman J, Blanke L (1991) Effects of crossmodal divided attention on late ERP components: II. Error processing in choice reaction tasks. Electroencephalogr Clin Neurophysiol 78:447–455
17. Festinger L (1943) Studies in decision: I. Decision-time, relative frequency of judgment and subjective confidence as related to physical stimulus difference. J Exp Psychol 32(4):291–306. doi:10.1037/h0056685
18. Galvin SJ, Podd JV, Drga V, Whitmore J (2003) Type 2 tasks in the theory of signal detectability: discrimination between correct and incorrect decisions. Psychon Bull Rev 10(4):843–876
19. Gehring WJ, Fencsik D (1999) Slamming on the brakes: an electrophysiological study of error response inhibition. In: Annual meeting of the Cognitive Neuroscience Society, Washington, DC, April 1999
20. Gehring WJ, Goss B, Coles MGH, Meyer DE, Donchin E (1993) A neural system for error detection and compensation. Psychol Sci 4(6):385–390
21. Gold JI, Shadlen MN (2007) The neural basis of decision making. Annu Rev Neurosci 30:535–574. doi:10.1146/annurev.neuro.29.051605.113038
22. Gratton G, Coles MGH, Sirevaag EJ, Eriksen CW, Donchin E (1988) Pre- and poststimulus activation of response channels: A psychophysiological analysis. J Exp Psychol Hum Percep Perform 14(3):331–344
23. Graziano M, Polosecki P, Shalom DE, Sigman M (2011) Parsing a perceptual decision into a sequence of moments of thought. Frontiers integr neurosci 5:45. doi:10.3389/fnint.2011.00045
24. Henmon VAC (1911) The relation of the time of a judgment to its accuracy. Psychol Rev 18(3):186–201. doi:10.1037/h0074579
25. Hester R, Barre N, Murphy K, Silk TJ, Mattingley JB (2008) Human medial frontal cortex activity predicts learning from errors. Cereb Cortex 18(8):1933–1940. doi:10.1093/cercor/bhm219
26. Holroyd CB, Coles MGH (2002) The neural basis of human error processing: Reinforcement learning, dopamine, and the error-related negativity. Psychol Rev 109(4):679–709
27. Hughes G, Yeung N (2011) Dissociable correlates of response conflict and error awareness in error-related brain activity. Neuropsychologia 49(3):405–415. doi:10.1016/j.neuropsychologia.2010.11.036
28. Jang Y, Wallsten TS, Huber DE (2012) A stochastic detection and retrieval model for the study of metacognition. Psychol Rev 119(1):186–200. doi:10.1037/a0025960
29. Jentzsch I, Dudschig C (2009) Why do we slow down after an error? Mechanisms underlying the effects of posterror slowing. Q J Exp Psychol 62(2):209–218
30. Joordens S, Piercey CD, Azarbehi R (2009) Modeling performance at the trial level within a diffusion framework: a simple yet powerful method for increasing efficiency via error detection and correction. Can J Exp Psychol 63(2):81–93. doi:10.1037/a0015385
31. Kepecs A, Uchida N, Zariwala HA, Mainen ZF (2008) Neural correlates, computation and behavioural impact of decision confidence. Nature 455(7210):227–231. doi:10.1038/nature07200
32. Kiani R, Shadlen MN (2009) Representation of confidence associated with a decision by neurons in the parietal cortex. Science 324(5928):759–764. doi:10.1126/science.1169405

33. Koechlin E, Summerfield C (2007) An information theoretical approach to prefrontal executive function. Trends Cogn Sci 11(6):229–235. doi:10.1016/j.tics.2007.04.005
34. Komura Y, Nikkuni A, Hirashima N, Uetake T, Miyamoto A (2013) Responses of pulvinar neurons reflect a subject's confidence in visual categorization. Nat Neurosci 16(6):749–755. doi:10.1038/nn.3393
35. Krajbich I, Armel C, Rangel A (2010) Visual fixations and the computation and comparison of value in simple choice. Nat Neurosci 13(10):1292–1298. doi:10.1038/nn.2635
36. Krigolson OE, Holroyd CB (2006) Evidence for hierarchical error processing in the human brain. Neuroscience 137(1):13–17. doi:10.1016/j.neuroscience.2005.10.064
37. Lashley KS (1951) The problem of serial order in behavior. In: Jeffress LA (ed) Cerebral mechanisms in behavior: the Hixon Symposium. Wiley, London, pp 112–136
38. Logan GD, Crump MJ (2010) Cognitive illusions of authorship reveal hierarchical error detection in skilled typists. Science 330(6004):683–686. doi:10.1126/science.1190483
39. Ma WJ, Beck JM, Latham PE, Pouget A (2006) Bayesian inference with probabilistic population codes. Nat Neurosci 9(11):1432–1438. doi:10.1038/nn1790
40. Maniscalco B, Lau H (2011) A signal detection theoretic approach for estimating metacognitive sensitivity from confidence ratings. Conscious Cogn. doi:10.1016/j.concog.2011.09.021
41. Middlebrooks PG, Sommer MA (2012) Neuronal correlates of metacognition in primate frontal cortex. Neuron 75(3):517–530. doi:10.1016/j.neuron.2012.05.028
42. Miller GA, Galanter E, Pribram KH (1960) Plans and the structure of behavior. Holt, Rinehart & Winston, New York
43. Moreno-Bote R (2010) Decision confidence and uncertainty in diffusion models with partially correlated neuronal integrators. Neural Comput 22(7):1786–1811. doi:10.1162/neco.2010.12-08-930
44. Nieuwenhuis S, Aston-Jones G, Cohen JD (2005) Decision making, the P3, and the locus coeruleus-norepinephrine system. Psychol Bull 131(4):510–532. doi:10.1037/0033-2909.131.4.510
45. Nieuwenhuis S, Ridderinkhof KR, Blom J, Band GPH, Kok A (2001) Error-related brain potentials are differentially related to awareness of response errors: evidence from an antisaccade task. Psychophysiol 38:752–760
46. Pleskac TJ, Busemeyer JR (2010) Two-stage dynamic signal detection: a theory of choice, decision time, and confidence. Psychol Rev 117(3):864–901. doi:10.1037/a0019737
47. Rabbitt PMA (1966) Errors and error correction in choice-response tasks. J Exp Psychol 71(2):264–272
48. Rabbitt PMA (2002) Consciousness is slower than you think. Q J Exp Psychol 55A(4):1081–1092
49. Rabbitt PMA, Cumming G, Vyas SM (1978) Some errors of perceptual analysis in visual search can be detected and corrected. Q J Exp Psychol 30:319–332
50. Rabbitt PMA, Rodgers B (1977) What does a man do after he makes an error? An analysis of response programming. Q J Exp Psychol 29:727–743
51. Rabbitt PMA, Vyas SM (1981) Processing a display even after you make a response to it. How perceptual errors can be corrected. Q J Exp Psychol 33A:223–239
52. Ratcliff R (1978) A theory of memory retrieval. Psychol Rev 85(2):59–108. doi:10.1037/0033-295x.85.2.59
53. Ratcliff R, Frank MJ (2012) Reinforcement-based decision making in corticostriatal circuits: mutual constraints by neurocomputational and diffusion models. Neural Comput (in press)
54. Ratcliff R, McKoon G (2008) The diffusion decision model: theory and data for two-choice decision tasks. Neural Comput 20(4):873–922. doi:10.1162/neco.2008.12-06-420
55. Resulaj A, Kiani R, Wolpert DM, Shadlen MN (2009) Changes of mind in decision-making. Nature 461 (7261):263–266. doi:10.1038/nature08275
56. Ridderinkhof KR, Ramautar JR, Wijnen JG (2009) To P(E) or not to P(E): a P3-like ERP component reflecting the processing of response errors. Psychophysiol 46(3):531–538. doi:10.1111/j.1469-8986.2009.00790.x

57. Rodriguez-Fornells A, Kurzbuch AR, Muente TF (2002) Time course of error detection and correction in humans: neurophysiological evidence. J Neurosci 22:9990–9996
58. Rushworth MFS, Behrens TE, Rudebeck PH, Walton ME (2007) Contrasting roles for cingulate and orbitofrontal cortex in decisions and social behaviour. Trends Cogn Sci 11(4):168–176
59. Scheffers MK, Coles MGH (2000) Performance monitoring in a confusing world: error-related brain activity, judgements of response accuracy, and types of errors. J Exp Psychol Hum Percep Perform 26(1):141–151
60. Schultz W, Dayan P, Montague PR (1997) A neural substrate of prediction and reward. Science 275:1593–1599
61. Shadlen MN, Newsome WT (2001) Neural basis of a perceptual decision in the parietal cortex (area LIP) of the rhesus monkey. J Neurophysiol 86(4):1916–1936
62. Smith PL, Ratcliff R (2004) Psychology and neurobiology of simple decisions. Trends Neurosci 27(3):161–168. doi:10.1016/j.tins.2004.01.006 (S0166223604000220 [pii])
63. Steinhauser M, Maier M, Hubner R (2008) Modeling behavioral measures of error detection in choice tasks: response monitoring versus conflict monitoring. J Exp Psychol Hum Percep Perform 34(1):158–176. doi:10.1037/0096-1523.34.1.158 2008-00937 010 [pii]
64. Steinhauser M, Yeung N (2010) Decision processes in human performance monitoring. J Neurosci 30(46):15643–15653
65. Usher M, McClelland JL (2001) The time course of perceptual choice: the leaky, competing accumulator model. Psychol Rev 108(3):550–592
66. Vickers D, Burt J, Smith P, Brown M (1985) Experimental paradigms emphasizing state or process limitations: 1. Effects on speed accuracy tradeoffs. Acta Psychol 59:129–161
67. Vickers D, Packer J (1982) Effects of alternating set for speed or accuracy on response time, accuracy and confidence in a unidimensional discrimination task. Acta Psychol (Amst) 50(2):179–197
68. Volkmann J (1934) The relation of the time of judgment to the certainty of judgment. Psychol Bull 31(672):673
69. Wald A, Wolfowitz J (1949) Bayes solutions of sequential decision problems. Proc Natl Acad Sci USA 35(2):99–102
70. Yeung N, Botvinick MM, Cohen JD (2004) The neural basis of error detection: conflict monitoring and the error-related negativity. Psychol Rev 111(4):931–959
71. Yu AJ, Dayan P (2005) Uncertainty, neuromodulation, and attention. Neuron 46(4):681–692
72. Zylberberg A, Barttfeld P, Sigman M (2012) The construction of confidence in a perceptual decision. Frontiers integr neurosci 6:79. doi:10.3389/fnint.2012.00079

Chapter 8
Metacognition and Confidence in Value-Based Choice

Stephen M. Fleming and Benedetto De Martino

Abstract Basic psychophysics tells us that decisions are rarely perfect: even with identical stimuli choice accuracy fluctuates and errors are often made. Metacognition allows appraisal of this uncertainty and correction of errors. For more complex value-based choices, however, metacognitive processes are poorly understood. In particular, how subjective confidence and valuation of choice options interact at the level of brain and behaviour is unknown. In this chapter, we summarise and discuss the results of a study designed to investigate this relationship. Subjects were asked to choose between pairs of snack items and subsequently provide a confidence rating in their choice. As predicted by a computational model of the decision process, confidence reflected the evolution of a decision variable over time, explaining the observed relation between confidence, value, accuracy and reaction time (RT). Furthermore, fMRI signal in human ventromedial prefrontal cortex (vmPFC) reflected both value comparison and confidence in the value comparison process. In contrast, individuals' metacognitive ability was predicted by a measure of functional connectivity between vmPFC and rostrolateral prefrontal cortex (rlPFC), a region that responded to changes in confidence but was

S. M. Fleming
Department of Experimental Psychology, University of Oxford, South Parks Road, Oxford OX1 3UD, UK

S. M. Fleming (✉)
Center for Neural Science, New York University, 4 Washington Place, New York, NY 10003, USA
e-mail: sf102@nyu.edu

B. De Martino
Department of Psychology, Royal Holloway University, London TW20 0EX, UK
e-mail: benedettodemartino@gmail.com

B. De Martino
Division of the Humanities and Social Sciences, California Institute of Technology, Pasadena, CA 91125, USA

S. M. Fleming and C. D. Frith (eds.), *The Cognitive Neuroscience of Metacognition*, DOI: 10.1007/978-3-642-45190-4_8,

not involved in representing the values used to guide choice. These results provide a novel link between noise in value comparison and metacognitive awareness of choice, extending the study of metacognition to value-based decision-making.

8.1 Introduction

The decision-making literature often draws a distinction between 'perceptual' and 'value-based' decisions. Perceptual decisions are those in which the aim of the decision-maker is to categorise ambiguous (or noisy) sensory information; for example, deciding whether a face in a crowd is a friend or stranger. Value-based decisions, on the other hand, require the selection of actions based on their subjective value, such as a choice between possible restaurants. In value-based decisions there is often little sensory ambiguity in the stimuli, with uncertainty instead arising from sources that are more difficult to measure experimentally. For example, a decision-maker faced with a choice between two restaurants needs to retrieve the memory traces associated with previous visits and integrate these traces with the current homeostatic state (e.g. level of hunger) in order to make an appropriate decision.

Notwithstanding this distinction, important commonalities exist between value-based and perceptual decisions [49]. For example, both are stochastic, with repeated presentation of the same stimulus set sometimes leading to different outcomes. Both show regularities in the relationship between response time and error rates (e.g. [34, 41, 51]), suggesting that common neural dynamics underlie both types of decision [19, 49]. Furthermore, whenever information is integrated or compared, approximate inferences [3] and computational constraints [47] will lead to additional behavioural variability regardless of the particular domain.

However, little is known about how metacognition, in particular decision confidence, operates during value-based decision-making. There are several reasons why value-based decision-making is an attractive model in which to study metacognition. First, the neural basis of value-based choice is well-established, constraining hypotheses about the system involved in choice confidence (see [40, 45], for reviews). Second, value-based choices provide a bridge between lower-level psychophysics and everyday decisions such as choosing a restaurant or accepting a job offer. Third, a framework for the study of confidence in value-based choice can be naturally extended to assess how confidence develops during learning [6, 7, 52]. Finally, the study of metacognition and confidence may shed light onto the fundamentals of the choice process itself. For example, the finding that different confidence levels provide a behavioural and neural correlate of the level of stochasticity in choice lends support to so-called 'random utility' developed in economics [33]. We will return to this issue in the Discussion.

The field of perceptual psychophysics has begun to examine in greater detail the psychological and neural underpinnings of confidence in perceptual decisions [1, 23, 24, 54]. In addition, there has been progress in understanding how perceptual confidence is appraised for metacognitive report [12, 13, 30, 39, 44]. In brief, these studies suggest that there is a second-order appraisal of confidence that relies on the rostrolateral prefrontal cortex (rlPFC) (see Fleming and Dolan, this volume, for a review). Prior work has established that the ventromedial prefrontal cortex (vmPFC) plays a central role in computing the value of potential choice options (e.g. [2, 11, 21, 46]), with activity in this region reflecting the dynamic evolution of a value comparison [19]. However, this work has largely focused on the choice process, without considering the subject's level of confidence in their decision. Consequently, it is unknown how a process of value comparison, instantiated in vmPFC, relates to subjective confidence.

Critically, dissociating confidence from other features of the decision process requires acquisition of separate measures of choice and confidence (Kepecs and Mainen, this volume). In this chapter, we review a recent study implementing such an approach to examine confidence and metacognition during value-based decision-making in humans [8]. We collected trial-by-trial estimates of decision confidence while healthy volunteers chose between pairs of snack items. We additionally measured the subjective value of each snack item via an incentive compatible bidding procedure often used in behavioural economics. This allowed us to behaviourally dissociate the subjective value of choice options presented on each trial from subjects' confidence levels following each choice.

To explore systematic relationships between confidence, accuracy, choice and reaction time (RT), we modelled our data using a variant of the race model [51]; part of a larger class of dynamic models of decision-making (see Yeung and Summerfield, this volume). This model predicts that subjective confidence reflects the stochastic accumulation of evidence during the value comparison process. Consistent with this prediction we show that the same anatomical region in vmPFC not only reflects a difference in value (DV) between available options, but also the confidence associated with a value comparison process. Finally, we show that individual differences in participants' ability to relate confidence to decision performance is linked to increased functional connectivity between vmPFC and rlPFC, a region previously shown to play a role in metacognitive appraisal in perceptual decision-making (see Fleming and Dolan, this volume).

8.2 Task Design

We scanned 20 hungry participants while they made choices between food items that they were able to consume later (Fig. 8.1a). After making each choice participants reported the degree of confidence in their decision (choice confidence). Note that confidence, or certainty, in the present study is conceptually distinct from risk, in that each choice determines a known outcome. Confidence here reflects the

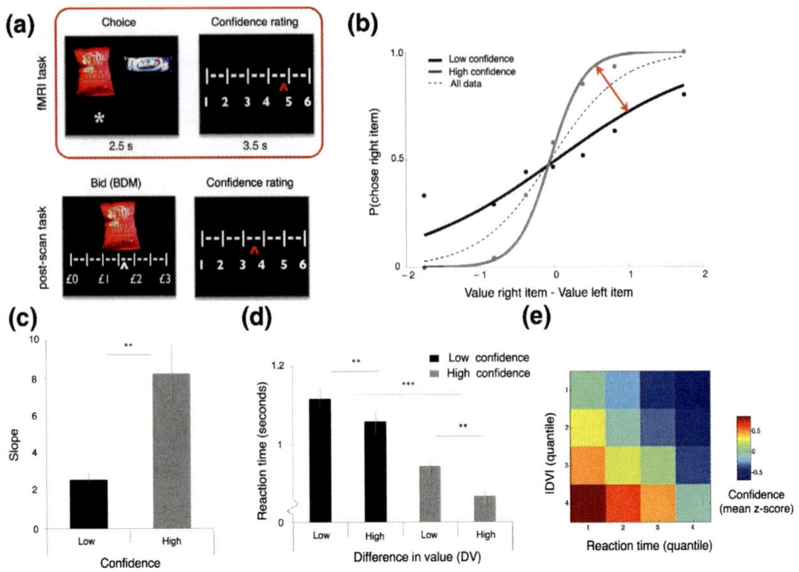

Fig. 8.1 Task and behavioural results. **a** fMRI task (*red box*): subjects were presented with a choice between 2 confectionary items and were then required to choose (2.5 s) one item to consume at the end of the experiment. After each choice, subjects indicated their level of confidence in having made a correct decision (choice confidence). Post-scanning task: subjects were presented with each item individually and had to submit a bid to buy each item. After each bid, they were asked to rate their level of confidence in having provided a correct bid price (bid confidence). **b** Probability of choosing the item on the right as a function of the difference in value (i.e. bid price) between the 2 items (logistic fit) for an exemplar subject. *Dotted line* = all choices; *black line* = *low* confidence choices; *grey line* = *high* confidence choices. The *red arrow* indicates the increase in choice accuracy (change in slope) for *high* versus *low* confidence trials used in the between subject analyses (Figs. 8.4b and 8.5b) **c** The slope of the logistic fit is systematically higher (sharper) in *high* compared to *low* confidence trials ($p < 0.0001$). **d** Average choice reaction time data as a function of confidence and |DV|. **e** Heatmap showing mean z-scored confidence (*colorbar*) across subjects, as a function of subject-specific |DV| and RT quantiles. *Error bars* represent the standard error of the mean (s.e.m.)

degree of subjective certainty in having made the best choice, which equates to choosing the higher valued item. To establish value for individual items we asked participants at the end of the scanning session to place a bid for each food item using a standard incentive compatible procedure, the Becker–DeGroot–Marschak (BDM) mechanism [4]. BDM is widely used in behavioural economics and neuroeconomics to elicit non-strategic reservation prices also known as willingness-to-pay (WTP). In this phase subjects were required to state their maximum willingness-to-pay for each food item (see [8] for further details). A number of studies have shown that the BDM mechanism reliably elicits goal values that are used to guide choice [9, 17, 38]. Participants' bids were then used to calculate a signed DV between each pair of items (V_right–V_left), which was then entered into a logistic regression to predict the probability that the subject chose the rightmost item on each trial (Fig. 8.1b—dotted line). In line with previous studies we show DV is a reliable predictor of participants' choices, with the slope of the logistic regression being a measure of choice accuracy, or noise in the choice process [48].

8.3 Results

8.3.1 Choice, Confidence and Reaction Time

At first glance, one might expect confidence to be trivially related to value. When choices are easy, and one item is valued much higher than another, then confidence in choosing the best option should also be high. Conversely, when there are only small differences in value between the options, the decision is difficult and confidence will be low. However, counter to this intuition, we observed that unsigned |DV| only accounted for an average of 17.7 % of the variance in participants' confidence ratings ($r = 0.42 \pm 0.19$ SD).

This partial independence between confidence and |DV| is reminiscent of findings in perceptual decision-making, where confidence fluctuates in tandem with accuracy despite stimulus evidence remaining constant [12]. By splitting our logistic regression fit into high and low confidence trials, we showed that higher confidence was consistently associated with increased choice accuracy (Fig. 8.1b, c). This effect of confidence on choice was also reflected in RT, with main effects of both |DV| and confidence (both $P < 0.001$), but no interaction (Fig. 8.1d). The three-way relationship between |DV|, confidence and RT is plotted in Fig. 8.1e. This plot shows that confidence is greatest when |DV| is high and response times are fast. Crucially, however, both factors influence reported confidence independently, suggesting confidence is dynamically constructed from choice values. We next turned to a computational model that could account for these effects.

8.3.2 Dynamic Model of Value Comparison

To predict how value, confidence and RT interact during decision-making, we simulated a dynamic model of decision process [51]. In the Vickers race model, separate decision variables accumulate evidence for distinct options, with the final decision determined by which accumulator reaches threshold first. On each time step during accumulation, a new evidence sample is drawn from a normally distributed random variable $s_t = \mathcal{N}(\mu_{\text{stim}}, \sigma_{\text{stim}})$. u_{stim} is positive if the correct choice (higher value item) is the right item; negative if the correct choice is the left item. Due to s_t being drawn from a normal distribution, the actual value of s_t at each time step may be positive or negative. The accumulators evolve according to the following equations:

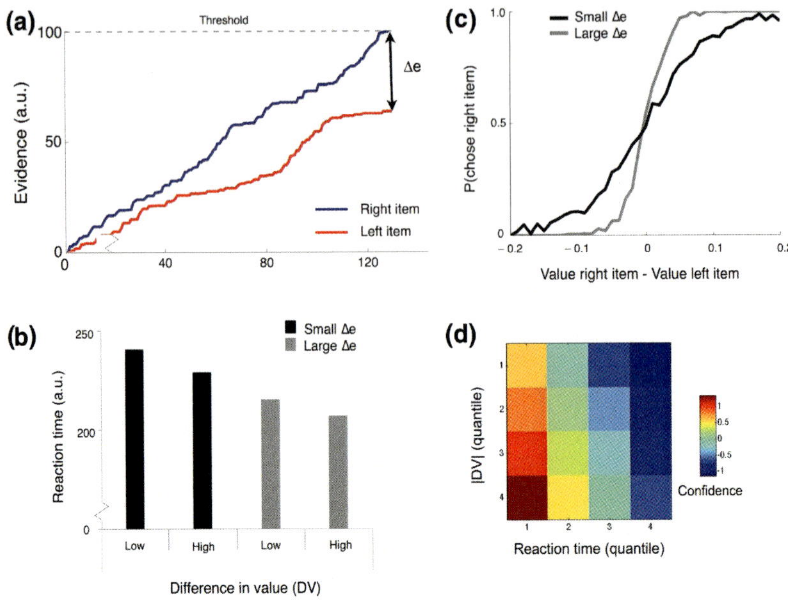

Fig. 8.2 Computational model. **a** Dynamic (race) model of value comparison. Evidence in favour of each option accumulates over time, with a choice in favour of one or other option being made when threshold is reached. In this model decision confidence is derived from the absolute difference between the two accumulators at the time of the decision (Δe). (b-d) Model predictions. **b** Reaction times are predicted to decrease when either |DV| or Δe increase. **c** When Δe is large (i.e. *high* confidence) choice accuracy is predicted to increase, reflected by a sharper curve in the logistic regression. **d** Matrix representing how model confidence changes across |DV| and RT quantiles. Note the close similarity between the model predictions and behaviour (Fig. 8.1c–e)

$$R_{t+1} = \begin{cases} R_t + s_t, & s_t > 0 \\ R_t, & s_t \leq 0 \end{cases}$$

$$L_{t+1} = \begin{cases} L_t, & s_t \geq 0 \\ L_t - s_t, & s_t < 0 \end{cases}$$

The race terminates when either R_t or L_t reach a predetermined threshold, θ, with the decision being determined by which accumulator reaches threshold first. Therefore at decision time, $t(\theta)$, either R_t or $L_t = \theta$. The finishing point of the losing accumulator depends on the values of u_{stim} and σ_{stim}.

An estimate of decision confidence, Δe, can be recovered from the race model as the distance between the two accumulators R_t and L_t at the time the race is terminated (Fig. 8.2a). We simulated the model using the same parameters as [23]. We simulated 1,000 trials at each level of u_{stim}, and recorded mean choice, confidence and RT. We display the simulation output in an identical manner to the behavioural data (Fig. 8.2b–d).

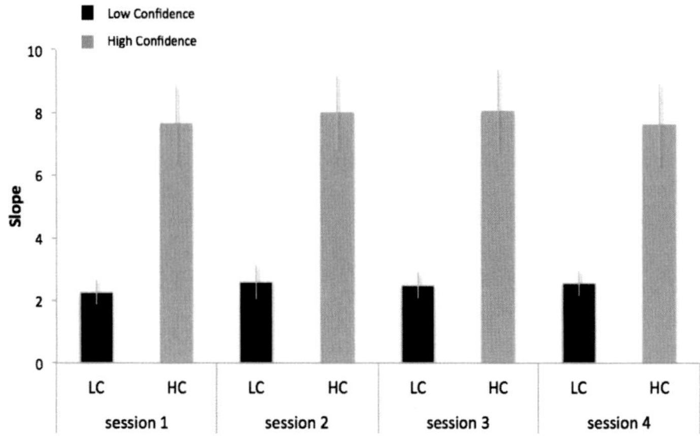

Fig. 8.3 Slope from a logistic regression model predicting choice from DV, separately for *high* and *low* confidence and split by session

Such a model predicts that when Δe is large then choice accuracy is increased, reflected by a sharper slope in the logistic regression (Fig. 8.2b). Thus, the race model neatly accounts for an increase in choice accuracy we observe behaviourally in the high confidence condition (Fig. 8.1b). Furthermore this model predicts a decrease in RT when either |DV| or Δe are increased (Fig. 8.2c), as seen in the behavioural data (Fig. 8.1d). The intuition is that, even within a particular level of *initial* DV, inter-trial noise in the value comparison process results in some trials having greater *final* DV's (higher confidence) than others. Such decisions will tend to be faster, more accurate and associated with higher confidence (Fig. 8.2d). Indeed, this predicted inter-relationship between RT, |DV| and confidence closely matches what is observed in the behavioural data (Fig. 8.1e). Finally, since the model predicts that confidence reflects the stochastic evolution of a value comparison process, it will only be weakly related to initial DV. This feature of the model provides a parsimonious explanation for why DV and confidence are dissociable in our behavioural data.

8.3.3 Stability of Confidence Over Time

We next examined whether the relationship between confidence and choice is stable over time. Splitting the logistic regression analysis into separate sessions revealed a robust main effect of confidence ($F(1, 19) = 39.75$; $P < 0.0001$) but a non-significant main effect of session ($F(3, 57) = 0.3$; $P = 0.7$) and a lack of interaction between session and confidence ($F(3, 57) = 0.13$; $P = 0.9$; Fig. 8.3). To examine whether local fluctuations in attention affected confidence, we constructed a serial autocorrelation regression model that predicted the current confidence rating from the confidence ratings given on the immediately preceding five trials, in addition to |DV|. None of the autocorrelation coefficients reached group-level significance (all $t < 1.2$, $P > 0.27$). Together these results indicate that confidence is a stable predictor of choice accuracy, and does not reflect local changes in attention.

As each item pairing was presented twice (once in each spatial configuration), it was also possible to examine the relationship between confidence ratings given for identical choice pairs. As confidence is partly determined by absolute DV (|DV|, which does not vary across choice pairs) we expected some stability purely driven by DV. Thus to address this question we computed the partial correlation between 1st and 2nd confidence ratings, controlling for DV. There was no significant difference between mean confidence ratings for the first and second presentations of the same item pairs ($t(19) = -0.64$, $P = 0.53$). For 19 out of 20 subjects, there was a significant partial correlation ($P < 0.05$) between confidence ratings for repeated item pairs after controlling for the influence of |DV|, indicating stability in confidence for judgments of particular item pairs that cannot be accounted for by |DV| alone.

Finally, we examined whether choices were stable over time. On average, 14.7 % of choices (±5.7 % SD) were reversed on the second presentation. Choices that were subsequently reversed were associated with significantly lower initial confidence than those that were subsequently repeated (reversal confidence (a.u.) = 210.6 ± 72.4 SD; repetition confidence = 340.2 ± 53.5 SD; $t(19) = 12.1$, $P < 10^{-10}$). In a logistic regression model predicting subsequent reversal from both |DV| and initial confidence, initial confidence was a significant negative predictor of choice reversal (mean standardised regression coefficient −0.0083 ± 0.0034 SD; one-sample t-test $t(19) = -10.9$, $P < 10^{-9}$). These data support an hypothesis that low confidence may be associated with subsequent changes of mind.

8.3.4 Other Factors Influencing Confidence

We recognise that aside from |DV| and RT other factors (internal and external) are likely to affect subjective confidence. In Table 8.1 we report analyses of a limited set of these factors for which we could exercise good experimental control. In these analyses, only familiarity of individual items explained significant variance in confidence ratings.

8.3.5 Confidence and Value in vmPFC

We turn next to the brain imaging data. If choice confidence is an emergent property of a value comparison process, the same brain regions involved in value-based decision-making should also represent subjective confidence in a value estimate. In other words, if a brain region involved in value comparison is

Table 8.1 Analysis of additional factors affecting confidence and value

Factor	Analysis	Results
Item familiarity[a]	Linear regression of familiarity and $\lvert DV \rvert$ on confidence	Mean familiarity was a significant predictor of choice confidence ($t(19) = 3.09$, $P < 0.01$, individually significant in 13 out of 20 subjects); as established previously, $\lvert DV \rvert$ also significantly predicted choice confidence in this model ($t(19) = 8.55$, $P < 0.001$).
Actual retail price[b]	Pearson correlation between price and BDM bid	1 out of the 20 participants showed a significant positive correlation between the actual retail price of each item and the bids they submitted for these items (group mean $r = 0.0064 \pm 0.24$).
Belief about retail price[a]	Pearson correlation between price and BDM bid	4 out of 20 subjects showed a significant correlation between their beliefs about retail prices and the bids they submitted for these items (group mean $r = 0.23 \pm 0.27$).
Taste[c]	One-way repeated measures ANOVA of mean confidence by factor sweet/salty/mixed	Confidence ratings were not affected by item type ($F(1.273, 24.182) = 1.001$, $P = 0.347$).
Calorie content[d]	One-way repeated measures ANOVA of mean confidence by factor high/low/mixed calorie	Confidence ratings were not affected by calorie level ($F(1.141, 21.671) = 0.681$, $P = 0.437$).

[a] Item familiarity and belief about retail price were collected in post-experiment questionnaires

[b] Actual retail prices were taken from a UK supermarket website

[c] Taste was determined by categorising each item as sweet or salty, and dividing post-choice confidence ratings into three groups: sweet (where both items in the choice pair comprised items categorised as sweet); salty (where both items in the choice pair comprised items categorised as salty); mixed (where a choice pair consisted of one item categorised as sweet and one item categorised as salty)

[d] Calorie content was determined by categorising items as high or low calorie (median split), and dividing post-choice confidence ratings into three groups: high calorie (where both items in the choice pair comprised items categorised as high calorie); low calorie (where both items in the choice pair comprised items categorised as low calorie); mixed calorie (where a choice pair consisted of one item categorised as high calorie and one item categorised as low calorie)

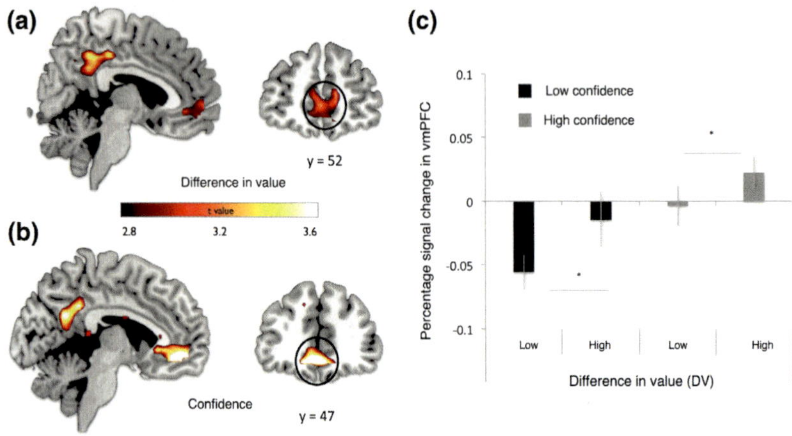

Fig. 8.4 vmPFC. **a** Brain activity in precuneus and vmPFC (MNI space coordinates (x, y, z) 12, 56, 4) correlating with increases in difference in value between the two items presented ($p < 0.05$ family wise error (*FWE*) corrected at cluster level). **b** Brain activity in precuneus and vmPFC (12, 47, −11) correlating with increases in subjective confidence ($p < 0.05$ FWE corrected at the cluster level). **c** Signal in vmPFC (6 mm *sphere* centred at 12, 56, 4) showing significant main effects of difference in value and level of confidence in the absence of an interaction. Note that the bar plot is shown only to clarify the signal pattern in vmPFC (i.e. lack of interaction between confidence and DV). *Error bars* represent the s.e.m

implementing a process akin to a race model, then activity here should be modulated by both initial |DV| and noise (confidence) on that trial. To test this hypothesis we constructed a general linear model (GLM) of our fMRI data in which each trial was modulated by two parametric regressors: |DV| and confidence orthogonalised with respect to |DV|. We found that activity in vmPFC is indeed modulated by both value and confidence (Fig. 8.4a, b; [12, 47, −11], $p < 0.05$ family wise error (FWE) corrected at cluster level). This pattern is consistent with the established role of this region in encoding goal values and with our novel hypothesis that this region also represents the confidence associated with a value comparison.

We next investigated whether |DV| and confidence interacted in vmPFC by splitting the model into high and low confidence trials, both parametrically modulated by |DV|. This analysis showed main effects of |DV| and confidence in vmPFC, but importantly no interaction (2 × 2 ANOVA with factors value, confidence: main effect of value: $F(1, 19) = 5.1, p < 0.05$; main effect of confidence: $F(1, 19) = 7.6$, $p < 0.05$; interaction: $F(1, 19) = 0.7$, $p > 0.5$) (Fig. 8.4c). The absence of an interaction at the neural level is consistent with a theoretical independence between value and noise in the choice process, such that subjects might have high confidence in a low value choice, and vice versa. Furthermore, the pattern across conditions closely resembles that seen for RTs (Fig. 8.1d) providing convergent evidence that vmPFC activity is tightly linked to behaviour.

Finally, we sought to rule out alternative explanations of the activity pattern in vmPFC. We first tested and confirmed that the response to confidence is not driven

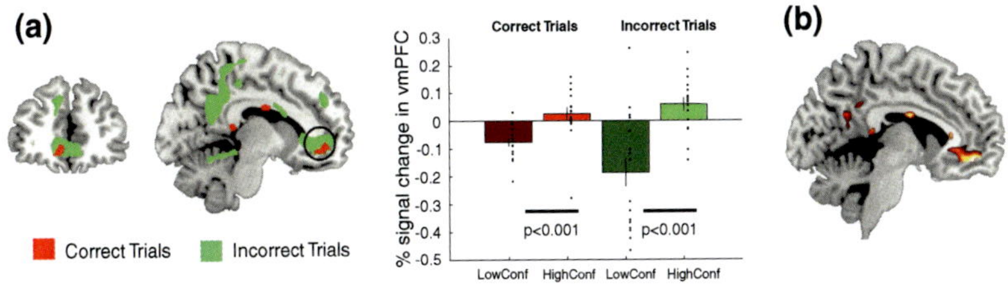

Fig. 8.5 **a** Analysis of the effect of confidence in vmPFC separately for correct and incorrect trials. Significant effects of confidence were seen for both trial types. **b** Effect of confidence in vmPFC after orthogonalising with respect to RT and |DV|

by a categorical response to errors [43] (Fig. 8.5a). Second, we included RTs as a covariate in the fMRI model to test whether an independent effect of confidence remained in vmPFC after controlling for (orthogonalising with respect to) |DV| and RT (Fig. 8.5b). This was the case, suggesting confidence in choice is an emergent property of the decision process that cannot be 'explained away' by other decision-related variables.

8.3.6 Confidence in rlPFC

A key question is how confidence-related information represented in vmPFC becomes available for self-report. One computationally plausible hypothesis is an hierarchical model where confidence in a comparison process is 'read out' by an anatomically distinct second-order network [20, 28, 37]. Right rlPFC is a likely candidate as this region is widely implicated in metacognitive assessments of perceptual decisions [12, 13, 32, 53]. Consequently, we tested whether this region plays a more general role in metacognitive appraisal by enabling explicit report of confidence in a value comparison.

We first established that right rlPFC tracks changes in reported confidence, but does not code for DV (Fig. 8.6a, b, [39, 41, 16], $p < 0.005$ SVC), as expected for a region providing a read-out of decision confidence. We next harnessed individual differences in metacognition to provide a more stringent test for the role of rlPFC. We defined an individual's metacognitive accuracy as the change in choice accuracy (slope of the logistic fit) between low and high confidence trials (Fig. 8.1b). We reasoned that if rlPFC plays a role in the metacognitive appraisal of confidence, activity in this region and/or its coupling with vmPFC should predict this change in slope across individuals. To test our first prediction, we entered change in slope as a between subjects covariate in the whole-brain analysis of confidence-related activity, finding that this parameter significantly modulated the response to confidence in right rlPFC ($p < 0.05$; SVC for multiple comparisons). In other words, participants manifest a neurometric-psychometric match between their behavioural and neural responses to change in confidence level (Fig. 8.6c).

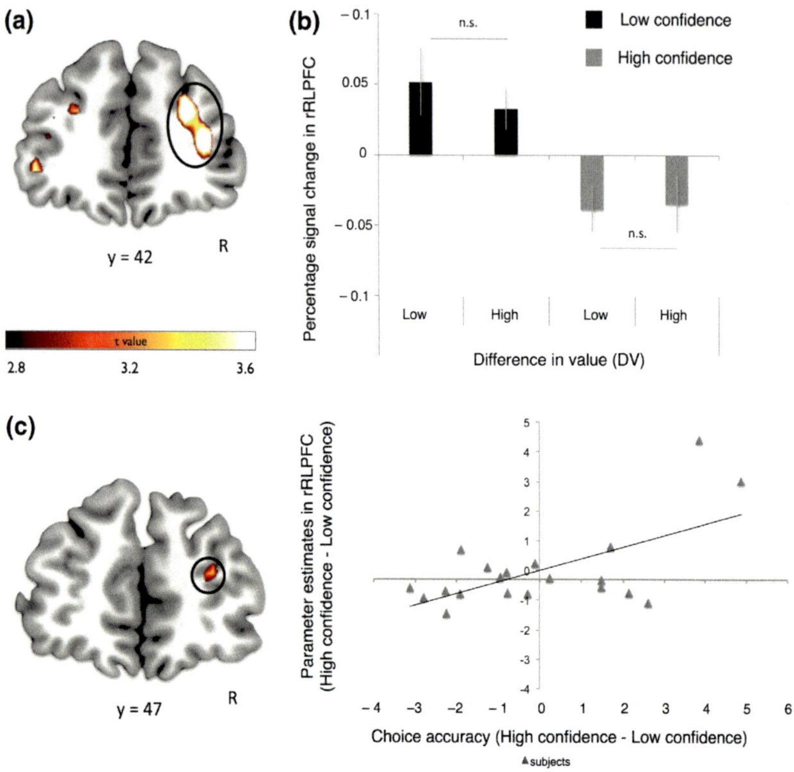

Fig. 8.6 rlPFC. **a** Brain activity in *right* rlPFC correlating with decreases in subjective confidence ($p < 0.005$ small volume FWE corrected). **b** Signal in *right* rlPFC (6 mm *sphere* MNI space coordinates (x, y, z) 39, 41, 16) showing a main effect of confidence but not difference in value. Note that the bar plot is shown only to clarify the signal pattern in rlPFC (i.e. absence of main effect of DV). **c** Between subject regression analysis entering the change in choice accuracy (slope of the logistic fit) between *low* and *high* confidence trials (see *red arrow* in Fig. 8.1b) as a covariate for confidence-related activity in *right* rlPFC (peak (x, y, z) 27, 44, 16; $p < 0.05$ small volume FWE corrected). Note that the scatter-plot is not used for statistical inference (which was carried out in the SPM framework), and is shown solely for illustrative purposes. *Error bars* represent the s.e.m

8.3.7 Metacognitive Access: Functional Interaction of vmPFC and rlPFC

To test our second prediction, that these two regions are part of the same functional network (in the context of our task), we performed a psychophysiological interaction (PPI) analysis using rlPFC as a seed (Fig. 8.7a—in blue). This analysis revealed a robust modulation of connectivity between rlPFC and vmPFC (peak (x, y, z) 9, 50, -11; $z = 3.05$; $p < 0.05$ small volume FWE corrected) by confidence level (Fig. 8.5a, b). Furthermore, the strength of connectivity between these two regions also predicted metacognitive accuracy across subjects (vmPFC 15, 56, -5;

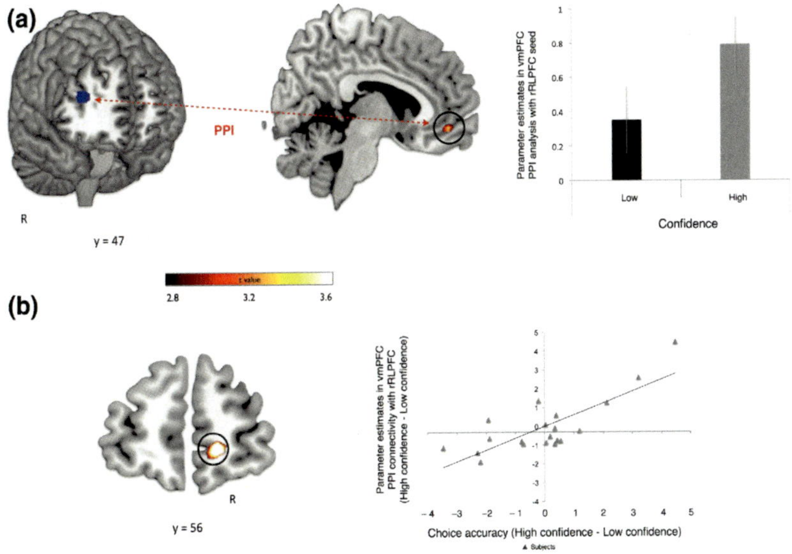

Fig. 8.7 Connectivity analysis. **a** PPI analysis: vmPFC (*circled* in *black*) shows increases in connectivity with a region of rlPFC (6 mm *sphere* (*x*, *y*, *z*) 39, 41, 16) (in *blue*) previously identified as being modulated by confidence (vmPFC peak (*x*, *y*, *z*) 9, 50, −11; $z = 3.05$; $p < 0.05$ small volume FWE corrected). **b** Between-subject regression analysis entering the increase in choice accuracy (see *red arrow* in Fig. 8.1b) between *high* confidence and *low* confidence conditions as a covariate for the modulation of connectivity (vmPFC peak (*x*, *y*, *z*) 15, 56, −5; $z = 3.91$; $p < 0.05$ small volume FWE corrected). Note that the scatter-plot is not used for statistical inference (which was carried out in the SPM framework), and is shown solely for illustrative purposes. *Error bars* represent the s.e.m

$p < 0.05$; SVC for multiple comparisons) (Fig. 8.7b). Thus, both the level of activity in rlPFC itself and its coupling strength with vmPFC influences the degree to which confidence is effectively 'read out' for metacognitive report.

How might this read-out process relate to our computational model of confidence? Intuitively, if reported confidence is a noisy facsimile of the confidence inherent to a decision process, the relationship between confidence and behaviour will weaken, and metacognitive accuracy will decrease (Maniscalco and Lau, this volume; [31]. We were able to modify the race model, introduced previously, to account for the inter-subject variability in metacognitive reports observed experimentally. We introduced an additional parameter (σ_{conf}) governing the noise in the read-out of Δe (i.e. decision confidence) computed during the value comparison. Variation in this parameter captured variability in the change in slope between high and low confidence conditions, despite overall choice accuracy remaining equal (Fig. 8.8b). Together with our imaging results, this analysis suggests that rlPFC may indeed mediate variability in reported confidence (see Fig. 8.8 and Discussion).

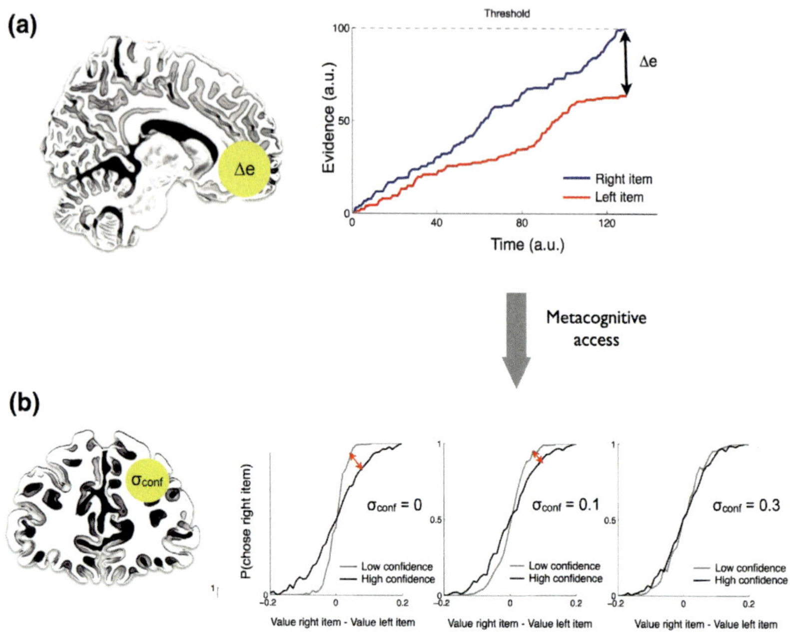

Fig. 8.8 Hypothesised relationship between our computational model and neuroimaging analyses. **a** Confidence in the decision (Δe) emerges from the value comparison process instantiated in vmPFC. **b** In order to reach metacognitive awareness (and be reported by the participant) this information is transferred to *right* rlPFC. An additional parameter (σ_{conf}) governs the noise in the read-out of Δe (i.e. decision confidence). If σ_{conf} is zero the information about confidence (Δe) is uncorrupted, resulting in a pronounced shift in the choice accuracy between *high* confidence and *low* confidence trials (*red arrow*). As the level of metacognitive noise increases (more positive values of σ_{conf}) the shift between the two *curves* (*low* and *high* confidence) diminishes. Differences in σ_{conf} account for the inter-subject variability in metacognitive reportability that we observed behaviorally

8.4 Discussion

The study outlined in this chapter extends the investigation of the neural basis of metacognition to value-based decision-making. Value is a particularly promising domain in which to understand metacognition, as the neural basis of value representation is relatively well-established in humans. We found that decision confidence emerges from a value comparison process in vmPFC, and that this region is in turn accessed by rlPFC to enable a metacognitive assessment of confidence. Our neural findings are consistent with previous evidence showing that choice difficulty is coded by vmPFC in humans and analogous OFC neurons in rodents [23, 43]. There is also an established body of work showing that this brain area represents the expected value of an outcome [40]. However, previous studies were unable to tease apart the relationship between value and confidence. Our design dissociates subjective confidence from DV on a trial-by-trial basis. In so doing we demonstrate that neural activity in the same anatomical region represents both variables, suggesting that confidence and DV are separate behavioural manifestations of the same underlying decision variable.

A central problem for computational models of metacognition is how confidence information is 'read out' for appraisal and communication to others. Insabato et al. proposed that such a computation can be achieved by a two-layer neural network architecture, in which the second-order network receives information about the performance of the first-order network, and uses this information to generate reports of confidence [20]; see also [37]. Our fMRI data can be interpreted in this framework and suggests that right rlPFC is a plausible locus for this second-order network. First, rlPFC represented confidence, but not DV, as predicted for a brain region that has access to information about confidence but is not directly involved in value comparison. Second, both confidence-related activity in rlPFC and coupling between rlPFC and vmPFC predicted the relationship between confidence and accuracy across individuals. This result can be explained if the coupling between vmPFC and rlPFC reflects the fidelity with which reported confidence tracks the evolution of a putative accumulator process in vmPFC (Figs. 8.2a and 8.8). Notably, confidence-related activity in rlPFC is also seen in perceptual decision-making [13], together with a modulation of connectivity with visual cortex. This pattern of findings suggests that rlPFC might play a domain-general role in metacognitive evaluation of decision-making, supporting the notion of segregated neural process governing metacognitive access. An alternative interpretation of our data is that information about choice confidence is coded elsewhere, perhaps in parallel to the construction of choice values, and is then communicated to vmPFC (possibly via rlPFC) where it is incorporated into the choice process. This mechanism would be analogous to a modulation of the vmPFC value signal during self-control by dorsolateral PFC [18]. Resolving this possibility is beyond the design of the current study and will require other techniques with high temporal resolution, such as MEG, that can track the evolution of confidence and valuation in the brain.

Psychological theories of metacognition emphasise the role of task-irrelevant heuristics driving an active process in the construction of metacognitive confidence [26]. Despite such heuristics, it is generally the case that metacognitive assessments are linked to underlying cognition, since there are reliable confidence-accuracy correlations in memory, perception and decision-making (Schwartz and Diaz, this volume). Here we provide a complementary perspective from computational neuroscience. By modelling confidence as the output of a simple accumulation process, we can accommodate several regularities between choice, confidence and response time in simple value-based decisions (see Yeung and Summerfield, this volume). This model provides a natural explanatory framework for why confidence and value activate the same system in vmPFC: they both reflect features of the same underlying decision variable. However, such a 'bottom-up', decisional perspective on confidence does not accommodate variability in metacognitive accuracy. Instead, this can be accommodated by the observation that other brain regions also represent confidence but not decision value, possibly reflecting higher-order metacognitive appraisal. We note that this view does not address how individuals with low metacognitive ability construct their confidence ratings, if not from decision-related activity. A future explanation may draw on notions of heuristics common in psychological models.

In keeping with recent research efforts that have incorporated dynamic models into the field of economic decision-making [49] we find that such a model captures several features of the relationship between choice, RT and confidence in a value-based choice paradigm. The separation between confidence and BDM values in this study provides a novel perspective on how an underlying decision variable can be fractionated into distinct behavioural components. Given that both DV and confidence had independent effects on vmPFC activity, this result provides convergent support for the notion that vmPFC acts as a dynamic accumulator of choice values [19]. However, in this study, we did not compare between variants of accumulation models that each have subtly different predictions for the inter-relationship between these measures of behaviour [5]. Different model variants, such as the drift–diffusion model, the race model, and so forth, can be summarised into one general class of race model with variable correlation between the accumulators [5]; non-linear interactions between the accumulators are also possible, e.g. [50]. The Vickers race model is actually an anomaly in this regard, accumulating noisy samples of the DV, rather than the value of each item separately. We chose this model as it generated the simplest predictions for choice confidence. An important area of future study is the modification of accumulation models to account for fluctuations in confidence in decision-making (e.g. [35, 39, 42]). However, this study, along with others in the evidence accumulation framework, demonstrate that any successful theory of metacognition and confidence should incorporate a detailed account of the functional dynamics of confidence (see also Middlebrooks et al., this volume; Yeung and Summerfield, this volume).

Our findings also accord with a theoretical Bayesian scheme in which uncertainty, or precision, is an inherent property of the neural code [14, 25]. This is in line with the theoretical notion that the estimates that guide value-based decisions are better described as samples from probability distributions (with variable degrees of uncertainty) rather than single values. In contrast to this theoretical view, most of the models currently used in neuroeconomics (such as expected utility theory or prospect theory; [22]) have a deterministic nature, and therefore treat values as single quantities and not as probability distributions with variable degrees of uncertainty or precision. It is only recently that random utility models [33] (which are probabilistic in nature) have gained popularity among decision theorists [16], partly on account of their ability to describe actual choices, including suboptimal ones [29]. The key idea behind these models is that utility (which is roughly equivalent to 'value' in economic jargon) is imperfectly observable by the experimenter; therefore, a random error term (usually a normal or logistic distribution) is added to the 'true' utility to account for the variability observed in the choice. However, these models are problematic for two main reasons: firstly, they focus only on the noise produced by errors in the experimental measurement while overlooking the noise inherent in the computational process itself [10, 15]; secondly, they ignore the causes underlying the error term (i.e. fluctuations in uncertainty). In this study we show that different confidence levels provide a behavioural and neural correlate of the level of stochasticity in choice. This relationship lends *prima facie* support to random utility models, and further suggests that (a) a large part of the noise is inherent to the neurocomputational process that leads to the construction of a value estimate, rather than experimental errors and (b) the noise term is amenable to behavioural measurement.

Our data show that humans have some degree of metacognitive access to noise in a value comparison, and that increased choice accuracy is associated with high subjective confidence. In other words, while choices often appear noisy from the point of view of the experimenter [15], subjective confidence ratings reveal systematic changes in the level of noise that are reflected by changes in choice accuracy. Most studies of the neural basis of metacognition, including the present one, have focused on the 'monitoring' aspect of Nelson and Narens' framework [36], leaving as an open question how metacognitive appraisal is used in guidance of subsequent behaviour (e.g. [27]). As an initial step towards addressing this question, we found that confidence was predictive of subsequent changes of mind when the same choice pair was repeated. We suggest that metacognitive confidence in value-based decision-making may be particularly important for guiding future behaviour in the absence of feedback on whether a choice was a good or bad one, as in the present case.

By integrating computational modelling with neural analysis, we provide evidence that subjective confidence is integral to the brain's representation of value in the vmPFC. Our work outlines a novel neural schema for how confidence-related information is computed and transferred to a distinct brain region (rlPFC), supporting metacognitive report. Far from being a blind process of selection corrupted by noise, value-based choices are accompanied by fluctuations in subjective confidence exquisitely sensitive to stochasticity in choice.

Acknowledgments The research reviewed in this chapter was supported by the Wellcome Trust. SMF is supported by a Sir Henry Wellcome Fellowship (WT096185). BDM is supported by a UCL early career fellowship.

References

1. Baranski JV, Petrusic WM (2001) Testing architectures of the decision-confidence relation. Can J Exp Psychol 55(3):195–206
2. Basten U et al (2010) How the brain integrates costs and benefits during decision making. Proc Natl Acad Sci 107(50):21767–21772
3. Beck JM et al (2012) Not noisy, just wrong: The role of suboptimal inference in behavioral variability. Neuron 74(1):30–39
4. Becker GM, DeGroot MH, Marschak J (1964) Measuring utility by a single-response sequential method. Behav Sci 9(3):226–232

5. Bogacz R et al (2006) The physics of optimal decision making: A formal analysis of models of performance in two-alternative forced-choice tasks. Psychol Rev 113(4):700–765
6. Daniel R, Pollmann S (2012) Striatal activations signal prediction errors on confidence in the absence of external feedback. NeuroImage 59(4):3457–3467
7. Daw N (2011) Trial-by-trial data analysis using computational models. In: Delgado MR, Phelps EA, Robbins TW (eds) Decision making, affect, and learning: Attention and performance XXIII. Oxford University Press, Oxford
8. De Martino B et al (2013) Confidence in value-based choice. Nat Neurosci 16(1):105–110
9. De Martino B et al (2009) The neurobiology of reference-dependent value computation. J Neurosci 29(12):3833–3842
10. Faisal AA, Selen LPJ, Wolpert DM (2008) Noise in the nervous system. Nat Rev Neurosci 9(4):292–303
11. FitzGerald THB, Seymour B, Dolan RJ (2009) The role of human orbitofrontal cortex in value comparison for incommensurable objects. J Neurosci 29(26):8388–8395
12. Fleming SM et al (2010) Relating introspective accuracy to individual differences in brain structure. Science 329(5998):1541–1543
13. Fleming SM, Huijgen J, Dolan RJ (2012) Prefrontal contributions to metacognition in perceptual decision making. J Neurosci 32(18):6117–6125
14. Friston K (2010) The free-energy principle: A unified brain theory? Nat Rev Neurosci 11(2):127–138
15. Glimcher PW (2005) Indeterminacy in brain and behavior. Annu Rev Psychol 56:25–56
16. Gul F, Pesendorfer W (2006) Random expected utility. Econometrica 74(1):121–146
17. Hare TA et al (2008) Dissociating the role of the orbitofrontal cortex and the striatum in the computation of goal values and prediction errors. J Neurosci 28(22):5623–5630
18. Hare TA, Malmaud J, Rangel A (2011) Focusing attention on the health aspects of foods changes value signals in vmPFC and improves dietary choice. J Neurosci 31(30):11077–11087
19. Hunt LT et al (2012) Mechanisms underlying cortical activity during value-guided choice. Nat Neurosci 15:470–476
20. Insabato A et al (2010) Confidence-related decision making. J Neurophysiol 104(1):539–547
21. Kable JW, Glimcher PW (2007) The neural correlates of subjective value during intertemporal choice. Nat Neurosci 10(12):1625–1633
22. Kahneman D, Tversky A (1979) Prospect theory—analysis of decision under risk. Econometrica 47(2):263–291
23. Kepecs A et al (2008) Neural correlates, computation and behavioural impact of decision confidence. Nature 455(7210):227–231
24. Kiani R, Shadlen M (2009) Representation of confidence associated with a decision by neurons in the parietal cortex. Science 324(5928):759
25. Knill DC, Pouget A (2004) The Bayesian brain: the role of uncertainty in neural coding and computation. Trends Neurosci 27(12):712–719
26. Koriat A (1997) Monitoring one's own knowledge during study: A cue-utilization approach to judgments of learning. J Exp Psychol Gen 126(4):349
27. Koriat A, Goldsmith M (1996) Monitoring and control processes in the strategic regulation of memory accuracy. Psychol Rev 103(3):490–517
28. Lau H, Rosenthal D (2011) Empirical support for higher-order theories of conscious awareness. Trends Cogn Sci 15(8):365–373
29. Louviere J et al (2002) Dissecting the random component of utility. Mark Lett 13(3):177–193
30. Maniscalco B, Lau H (2012) A signal detection theoretic approach for estimating metacognitive sensitivity from confidence ratings. Conscious Cogn 21(1):422–430
31. Maniscalco B, Lau H (2010) Comparing signal detection models of perceptual decision confidence. J Vis 10(7):213
32. McCurdy LY et al (2013) Anatomical coupling between distinct metacognitive systems for memory and visual perception. J Neurosci 33(5):1897–1906

33. McFadden D (1980) Econometric models for probabilistic choice among products. J Bus pp S13–S29
34. Milosavljevic M et al (2010) The drift diffusion model can account for value-based choice response times under high and low time pressure. Judgement Decis Making 5:437–449
35. Moreno-Bote R (2010) Decision confidence and uncertainty in diffusion models with partially correlated neuronal integrators. Neural Comput 22(7):1786–1811
36. Nelson TO, Narens L (1990) Metamemory: A theoretical framework and new findings. Psychol Learn Motiv 26, 125–322
37. Pasquali A, Timmermans B, Cleeremans A (2010) Know thyself: Metacognitive networks and measures of consciousness. Cognition 117(2):182–190
38. Plassmann H, O'Doherty J, Rangel A (2007) Orbitofrontal cortex encodes willingness to pay in everyday economic transactions. J Neurosci 27(37):9984–9988
39. Pleskac TJ, Busemeyer JR (2010) Two-stage dynamic signal detection: A theory of choice, decision time, and confidence. Psychol Rev 117(3):864–901
40. Rangel A, Hare T (2010) Neural computations associated with goal-directed choice. Curr Opin Neurobiol 20(2):262–270
41. Ratcliff R, Rouder JN (1998) Modeling response times for two-choice decisions. Psychol Sci 9(5):347–356
42. Ratcliff R, Starns JJ (2009) Modeling confidence and response time in recognition memory. Psychol Rev 116(1):59–83
43. Rolls ET, Grabenhorst F, Deco G (2010) Choice, difficulty, and confidence in the brain. NeuroImage 53(2):694–706
44. Rounis E et al (2010) Theta-burst transcranial magnetic stimulation to the prefrontal cortex impairs metacognitive visual awareness. Cogn Neurosci 1(3):165–175
45. Rushworth MFS et al (2011) Frontal cortex and reward-guided learning and decision-making. Neuron 70(6):1054–1069
46. Smith DV et al (2010) Distinct value signals in anterior and posterior ventromedial prefrontal cortex. J Neurosci 30(7):2490–2495
47. Soltani A, De Martino B, Camerer C (2012) A range-normalization model of context-dependent choice: A new model and evidence. PLoS Comput Biol 8(7):e1002607
48. Sugrue LP, Corrado GS, Newsome WT (2005) Choosing the greater of two goods: Neural currencies for valuation and decision making. Nat Rev Neurosci 6(5):363–375
49. Summerfield C, Tsetsos K (2012) Building bridges between perceptual and economic decision-making: Neural and computational mechanisms. Frontiers Neurosci 6:70
50. Usher M, McClelland JL (2001) The time course of perceptual choice: The leaky, competing accumulator model. Psychol Rev 108(3):550–592
51. Vickers D (1979) Decision processes in visual perception. Academic Press, New York
52. Wang S et al (2012) The role of risk aversion in non-conscious decision making. Frontiers Psychol 3
53. Yokoyama O et al (2010) Right frontopolar cortex activity correlates with reliability of retrospective rating of confidence in short-term recognition memory performance. Neurosci Res 68(3):199–206
54. Zylberberg A, Barttfeld P, Sigman M (2012) The construction of confidence in a perceptual decision. Frontiers Integr Neurosci 6

Chapter 9
What Failure in Collective Decision-Making Tells Us About Metacognition

Collective Failure and Metacognition

Dan Bang, Ali Mahmoodi, Karsten Olsen, Andreas Roepstorff, Geraint Rees, Chris Frith and Bahador Bahrami

Abstract Condorcet [2] proposed that a majority vote drawn from individual, independent and fallible (but not totally uninformed) opinions provides near-perfect accuracy if the number of voters is adequately large. Research in social

This chapter is adapted from: Bahrami B, Olsen K, Bang D, Roepstorff A, Rees G, Frith C (2012) What failure in collective decision-making tells us about metacognition. Phil Trans R Soc B 367:1350–1365

D. Bang (✉)
Department of Experimental Psychology, University of Oxford, South Parks Road, Oxford OX1 3UD, UK
e-mail: danbang.db@gmail.com

D. Bang
Calleva Research Centre for Evolution and Human Sciences, Magdalen College, High Street, Oxford OX1 4AU, UK

D. Bang · K. Olsen · A. Roepstorff · C. Frith · B. Bahrami
The Interacting Minds Centre, Aarhus University, Jens Chr. Skous Vej 4, Building 1483, 8000 Aarhus, Denmark

A. Mahmoodi
Control and Intelligent Processing Centre of Excellence, School of Electrical and Computer Engineering, College of Engineering, University of Tehran, North Kargar Avenue, Tehran, Iran

A. Mahmoodi
School of Cognitive Science, Institute for Research in Fundamental Sciences (IPM), Bahonar Square, Tehran, Iran

G. Rees · B. Bahrami
UCL Institute of Cognitive Neuroscience, University College London, Alexandra House, London WC1N 3AR, UK

G. Rees · C. Frith · B. Bahrami
Wellcome Trust Centre for Neuroimaging, Institute of Neurology, University College London, London WC1N 3BG, UK

S. M. Fleming and C. D. Frith (eds.), *The Cognitive Neuroscience of Metacognition*, DOI: 10.1007/978-3-642-45190-4_9,

psychology has since then repeatedly demonstrated that collectives can and do fail more often than expected by Condorcet. Since human collective decisions often follow from exchange of opinions, these failures provide an exquisite opportunity to understand human communication of *metacognitive confidence*. This question can be addressed by recasting collective decision-making as an information integration problem similar to multisensory (cross-modal) perception. Previous research in systems neuroscience shows that one brain can integrate information from multiple senses nearly optimally. Inverting the question, we ask: under what conditions can two brains integrate information about one sensory modality optimally? We review recent work that has taken this approach and report discoveries about the quantitative limits of collective perceptual decision-making, and the role of the mode of communication and feedback in collective decision-making. We propose that shared metacognitive confidence conveys the strength of an individual's opinion and its reliability inseparably. We further suggest that a functional role of shared metacognition is to provide substitute signals in situations where outcome is necessary for learning but unavailable or impossible to establish.

Keywords Metacognition · Collective decision-making · Signal detection · Communication · Cooperative behaviour · Feedback · Confidence

9.1 Introduction

In *The extraordinary and popular delusions and madness of crowds*, Charles Mackay chronicled a colourful and prolific history of humankind's collective follies [1].[1] Mackay's decision to doubt and re-examine the popular belief that 'two heads are better than one' has since then guided numerous disciplines interested in human collective decision-making from political sciences to economics and social psychology. Mackay's negative revisionism was preceded by a wave of optimistic trust in mass decisions initiated by Marquis de Condorcet [2], a mathematician and political philosopher of the French revolution. Condorcet's jury theorem elegantly proved that a simple 'democratic' majority vote drawn from the aggregated opinions of individual, independent and fallible (but not totally uninformed) lay people provides near-perfect accuracy if the number of voters is adequately large [2].

At a local livestock fair in Plymouth, early in the twentieth century Galton [3] found strong empirical support for Condorcet's theoretical proposition. At a weight-judging contest, participants estimated the weight of a chosen live ox after it had been slaughtered and dressed. Participants entered the competition by

[1] Interestingly, his major case studies, financial bubbles and religious conflicts over Jerusalem, do not show any signs of running out of steam yet.

privately writing their estimate on a ticket and submitting it to the fair organisers. The winner was the one who submitted the most accurate estimate. After the competition, Galton collected the $\sim$800 submitted tickets and demonstrated in a paper [3] that, indeed, the simple average of the estimates of the entire crowd was even more accurate than the winner. The most striking aspect of this finding was that the majority of participants had very little specialised knowledge of butchery; yet their contribution to the average opinion outperformed the best expert opinion. Theoretically and empirically, masses ruled supreme. So, can we dismiss Mackay's [1] worries? Definitely not!

A large body of work in political sciences and social psychology has examined collective decision-making and, indeed, numerous examples of collective failure have been discovered [4]. Indeed, this research clearly shows that Condorcet's assumption about the independence of individual opinions—which was neatly satisfied in the weight-judging contest—is often not applicable to real-world situations of collective behaviour [5]. However, by carefully identifying the determinants of collective failure, we can use Mackay's [1] insight, that collective benefit is the exception rather than the rule, to better understand the nature of human social interaction. To rephrase, one could ask which features of interpersonal communication and/or interaction contribute to collective failures.

In this chapter, we first review previous work that has addressed this question by recasting collective decision-making as an 'information integration' problem similar to multisensory (cross-modal) perception. In multisensory perception, the participant combines information from different sensory modalities (e.g. vision and touch) taking into account the reliability (or variance) of each modality such that the multisensory decision is more strongly influenced by the sensory modality with the higher reliability (i.e. lower variance) [6–8]. By analogy, collective decision-making also requires combining information, but from different participants. We argue that establishing the reliability of this information constitutes an integral part of information integration at the collective level.

Using collective decision-making in the perceptual domain as a framework, we describe and compare two models of how we communicate and integrate our individual perceptions and their reliability [9, 10]. Both models posit that participants convey the reliability of their individual perceptions by communicating their *confidence* in their perceptual decisions, i.e. their metacognitive awareness of their perceptual decisions. However, the models make very different assumptions about the exact content of the communicated confidence and the computational strategy by which participants combine them to arrive at a collective decision. We will review the predictions of each model and assess them in the light of the existing literature. We will also present new empirical data revealing further features of collective perceptual decision-making by confidence sharing. We will place the findings from collective perceptual decision-making in the wider context of group decision-making [4] and discuss a possible functional social role for metacognitive awareness. Finally, we will briefly compare and attempt to connect our current understanding of metacognition at the levels of brain mechanism, individual behaviour and social interaction.

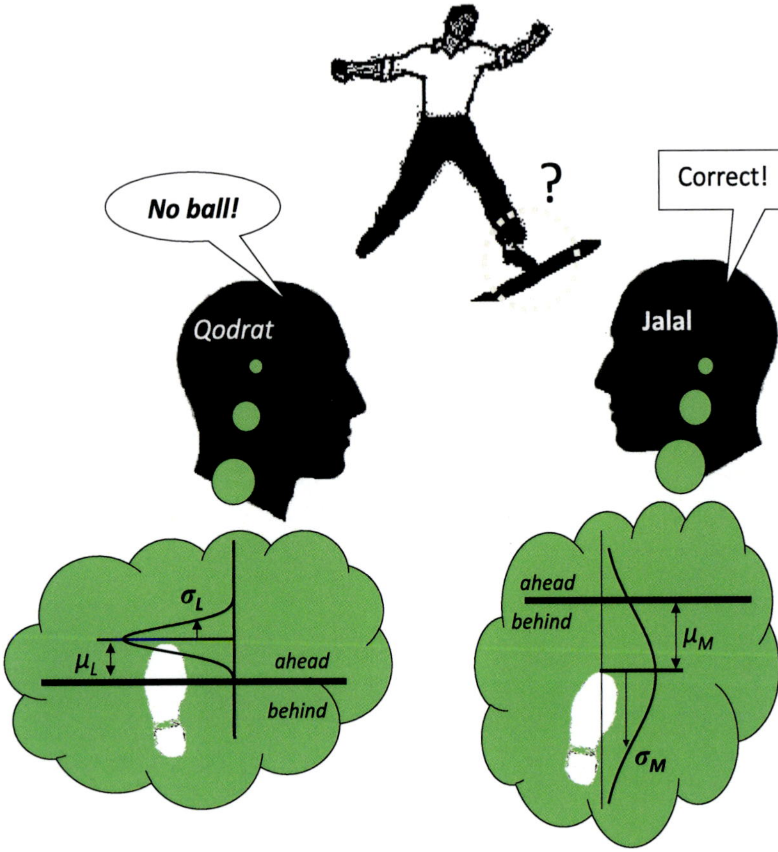

Fig. 9.1 Two cricket umpires, Qodrat and Jalal, disagree about whether the bowler crossed the line. The *low-quality image* depicting the bowler was intentionally constructed to indicate perceptual noise. Each umpire's individual decisions are based on his respective noisy perceptual representation, which we model as a Gaussian distribution. The figure is inspired by [52]

9.2 Collective Perceptual Decision-Making

Qodrat and Jalal (Fig. 9.1) are two cricket umpires [2] The bowler (Fig. 9.1, top) has just made his run and bowled the ball; but the two umpires disagree about whether his foot crossed the line or not. Whereas Qodrat has announced a 'no ball', Jalal contends that there was no such error. Let us stop here and examine the situation.

[2] The names chosen for the cricket umpires are inspired by Graham Greene's (1940) *The Power and the Glory* (Qodrat o Jalal).

We can think of each umpire's visual perception of the events as represented in the brain by a normal distribution with a mean (μ_Q for Qodrat and μ_J for Jalal) and a standard deviation (σ_Q and σ_J). This normal distribution could correspond to, for example, the firing pattern of neurons in each umpire's early visual cortex. The umpire's decision about whether the bowler's foot landed ahead of (e.g. Qodrat, $\mu_Q > 0$) or behind (e.g. Jalal, $\mu_J < 0$) the line is given by the signed mean of the distribution. The standard deviation of each distribution relates to how noisy the umpire's perception is. As such, a reliable percept would be characterised by a large mean (e.g. Jalal) and a small standard deviation (e.g. Qodrat). But how do Qodrat and Jalal resolve their disagreement and come to a joint decision? The simple formulation of the situation given above is the basis of two recent models [9, 10] of collective perceptual decision-making.

Sorkin et al. [10] proposed that, by communicating their confidence in their perceptual decision, the umpires are in fact communicating their respective μ and σ *separately* and *distinctly* to one another. As we will see further below, the distinctness of these two pieces of information is a critical feature of this model. To make an optimal collective decision (i.e. to minimise the chances of error given each umpires' decision noise), the two umpires (i.e. the group) somehow evaluate the term $\left(\frac{\mu_Q}{\sigma_Q^2} + \frac{\mu_J}{\sigma_J^2}\right)$ and take its sign as their joint decision. Defining perceptual sensitivity (s) as inversely proportional to standard deviation (such that $s = k/\sigma$; where k represents a constant term. See Eq. 9.6 below for the exact definition of slope), the group's sensitivity, S_{group} is then expected to be

$$S_{group} = \sqrt{S_Q^2 + S_J^2} \tag{9.1}$$

In a standard sensory signal detection task performed by individuals and groups in separate experiments, Sorkin et al. [10] showed that groups achieved a robust collective benefit over and above the sensitivities of the constituent individuals as measured when these individuals performed the task in isolation. Their model was able to predict the collective benefits accrued by the groups. Interestingly, their model could readily be extended to groups larger than two people. However, as group size expanded, group performance did not improve as fast as predicted by the model, indicating that, perhaps, different group dynamics may be at work as group size increases.

Sorkin et al. [10] model is conceptually identical to the model used in multisensory perception research to describe how information from different sensory modalities, such as touch and vision, are combined within the brain of one participant [6, 8, 11]. That dyads performed as well as Eq. 9.1 would lead to the uncomfortable conclusion that communication between brains is as reliable and high-fidelity as communication within the same brain. Moreover, this formulation implies that groups would *never* do worse than individuals. Recalling the case of Condorcet, Mackay and Galton, once again, groups seemed to be doing much better (theoretically and empirically) than common sense would suggest.

Nearly a decade later, Bahrami et al. [9] performed an experiment almost identical to that of Sorkin et al. [10], but they made a different assumption about the content of the information communicated between individuals. Noting that a reliable decision (Fig. 9.1) is one based on a large mean and a small standard deviation, they suggested that Qodrat's communicated confidence in his decision could be defined as the ratio μ_Q/σ_Q. The magnitude of this signed ratio indicates the probability that Qodrat has made the right decision.[3] The collective decision could then simply be defined as the sign of the sum of shared confidences ($\mu_Q/\sigma_Q + \mu_J/\sigma_J$), giving the group sensitivity by

$$s_{\text{group}} = \frac{s_Q + s_J}{\sqrt{2}} \tag{9.2}$$

Bahrami et al. [9] dubbed this model the Weighted Confidence Sharing (WCS) model. Comparison with the Sorkin et al. [10] model shows that when sensitivities are similar (i.e. $\sigma_L = \sigma_Q$), the two decision boundaries $\left(\frac{\mu_Q}{\sigma_Q^2} + \frac{\mu_J}{\sigma_J^2} = 0 \text{ and } \frac{\mu_Q}{\sigma_Q} + \frac{\mu_J}{\sigma_J} = 0\right)$ become identical and the outcome is equivalent to that seen in multisensory perception. When the individual sensitivities are different (say, $s_Q > s_J$), however, the two models diverge in their predictions. To demonstrate this, if we rewrite Eq. 9.2 as

$$\frac{s_{\text{group}}}{s_Q} = \frac{1}{\sqrt{2}} + \frac{1}{\sqrt{2}}\left(\frac{s_J}{s_Q}\right) \tag{9.3}$$

we can see that the expected collective benefit (s_{group}/s_Q; i.e. group sensitivity relative to its more sensitive member) is a linear function of the similarity between group members' sensitivities (s_J/s_Q). This linear relationship means that if Jalal's sensitivity is no better than ~40 % of Qodrat's ($s_J/s_Q < 2^{1/2}-1 \approx 0.4$), then—in sharp contrast to Sorkin et al. [10] model—this model predicts that Qodrat and Jalal together should do worse than the more sensitive participant (Qodrat) alone ($s_{\text{group}} < s_Q$).

Bahrami et al. [9] tested dyads in a simple perceptual decision-making task that involved visual contrast discrimination (Fig. 9.2). In every trial, individuals first made a private decision about a briefly viewed stimulus. If their private decisions disagreed, they were asked to negotiate a joint decision. When dyad members had similar sensitivities, dyad decisions were more accurate than those of the better individual. However, if one participant was much less sensitive than the other, the dyad failed to outperform the better member; c.f. Experiment 2 in [9]. Importantly, group performance in these latter situations was markedly worse than expected

[3] The confidence ratio μ_i/σ_i can be monotonically transformed into $P(x > 0)$, which gives the probability of being correct given perceptual sample x. The decision rule ($\mu_Q/\sigma_Q + \mu_J/\sigma_J$) is thus equivalent to accepting the decision of the person with the higher probability of being correct (i.e. the higher confidence μ_i/σ_i). See the Supplementary Materials to Bahrami et al. [9] for further mathematical details.

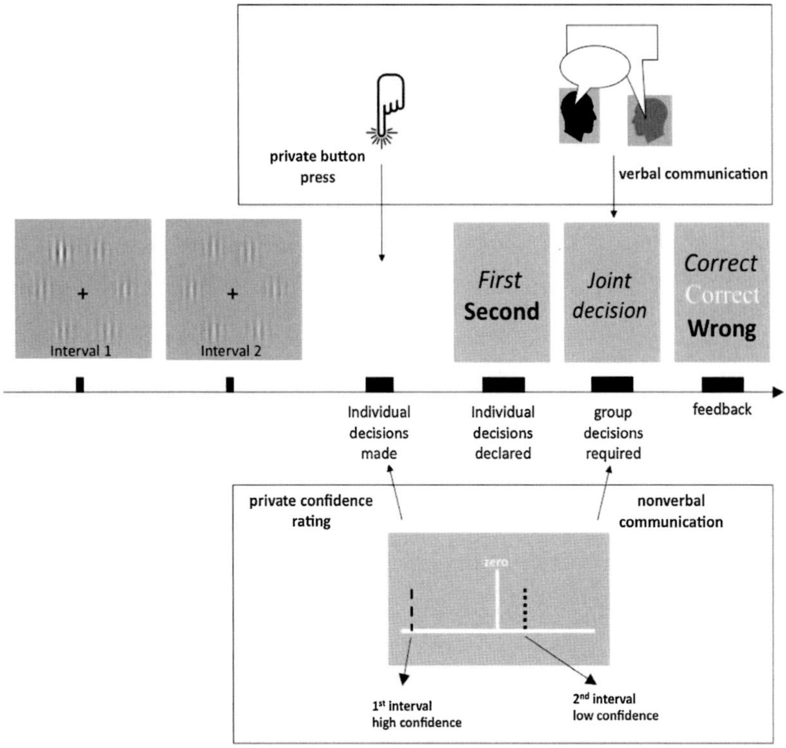

Fig. 9.2 Stimuli, task and modes of communication. Each trial consisted of two observation intervals followed by private decisions by each participant. In the verbal communication mode (*top box*), participants indicated their individual decision by a button press. In the nonverbal mode (*bottom box*), participants reported the target interval by dragging a marker to the *left* (1st) or *right* (2nd) of the centre and indicated their confidence by the distance of the line from the centre. Individual decisions were then announced, and in cases of disagreement, participants either talked to each other (*top*) or saw each others' confidence rating (*bottom*) in order to reach a joint decision. Then one of the participants (indicated by the colour of the sentence 'joint decision?') announced the dyad decision. *Italics*, *boldface* and *white* shades correspond to *blue*, *yellow* and *white* colour codes that were used in the experiments to indicate the participant using the keyboard, the one using the mouse and the dyad, respectively

from Eq. 9.1 but not statistically different from the predictions of Eq. 9.2. Mackay's intuition had once again proved useful: examination of collective failures suggested that perceptual decisions and their reliability are not spontaneously communicated separately but instead together in the form of a ratio. If what Qodrat communicates is the ratio (μ_Q/σ_Q), then Jalal (i.e. the recipient) will be unable to resolve (μ_Q) and (σ_Q) from one another. Consequently, the process of interpersonal communication involves information loss. But is there a way to avert or reduce this loss of information?

Previous research suggests that failures of communication are not solely due to noise and random errors [12]. A number of systematic egocentric biases that impair communication have been identified on both sides of a verbal exchange. When communicating their internal intentions verbally, people often overestimate the clarity of their communicated message, as if their internal states were readily evident, indeed *transparent*, to their addressee [13]. Similarly, an egocentric bias afflicts the addressee: listeners often interpret the meaning of what they are told from their own (rather than the speaker's) perspective [14].

We hypothesised that if the collective failures observed by Bahrami et al. [9] were a consequence of egocentric biases that plague face-to-face verbal communication [12], then providing a nonverbal, scalar system for participants to share information may remove or reduce these egocentric biases and in turn improve collective decision-making; especially when participants have vastly different sensitivities. To test this hypothesis, we first replicated the collective failure reported by Bahrami et al. [9] (Experiment 1: Verbal condition). Then we devised a nonverbal confidence rating/sharing schema to replace face-to-face verbal communication of decisions while keeping all other aspects of the experiment constant (Experiment 1: NonVerbal condition). If egocentric biases in face-to-face verbal communication were at least partially responsible for the collective failures in the Verbal condition, then the nonverbal confidence rating/sharing schema employed in the nonverbal condition should improve collective performance under conditions of asymmetric sensitivity.

9.3 Experiment 1

9.3.1 Methods

9.3.1.1 Participants

Participants were recruited from undergraduate, graduate and faculty members of Aarhus University, Denmark. Verbal (V) condition: $N = 30$; mean age ± sd: 23 ± 2.5; NonVerbal (NV) condition: $N = 30$; mean age ± sd: 23.9 ± 2.5). All participants were healthy male adults with normal or corrected-to-normal visual acuity. Members of each dyad knew each other. No participant was recruited for more than one experiment. The local ethics committee approved all experiments; and written informed consent was obtained from all participants.

9.3.1.2 Display Parameters and Response Mode

In all experiments, both dyad members sat in the same testing room. Each viewed his own display. Display screens were placed on separate tables at a right angle to each other. The two displays were connected to the same graphic card via a video amplifier splitter and controlled by the Cogent toolbox (http://www.vislab.ucl.ac.uk/cogent.php/) for MATLAB (Mathworks Inc).

Each participant viewed an LCD display at a distance of $\sim$57 cm (resolution = 800 × 600—Fujitsu Siemens AMILO SL 3220 W, 22″) for which a look-up table linearised the output luminance. Background luminance was 62.5 Cd/m^2 in both displays. The displays were connected to a personal computer through an output splitter that sent identical outputs to both of them. Within each session of the experiment, one participant responded with the keyboard and the other with the mouse. Both participants used their right hand to respond.

Each participant viewed one-half of their screen: the left half of one display for the participant responding with the keyboard, and the right half of the other display for the participant responding with the mouse. A piece of thick black cardboard placed on each display was used to occlude the half not viewed by each participant. Two stimulus arrays were presented on both displays simultaneously, each on one-half of the display. Control over which one the participants saw was achieved by using the occluding cardboard. Stimulus eccentricity and retinal size were identical for both experiments. This configuration permitted us to display stimuli with different levels of noise to participants in the same dyad. Other stimulus characteristics (retinal size, luminance, contrast, duration) were identical for both participants. Moreover, in the NV condition, the use of a bipartite display allowed us to assess the participants' confidence privately (i.e. each participants only saw his own confidence bar) at the individual decision stage (see Procedure).

9.3.1.3 Design and Task

In all experiments, a 2-Alternative Forced Choice (2AFC) design was employed (Fig. 9.2). Two observation intervals were provided. A target stimulus always occurred either in the first or the second interval. Participants were instructed to choose the interval most likely to have contained the target. In the NV condition, participants rated their confidence in their decision on a scale from 1 (indicating 'very doubtful') to 5 (indicating 'absolutely sure') (see below for a description of the confidence rating procedure and display).

9.3.1.4 Stimuli

The stimulus set displayed in each interval consisted of six vertically oriented Gabor patches (standard deviation of the Gaussian envelope: 0.45°; spatial frequency: 1.5 cycles/degree; contrast: 10 %) organised around an imaginary circle (radius: 8°) at equal distances from each other. The target stimulus was generated by elevating the contrast of one of the six patches, which produced a contrast oddball. The target location and interval were randomised across the experimental session. The stimulus duration in each interval was 85 ms. Target contrast was obtained by adding one of four possible values (1.5, 3.5, 7.0 or 15 %) to the 10 % contrast of the non-target items.

For one participant, in each trial and for each item in the stimulus array, freshly generated white noise was added to the grey value of each pixel in each Gabor patch. The additional white noise was drawn, on each update, from a random uniform distribution ranging from 0 to 30 % of the monitor's maximum luminance. The participants did not know about the addition of noise. The choice of which participant would receive the noise was determined by a preliminary test before the experiment (see below).

9.3.1.5 Procedure

Each trial was initiated by the participant responding with the keyboard (see Fig. 9.2). A black central fixation cross (width: 0.75° visual angle) appeared on the screen for a variable period, drawn uniformly from the range 500 to 1,000 ms. The two observation intervals were separated by a blank display lasting 1,000 ms. The fixation cross turned into a question mark after the second interval to prompt the participants to respond. The question mark stayed on the screen until both participants had responded. Each participant initially responded without consulting the other.

In the V condition, participants communicated by talking to each other. Participant who used the keyboard responded by pressing 'N' and 'M' for the first and second interval, respectively; the participant who used the mouse responded with a left and right click for the first and second interval, respectively. Individual decisions were then displayed on the monitor (Fig. 9.2), so both participants were informed about their own and their partner's choice of the target interval. Colour codes were used to denote keyboard (blue—illustrated in Fig. 9.2 by italics) and mouse (yellow—illustrated in Fig. 9.2 by boldface) responses. Vertical locations of the blue and yellow text were randomised to avoid spatial biasing. If the private decisions disagreed, a joint decision was requested. The request was made in blue if the keyboard participant was to announce the decision and in yellow if the mouse participant was to announce the decision. The keyboard participant announced the joint decision in odd trials; the mouse participant on even trials. Participants were free to verbally discuss their choice with each other as long as they wanted. They were also free to choose any strategy that they wished. The experimenter was present in the testing room throughout all experiments to make sure that the instructions were observed.

In the NV condition, participants did not talk to each other but instead used a visual schema (Fig. 9.2, lower panel) to communicate their confidence in their private decisions. After the two observation intervals, a horizontal line appeared on the screen with a fixed midpoint. The left side of the line represented the first interval, the right side of the line represented the second interval. An additional vertical 'confidence marker' (colour coded for keyboard and mouse responses—see above) was displayed in each participant's panel. By dragging the confidence marker to the left or right from centre, the participant reported his choice about whether the target was in the first or second interval, respectively. The confidence marker could be moved along the line by up to five steps on either side. Each step farther from the centre indicated higher confidence. We chose this method for obtaining the decision (left or right side of centre) and the confidence in the decision (distance from the centre) all in one step rather than having the participants report them serially. This ensured that the participants' private task involved only one step in all conditions here and in Experiment 2. The participant who used the keyboard navigated the marker on the confidence rating scale by pressing 'N' or 'M' to move the marker left or right, respectively. He would then confirm his decision by pressing 'B' when he thought the marker correctly indicated his confidence. The participant who used the mouse moved the confidence marker by pressing left or right button to move the marker left or right, respectively. He would then press the middle button when he thought the marker correctly indicated his confidence. Participants did not see each other's confidence rating at this stage. After the private confidence ratings were made, confidence values were announced by displaying both participants' confidence markers along the horizontal line. In the case of disagreement, a joint decision was requested. Here, the keyboard participant announced the joint decision in odd trials, and the mouse participant on even trials. For the joint decision, a white confidence marker was used with the same five levels as private decisions; the marker was not visible to the other participant until a joint decision had been made. Participants did not talk to each other. They were given earphones to eliminate any meaningful auditory communication. In addition, a screen was placed between them to prevent them from seeing each other if they turned around. The experimenter was present in the testing room throughout all experiments to make sure that the instructions were observed.

In all conditions, participants received feedback either immediately after they made their private decision, in cases where their private decisions agreed, or after the joint decision had been made in cases where their private decisions disagreed. The feedback either said 'CORRECT' or 'WRONG'. Feedback was given for each participant (keyboard: blue—illustrated in Fig. 9.2 by italics; mouse: yellow—illustrated in Fig. 9.2 by boldface) and for the dyad (white). Feedback remained on the screen until the participant using the keyboard initiated the next trial (Fig. 9.2). Vertical order of the blue and yellow was randomised and the dyad feedback always appeared in the centre.

In both conditions, participants started the experiment with a preliminary, non-interactive session (8 blocks of 16 trials) that was conducted in order to identify the participant who would receive noise in the subsequent main session (see Assignment of noise). Then, the main experimental session (8 blocks of 16 trials) was conducted.

9.3.1.6 Assignment of Noise

We determined which participant would receive the noisy stimuli by first testing the participants in an isolated version of the task. In each trial, participants made a private decision about the target interval and then received private feedback (i.e. there was no sharing of private decisions and feedback). At the end of this session, the two participants' sensitivity (i.e. the slope of the psychometric function, see Data Analysis) was assessed, and the less sensitive participant was chosen to receive the noisy stimuli in the experiment proper. The participants were not informed about this procedure and were told that the preliminary test served as practice.

9.3.1.7 Data Analysis

Psychometric functions were constructed for each participant and for the dyad by plotting the proportion of trials in which the oddball was seen in the second interval against the contrast difference at the oddball location (the contrast in the second interval minus the contrast in the first; see Fig. 9.3a).

The psychometric curves were fit to a cumulative Gaussian function whose parameters were bias, b, and variance, σ^2. To estimate these parameters, a probit regression model was employed using the *glmfit* function in MATLAB (Mathworks Inc). A participant with bias b and variance σ^2 would have a psychometric curve, denoted $P(\Delta c)$ where Δc is the contrast difference between the second and first presentations, given by

$$P(\Delta c) = H\left(\frac{\Delta c + b}{\sigma}\right), \tag{9.4}$$

where $H(z)$ is the cumulative Normal function,

$$H(z) \equiv \int_{-\infty}^{z} \frac{\mathrm{d}t}{(2\pi)^{1/2}} \exp\left[-t^2/2\right]. \tag{9.5}$$

As usual, the psychometric curve, $P(\Delta c)$, corresponds to the probability of saying that the second interval had the higher contrast. Thus, a positive bias indicates an increased probability of saying that the second interval had higher contrast (and thus corresponds to a negative mean for the underlying Gaussian distribution).

Given the above definitions for $P(\Delta c)$, we see that the variance is related to the maximum slope of the psychometric curve, denoted s, via

$$s = \frac{1}{(2\pi\sigma^2)^{1/2}}. \tag{9.6}$$

A large slope indicates small variance and thus highly sensitive performance. Using this measure, we quantified individual participants' as well as the dyad's sensitivity. We defined 'collective benefit' as the ratio of the dyad's slope (s_{dyad}) to that of the more sensitive participant (i.e. the one with higher slope, $s_{\max}$). A collective

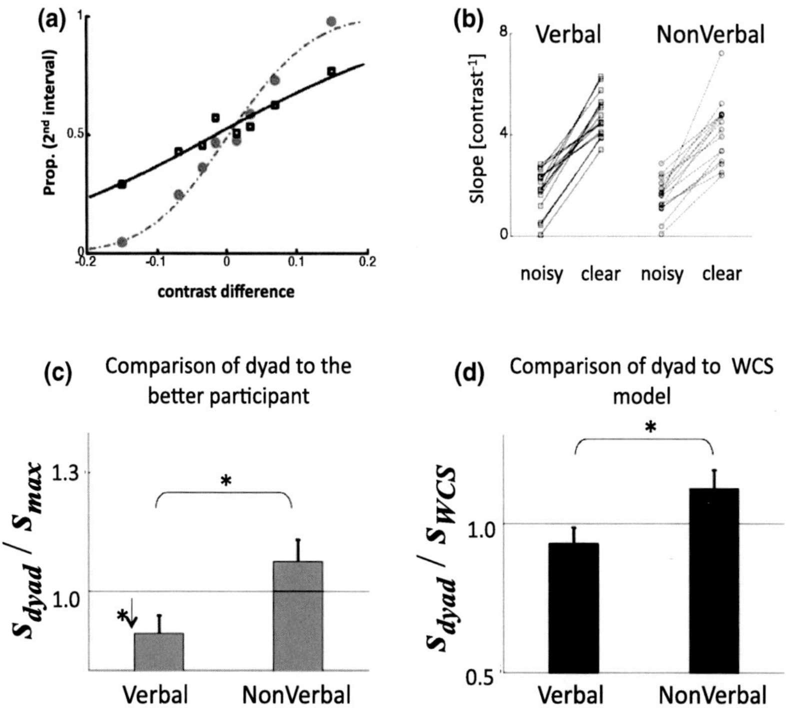

Fig. 9.3 Experiment 1 results. **a** Psychometric function relating performance contrast. Data are from the Verbal condition of Experiment 1 averaged across $N = 15$ participants for each curve. The proportion of trials in which the target was reported in the 2nd interval is plotted against the contrast difference at the target location (i.e. contrast in the second interval minus contrast in the first). Participants who received clear stimuli (*grey*) produced steeply rising psychometric functions with large slopes. Participants who received noise (*black*) had a much shallower slope. **b** The slope of psychometric functions of the dyad members in the Verbal and the NonVerbal conditions of Experiment 1. Each line corresponds to a dyad. Addition of noise was clearly effective at reducing the slope in both experiments. **c** Collective benefit (s_{dyad}/s_{max}; see Methods) accrued in the Verbal and the NonVerbal conditions of Experiment 1. *Horizontal line* indicates that dyad slope was equal to the more sensitive participant's. **d** Optimality of group performance in the Verbal and the NonVerbal conditions of Experiment 1. *Horizontal line* indicates that group performance was as good as predicted by the WCS model (cf. Bahrami et al. [9]). * : $p < 0.05$

benefit value above one would indicate that the dyad managed to gain an advantage over its better participant. Values below one would indicate that collaboration was counterproductive and that the dyad did worse than its more sensitive member.

The WCS model expressed in Eq. 9.2 [9] identifies the dyad's *potential* for collective achievement under the assumption that the members can communicate their confidence to each other accurately. We compared the empirically obtained data to this potential upper bound to see whether and how different modes of communication helped or hindered collective decision-making. We defined an 'optimality index' as the ratio of the dyad's slope to that predicted by the WCS model (Eq. 9.2).

9.4 Results

As demonstrated in Fig. 9.3b, all participants who received noise showed lower sensitivity (as measured by the slope of their psychometric function) compared to their partner who received noise-free stimuli. This result showed that our noise manipulation effectively rendered one participant's perceptual decisions much less reliable than those of the other. Under such conditions, the WCS model predicts that the dyad will do worse than the better participant.

9.4.1 Comparison of Dyad to the Better Participant

In the V condition dyad sensitivity was significantly worse than that of the better participant (one sample t test comparing collective benefit to baseline; $t(14) = -2.34$, $p = 0.03$; see Fig. 9.3c). This result is consistent with the predictions of the WCS model, which predicted that collective decision-making will be counterproductive when dyad members have very different sensitivities [9]. In the NV condition, on the other hand, dyad sensitivity was no worse than the more sensitive participant (one sample t-test comparing collective benefit to baseline; $t(14) = 1.42$, $p = 0.17$). Figure 9.3c showing that groups had been at least as good as the better participant. Importantly, direct comparison of the two conditions showed that collective benefit was significantly greater in the NV condition (independent samples t-test, $t(28) = 2.61$, $p = 0.014$).

9.4.2 Comparison of Dyad to the WCS Model

The WCS model (Eq. 9.2) slightly (but not significantly) overestimated dyad performance in the V condition (one sample t-test; $t(14) = 1.25$, $p = 0.23$. Figure 9.3d). In the NV condition, the WCS model showed a trend to underestimate the dyad performance (one sample t-test; $t(14) = 1.87$, $p = 0.08$. Figure 9.3d). Direct comparison of the two conditions showed that the optimality index was significantly higher in the NV condition (independent sample t-test; $t(28) = 2.24$, $p = 0.03$).

9.5 Data Summary

The impact of verbal and nonverbal communication on collective decision-making was compared in an experimental situation where previous work had shown that dyads would perform no better than their constituting individuals [9]. The results replicate the previous findings, but go beyond them in several respects:

Bahrami et al. [9] had assigned noise to either one of the dyad members *at random*. This trial-by-trial random noise assignment made it impossible for the participants to form any stable idea of which dyad member was the less reliable one in any trial. Here we used a block design and assigned noise consistently to one member of the dyad. Thus, the results of the V condition (Fig. 9.3c, d) show an even more impressive collective failure: dyad sensitivity was significantly *worse* than the more sensitive dyad member. A conspicuous difference in performance did not protect the groups from suffering counterproductive collaboration.

The results of the NV condition showed that a nonverbal schema for reporting and sharing decision confidence (Fig. 9.1, lower panel), to some extent, could remedy the defective collective decision-making process and make it more productive; even though all the low-level conditions, especially the asymmetric administration of noise, were retained. This result is consistent with the suggestion that the collective failure observed in the V condition is not due to random errors caused by asymmetric noise but, rather, that direct, verbal communication and its associated underlying cognitive biases cause the collective failure.

Having demonstrated the beneficial impact of nonverbal communication on defective collective decisions, we asked whether this benefit is general. Experiment 1 tested collective decisions under asymmetric administration of noise where we expected no collective benefit to start with. When dyad members have access to similarly reliable perceptual information, however, direct verbal communication can indeed confer a robust group benefit that is no less than expected from the optimal combination of individuals' decisions [9, 10]. We next asked whether the nonverbal communication of confidence could provide any additional benefit over and above direct verbal communication. Indeed, we do not know whether and how different modes of communication interact with one another towards collective decisions. In order to address this question, we used a 2×2 design where collective decision-making was tested under all four possible combinations of the two modes of communication (see Fig. 9.4), without the addition of noise.

9.6 Experiment 2

9.6.1 Methods

9.6.1.1 Participants

Participants were recruited from undergraduate, graduate and faculty members of Aarhus University, Denmark. Verbal and NonVerbal (V&NV) condition: $N = 30$; mean age $\pm$ sd: 24.8 ± 3.5; Verbal (V) condition: $N = 30$; mean age $\pm$ sd: 28.30 ± 6.27; NonVerbal (NV) Condition: $N = 30$; mean age $\pm$ sd: 22.2 ± 2; None (N) condition: $N = 28$; mean age $\pm$ sd: 23.2 ± 2. All participants were healthy male adults with normal or corrected-to-normal visual acuity. Members of each dyad knew each other. No participant was recruited for more than one

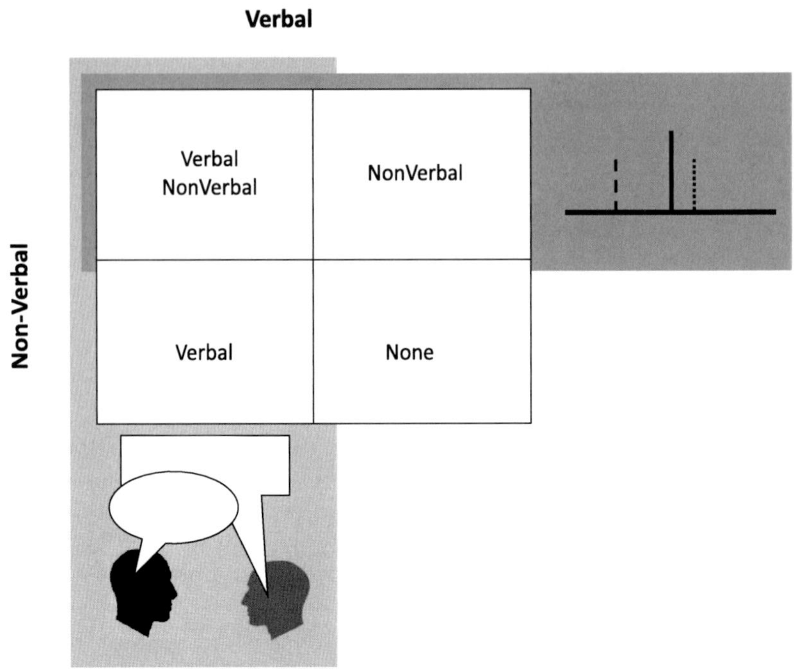

Fig. 9.4 Two by two design employed in Experiment 2

experiment. The local ethics committee approved all experiments; and written informed consent was obtained from all participants. Data from the V and N conditions have been reported elsewhere [9].

9.6.1.2 Task and Design

We employed a 2 × 2 design to investigate the impact of verbal communication (two levels: with and without) and nonverbal confidence sharing (two levels: with and without) on collective decision-making (Fig. 9.4). In the V and V&NV conditions participants communicated verbally. In the NV and V&NV conditions, participants communicated using the confidence marker (as in the NV condition of Experiment 1). In the None (N) condition, participants were not allowed to communicate anything but their decision (first or second interval). The task was identical to Experiment 1 in all other aspects.

9.6.1.3 Display and Stimuli

Participants received identical visual stimuli and no participant was given any additional noise. All stimulus characteristics were identical to the noise-free stimuli in Experiment 1. In conditions that did not involve nonverbal communication (i.e. V and N conditions) the bipartite display was not used and both participants viewed a single stimulus set displayed at the centre of the entire screen. All other display and stimulus characteristics were identical to Experiment 1.

9.6.1.4 Procedure

In all conditions, after one practice block of 16 trials, two main experimental sessions were conducted. Each main session consisted of 8 blocks of 16 trials. Participants switched places (and thereby response device) at the end of session one. The participants set the pace of the experiment's progress. All other aspects of the procedure were identical to Experiment 1.

9.7 Results

9.7.1 Comparison of Dyad to the Better Participant

We first looked at the impact of the mode of communication on the collective benefit. Following our design (Fig. 9.4), we employed a two (with and without verbal communication) by two (with and without nonverbal confidence sharing) between-subject ANOVA with collective benefit (s_{dyad}/s_{max}; see Experiment 1) as the dependent variable (Fig. 9.5). The main effect of verbal communication was highly significant ($F(1, 59) = 8.4$; $p = 0.005$). *Post hoc* comparison showed that collective benefit was significantly higher when verbal communication was allowed (i.e. Conditions (V and V&NV) versus (NV and N); independent sample t-test, $t(57) = 2.7$, $p = 0.008$). Comparison to baseline (see horizontal lines in Fig. 9.5) showed that a robust collective benefit (i.e. group performance advantage over and above the better participant) was observed only where verbal communication allowed, i.e. V&NV (Fig. 9.5a, one sample t-test, $t(14) = 2.47$, $p = 0.026$) and V conditions (Fig. 9.5b; one sample t-test, $t(14) = 5.38$, $p < 0.0001$). When communication was strictly nonverbal (NV condition), collective benefit marginally approached significance (Fig. 9.5c; one sample t-test, $t(14) = 2.00$, $p = 0.064$). The main effect of nonverbal communication was not significant ($F = 0.8$). Finally, a significant interaction was found between verbal and nonverbal communication ($F(1, 59) = 6.56$; $p = 0.013$). *Post hoc* comparison showed that the interaction was driven by a significantly higher collective benefit in the V condition where participants communicated *only* verbally: collective decision-making was significantly *less* successful when participants were required to use *both* verbal and nonverbal communication (i.e. V&NV vs. V condition;

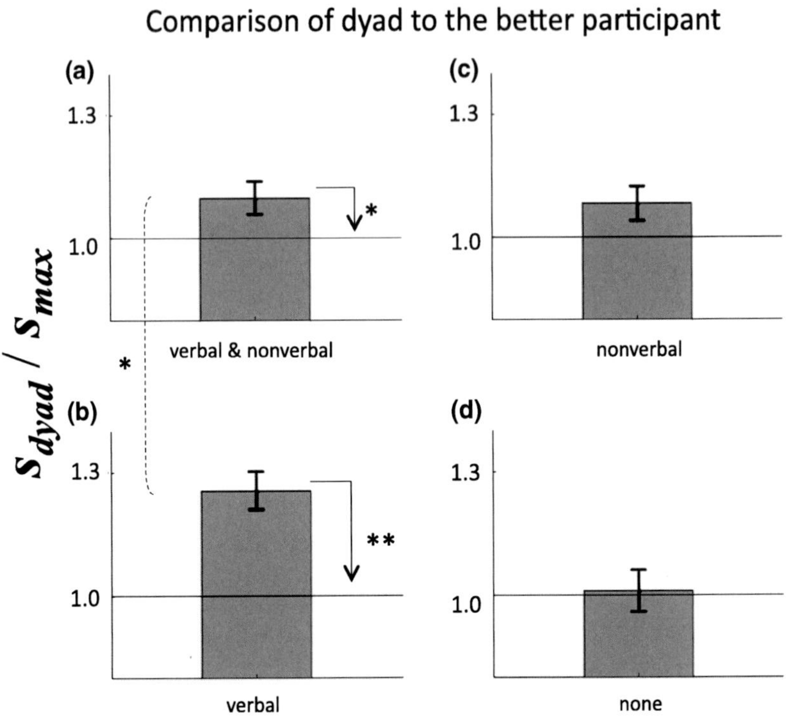

Fig. 9.5 Collective benefit accrued in each condition of Experiment 2. Panels correspond to the conditions illustrated in Fig. 9.3. *Horizontal line* indicates that dyad slope was equal to the more sensitive participant's. * : $p < 0.05$; ** : $p < 0.01$

independent sample *t*-test, $t(28) = 2.54$, $p = 0.016$). Compared to no communication (N condition), the collective benefit accrued from nonverbal communication (NV) was not significant.

9.7.2 Comparison of Dyad to the WCS Model

To compare the dyads' collective performance to the upper bound set by the WCS model under different modes of communication, we applied a similar two by two repeated measure ANOVA to the optimality index (Fig. 9.6). Similar to the collective benefit analysis, the main effect of verbal communication ($F(1, 59) = 8.745$; $p = 0.004$) and the interaction between verbal communication and nonverbal communication $F(1, 59) = 5.4$; $p = 0.024$) were significant.

Post hoc comparison showed that the interaction was driven by the fact that sharing confidence nonverbally (NV condition, Fig. 9.6c) allowed the dyads to approach the WCS model significantly better than without any communication

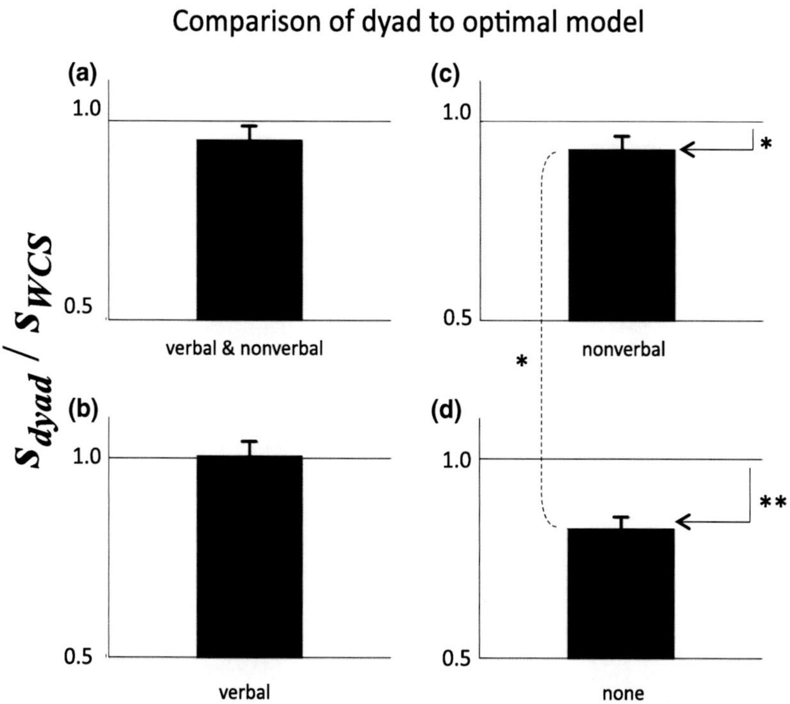

Fig. 9.6 Optimality of dyad performance in each condition of Experiment 2. Panels correspond to the conditions illustrated in Fig. 9.3. *Horizontal line* indicates that group performance was at the level predicted by the WCS model. * : $p < 0.05$; ** : $p < 0.01$

(N condition; Fig. 9.6d) (independent sample *t*-test, $t(27) = 2.41, p = 0.02$). This result demonstrated that although nonverbal confidence sharing is not an ideal mode of communication for collective decision-making (recall that the difference in collective benefit between the N and NV conditions was not significant), communicating 'something' is still better than 'nothing' for collective decision-making.

Comparison to baseline showed that when verbal communication was possible (V and V&NV conditions) dyads fulfilled the WCS model's expectations. With nonverbal communication only (Fig. 9.6c; $t(14) = 2.19$, $p = 0.04$) and without any communication whatsoever, the model prediction exceeded the empirical dyad performance significantly (Fig. 9.6d; $t(14) = 5.91$; $p < 10^{-4}$, paired *t*-test).

9.7.3 Meta-d' Analysis

We used Maniscalco and Lau's (2011) 'meta-*d*' analysis' to compare the reliability of participants' confidence estimates in the NV and the NV&V conditions.

Crucially, the analysis filters out confounds due to sensitivity (i.e. ability to detect contrast target) and response bias (i.e. tendency to endorse low/high confidence) and thus provides a 'pure' measure of the efficacy by which participants' use the confidence scale to discriminate between their own incorrect and correct decisions (see Maniscalco and Lau 2011, for mathematical details and freely available Matlab code). We found no significant difference in meta-*d*' between the NV and the NV&V conditions ($t(58) = 0.890$, $p > 0.37$). We then divided the data into two sessions (i.e. two subsets of 128 trials for each dyad) to evaluate whether meta-*d*' improved with time in each condition. While we found no evidence for improvement in the NV condition (paired *t*-test between first and second subset, $t(29) = -0.24$, $p > 0.43$), there was a marginally significant improvement in the V&NV condition (paired *t*-test between first and second subset, $t(29) = -1.89$, $p = 0.06$). Unfortunately, these results do not provide adequate power to test the question whether metacognition could be affected by social interaction but the trend does point to such possibility of social modulation of metacognition, which could be pursued in future studies.

9.8 Data Summary

Because dyad members received identical visual stimuli without any asymmetric noise, collective benefit was expected in all communicative conditions (except the N condition). Collective benefit was robustly obtained when participants communicated only verbally. Nonverbal communication alone (NV condition) also showed some benefit: dyad performance was closer to the optimal upper bound (predicted by the model) than no communication (N condition) (Fig. 9.6) and a trend was observed for collective benefit (Fig. 9.5). Surprisingly, when dyad members communicated by both means, they obtained less benefit than when they communicated only verbally (Fig. 9.5a vs. b). The benefits of verbal and nonverbal communication were, so to speak, sub-additive.

9.9 Discussion

'How can we aggregate information possessed by individuals to make the best decisions?' Condorcet, Galton and Mackay would have been pleased (or disappointed?) to know that a recent survey (http://bit.ly/hR3hcS) of current academic opinions has listed this question as one of the 10 most important issues facing social sciences in the twenty first century. The data presented here directly address this question and the results provide recommendations for enhancing the accuracy of collective decisions under different circumstances.

Experiment 1 showed that the success of collective decision-making is severely compromised if the quality of evidence available to verbally communicating collaborators is very different. When participants could communicate verbally, asymmetric sensitivity of the team members led to counterproductive collaboration, even though block design administration of noise to one member, but not the other, caused a striking and persistent difference in outcome accuracy between the two collaborators. These results delineate a critical danger facing collective decisions: too wide a competence (i.e. in our case, perceptual sensitivity) gap among interacting agents leads to collaborative failure even if the gap is conspicuously obvious. If we needed any quantitative evidence for the 'madness of the crowds', this could be it. However, when participants could only share their confidence nonverbally, dyads did significantly better than those who had talked to each other directly, even though the competence gap was still firmly in place. The latter findings suggest that groups composed of members with very different competences could avoid major losses (and perhaps even accrue some collaborative benefit) if a suitable mode of communication was adopted.

This result is consistent with the 'egocentric bias' hypothesis from earlier work [12] suggesting that verbally interacting human agents operate under the assumption that their collaborators' decisions and opinions share the same level of reliability. As long as this assumption holds (i.e. $s_Q \approx s_J$), verbal communication provides an efficient strategy for aggregating information across individuals and making decisions that are as good as if the individuals had direct access to each other's mental representations (c.f. comparison of Eqs. 9.1 and 9.2). However, verbal communication backfires when the egocentric assumption does not hold (e.g. $s_Q \gg s_J$). What aspects of verbal communication might be responsible for upholding the egocentric bias?

One critical aspect might be the urge to contribute (or *make a difference*) to the group despite objectively being less competent. To explore this hypothesis, we compared the percentage of trials in which the less sensitive dyad member announced his own decision as the joint decision in the NV and the V conditions of Experiment 1. If the urge to *make a difference* were the cause of collective failure in the V versus NV condition of Experiment 1, we would expect the less sensitive participants to confirm his own decision more often in the Verbal condition. While the less sensitive dyad member did tend to announce his own decision as the joint decision more often in the V than in the NV condition, this difference was not significant ($p > 0.2$). Another critical aspect might be a *social obligation* to treat others as equal to oneself despite their objective incompetence. To explore this hypothesis, we compared the percentage of trials in which the more sensitive dyad member announced his partner's decision as the joint decision in the NV and the V conditions of Experiment 1. Again, while the more sensitive dyad member tended to announce his partner's decision as the joint decision more often in the V than in the V condition, this difference was not significance ($p > 0.2$). Future research is needed to identify the aspects of verbal communication that are responsible for upholding the egocentric bias despite recurring collective failure.

In Experiment 2, in a 2 × 2 design (Fig. 9.3), we systematically investigated the impact of verbal communication and nonverbal confidence sharing on collective decision-making. The results showed that combining the two modes of communication was counterproductive. When group members had similar sensitivity and made decisions based on similarly reliable information, best (near optimal) performance was achieved with direct verbal communication. Imposing an additional nonverbal communication tool significantly reduced the group performance. Collective benefit was, so to speak, *crowded out*.

A plausible explanation for this observation may be found in the literature on introspection and metacognition. The confidence rating schema required participants to actively introspect about their perceptual experience and then graphically indicate their internal, metacognitive estimate of the reliability of their decision (i.e. their confidence). This is no mean feat and indeed a costly cognitive task that requires allocation of top-down attention [15, 16]. In the condition where participants both communicated verbally and used the confidence rating schema, it is conceivable that the cognitive load introduced by active introspection may have interfered with the verbal communication and the collective decision-making process. Unconstrained verbal communication is perhaps more automatic and humans may be thought of as 'natural experts' in it. Indeed, a recent study [17] (see also Overgaard and Sandberg, this volume) has suggested that asking people directly about their perceptual experience, rather than having them rate their confidence, may give a more accurate measure of their metacognitive awareness. Future research could show if more practice with active confidence rating could lead to more automatic, effortless introspection, which in turn might contribute to enhanced collective decision-making beyond what is achievable by direct verbal communication.

When we consider Experiments 1 and 2 together, an intriguing crossover effect is observed: in Experiment 1 (with unequal external noise) collective benefit was higher when participants only share confidence nonverbally, whereas in Experiment 2 (with equal external noise), verbal communication was clearly superior. The crossover is not consistent with any explanation relying on a single mechanism to determine the success of collective benefit. Egocentric bias inherent in verbal communication (Experiment 1) and cognitive load of introspection (Experiment 2) may be interacting with one another to give rise to this crossover. This suggests that the preferred mode of communication of confidence for collective decisions depends on the *similarity* of dyad members' sensitivity. This idea is illustrated in Fig. 9.7. Each panel shows the relationship between similarity (s_{min}/s_{max}) and collective benefit (s_{dyad}/s_{max}). Data are from parts of Experiments 1 and 2 in which communication was exclusively verbal or exclusively nonverbal. The instructive conclusions for how to maximise collective benefit are clear: when dyad members are highly similar ($s_{min}/s_{max} > 0.6$) direct verbal communication should be used (squares in Fig. 9.7a). But the substantial benefit from verbal, direct engagement strongly depends on the similarity of dyad members' competence. When participants have very dissimilar sensitivities—$s_{min}/s_{max} < 0.6$—direct communication is disastrous (x symbols in Fig. 9.7a). In such situations,

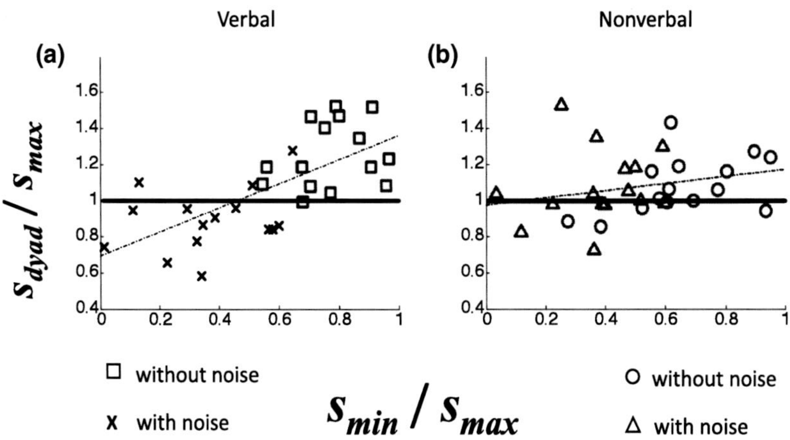

Fig. 9.7 The relationship between collective decisions and similarity of dyad members' sensitivity. Each panel shows the relationship between collective benefit ($s_{\mathrm{dyad}}/s_{\max}$) and similarity ($s_{\min}/s_{\max}$, see Methods). **a** Data from experiments with exclusively verbal communication mode (*cross symbol* Verbal condition in Experiment 1; *squares* Verbal condition in Experiment 2). **b** Data from experiments with exclusively nonverbal communication mode (*triangles* NonVerbal condition in Experiment 1; *circles* NonVerbal condition in Experiment 2)

nonverbal confidence sharing communication (triangles in Fig. 9.7b) is recommended: it could save the dyad by avoiding the counterproductive collaboration that is observed with direct verbal communication.

These results provide algorithmic guidelines for Qodrat and Jalal (Fig. 9.1) in their effort to maximise the accuracy of their perceptual judgement as a group of umpires. However, perhaps with the exception of refereeing in sports games, collective decisions are rarely about purely perceptual events amid uncertainty and noise. In the next section, we will discuss other domains of social interaction where collective failures have been reported and compare them to these findings.

9.10 Collective Failures in Non-perceptual Domains

Numerous studies in social psychology have documented instances where group performance is worse than the performance of the best member. In social loafing [18], individuals exert less effort in the presence of others leading to reduced overall group performance. Thus social loafing refers to the difference in individual performance when individuals act in isolation versus when they act together as a group. An important feature of collective situations in which social loafing has been observed (e.g. the 'tug of war' game) is that group members share the responsibility for possible failures such that no specific member could be singled out and held directly responsible for the group's misfortune [19].

The collective failures described here and by Bahrami et al. [9] are different from social loafing for two reasons. First, here dyad members were always tested in the presence of their partner. The social setup of the task was identical for dyads that consisted of similarly sensitive members (who achieved a collective benefit) and those with dissimilarly sensitive members (who incurred a collective loss). In none of the conditions discussed above did the participants perform the task 'in isolation'. Even when participants did not communicate either verbally or nonverbally (Figs. 9.3, 9.5d and 9.6d), they were still sitting in the same room and shared decisions and made joint decisions when in disagreement. Interestingly, a recent finding has suggested that individual sensitivity assessed in collaborative settings (i.e. private decision stage; Fig. 9.2) was superior to individual sensitivity assessed in non-collaborating setting where two participants were independently tested simultaneously in the same room [20]. This individual sensitivity advantage required the dyad to actively engage in the joint decision making and was therefore different from social facilitation induced by the mere inactive presence of another person [21]. Second, in all experiments described here, decision outcomes were clearly stated for the group as well as both participants leaving little room for sharing the responsibility for group failures. The participant who led the group to the wrong decision had, so to speak, nowhere to hide. This is an important feature of these experiments which shields the group performance against motivation loss [22].

Groupthink [23] is another case of collective failure. When individuals are not given the opportunity to make their own decisions privately, they subsequently fail to develop and voice their disagreeing opinions. Interdependence of individual decisions leads to groupthink [24]. This phenomenon cannot account for the results reported by Bahrami et al. [9] because individual decisions were always first made privately and independently.

Interpersonal competition [25] is also ruled out since the participants were not differentially rewarded for their decisions and there was no incentive for competition.

Finally, the 'hidden profile paradigm' [26–28] is another extensively studied case of collective failure with interesting similarities with and differences from the cases discussed so far. In 1985, Stasser and Titus discovered that group interactions tend to focus on information shared by everybody. This even happens when some of the interacting individuals have access to unshared information that is fundamentally relevant—and provides the best solution—to the joint decision problem and it is in the interest of all individuals to share that exclusive information. In other words, group interactions are biased away from hidden profiles. Groups composed of members with dissimilar knowledge profiles thus tend to under-exploit their unshared but available and relevant information.[4] The pattern of collective behaviour in the hidden profile paradigm is consistent with the illusion of transparency [13] and the egocentric biases [14] in interpersonal communication. Indeed, the verbal condition in Experiment 1—where one person is much better than the other ($S_{max} \gg S_{min}$)—may involve a similar situation: the better person might be seen as having some implicit knowledge (e.g. less noisy stimulus) that the other does not have. However, it is difficult to explain the collective failures that were exposed here based on the hidden profile paradigm *per se*. The marked difference in participants' accuracy on a trial-by-trial basis was common knowledge because the feedback was given to both individuals at the same time. As such, after a few trials, the asymmetric reliability of the participants in Experiment 1 was not exclusive knowledge at all. Moreover, the detrimental impact of asymmetric noise on collective performance was only observed when dyads communicated directly rather than when they shared confidence using the visual schema. If the hidden profile paradigm were responsible for the collective failures, one would expect not less but, rather, maybe even more collective failure when communication was minimised and only nonverbal confidence sharing was used. Nonetheless, it is possible that verbal communication masks the sensitivity gap, whereas nonverbal communication strips away the social interaction and magnifies the gap, making it easier to discard the less sensitive participant's opinion. At present, the only firm conclusion on this issue would be that more research is needed to address these possibilities.

9.11 The Impact of Interaction on Alignment of Metacognition

What are the qualitative features of sharing and discussing metacognitive awareness when Qodrat and Jalal (Fig. 9.1) discuss their opinions? Recently we have undertaken linguistic analysis of the conversations leading to the collective decisions in the Verbal condition of Experiment 2 [29]. The results (not reported here) showed that dyadic conversations often focus on participants' confidence in their decisions. Most groups used more everyday expressions such as 'I was not so sure'

---

[4] Thomas Bayes (http://bit.ly/f0uTBk) would perhaps have found the bias for favouring redundant and frequent information only wise and sensible. In the words of Bellman in Lewis Carrol's brilliant *The Hunting of the Snark*,

'JUST the place for a Snark!' the Bellman cried,
As he landed his crew with care;
Supporting each man on the top of the tide
By a finger entwined in his hair.
'Just the place for a Snark! I have said it twice:
That alone should encourage the crew.
Just the place for a Snark! I have said it thrice:
What I tell you three times is true.'

and 'I saw it clearly'. The conversations rarely (i.e. 1 out of 30 sessions studied) led to spontaneous use of explicit numerical scales to express and compare confidence. On a trial-by-trial basis, interacting participants tended to align with each other's confidence expressions. For example, if one started the conversation with 'I did not see anything', the other person would most likely respond with some expression using 'see'. Over time, the content of conversations tended to diminish with practice such that by the end of the experiment, dyad members had converged to a small, repeatedly used set of expressions.

These qualitative and quantitative observations [29] prompted us to wonder if a similar practice-dependent alignment of confidence could be observed in the conditions of Experiment 2 where the confidence rating schema was employed. To test this hypothesis, we revisited the data from the NV and V&NV conditions of Experiment 2. For each trial, we calculated the absolute difference in signed confidence rating (see methods) between the two participants and defined alignment as the inverse of this difference. The results (Fig. 9.8) showed robust evidence for increased alignment in the NV condition where participants only used nonverbal confidence sharing (Fig. 9.8 left panel, grey; One-Way ANOVA with 3 levels for first, middle and last 1/3 of trials, $F(2, 28) = 5.38$, $p < 0.012$). These results are in line with the qualitative findings about linguistic alignment of confidence [29] in the verbal condition of Experiment 2 and show that as dyad members gain experience from their interactions, they tend to 'describe' their confidence more similarly using the confidence rating bar.

However, in the V&NV condition of Experiment 2 where participants used both verbal and nonverbal communication, confidence alignment decreased (Fig. 9.8, left panel, black; One-Way ANOVA, $F(2, 28) = 5.16, p < 0.013$). In other words, confidence judgements diverged from one another over time. Whereas use of the confidence rating schema alone led to alignment of participants' metacognitive reports, combining verbal communication and nonverbal confidence ratings led to a divergence of confidence ratings.

Direct comparison of the NV and V&NV conditions using a mixed ANOVA (with 2 levels for conditions and 3 levels for trial bins) supported this conclusion with a significant interaction ($F(2, 56) = 9.34$, $p < 0.0001$). These results corroborate the idea suggested earlier that combining both modes of communication—as in the V&NV condition—leads to an interference in task performance both at the individual (as we see here) and collective level (as we saw earlier in the results of Experiment 2).

Does alignment of metacognition have any relevance for collective decision-making? In the conditions of Experiment 2 where participants rated their confidence, for each dyad, we calculated the collective benefit accrued within each one third of the experiment (Fig. 9.8, middle panel). We then tested whether there was any correlation between alignment (Fig. 9.8, left panel) and collective benefit (Fig. 9.8 middle panels) across the dyads. The results showed that a significant correlation (Pearson $r = 0.6, p = 0.01, N = 15$) emerged in the last third of the NV condition (Fig. 9.8, upper right panel). When participants only shared confidence rating but did not talk, metacognitive alignment was associated with collective

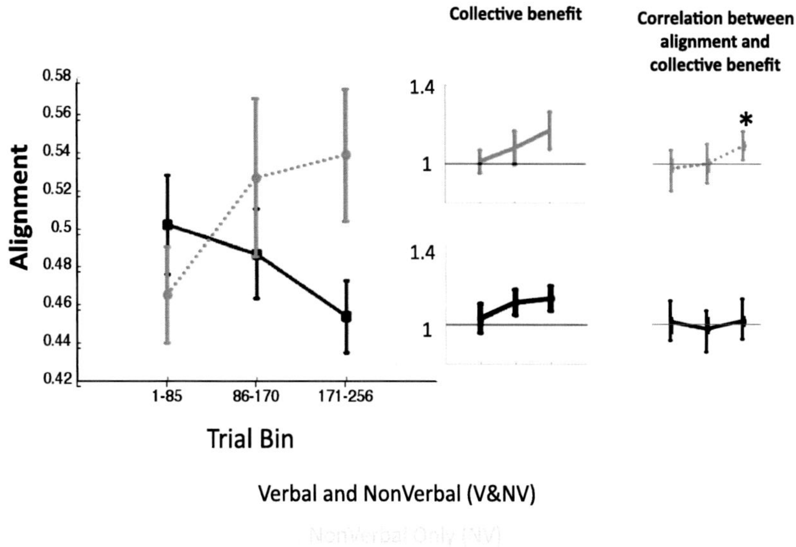

Fig. 9.8 *Left* Alignment of confidence plotted for each time bin consisting of 1/3 of the trials. *Black symbols* and *curve* correspond to the Verbal & NonVerbal condition in Experiment 2 where participants both communicated verbally and used the confidence rating schema. *Grey symbols* and *curve* correspond to the NonVerbal condition in Experiment 2 where participants only communicated via the confidence rating schema. Error bars are 1 SEM across dyads ($N = 15$). *Middle* Collective benefit (s_{dyad}/s_{max}) is plotted for each bin. *Horizontal line* indicates no benefit (i.e. $s_{dyad} = s_{max}$). Error bars are 1 SEM across dyads ($N = 15$). *Right* Correlation coefficients between alignment and collective benefit across dyads for each time bin. For V&NV condition Pearson $r = [0.08, -0.15, 0.11]$ and all $p > 0.55$. For NV condition, Pearson $r = [0.1, -0.01, 0.6]$ and $p = [0.7, 0.9, 0.01]$. Horizontal line indicated zero. Departure from null hypothesis ($p = 0.01$) is marked by *. Error bars are 95 % CIs for Pearson correlation using Fisher transformation [53]

benefit suggesting that with enough practice, dyad members may arrive at and utilize the alignment to make better group decisions. In line with our previous observations, in the V&NV condition (Fig. 9.8, lower right panel) where participants both talked and used the confidence rating schema no such relationship ever emerged. This finding once again underscores our conclusion that the combination of two modes of communication was not productive. Moreover, the fact that the overall collective benefit was not statistically different between the V&NV and the NV conditions (Fig. 9.5a, c) suggests that dyads in these two conditions achieved the same level of performance employing different strategies for communication and decision rules. More research is needed to understand the nature of these different strategies and decision rules.

We also conducted a similar alignment analysis on the data from the NV condition of Experiment 1. Our results did not show any significant findings either for the alignment of confidence ratings or for a correlation between alignment and collective benefit. We believe that this is most likely due to the fact that the number of trials in the interactive phase of Experiment 1 (i.e. 128) was half that of Experiment 2; remember that the effects that we see in Fig. 9.8 did not emerge before the final third of the trials (171–256). Aside from statistical power issues, however, we are also reluctant to make a strong prediction about alignment of confidences in Experiment 1 since the main manipulation in that experiment was to deliver different and uncorrelated levels of independently generated visual input noise to the two participants in each trial which is expected to weaken any developing/existing correlation among any outputs from the two participants including confidence ratings.

9.12 The Role of Feedback and the Contribution of Shared Metacognition to Social Learning

Could Qodrat and Jalal (Fig. 9.1) achieve any collective benefit from sharing their opinions and discussing their disagreements if they never found out who was actually right and who was wrong? This is an important question because none of the information integration models that we have discussed here [9, 10] assume any role for decision outcomes. Moreover, and perhaps more importantly, both models attempt to explain the dyad behaviour as a stable, stationary phenomenon with little variability over time. As useful as these assumptions may be for simplifying the problem, few would agree that human group behaviour is—in general—independent of outcomes and unaffected by social learning.

A recent study [30] examined the development of collective benefit when feedback was withdrawn from the dyads. Without feedback, dyads did not initially achieve any collective benefit. However, with practice dyads started to exceed their more sensitive member such that by the end of the experiment, the collective benefit of interacting dyads with and without feedback were statistically indistinguishable. Thus, knowledge about outcomes only seemed to accelerate the process of social learning required for efficient confidence sharing.

Interestingly, feedback is not necessary for optimal multisensory integration of visual and haptic information [31]. Following the standard practice in psychophysics, those results were obtained from several thousands of experimental trials for each participant to make sure that performance is measured long after any learning process is finished. It is, therefore, likely that feedback plays a similar accelerating role in achieving optimality in multisensory integration. To our knowledge, previous research in perceptual learning of multisensory integration has not addressed the role of outcome information on the speed of learning.

Once again, collective failure is instructive in helping us to phrase the right research question. The initial failure of the no-feedback groups to exceed their best member and their subsequent improvements to the same level as feedback groups pose serious problems for models of collective decision-making that assume no social learning [9, 10]. An important question for future research is to explain the dynamics of social learning needed to achieve effective collective behaviour over the course of repeated interactions in the absence of feedback.

Currently, a number of computational models have been proposed for social learning based on principles of associative reinforcement learning [32–34]. The critical question here is: how could dyad members in the no-feedback experiment [30] have accomplished reinforcement learning without any reinforcement (i.e. without knowing the outcome of their decisions)? It has been suggested [30] that sharing metacognitive awareness may provide sufficient information to replace feedback and reinforce social learning. On this account, when participants are sincere in their opinions, the shared metacognitive awareness that informs the joint decision provides a noisy but still informative estimate of the true state of the world which can be used as a substitute for the missing feedback about decision outcomes [35]. With a noisy substitute, the reinforcement learning process could still happen but would take longer to develop. With enough practice, learning with and without feedback would eventually stabilise at similar performance levels. This account [30] has an interesting, if unexpected corollary: a functional role of shared metacognitive awareness may be to replace missing reinforcement signals when decision outcomes are not available (e.g. too complex to estimate or too far in the future to wait for). Given the abundance of situations in everyday life where immediate outcomes are difficult, sometimes even impossible to establish, the hypothesis proposed by Bahrami et al. [9] offers an ecologically relevant role for metacognition.

9.13 Neuronal, Behavioural and Social Metacognition

Historically, decision science has focused on three aspects of every decision: accuracy, reaction time and confidence [36, 37] often assuming that all three originate from the same underlying process [38]. The *sequential sampling* family of models [39] was developed to account for speed-accuracy trade-offs observed in two-alternative choice tasks (for a review see Kepecs and Mainen, this volume). The idea in sequential sampling is that when a participant is presented with some sensory signal and asked to categorise it as A or B, s/he keeps sampling the signal and accumulating the evidence for each alternative. The race between the two accumulators goes on until evidence collected for one category hits a predefined boundary determining the chosen category for the signal. These three components, a sensory receptor, an accumulator and a boundary, are the backbone of perhaps the most widely popular decision-making models in today's system neuroscience [40, 41].

Sequential sampling models have been extended to account for decision confidence as the difference in accumulated evidence supporting each category at the decision time [42]. Heath [43] showed that such a 'balance of evidence' concept can account for a number of qualitative features of decision confidence [43].

Recent works have found neuronal substrates in rodent [44] and non-human primate [45] brains for decision confidence that closely overlap with the known neural machinery involved in decision accuracy and speed. Moreover, the firing patterns of these confidence neurons closely match the predictions of the sequential sampling models. These latter findings thus provide evidence for the earlier intuition that decision accuracy, reaction time and confidence arise from the same latent neuronal process [36]. As such, neuronal encoding of confidence seems to be the cost-free, automatic by-product of the decision process.

But this view is hardly consistent with what is known about metacognition at the level of behaviour. Introspection is cognitively demanding [15] and therefore neither automatic nor cost-free. Moreover, if the confidence and choice processes were one and the same, then restriction of choice time should systematically reduce metacognitive accuracy in a manner parallel to standard speed-accuracy trade-offs. However, when speed is stressed in a choice reaction time task, choice accuracy decreases as expected but, paradoxically, metacognitive accuracy increases [46]. This suggests that rating confidence may involve some post-decisional processing distinct from the race-to-boundary stage (also see Yeung and Summerfield, this volume). A tantalising prediction arising from this notion is that, if one repeats our experiments (reported above) with an emphasis on speed (rather than accuracy) in the initial perceptual task, then sharing (supposedly) more accurate metacognitive awareness should enhance the collective benefit.

But the data we have reported here (Experiments 1–2) caution against tightly connecting behavioural metacognition with shared, social metacognition. Our results showed that effective sharing of metacognitive awareness depends on some form of social heuristics (e.g. egocentric bias) and that the sharing process seems to be dissociable from and interacts with the cognitive demands of introspection (i.e. behavioural metacognition). As such, our understanding of metacognition at the levels of neuronal representation, behaviour and social interaction seem to be disconnected at present, calling for future research to see if it is possible or meaningful to bring them together.

The neuronal and behavioural interpretations of confidence diverge from each other in another important conceptual dimension. Confidence in the perceptual sciences and 'uncertainty' in the decision sciences both concern the subjective probability of choice outcomes. Uncertainty is typically decomposed into 'risk' and 'ambiguity' and neuroeconomic studies [47] have demonstrated the behavioural and neuronal correlates of each component [48–50]. When the possible outcomes and their respective probabilities are known, the decision is said to be 'risky'. 'Ambiguity', on the other hand, refers to situations in which the outcome alternatives and/or their respective probabilities are unknown. At first glance, the process of continuously estimating confidence in a perceptual task could be thought of as—through learning—minimising the ambiguity associated with one's choices so as to reliably estimate their associated risk. However, an important distinction seems to be that the notions of risk and ambiguity both refer to subjective probabilities *prior* to choice whereas the concept of confidence refers to subjective probabilities that arise *during* evidence accumulation and *after* the choice has been made. Furthermore, while models of perceptual confidence assume that it depends on internal (neural) and external (environmental) sources of noise, models of economic uncertainty appear to only address the latter. In sum, the connection between perceptual confidence and economic uncertainty is at present unclear but indeed a very interesting topic that is just beginning to be investigated [51].

One final word of caution on the scopes and limits of our interpretation of the data that we have presented here is due. Our results were obtained from male-only groups of individuals and were geographically restricted to one country, Denmark. Whether our models of social interaction would generalise to explaining and predicting the behaviour of female or non-Danish dyads is an empirical question that only future research can inform us about.

9.14 Closing Remarks

Two heads are not always better than one. This paper focused on recent models and empirical findings that explored collective failures. These models are inspired by thinking of collective decision-making as an 'information integration' problem similar to that of multisensory perception. The intuition obtained from these theoretical and empirical findings is that shared metacognitive awareness (socially communicated confidence in one's own perceptual decisions that contributes to collective perceptual decisions) conveys the strength of the sensory experience and its reliability inseparably. An important functional role of such metacognitive awareness may be to substitute missing outcomes in situations where outcome is necessary for learning but unavailable or impossible to establish.

Acknowledgments This work was supported by a British Academy postdoctoral fellowship (BB), European Research Council (NeuroCoDec, grant number 309865), the Calleva Research Centre for Evolution and Human Sciences (DB), the Danish National Research Foundation and the Danish Research Council for Culture and Communication (BB, KO, AR, CF) and by the Wellcome Trust (GR). Support from the MIND*Lab* UNIK initiative at Aarhus University was funded by the Danish Ministry of Science, Technology and Innovation.

References

1. Mackay C (1841) The extraordinary and popular delusions and madness of crowds, 4th edn. Wordsworth Editions Limited, Ware
2. Condorcet M (1785) Essai sur l'application de l'analyse á la probabilité des décisions rendues á la pluralité des voix. de l'Impr. royale, Paris
3. Galton F (1907) Vox populi. Nature 1949(75):450–451
4. Kerr NL, Tindale RS (2004) Group performance and decision making. Annu Rev Psychol 55:623–655

5. Lorenz J, Rauhut H, Schweitzer F, Helbing D (2011) How social influence can undermine the wisdom of crowd effect. Proc Natl Acad Sci USA 108(22):9020–9025
6. Alais D, Burr D (2004) The ventriloquist effect results from near-optimal bimodal integration. Curr Biol 14(3):257–262
7. Deneve S, Latham PE, Pouget A (2001) Efficient computation and cue integration with noisy population codes. Nat Neurosci 4(8):826–831
8. Ernst MO, Banks MS (2002) Humans integrate visual and haptic information in a statistically optimal fashion. Nature 415(6870):429–433
9. Bahrami B, Olsen K, Latham PE, Roepstorff A, Rees G, Frith CD (2010) Optimally interacting minds. Science 329(5995):1081–1085
10. Sorkin RD, Hays CJ, West R (2001) Signal-detection analysis of group decision making. Psychol Rev 108(1):183–203
11. Ma WJ, Beck JM, Latham PE, Pouget A (2006) Bayesian inference with probabilistic population codes. Nat Neurosci 9(11):1432–1438
12. Keysar B (2007) Communication and miscommunication: The role of egocentric processes. Intercultral Pragmatics 4(1):71–84
13. Gilovich T, Savitsky K, Medvec VH (1998) The illusion of transparency: Biased assessments of others' ability to read one's emotional states. J Pers Soc Psychol 75(2):332–346
14. Keysar B, Lin S, Barr DJ (2003) Limits on theory of mind use in adults. Cognition 89(1):25–41
15. Corallo G, Sackur J, Dehaene S, Sigman M (2008) Limits on introspection: distorted subjective time during the dual-task bottleneck. Psychol Sci 19(11):1110–1117
16. Ericsson KA, Simon HA (1980) Verbal reports as data. Psychol Rev 87(3):215–251
17. Sandberg K, Timmermans B, Overgaard M, Cleeremans A (2010) Measuring consciousness: is one measure better than the other? Conscious Cogn 19(4):1069–1078
18. Latane B, Williams K, Harkins S (1979) Many hands make light the work—causes and consequences of social loafing. J Pers Soc Psychol 37(6):822–832
19. Karau SJ, Williams KD (1993) Social loafing—a metaanalytic review and theoretical integration. J Pers Soc Psychol 65(4):681–706
20. Olsen K, Christensen P, Bang D, Roepstorff A, Rees G, Frith C, Bahrami B (In preperation) Interaction accelerates the rate of visual perceptual learning in humans
21. Zajonc RB (1965) Social facilitation. Science 16(149):269–274
22. Williams K, Harkins S, Latane B (1981) Identifiability as a deterrent to social loafing: two cheering experiments. J Pers Soc Psychol 40(2):303–311
23. Turner ME, Pratkanis AR (1998) Twenty-five years of groupthink theory and research: Lessons from the evaluation of a theory. Organ Behav Hum Decis Process 73(2–3):105–115
24. Raafat RM, Chater N, Frith C (2009) Herding in humans. Trends Cogn Sci 13(10):420–428
25. Hastie R, Kameda T (2005) The robust beauty of majority rules in group decisions. Psychol Rev 112(2):494–508
26. Stasser G, Titus W (1985) Pooling of unshared information in group decision-making: biased information sampling during discussion. J Pers Soc Psychol 48(6):1467–1478
27. Stasser G, Titus W (1987) Effects of information load and percentage of shared information on the dissemination of unshared information during group discussion. J Pers Soc Psychol 53(1):81–93
28. Stasser G, Titus W (2003) Hidden profiles: a brief history. Psychol Inq 14(3–4):304–313
29. Fusaroli R, Bahrami B, P, Olsen K, Roepstorff A, Frith C, et al. (2011) Coming to terms: an experimental quantification of the coordinative benefits of linguistic interaction. Under review 2011
30. Bahrami B, Olsen K, Bang D, Roepstorff A, Rees G, Frith C (2011) Together, slowly but surely: the role of social interaction and feedback on the build-up of benefit in collective decision-making. J Exp Psychol Human Percept Perform 38:3
31. Burge J, Girshick AR, Banks MS (2010) Visual-haptic adaptation is determined by relative reliability. J Neurosci 30(22):7714–7721

32. Behrens TE, Hunt LT, Woolrich MW, Rushworth MF (2008) Associative learning of social value. Nature 456(7219):245–249
33. Behrens TE, Hunt LT, Rushworth MF (2009) The computation of social behavior. Science 324(5931):1160–1164
34. Hampton AN, Bossaerts P, O'Doherty JP (2008) Neural correlates of mentalizing-related computations during strategic interactions in humans. Proc Natl Acad Sci USA 105(18):6741–6746
35. Austen-Smith D, Banks J (1996) Information aggregation, rationality, and the condorcet jury theorem. Am Political Sci Rev 90(1):34–45
36. Blackwell D, Girschick MA (1954) Theory of games and statistical decisions. Wiley, New York
37. Maloney LT (2002) Statistical decison theory and biological vision. In: Heyer D, Mausfeld R (eds) Perception and physical world. Wiley, New York, pp 145–189
38. Pierce CS, Jastrow J (1884) On small differences in sensation. Mem Natl Acad Sci 3:75–83
39. Wald A (1947) Sequential analysis. Wiley, New York
40. Gold JI, Shadlen MN (2007) The neural basis of decision making. Annu Rev Neurosci 30:535–574
41. Ratcliff R (1978) Theory of memory retrieval. Psychol Rev 85(2):59–108
42. Vickers D (1979) Decision processes in visual perception. Academic, New York
43. Heath RA (1984) Random-walk and accumulator models of psychophysical discrimination: a critical evaluation. Perception 13(1):57–65
44. Kepecs A, Uchida N, Zariwala HA, Mainen ZF (2008) Neural correlates, computation and behavioural impact of decision confidence. Nature 455(7210):227–231
45. Kiani R, Shadlen MN (2009) Representation of confidence associated with a decision by neurons in the parietal cortex. Science 324(5928):759–764
46. Pleskac TJ, Busemeyer JR (2010) Two-stage dynamic signal detection: a theory of choice, decision time, and confidence. Psychol Rev 117(3):864–901
47. Glimcher PW, Rustichini A (2004) Neuroeconomics: the consilience of brain and decision. Science 306(5695):447–452
48. Hsu M, Bhatt M, Adolphs R, Tranel D, Camerer CF (2005) Neural systems responding to degrees of uncertainty in human decision-making. Science 310(5754):1680–1683
49. Huettel SA, Stowe CJ, Gordon EM, Warner BT, Platt ML (2006) Neural signatures of economic preferences for risk and ambiguity. Neuron 49(5):765–775
50. Platt ML, Huettel SA (2008) Risky business: the neuroeconomics of decision making under uncertainty. Nat Neurosci 11(4):398–403
51. Fleming SM, Dolan RJ (2010) Effects of loss aversion on post-decision wagering: implications for measures of awareness. Conscious Cogn 19(1):352–363
52. Ernst MO (2010) Behavior. Decisions made better. Science 329(5995):1022–1023
53. Fisher RA (1921) On the probable error of a coefficient of correlation deduced from a small sample. Metron 1(4):3–32

Part III
Cognitive Neuroscience of Metacognition

Chapter 10 Studying Metacognitive Processes at the Single Neuron Level

Paul G. Middlebrooks, Zachary M. Abzug and Marc A. Sommer

10.1 Introduction

Over the past few decades, strides have been made toward understanding how higher level cognitive processes are mediated by neuronal spiking activity. Neuronal correlates of functions such as attention, executive control, working memory, decision-making, and reward processing have all been elucidated, to an impressive level of detail, at the single cell and circuit levels. This explosion in neuroscience-based discovery has depended crucially on nonhuman animal (*animal*, hereafter) models of the behaviors and processes under question. Developing animal models becomes a greater challenge for cognitive functions that approach the complexity of those arguably unique to humans.

A prime example is metacognition. As reviewed in the *Foundations of metacognition* section of this volume, we know that humans engage in complex metacognitive behaviors. A metacognitive process is by definition *about* one of our own cognitive process, and is often referred to as "thinking about thinking." Hence it is not surprising that metacognition is often associated with our subjective or conscious sense of self (e.g., [38]). Beyond the human brain, evidence for metacognition is less clear. There is, as yet, no definitive evidence that animals experience a subjective awareness, or a continuity of mental experience, similar to our own. Consequently, many investigators conclude that animals must not possess metacognition as humans do. Recent behavioral evidence, however, makes a case for some degree of metacognitive capability in a variety of animal species.

P. G. Middlebrooks (✉)
Department of Psychology, Vanderbilt University, 301 Wilson Hall, Nashville, TN 37240, USA
e-mail: paul.g.middlebrooks@vanderbilt.edu

Z. M. Abzug · M. A. Sommer
Department of Biomedical Engineering, the Center for Cognitive Neuroscience, and the Duke Institute for Brain Sciences, Duke University, Durham, NC 27708, USA

S. M. Fleming and C. D. Frith (eds.), *The Cognitive Neuroscience of Metacognition*, DOI: 10.1007/978-3-642-45190-4_10,

Early attempts to test animals' metacognitive skills used paradigms that analyzed relatively simple metacognitive behaviors. Subsequent single neuron studies have followed suit by developing streamlined tasks that are quick in duration, austere in terms of sensory stimulation and motor response, and balanced as much as possible by control conditions. A subtle issue is that animals, primates in particular, are notorious for finding the simplest strategy for accomplishing a task, rather than the strategy desired by the experimenter. It is important to verify that subjects are not "cheating" at metacognitive tasks by using external cues (e.g., visual differences between conditions or motor differences between responses) instead of internal perceptions and memories (see [30] for a review). Taking all of these considerations into account, investigators have designed a variety of tasks for evaluating the association between neuronal activity and metacognition. All of the tasks to date consist of a "cognitive" period followed immediately by a "metacognitive" period. Likewise, all of them use a confidence response or surrogate thereof. Though the single neuron studies we describe in this chapter do not attempt to investigate the richness of metacognitive skills that we take for granted as humans, they serve as a starting point for what hopefully will continue to develop into a mechanistic neuronal account of metacognition in general.

We begin by describing the behavioral tasks used to test metacognition in animals, with a focus on those used in single neuron studies. Next we discuss a few possible ways neuronal firing rates might encode metacognitive processes. The bulk of the chapter is then devoted to describing and critiquing three studies that examined metacognitive processes at the single neuron level. Finally, we discuss the implications and limitations of these and future single neuron studies of metacognition in animals.

10.2 Streamlined Metacognitive Paradigms: Opt-Out and Betting Tasks

Before describing the tasks and experiments for studying the neuronal basis of metacognition, it is important to consider how the field arrived at this point. Before the term "metacognition" was used, experiments were performed in which subjects were asked to assess their own "feeling of knowing" whether an item was in their memory even though they could not presently recall it [23]. In the 1970s John Flavell coined the terms "metamemory" [12] and "metacognition" [13] in his studies of child development. Subsequently, Nelson and Narens [39] developed a systematic framework for the study of metacognition that has been widely used since. In their framework, a distinction was made between two types of metacognitive processes. A "monitoring" process *receives* information about ongoing cognitive operations. For example, a student might experience a sense of whether she is correctly recalling a list of memorized words. A "control" process *provides* information to ongoing cognitive processes and allows a subject to strategically plan. For example, a student can estimate the effort it will take to memorize a list

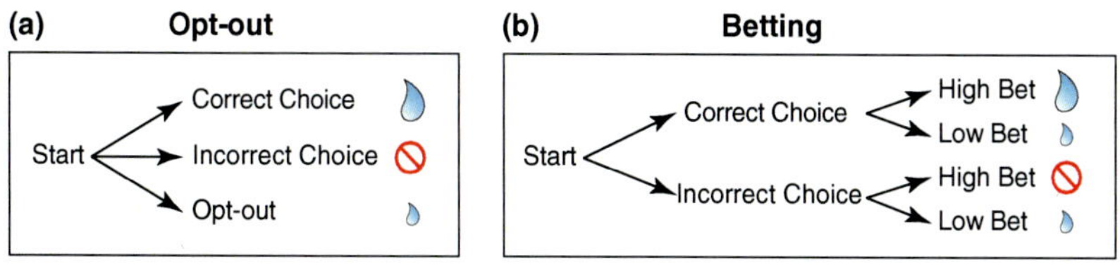

Fig. 10.1 Metacognitive monitoring tasks. **a** Schematic of opt-out task paradigms. Opt-out tasks generally involve a two-choice perceptual discrimination. On some proportion of trials, a third opt-out target appears. Selection of the opt-out target results in a small but ensured reward. Participants utilizing a metacognitive strategy should select the opt-out target more often on more difficult trials, and make more accurate responses on trials the opt-out is offered. **b** Schematic of betting task paradigms. Betting tasks generally involve a choice stage followed by a betting stage. Selection of the high bet target results in a large reward after a correct response, and no reward after an incorrect response. Selection of the low bet target results in a small but ensured reward. Participants using a metacognitive strategy should select the high bet target more often after correct decisions. Unlike opt-out tasks, betting tasks require a primary task decision on every trial

of words. Within these two main divisions, monitoring and control processes, Nelson and Narens' framework provides multiple subcategories that classify metacognitive processes according to factors such as which facet of a cognitive process is interacting with the metacognitive process, the responses required by the subject, and whether the response occurs while a subject is learning or recalling material. A main goal for the neuroscientific study of metacognition is to embrace these psychological principles while adapting the tasks for use in nonverbal subjects (animals) in settings that demand speed and efficiency (single neuron recordings).

Experiments to test animals' metacognitive abilities have almost all focused on monitoring processes. In the 1990s David Smith et al. tested whether animals (dolphins in particular) could monitor their own uncertainty [56] during decision making. Variously referred to as "uncertainty monitoring," "decline," or "escape" tasks, we will refer to this general class as "opt-out" tasks (Fig. 10.1a). Animals are required to perform a primary decision task, such as making a two-choice perceptual discrimination. Reward is earned for correct responses. On some trials, an additional "opt-out" response choice is offered that, when selected, always delivers minimal reward (most studies have offered the opt-out concurrent with the primary task response targets, but see [20] for an important innovation in which the opt-out is presented before the animal responds to the primary task). The animal thus can opt-out of the primary task, which will earn either a large reward if correct or no reward if incorrect, and instead receive an ensured small reward. The basic premise is that an animal capable of monitoring its own uncertainty will select the opt-out response more often during difficult trials. Likewise, when the animal does make a response to the primary task, accuracy will be higher on trials in which the opt-out response was offered than those when the animal was forced to perform the primary task. Multiple species have been shown to opt-out in a manner consistent with the ability to monitor their uncertainty, including dolphins [56], rats [14, 27], rhesus macaques [2, 20, 28, 53, 57, 63], orangutans [61], and gorillas [62].

Another line of studies used what we refer to as "betting" tasks (Fig. 10.1b). Like opt-out tasks, betting tasks require performance of a primary task, such as making a two-choice perceptual discrimination. However, a response to the primary task is required on every trial. Reward is not earned immediately. Instead, once a response is made, there is an option to make either a high bet or a low bet. High bets earn a large reward after correct decisions and no reward (or a timeout punishment) after incorrect decisions. Low bets earn a small but ensured reward following either correct or incorrect decisions. The premise is that animals able to monitor their own decisions will bet high more often after correct responses and bet low more often after incorrect responses. Rhesus macaques perform betting tasks in a manner that suggests they are able to monitor their ongoing cognitive operations [31, 36, 54, 59].

Comparative studies of metacognition sometimes use other tasks as well. For example, a series of experiments showed that gorillas, chimpanzees, bonobos, orangutans, and rhesus monkeys will seek information when it is needed to perform better on a task [1, 5, 6, 21], a type of metacognitive control behavior. The field of comparative metacognition continues to grow and improve methodologically. But from a neuroscientific point of view, the relative simplicity of opt-out and betting tasks is attractive. Both tasks are rooted in one of the most successful fields of neuroscience: decision making. It is not surprising, then, that the single neurons studies described below employed tasks adapted from previous behavioral opt-out and betting task studies.

10.3 Mechanisms of Metacognition

Before we highlight the single neuron studies related to metacognition, it is worthwhile to consider some theoretical perspectives that propose what a metacognitive neuronal signal might look like. In what follows, we outline a few neuronal coding schemes that plausibly underlie metacognition. In keeping with the rest of this chapter, we frame our discussion within the context of opt-out and/or betting tasks, in which a metacognitive judgment is in temporal proximity to its referent cognitive behavior (a decision). In principle, though, the mechanisms could apply to other metacognitive tasks with some modification.

Researchers have approached the study of metacognition from decision-making sciences. The framework of decision making can be extended to include how information in the signals that encode decisions could be used and/or further processed to encode related behavior, i.e., a metacognitive signal. Much of what we understand about how decisions are made, and especially how perceptual decisions are made, is encapsulated by a family of cognitive models known as sequential sampling models [17, 55, 68]. Rooted in signal detection theory [19], sequential sampling models posit that available perceptual evidence is repeatedly sampled until the amount of evidence reaches a criterion threshold. At that point an appropriate response is executed. Neuronal firing rates in many regions of the

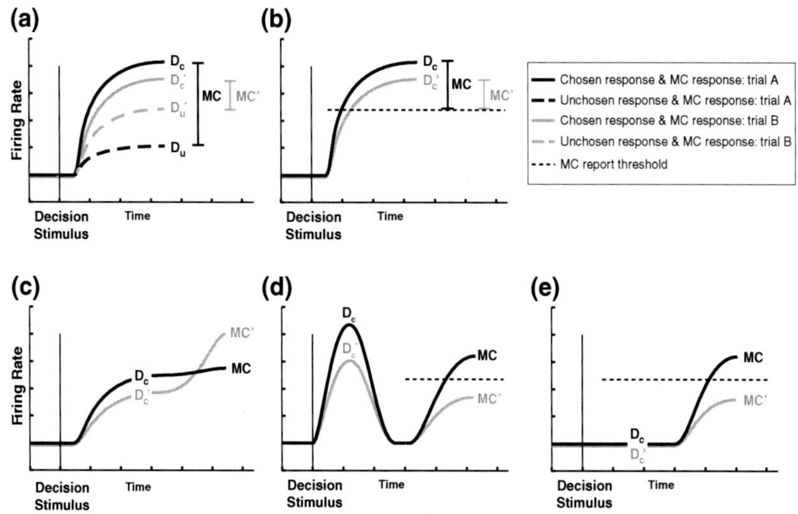

Fig. 10.2 Possible mechanisms of metacognition. **a** Comparison of decision-related activity between response alternatives. In this model, confidence (*MC*) is a function of the difference between neural activity for the chosen (D_c) and unchosen (D_u) responses. **b** Comparison of decision-related activity to an independent threshold. In this model, confidence is a function of the difference between neural activity for the chosen response and the threshold. It is not dependent on activity for the unchosen response. **c** Sequential coding of decision and confidence. In this model, evidence continues to accumulate after the decision is made (D_c or D_c') to subsequently produce a confidence response (MC or MC', respectively). **d**, **e** Distinct coding of decision and confidence. In these models, confidence is encoded separately from the decision. Confidence can be encoded by the same neurons involved in the decision (**d**) or by distinct neurons (**e**)

brain resemble what sequential sampling models predict. Neurons' firing rates increase stochastically, at a rate proportional to available sensory evidence, until a consistent threshold is reached and a response is made. These regions include superior colliculus [32, 47, 48], lateral intraparietal (LIP) cortex [49, 52], dorsolateral prefrontal cortex (PFC) [29], the caudate nucleus [10], and the frontal eye field (FEF) [11, 22, 46].

Psychologists have long thought confidence may be encoded simultaneously with decisions (e.g., [24, 33, 43, 71]). After all, we usually experience a sense of how well we're performing some task while in the middle of performing it. Building on that theme, various proposals have been made by which confidence in a decision could be encoded using the same mechanisms underlying the decision, at the same time the decision process occurs. In that case, brain regions encoding cognitive decisions could concurrently encode metacognitive decisions.

One example proposes that a metacognitive signal is encoded by comparing the firing rates of neurons selective for the alternative responses in a decision task (Fig. 10.2a). Consider two hypothetical neurons, each neuron selective for one of the two alternative responses. During any given trial, the firing rate of the neuron selective for the response that was chosen (D_c, solid black) will likely be different than (and usually exceed) the firing rate of the neuron selective for the unchosen response (D_u, dashed black). As the signals develop over the course of the trial, a metacognitive signal could be computed at any time by taking the difference between the two decision signals (MC). The magnitude of the difference could guide the metacognitive behavior. A large difference would correlate with a high level of confidence, for example. During a different, more difficult trial (grey lines), the firing rates of the two neurons may differ less and thus lead to a lower confidence rating (MC'). Models of this nature have accounted for human confidence ratings (e.g., [9, 35, 69]).

A similar but alternate mechanism (Fig. 10.2b) would compare firing rates of response selective neurons not with the firing rate of neurons selective for the other response, but with a threshold level representing the boundary between the alternative choices. During one trial (black line), a decision response (D_c) is made and the magnitude of the metacognitive signal (MC, black) is proportional to the difference between the neuron's firing rate and the threshold (black dashed line). During a more difficult trial (grey line), the same decision response (D_c') is made but with less confidence (MC'). A model of this kind was used to describe rat behavior in one of the single neurons studies described below [27].

The two mechanisms described assume that cognitive and metacognitive processes are encoded simultaneously within a brain area. An alternative proposal entails a sequence of processing stages within a single brain region, in which the metacognitive follows the cognitive process (Fig. 10.2c). Like the previous models, this model exploits the sequential sampling framework. During a given trial (black line), evidence accumulates to a decision (D_c). The metacognitive process, however, depends on evidence continuing to accumulate until a metacognitive response is made (MC). During a different trial (grey line), the same decision may be made (D_c'), but further processing could lead to a higher confidence response (MC'). The stage processing mechanism accounts for human confidence responses [44] and for changes of mind after a decision has been made [70]. It should be noted that simultaneous and multistage models of metacognition are not mutually exclusive. A metacognitive process could be encoded in parallel with a cognitive process *and* after the cognitive process, and there is some evidence for such a scenario [42]. In that study, humans' decision response times (RTs) increased when the task required confidence responses, suggesting the metacognitive process interacted with the cognitive process. In addition, confidence response RTs varied with the confidence level reported, suggesting some post-decisional processing took place as well.

So far we've considered extensions of the sequential sampling models that have enjoyed much success describing decision making. The framework developed by Nelson and Narens [39] suggests separate cognitive and metacognitive processes that interact via information flow, as described earlier. For opt-out and betting tasks, confidence in a given decision would be encoded separately from the decision. It is possible this could occur in one brain region, as shown in Fig. 10.2d. The hypothetical neuronal firing rates encode the decision (D_c vs. D_c') and later the metacognitive signal (MC vs. MC'). Alternatively, perhaps most closely aligned with the Nelson and Narens framework, the metacognitive signal could be encoded in a separate brain region than the cognitive signal (Fig. 10.2e). If that were the case, one might observe little or no decision-related activity (D_c vs. D_c'). Instead, information about the decision would arrive from an external source, as for example a corollary discharge from the brain region encoding the decision [8, 58]. This copy of the information could be used to encode the metacognitive signal (MC vs. MC').

The mechanisms discussed are by no means exhaustive. Perhaps the most obvious alternative is to posit that metacognitive signals are not encoded by firing rates, but by a different signal. For example, neural oscillations could be used as a coding principle, affecting the correlated timing of spikes within a brain region and/or within a brain circuits across regions [4]. Another possibility is that metacognitive signals are encoded by reading out some function of the variance of decision-related spiking neurons during a task [73]. Finally, the worst case scenario (or most interesting scenario, depending on one's viewpoint) is that metacognition is represented along multiple dimensions of neuronal activity, including one, more than one, or all of the possibilities listed in this section. This is one reason that single neuron studies are so important. Different neurons within a brain region or between brain regions may in fact be encoding similar cognitive attributes in different ways. Methods that sample aggregate activity (e.g., fMRI, EEG) are unable to tease apart such variegated strategies for neuronal encoding. While single neuron recordings suffer from their own limitations (e.g., small sample sizes), they are exquisitely appropriate for discovering the coding mechanisms exploited by the brain for sensory, motor, or cognitive functions [72].

10.4 Single Neuron Studies

Metacognition has been studied only recently at the neuronal level in animals. Though many studies allow for the possibility of metacognition within their design, only three thus far have tested metacognitive processes specifically. By a metacognitive task, we mean one in which the activity of single neurons is correlated with, and therefore could be used for, monitoring a cognitive process and acting with respect to that process. A related field of study in neuroscience is so-called "performance monitoring", which correlates neuronal activity with trial outcomes and rewards [60]. Performance monitoring signals have been shown to correlate with adjustments in performance, like changes in trial RTs that depend on previous trial outcomes (e.g. [45]). But previous performance monitoring tasks were not designed to test whether information in the signals could be use to directly affect the outcome within a concurrent trial. Here we focus on studies in which animals were encouraged to use the monitoring information functionally.

Orbitofrontal cortex Kepecs et al. [27] examined neuronal correlates of confidence in rat orbitofrontal cortex (OFC), an area associated with reward, risk, and uncertainty (e.g., [25, 41, 67]). The rats performed an odor discrimination task, the goal of which was to report the majority component odor within a mixture of two odors (Fig. 10.3a). Decisions were reported by poking their nose into one of two ports, one port for each odor. Reward was delivered after a brief delay if the decision was correct. As expected, rats made more correct decisions on easy trials, i.e., when the proportion of one odor dominated the other (Fig. 10.3b).

OFC neuronal activity was analyzed during a time when the rats would likely experience confidence in their decisions: after the decision had been made, while the rats waited for reward delivery. Firing rates varied as a function of trial difficulty (Fig. 10.3c) and choice accuracy (Fig. 10.3d). In the population of recorded neurons, many (21 %, or 120/563) had higher firing rates during more difficult trials, like the example neuron in Fig. 10.3c, d. Some had the opposite pattern, higher firing rates during easier trials (12 %, 66/563). Further, many neurons differentiated between correct and incorrect decisions within a single level of difficulty. Most had higher firing rates during incorrect than correct trials, like an example neuron during relatively difficult trials (Fig. 10.3e) and during easier trials (Fig. 10.3f). The neuron's firing rate began to distinguish correct from incorrect choices before the decision was made, and sustained the signal throughout reward anticipation and reward delivery. This pattern of activity was evident across the subpopulation of neurons with higher firing rates for incorrect choices (Fig. 10.3g, h). Another subpopulation of neurons had the opposite pattern—higher firing rates during correct choices (not shown).

The OFC neuron signals could encode confidence in the decisions, and are consistent with mechanisms in Fig. 10.2a–c. However, the rats were not required to behave in a metacognitive fashion. To assess the rats' confidence in their decisions, the authors added a manipulation to the experiment. Once a rat poked its nose into a port, a random delay was imposed before reward delivery after a correct decision (as before no reward was delivered after an incorrect decision). The rats could endure the wait and earn reward (or risk waiting longer for no reward), or they could abort the trial and immediately start the next trial. Thus, the task was a hybrid between an opt-out and a betting task. The rats' behaved as if they experienced varying levels of confidence. They waited longer for reward after an easy correct trial than a difficult correct trial, and conversely after errors they aborted more often when the error was made on an easy trial than on a difficult trial. OFC neurons were not recorded during the delayed-reward trials, so we must cautiously assume the neuronal activity during the modified task was similar to that during the original task (which is not a fail-safe assumption; see [42]). Instead Kepecs et al. [27] offered two models, like Fig. 10.2a, b, in which confidence was encoded simultaneously with the decision. The models correctly predicted the animals' behavior during the delayed-reward (opt-out version) trials and matched the pattern of OFC firing rates from recordings made during the initial discrimination task.

In sum, rat OFC neurons recorded during an odor discrimination task carried signals that could be used to make metacognitive judgments about the decisions. When the rats were subjected to a modified version of the task that encouraged metacognitive behavior, their performance was consistent with experiencing

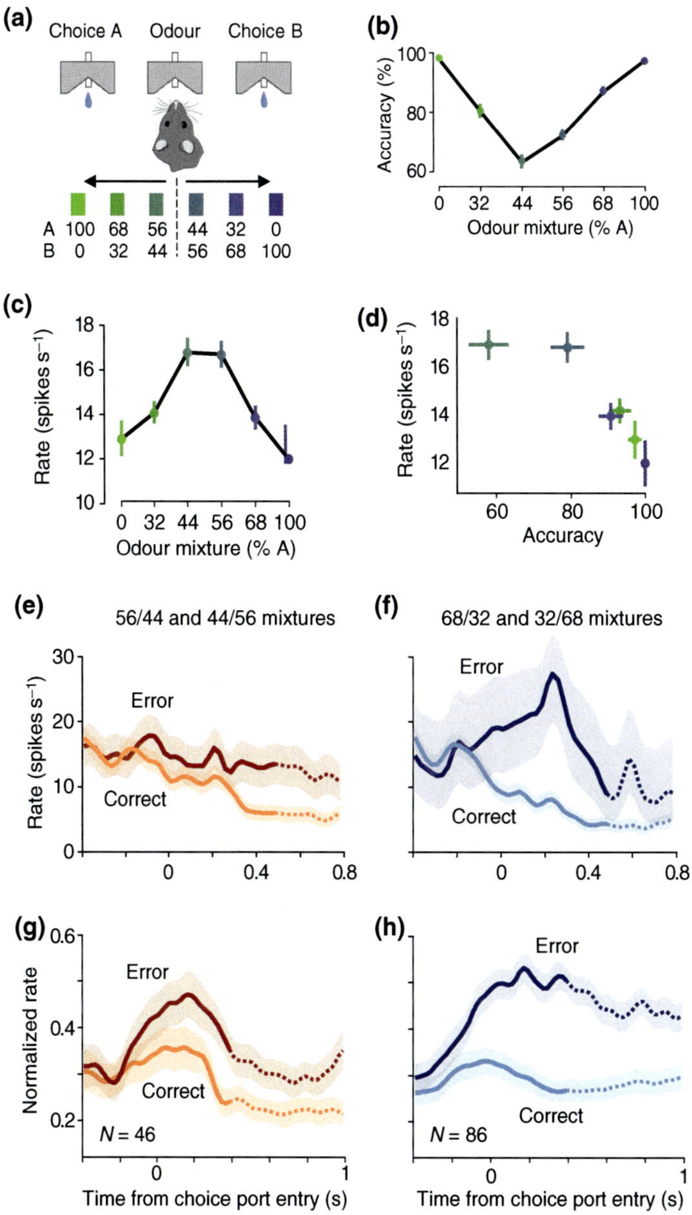

Fig. 10.3 Confidence-related neural activity in rodent orbitofrontal cortex. **a** Rodents discriminated the majority odor component in a two-choice odor discrimination task. Decisions were reported by a nose-poke into one of two adjacent ports. **b** Rodents performed better when the one odor component dominated the other. **c** An example OFC neuron that had higher firing rates

◀ during more difficult trials. Firing rates were measured after the decision, while the rat waited for reward. **d** The same neuron had higher firing rates after less accurate decisions. **e, f** Activity of an example OFC neuron differentiated between correct decisions and errors. This difference was greater and appeared sooner on easier trials (**f**) than harder trials (**e**). **g, h** Population activity differentiated between correct decisions and errors. The patterns seen in (**e, f**) are conserved in a subpopulation of neurons (66/563). Adapted with permission from Kepecs et al. (2007)

varying degrees of confidence. Models provided a link between neuronal activity during the discrimination task and performance during the metacognitive task.

Lateral intraparietal cortex Kiani and Shadlen [28] used an opt-out task and recorded single neurons in rhesus macaque LIP cortex, an area implicated in visuospatial cognition, attention, and decision making [7, 18, 52]. Monkeys were trained to discriminate the motion direction of a visual display of randomly moving dots that had overall coherence in one direction (Fig. 10.4a). During an initial fixation period, targets appeared in the periphery. A patch of moving dots appeared briefly then disappeared, followed by a delay, and then a cue to make a response. Trial difficulty varied with the overall motion strength of the moving dots and with the duration that the moving dots appeared. On half of the trials, a response to the stimulus was required, by making an eye movement in the same direction the dots appeared to be moving (Fig. 10.4a, lower panels). Reward was delivered after correct decisions. On the other half of trials, an opt-out response was offered after the moving dots stimulus disappeared, called the "sure target." If chosen it ensured a small reward (Fig. 10.4a, upper panels).

When offered the sure target, the likelihood of choosing it increased with trial difficulty (Fig. 10.4b). In addition, more accurate responses were made on trials when the sure target was offered but a motion stimulus target was chosen than on trials when the monkey was forced to choose a motion stimulus target (Fig. 10.4c, closed circles are trials with sure target present, open circles are forced-choice trials). Thus, the monkeys optimized reward by choosing the sure target when the probability of being correct was low, performance consistent with experiencing less confidence (more uncertainty) on those trials.

LIP activity varied as a function of choosing the sure target or one of the motion targets, illustrated by an example neuron (Fig. 10.4d). During forced-choice trials (left), while the monkey viewed the moving dots, the neuron's firing rate increased on trials when the target in the response field was the correct motion stimulus target (black line) relative to the incorrect target (grey line). These signals were maintained until a saccade was made to a target, similar to many previous reports of LIP neurons during the dots task (e.g., [49]). In trials when the sure target was offered, firing rates were again high or low when one of the motion targets was chosen (Fig. 10.4d right panel, solid black, and gray lines). When the sure target was chosen, however, the neuron's firing rates were intermediate (dashed black and gray lines). Thus, varying levels of firing rates were suggested to correlate with the monkeys' confidence. This same pattern of activity was evident across the population of 70 LIP neurons (Fig. 10.4e).

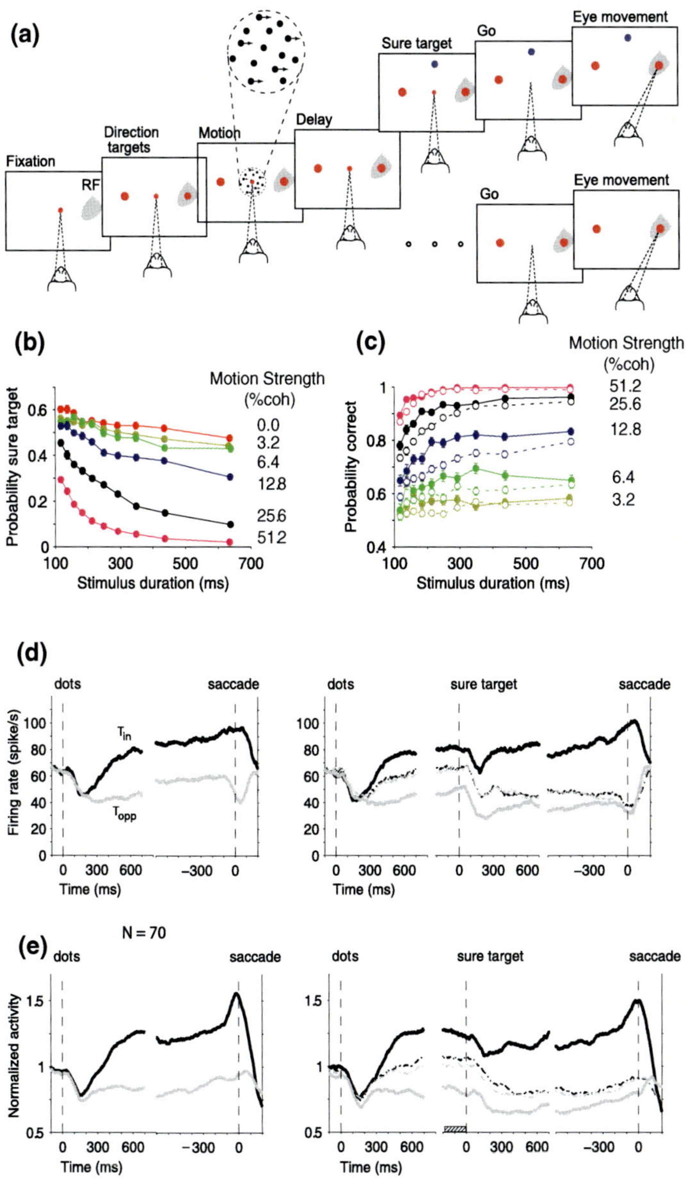

Fig. 10.4 Confidence-related neural activity in macaque lateral intraparietal cortex (*LIP*). **a** Opt-out task schematic. Monkeys had to discriminate the motion direction of a random dot-motion stimulus. Decisions were reported by a saccade to one of two peripheral targets. On some trials, a third "sure target" appeared after the motion stimulus but before the animal was permitted to respond. **b** Subjects were more likely to select the sure target when stimulus presentation time was shorter and overall motion coherence was lower. **c** Subjects also performed better on trials in which the sure target was offered than on trials in which there was no sure target. **d** Activity of an

◀ example LIP neuron varied with the decision to choose one of the direction targets or the sure target. In trials without sure target (*left*), the cell was more active when the direction target in its receptive field was selected (*black line*) than when the alternate target was selected (*grey line*). In trials with sure target (*right*), activity corresponding to its selection was intermediate (*dashed lines*). **e** This same pattern of activity was found in a population of 70 LIP neurons. Adapted with permission from Kiani and Shadlen [28]

Kiani and Shadlen [28] concluded, as did Kepecs et al. [27], that confidence was encoded along with the decision-related signal, manifested as graded levels of that signal. The authors likewise modeled their data using a sequential sampling framework. Response to the sure target depended on a dynamic threshold of neuronal activity throughout the trial, the level of which was set as a function of prior likelihoods of choosing the correct motion target. The LIP neuronal data and model therefore, like the rat OFC activity, are consistent with the mechanisms proposed in Fig. 10.2a, b.

Frontal eye field, dorsolateral prefrontal cortex, and supplementary eye field Middlebrooks and Sommer [36] carried out the most recent single neuron study of metacognition. A betting task (Fig. 10.1b) was used, inspired by previous behavioral experiments that tested monkeys' metacognitive skills [31, 54].

Each trial consisted of a decision stage and a subsequent bet stage (Fig. 10.5a). The goal of the decision stage was to detect the location of a red target square. The trial began by fixating a central spot. A red target appeared randomly at one of four possible locations, then after a varying delay white mask stimuli appeared at all four locations. A correct decision was reported by making a saccade to the location where the target appeared, and an incorrect decision was a saccade to one of the other locations. Difficulty varied as a function of the delay between the target and mask appearance, known as the stimulus onset asynchrony (SOA). Immediately after a decision was made, a new fixation spot appeared in the center of the screen to begin the bet stage. The goal of the bet stage was to make a bet regarding whether the decision was correct. Once the new fixation spot was obtained, two bet targets appeared in the periphery—a red high bet target and a green low bet target. The monkey placed a bet by making a saccade to one of the bet targets. Reward was earned based on the conjunction of decision responses and bets. A correct decision followed by a high bet (CH: correct-high) earned maximum reward, and an incorrect decision followed by a high bet (IH: incorrect-high) earned a brief timeout punishment. Low bets earned minimal juice rewards regardless of the decision (CL and IL: correct- and incorrect-low). Thus, a metacognitive strategy would maximize reward: bet high after correct decisions and bet low after incorrect decisions.

There are a few noteworthy differences between the betting task and the tasks described above. First, each trial requires both a decision and a bet. The moving dots opt-out task (Fig. 10.4a) required a single response (a decision or an opt-out), so neuronal activity related to a decision is potentially complicated during trials in which the animal opted out. The odor discrimination task (Fig. 10.3a) is more similar to the betting task by requiring a decision response each trial followed by either an action (abort trial to restart) or no action (wait for reward). However,

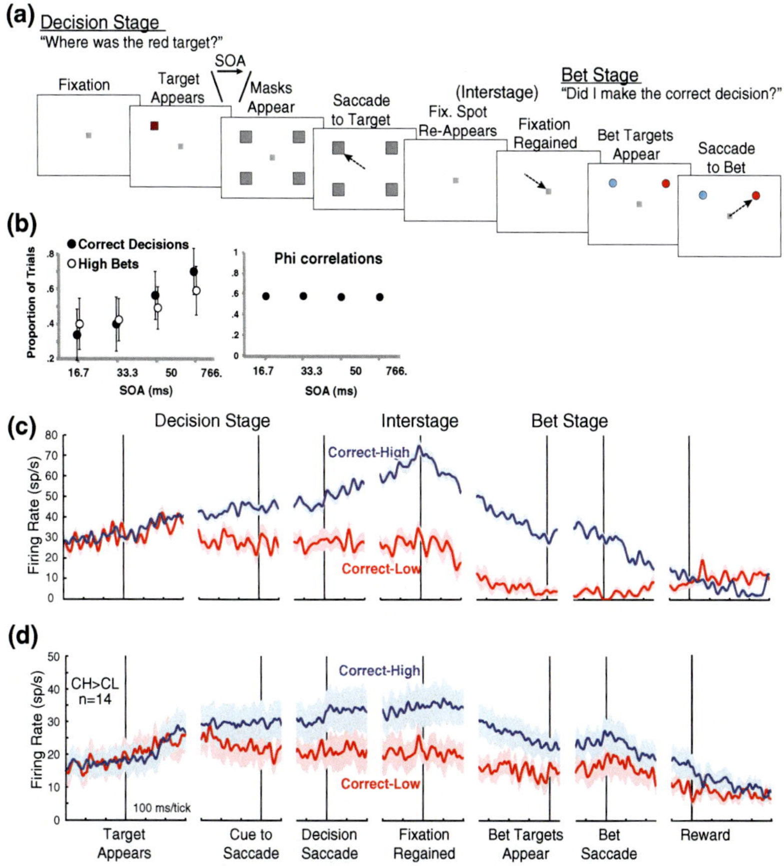

Fig. 10.5 Confidence-related neural activity in macaque supplementary eye field (*SEF*). **a** Betting task schematic. During the decision stage of the task, the goal was to detect the location of a *red* target. The *red* target appeared at one of *four* locations, and *white* mask stimuli appeared at all four locations after a brief delay (SOA). Decisions were reported by a saccade to one of the *four masks*. After a decision, the animal regained fixation to begin the bet stage. During the bet stage a high bet target and a low bet target appeared in the periphery, and the monkey made a saccade to one of the bet targets. **b** Decision accuracy and the proportion of high bets increased as task difficulty decreased (*left* panel, greater SOA values correspond to easier target detection). On trial-by-trial basis, high bets were correlated with correct decisions regardless of trial difficulty (*right* panel, phi correlations greater than zero indicated correlated decisions and bets) (**c**) Activity in an example SEF neuron varied with the likelihood of choosing the high bet after correct decisions. **d** This pattern of activity was conserved in a population of 14 SEF neurons. Adapted with permission from Middlebrooks and Sommer [36]

by requiring a saccadic bet on each trial, trials can be compared in which identical behaviors can result in alternative outcomes, controlling for behavior as a possible explanation of neuronal activity.

Another difference between tasks involves the type of perceptual decisions required. The decision stage of the betting task required *detection* of a stimulus. In contrast, the odor discrimination and the moving dots tasks both required *discrimination* of a stimulus. The subtle difference may serve better to separate decision-related signals from metacognitive signals. During discrimination tasks, perceptual evidence is thought to accumulate in neuronal activity over time until a decision is reached. This approach has provided rich contributions to understanding how decisions are made in the brain [17, 55]. It is possible though that signals related to the evolving perceptual evidence overlap with signals related to metacognition. Although this complication is not an issue when analyzing neuronal activity well after a decision (like the OFC activity during the odor discrimination task), it may affect interpretation of signals early during the task (like the LIP activity of the moving dots task). Using a detection task, involving a brief pulse of sensory information on which to base a decision, provides separation between perceptual and metacognitive signals, thus untangling them.

The monkeys' performance during the betting task indicated they used a metacognitive strategy. During the decision stage, target detection varied as expected with trial difficulty—correct decisions increased as a function of SOAs (Fig. 10.5b, left panel). Bets also varied with trial difficulty—high bets increase as a function of SOA. This overall pattern, the tendency to bet high on trials more likely to be correct was expected if the animals monitored their decisions. But it also could result from a probabilistic betting strategy based solely on the difficulty of the decisions. If so, high bets on average would parallel correct decisions (and low bets would parallel incorrect decisions), but on a trial-by-trial basis high (low) bets might not follow correct (incorrect) decisions. To ensure the monkeys adopted a metacognitive betting strategy, a trial-by-trial analysis confirmed that high bets mostly followed correct decisions and low bets mostly followed incorrect decisions, regardless of trial difficulty (Fig. 10.5b, right panel). Thus, monkeys accurately monitored their decisions to make appropriate bets.

Neurons were recorded in three separate cortical regions: the FEF, dorsolateral PFC, and the supplementary eye field (SEF). The decision stage of the task was inspired by previous reverse masking tasks in which FEF neuron firing rates varied with monkeys' ability to detect the target [65, 66]. FEF is involved in oculomotor behavior [3], higher level processes like attention [37], and is known to send copies of eye movement signals to other brain areas [8, 58]. Middlebrooks and Sommer [36] reasoned that FEF activity might also vary with monkeys' processing of the decision stage to guide a subsequent metacognitive bet. For similar reasons, neurons in PFC and SEF were recorded. PFC has been implicated in a range of high-level cognition, like working memory [15], decision-making [29], and goal-driven behavior [64]. SEF, in addition to having activity related to visual processing and oculomotor behavior [50, 51], has a known role in so-called performance monitoring—signals related to errors, response conflicts, and rewards [60]. Performance monitoring signals produced during the decision stage could be used to encode an upcoming bet.

Neuronal firing rates were first analyzed with respect to decision outcomes during the task, regardless of subsequent bets. Early during the decision stage, when sensory evidence about target location might be encoded, all three cortical regions' neuronal firing rates were modulated with decision accuracy. During the planning, execution, and immediate aftermath of the saccadic response, only SEF firing rates were modulated. FEF and PFC neurons were active and task-related, but did not differentiate correct and incorrect decisions (Schall and Hanes 1996, e.g.). In short, as expected, neuronal activity in each brain region varied with decision accuracy.

To test whether neurons in these regions were involved in monitoring decisions, neuronal firing rates were compared between the conjunctions of decision and bet outcomes—the metacognitive processes. If neurons encoded the accuracy of monitoring decisions, a prediction would be that their firing rates would be modulated between trials in which different bets were made after having made the same (correct or incorrect) decision. Trial outcomes were thus divided to compare CH versus CL and to compare IH versus IL trials.

Of the three cortical regions tested, SEF seemed most involved in metacognitive processing. There were neurons in SEF that differentiated CH and CL trials (15 %, or 20/133) and neurons that differentiated IH and IL trials (8 %, or 10/133). An example neuron that had higher firing rates for CH than CL trials is shown in Fig. 10.5c. Firing rates for CH and CL outcomes are shown throughout the trial. The signals diverge quickly after the target appears (before the decision has been made), reach a peak difference between the decision stage and the bet stage, and maintain a difference through the betting stage. The example neuron was typical of the population that had CH firing rates greater than CL (Fig. 10.5d). In general, SEF activity during the betting task provided more support that metacognitive processes could be encoded concomitant with and in the same brain region as cognitive processes.

10.5 Discussion

What do we know about the neuronal basis of metacognition? As attested by the studies described above, it is too early in this burgeoning field to make definitive claims about how neuronal activity translates into metacognitive behavior. Neurons in LIP, OFC, and SEF all had firing rates that varied with metacognitive behavior. There is no way to tell whether the neuronal activity was necessary for the metacognitive behavior, however, because none of the studies used causal manipulations. Microstimulation techniques and reversible inactivation or lesions of brain regions are needed to provide evidence that any region plays a causal role. A caveat to such approach however, is that it may be difficult to ascribe effects solely to metacognitive processing if the same brain regions are encoding the cognitive processes.

A major challenge facing single neuron metacognition research is the extent to which animal models of metacognition apply to human metacognition. It seems likely, based on the success of opt-out tasks, that many animals experience some measure of confidence along with the decisions they make. It also seems likely, based on betting tasks, that some animals keep track of the accuracy of their decisions, at least over short period of times. It is an open question whether these behaviors occur naturally in the environment or are a product of nurturing rudimentary metacognitive abilities by extensive laboratory training.

The relative simplicity and streamlined design of the tasks described above has advantages and disadvantages. Each task used a metacognitive component temporally yoked to the cognitive component. Notable advantages of this task design are the abilities to observe the dynamics of neuronal activity within a single trial, and to interpret the signals within the context of the large body of knowledge in decision-making neuroscience. A disadvantage is that they do not capture the complexity we traditionally associate with human metacognitive processes, which can refer to events many years in the past or even potential events years in the future. It will be a challenge for future animal studies to tap into more complex forms of metacognition.

All three studies in this chapter reported neuronal signals consistent with an account of metacognition being encoded in near simultaneity and in the same brain region as the referent cognitive process (Fig. 10.2a–b). None reported signals that clearly support Nelson and Narens' [39] framework, in which a metacognitive process is distinct from and monitors or controls a cognitive process. One explanation is that the limited scope of metacognitive behaviors tested, confidence and uncertainty in perceptual processes, falls short of complexity that would require distinct circuits. Though we generally refer to metacognition as if it were a single process, it is more likely to encompass multiple functions that require various brain circuits (e.g., [16, 26]), depending on the cognitive processes involved and the nature of the task. Thus, there may be systems yet discovered that encode metacognitive processes in a way more compatible with Nelson and Narens' framework.

It should also be noted that even if metacognitive signals reported are directly available from the cognitive signals, they are not instantaneously available. Instead, most proposed mechanisms require *some* computation to read out the metacognitive signal, whether it's a comparison between two neurons' firing rates (Fig. 10.2a), or a comparison between one neuron's firing rate and a signal representing a threshold from memory, etc. Therefore, the results of the single neuron studies do not rule out separate cognitive and metacognitive systems. In Nelson and Narens' framework, information is proposed to flow between the metacognitive and cognitive processors. Information in the brain, in the form of actions potential patterns, flows at the millisecond time scale. Hence cognitive and metacognitive processes could easily overlap in time and location.

An important point to consider is that metacognitive judgments may dissociate from cognitive performance. In other words, the monitoring or control of cognitive information (metacognition) is likely based on reduced-fidelity versions of that information. This occurs in healthy individuals but is worsened in some neuropsychiatric disorders (see the chapters in Part IV). One possible source for these metacognitive "errors" is inaccurate transformations/computations during the readout of decision-related signals. This is consistent with an account of metacognitive judgments that depend primarily on accurate translation of cognitive signals. Another potential source of error is misinterpretation of external cues like familiarity with task stimuli, consistent with metacognitive judgments derived from sources outside the cognitive signals [30]. These potential sources of error are not mutually exclusive, as metacognitive judgments could be affected by both factors.

A related issue is that studying high-level processes at the single neuron level presents the inherent difficulty of interpreting what is actually represented in the neuronal signals. Because metacognition can involve so many other cognitive processes, one might expect multiplexed information in the neuronal firing rates. Each of the three described studies addressed this issue and ruled out some alternative accounts of the neuronal signals. Thus, cognitive functions like risk assessment and reward-related processing did not explain the neuronal activity overall. However, it is unknown how much these and other processes, like attention, might contribute from trial to trial. It is important to consider these issues moving forward.

As interesting as it is that some cortical regions were involved in metacognitive processes, it is also interesting that others were not. Specifically, neither FEF nor PFC neurons varied with metacognitive performance. The simplest interpretation is that these regions are not part of the circuit that mediates metacognition. Another possibility is that metacognition is implemented by some other coding scheme than firing rates. For example, variation in coherence of action potential timing among pools of neurons may contribute to metacognitive processes (e.g., [34, 40]). Lastly, perhaps FEF and PFC do not contribute to the specific type of task used but may contribute when other facets of metacognition are tested.

In conclusion, the study of metacognition at the level of single neurons has been productive. With further refinement of animal-specific tasks and more detailed surveys of task-related signals across brain areas and species, single neuron data should continue to complement and inform the growing body of research on human metacognition.

Acknowledgments Supported by the NIH (NIMH Kirschstein NRSA F31 MH087094 to P.G.M. and NEI R01 EY017592 to M.A.S.) and the NSF (Graduate Research Fellowship to Z.M.A.).

References

1. Beran MJ, Smith JD (2011) Information seeking by rhesus monkeys (*Macaca mulatta*) and capuchin monkeys (*Cebus apella*). Cognition 120:90–105
2. Beran MJ, Smith JD, Redford JS et al (2006) Rhesus macaques (*Macaca mulatta*) monitor uncertainty during numerosity judgments. J Expt Pscyhol Anim Behav Proc 32:111–119
3. Bruce CJ, Goldberg ME (1985) Primate frontal eye fields. I. Single neurons discharging before saccades. J Neurophysiol 53:603–635

4. Buzsaki G (2006) Rhythms of the brain. Oxford, New York
5. Call J (2010) Do apes know that they could be wrong? Anim Cogn 13:689–700
6. Call J, Carpenter M (2001) Do chimpanzees and children know what they have seen? Anim Cogn 4:207–220
7. Colby CL, Duhamel JR, Goldberg ME (1996) Visual, presaccadic, and cognitive activation of single neurons in monkey lateral intraparietal area. J Neurophysiol 76:2841–2852
8. Crapse TB, Sommer MA (2008) Corollary discharge across the animal kingdom. Nat Rev Neurosci 9:587–600
9. De Martino B, Fleming SM, Garrett N et al (2013) Confidence in value-based choice. Nat Neurosci 16:105–110
10. Ding L, Gold J (2010) Caudate encodes multiple computations for perceptual decisions. J Neurosci 30:15747–15759
11. Ding L, Gold J (2012) Neural correlates of perceptual decision making before, during, and after decision commitment in monkey frontal eye field. Cereb Ctx 22:1052–1067
12. Flavell JH (1971) First discussant's comments: what is memory development the development of? Human Dev 14:272–278
13. Flavell JH (1976) Metacognitive aspects of problem solving. In: Resnick LB (ed) The nature of intelligence. Erlbaum, Hillsdale, pp 231–236
14. Foote AL, Crystal JD (2007) Metacognition in the rat. Curr Biol 17:551–555
15. Funahashi S, Bruce CJ, Goldman-Rakic PS (1989) Mnemonic coding of visual space in the monkey's dorsolateral prefrontal cortex. J Neurophysiol 61:331–349
16. Gigerenzer G, Hoffrage U, Kleinbölting H (1991) Probabilistic mental models: a Brunswikian theory of confidence. Psychol Rev 98:506–528
17. Gold JI, Shadlen MN (2007) The neural basis of decision making. Ann Rev Neurosci 30:535–574
18. Gottlieb JP, Kusunoki M, Goldberg ME (1998) The representation of visual salience in monkey parietal cortex. Nature 391:481–484
19. Green DM, Swets JA (1966) Signal detection theory and psychophysics. Wiley, New York
20. Hampton RR (2001) Rhesus monkeys know when they remember. PNAS 98:5359–5362
21. Hampton RR, Zivin A, Murray EA (2004) Rhesus monkeys (*Macaca mulatta*) discriminate between knowing and not knowing and collect information as needed before acting. Anim Cogn 7:239–246
22. Hanes DP, Schall JD (1996) Neural control of voluntary movement initiation. Science 274:427–430
23. Hart JT (1965) Memory and the feeling-of-knowing experience. J Educ Psychol 56:208–216
24. Heath RA (1984) Random-walk and accumulator models of psychophysical discrimination: a critical evaluation. Perception 13:57–65
25. Hsu M, Bhatt M, Adolphs R et al (2005) Neural systems responding to degrees of uncertainty in human decision-making. Science 310:1680–1683
26. Juslin P, Olsson H (1997) Thurstonian and Brunswikian origins of uncertainty in judgment: a sampling model of confidence in sensory discrimination. Psychol Rev 104:344–366
27. Kepecs A, Uchida N, Zariwala HA et al (2008) Neural correlates, computation and behavioural impact of decision confidence. Nature 455:227–231
28. Kiani R, Shadlen MN (2009) Representation of confidence associated with a decision by neurons in the parietal cortex. Science 324:759–764
29. Kim JN, Shadlen MN (1999) Neural correlates of a decision in the dorsolateral prefrontal cortex of the macaque. Nat Neuro 2:176–185
30. Kornell N (2013) Where is the "meta" in metacognition? J Comp Psychol 1–27
31. Kornell N, Son LK, Terrace HS (2007) Transfer of metacognitive skills and hint seeking in monkeys. Psychol Sci 18:64–71
32. Krauzlis R, Dill N (2002) Neural correlates of target choice for pursuit and saccades in the primate superior colliculus. Neuron 35:355–363
33. Link SW (1992) The wave theory of difference and similarity. Erlbaum, Hillsdale
34. Lisman JE, Jensen O (2013) The theta-gamma neural code. Neuron 77:1002–1016

35. Merkle EC, van Zandt T (2006) An application of the Poisson race model to confidence calibration. J Exp Psychol Gen 135:391–408
36. Middlebrooks PG, Sommer MA (2011) Metacognition in monkeys during an oculomotor task. J Exp Psychol Learn Mem Cogn 37:325–337
37. Moore T, Fallah M (2001) Control of eye movements and spatial attention. PNAS 98:1273–1276
38. Nelson TO (1996) Consciousness and metacognition. Am Psychol 51:102–116
39. Nelson TO, Narens L (1990) Metamemory: a theoretical framework and new findings. Psychol Learn Motiv 26:125–141
40. Nikolić D, Fries P, Singer W (2013) Gamma oscillations: precise temporal coordination without a metronome. Trends Cogn Sci 17:54–55
41. O'Neill M, Schultz W (2010) Coding of reward risk by orbitofrontal neurons is mostly distinct from coding of reward value. Neuron 68:789–800
42. Petrusic WM, Baranski JV (2003) Judging confidence influences decision processing in comparative judgments. Psychon Bull Rev 10:177–183
43. Pierce CS, Jastrow J (1884) On small differences in perception. Mem Natl AcadSci 3:75–83
44. Pleskac TJ, Busemeyer JR (2010) Two-stage dynamic signal detection: a theory of choice, decision time, and confidence. Psychol Rev 116:864–901
45. Pouget P, Logan GD, Palmeri TJ et al (2011) Neural basis of adaptive response time adjustment during saccade countermanding. J Neurosci 31:12604–12612
46. Purcell BA, Heitz RP, Cohen JY et al (2010) Neurally constrained modeling of perceptual decision making. Psych Rev 117:1113–1143
47. Ratcliff R, Cherian A, Segraves M (2003) A comparison of macaque behavior and superior colliculus neuronal activity to predictions from models of two-choice decisions. J Neurophysiol 90:1392–1407
48. Ratcliff R, Hasegawa YT, Hasegawa RP et al (2007) Dual diffusion model for single-cell recording data from the superior colliculus in a brightness-discrimination task. J Neurophysiol 97:1756–1774
49. Roitman JD, Shadlen MN (2002) Response of neurons in the lateral intraparietal area during a combined visual discrimination reaction time task. J Neurosci 22:9475–9489
50. Schall JD (1991) Neuronal activity related to visually guided saccadic eye movements in the supplementary motor area of rhesus monkeys. J Neurophysiol 66:530–558
51. Schlag J, Schlag-Rey M (1987) Evidence for a supplementary eye field. J Neurophysiol 57:179–200
52. Shadlen MN, Newsome WT (2001) Neural basis of a perceptual decision in the parietal cortex (area LIP) of the rhesus monkey. J Neurophysiol 86:1916–1936
53. Shields WE, Smith JD, Washburn DA (1997) Uncertain responses by humans and rhesus monkeys (*Macaca mulatta*) in a psychophysical same-different task. J Exp Psychol Gen 126:147–164
54. Shields WE, Smith JD, Guttmannova K et al (2005) Confidence judgments by humans and rhesus monkeys. J Gen Psychol 132:165–186
55. Smith PL, Ratcliff R (2009) An integrated theory of attention and decision making in visual signal detection. Psych Rev 116:283–317
56. Smith JD, Schull J, Strote J et al (1995) The uncertain response in the bottlenosed dolphin (*Tursiops truncatus*). J Exp Psychol Gen 124:391–408
57. Smith JD, Shields WE, Allendoerfer KR et al (1998) Memory monitoring by animals and humans. J Exp Psychol Gen 127:227–250
58. Sommer MA, Wurtz RH (2008) Brain circuits for the internal monitoring of movements. Ann Rev Neurosci 31:317–338
59. Son LK, Kornell N (2005) Meta-confidence judgments in rhesus macaques: explicit versus implicit mechanisms. In: Terrace HS, Metcalfe J (eds) The missing link in cognition: origins of self-reflective consciousness. Oxford Press, Oxford, pp 296–320
60. Stuphorn V, Taylor TL, Schall JD (2000) Performance monitoring by the supplementary eye field. Nature 408:857–860

61. Suda-King C (2008) Do orangutans (*Pongo pygmaeus*) know when they do not remember? Anim Cogn 11:21–42
62. Suda-King C, Bania AE, Stromberg EE et al (2013) Gorillas' use of the escape response in object choice memory tests. Anim Cogn 16:65–84
63. Tanaka A, Funahashi S (2012) Macaque monkeys exhibit behavioral signs of metamemory in an oculomotor working memory task. Behav Brain Res 233:256–270
64. Tanji J, Hoshi E (2008) Role of the lateral prefrontal cortex in executive behavioral control. Physiol Rev 88:37–57
65. Thompson KG, Schall JD (1999) The detection of visual signals by macaque frontal eye field during masking. Nat Neurosci 2:283–288
66. Thompson KG, Schall JD (2000) Antecedents and correlates of visual detection and awareness in macaque prefrontal cortex. Vision Res 40:1523–1538
67. Tobler PN, O'Doherty JP, Dolan RJ et al (2007) Reward value coding distinct from risk attitude-related uncertainty coding in human reward systems. J Neurophysiol 97:1621–1632
68. Usher M, McClelland JL (2001) The time course of perceptual choice: the leaky, competing accumulator model. Psych Rev 108:550–592
69. van Zandt T (2000) ROC curves and confidence judgments in recognition memory. J Expt Psych Learn Mem Cogn 26:582–600
70. van Zandt T, Maldonado-Molina MM (2004) Response reversals in recognition memory. J Expt Psych Learn Mem Cogn 30:1147–1166
71. Vickers D (1979) Decision processes in visual perception. Academic Press, New York
72. Wurtz RH, Sommer MA (2006) Single neurons and primate behavior. In: Senior C, Russell T, Gazzaniga M (eds) Methods in mind. MIT, Cambridge, pp 123–139
73. Yeung N, Summerfield C (2012) Metacognition in human decision-making: confidence and error monitoring. Philos Trans R Soc Lond B Biol Sci 367:1310–1321

Chapter 11
The Neural Basis of Metacognitive Ability

Stephen M. Fleming and Raymond J. Dolan

Abstract Ability in cognitive domains is usually assessed by measuring task performance, such as decision accuracy. A similar analysis can be applied to metacognitive reports about a task to quantify the degree to which an individual is aware of his or her success or failure. Here, we review the psychological and neural underpinnings of metacognitive accuracy, drawing primarily on research in memory and decision-making. These data show that metacognitive accuracy is dissociable from task performance and varies across individuals. Convergent evidence indicates that the function of rostral and dorsal aspects of lateral prefrontal cortex is important for the accuracy of retrospective judgements of performance. In contrast, prospective judgements of performance may depend upon medial prefrontal cortex. We close by considering how metacognitive processes relate to concepts of cognitive control, and propose a neural synthesis in which dorsolateral and anterior prefrontal cortical subregions interact with interoceptive cortices (cingulate and insula) to promote accurate judgements of performance.

Keywords Metacognition · Confidence · Conflict · Prefrontal cortex · fMRI · Individual differences

---

This chapter is adapted from: Fleming SM, Dolan RJ (2012) The neural basis of metacognitive ability. Phil Trans R Soc B 367:1338–1349

S. M. Fleming (✉)
Center for Neural Science, New York University, 4 Washington Place, Room 809,
New York, NY 10003, USA
e-mail: fleming.sm@gmail.com

S. M. Fleming
Department of Experimental Psychology, University of Oxford, South Parks Road,
Oxford OX1 3UD, UK

R. J. Dolan
Wellcome Trust Centre for Neuroimaging, University College London,
12 Queen Square, London WC1N 3BG, UK

S. M. Fleming and C. D. Frith (eds.), *The Cognitive Neuroscience of Metacognition*,
DOI: 10.1007/978-3-642-45190-4_11,

I am not yet able, as the Delphic inscription has it, to know myself, so it seems to me ridiculous, when I do not yet know that, to investigate irrelevant things.

Plato's *Phaedrus*

The notion that accurate self-knowledge has value, and is something to strive for, has preoccupied thinkers since Socrates. As the above quotation from Plato illustrates, self-knowledge is not always (or even often) transparent and at best tends to be a noisy and inaccurate impression of one's mental milieu [1]. Moreover, empirical data in the psychological sciences have thrown up counterintuitive examples of intentions being confabulated, dissociated from reality or otherwise inaccurate [2, 3]. To take one striking case, when decisions about facial attractiveness or supermarket goods are surreptitiously reversed, subjects are often unaware of these reversals, confabulating explanations of why they chose options they had in fact just rejected [4, 5]. Furthermore, self-assessments of personality and cognitive biases tend to be poorer than similar assessments applied to others, leading to an 'introspection illusion' [6]. Such subjective inaccuracy can account for the demise of an introspectionist method in the late nineteenth century; if verbal reports vary from setting to setting, and can be contradicted from trial to trial, then what hope is there for an objective science of the subjective? [7].

The very notion that an individual can turn his or her mental faculties inward was thought logically incoherent by Comte, who considered the idea that the mind might divide into two to permit self-observation as absurd (cited in [8]). We now understand the brain as a network of regions working in concert, and it is unsurprising that one set of regions (such as the prefrontal cortex) might process, hierarchically, information arising from lower levels (such as primary sensory regions). Indeed, several recent models of local and large-scale brain function rely on hierarchy as a principal organising factor [9, 10]. A view that self-knowledge, and its accuracy, is under neural control is now supported by mounting evidence in the neuropsychological literature, some of which will be reviewed later in this chapter. For example, in cases of traumatic injury to the frontal lobes, individuals may have deficits in self-knowledge of altered cognition and personality, as measured by the discrepancy between reports from the patient and family members [11]. Such studies have focussed on alterations in self-related, or autonoetic, metacognition (see [12]), but analogous discrepancies can be measured in assessments of task performance in healthy individuals.

By focussing on self-reports about memory performance—metacognitive reports—Flavell provided a systematic framework for the study of self-knowledge in healthy individuals [13]. Here, metacognitive report is treated as an object of study in its own right, and the accuracy of such reports (as dissociated from accuracy, or performance, on the task itself) provides an empirical scaffold upon which to build studies of self-knowledge [14, 15]. An influential model of metacognition was developed to account for behavioural dissociations between the 'object' level—cognition, or, more correctly, task performance—and the 'meta'

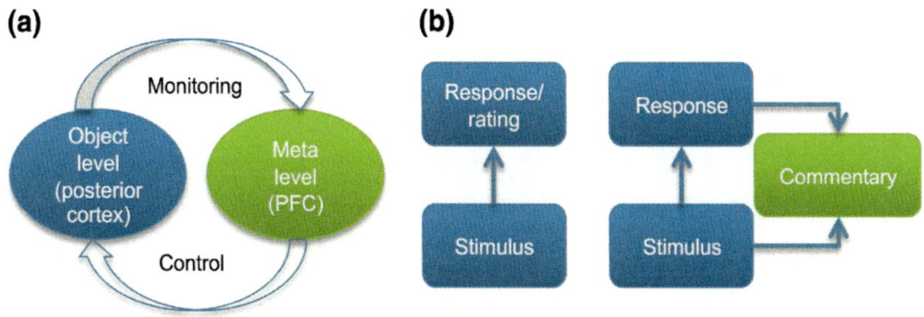

Fig. 11.1 **a** A schematic adapted from Ref. [26] showing how the levels of Nelson and Narens' cognitive psychology model of metacognition can be naturally mapped onto a hierarchical brain structure. **b** The *left panel* shows a first-order process, such as a simple visual discrimination, that may occur in the absence of metacognitive report. The *right panel* shows the same discrimination, this time with the information available for a second-order commentary about the decision

level, conceptualised as both monitoring and controlling the object level (Fig. 11.1; [16]). This approach shares similarities with an influential model of executive function [17]. This two-level framework has also been extended to study monitoring of perception [18, 19], decision-making [20, 21], sense of agency [22] and learning [23]. To the extent that the meta-level imperfectly monitors the object level, self-reports about cognition will be inaccurate, perhaps manifesting as a lack of awareness of the object level [24].

Despite progress in the definition and measurement of metacognition, the psychological and neural underpinnings of metacognitive accuracy remain ill understood [25, 26]. In this chapter, we review different approaches to both eliciting metacognitive reports and quantifying their accuracy. We also consider psychological and computational explanations for dissociations between metacognitive accuracy and task performance. We go on to consider recent studies that apply convergent neuroscience methodologies—functional and structural magnetic resonance imaging (MRI), transcranial magnetic stimulation (TMS) and neuropsychological approaches—to reveal cortical substrates mediating differences in metacognitive accuracy both between and within individuals. We end with a discussion of how metacognitive processes relate to notions of cognitive control.

11.1 Measurement of Metacognition

There are several flavours of metacognitive report, but all involve the elicitation of subjective beliefs about cognition—in other words how much do I know (viz. what can I report) about ongoing task performance? In this section, we review the behavioural methods available to the researcher interested in metacognition, focussing primarily on measures employed in the cognitive neuroscience studies that are discussed in subsequent sections.

Table 11.1 Summary of metacognitive measures classified by domain and time of elicitation. We note that a more general class of prospective judgements is also possible that refers to cognitive abilities not tied to a particular task

Timing	Object-level domain		
	Memory	Decision-making	Sensory
Prospective	Judgement of learning; feeling of knowing	Performance estimate	N/A
Retrospective	Confidence	Confidence, wager	Rating of sensory quality

A key distinction is that judgements can either be prospective, occurring prior to performance of a task, or retrospective, occurring after task completion (Table 11.1). In metamemory research, prospective judgements include feelings of knowing (FOK) and judgements of learning (JOL). A judgement of learning elicits a belief during learning about how successful recall will be for a particular item on subsequent testing [27]. In contrast, a FOK is a judgement about a different aspect of memory, namely that of knowing the answer to a particular question despite being unable to explicitly recall it [28]. FOKs are usually studied by first asking participants to recall answers to general knowledge questions, and, for answers they cannot recall, to predict whether they might be able to recognise the answer from a list of alternatives. A related phenomenon to FOKs is tip-of-the-tongue states, in which an item cannot be recalled despite a feeling that retrieval is possible [29].

Retrospective reports can be similarly elicited by asking a subject to give an additional report or commentary over and above their initial forced-choice response. For example, Peirce and Jastrow asked observers to rate their degree of confidence in a perceptual judgement using the following scale [30]:

> 0 denoted absence of any preference for one answer over its opposite, so that it seemed nonsensical to answer at all. '1' denoted a distinct leaning to one alternative. '2' denoted some little confidence of being right. '3' denoted as strong a confidence as one would have about such sensations.

Since this seminal work, asking for confidence in accuracy has become a standard tool for eliciting judgements of performance in a variety of settings (e.g. [23, 31]). One potential problem with eliciting subjective confidence is that of reliability: why should the subject be motivated to reveal his or her true confidence when there is little incentive to do so [32]? In addition, the necessarily subjective instructions given when eliciting reports of confidence preclude the use of these measures in non-human animal species.

To address these concerns, Kunimoto and colleagues [33] introduced wagers contingent on the correctness of the decision as an intuitive measure of retrospective confidence (see also [34]). In the simplest form of post-decision wagering (PDW), a participant is asked to gamble on whether their response was correct. If the decision is correct, the wager amount is kept; if it is incorrect, the amount is lost. The size of the chosen gamble is assumed to reflect a subject's confidence in his or her decision. In the same spirit as PDW, the Lottery Rule aims to elicit true underlying decision confidence [35], and is similar to the Becker-DeGroot-Marschak procedure used to elicit item values in behavioural economics [36].

Once a metacognitive judgement is elicited, how might we assess its accuracy? Again, several, often complementary, methods are available. Metacognitive accuracy is defined by how closely metacognitive judgements track ongoing task performance. Crucially, therefore, all measures require that an independent measure of the object level—task performance—is acquired, in order to quantify the relationship between the meta and object levels (Fig. 11.1). For example, after asking for a FOK judgement, we might assess the proportion of times a participant is indeed able to recognise the correct, but hitherto unrecalled, item from a list of alternatives. Then, by plotting the strength of the JOL or FOK against objective memory performance (actual recall success for JOLs, and recognition performance for FOKs), a measure of metacognitive accuracy can be derived from the associated correlation [15]. Similar confidence-accuracy correlations can be computed for retrospective confidence judgements. If the metacognitive report bears some relation to task performance, then these correlation coefficients will be significantly non-zero [37].

A related approach quantifies the accuracy of metacognitive assessments using the logic of signal detection theory (SDT), which assesses how faithfully an organism separates signal from noise [38, 39]. In standard applications of SDT (Type 1), sensitivity is defined by how well an observer can discriminate an objective state of the world (e.g. the presence or absence of a stimulus; Fig. 11.2a). By applying similar logic to metacognitive reports, the objective state of the world becomes the subject's trial-by-trial task performance (correct or incorrect; Fig. 11.2a) and the subjective report is now a judgement of that performance [40, 41]. An advantage of the SDT approach is that it dissociates bias from sensitivity; in other words, measures of metacognitive accuracy are relatively unaffected by an observer's overall tendency to use higher or lower confidence ratings (Fig. 11.2b; although see [42, 43]). Further, it naturally connects a process-level characterisation of the relationship between the object (Type 1) and meta-level (Type 2) to measures of behaviour, and this relationship can be taken into account to provide an unbiased measure of metacognitive accuracy [44]. This generative aspect of SDT will be discussed further in the following section.

Before closing our discussion on measures of metacognition, we note that a separate line of research has assessed the extent to which humans and other species use, or represent, uncertainty about the consequences of their actions to optimise decision-making (see [45, 46] for reviews). To highlight one example, Barthelme and Mamassian [47] showed that when human observers are allowed to choose between pairs of visual stimuli upon which to carry out a task, they systematically chose the less uncertain, thus improving their performance. Related work has demonstrated that subjects use knowledge of uncertainty to optimally bias decision-making in perceptual [48, 49] and motor [50] tasks, and that species as diverse as dolphins, pigeons and monkeys can use an 'opt-out' response to improve their reward rate when decisions are uncertain [51]. Recent single-neuron recording studies have begun to outline candidate mechanisms for a representation of uncertainty in the decision system [52, 53]. However, and crucially for the purposes

(a)

Type 1	"A"	"B"
S = A	Hit	Miss
S = B	FA	CR

Type 2	"High"	"Low"
Correct	Hit	Miss
Incorrect	FA	CR

(b)

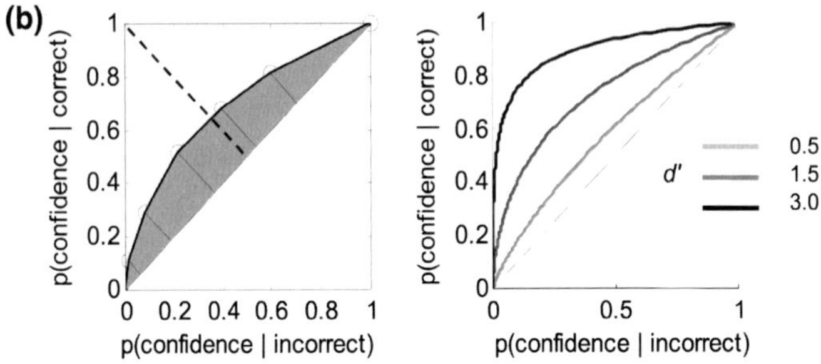

Fig. 11.2 **a** Contingency tables for (*left*) Type 1 SDT, and (*right*) Type 2 SDT. Rows correspond to objective states of the world; columns correspond to subjects' reports about the world. In the Type 2 table, *High* and *Low* refer to decision confidence. The *linking arrow* and *colour scheme* indicates that 'correct' and 'incorrect' states of the world for the Type 2 analysis are derived from averaging particular Type 1 outcomes. **b** *Left panel*—example Type 2 ROC function for a single subject in a perceptual decision task where performance is held constant using a staircase procedure. The *shaded area* indicates the strength of the relationship between performance and confidence. *Right panel*—theoretical Type 2 ROC functions for different levels of Type 1 *d'* (assuming neutral Type 1 response criteria), demonstrating that metacognitive accuracy is predicted to increase as task performance increases

of this chapter, use of uncertainty measures do not dissociate metacognition from task performance on a trial-by-trial basis, and thus cannot be used to study mechanisms underlying beliefs about performance. For example, on each trial of the 'opt-out' paradigm, the animal either chooses to complete the task or opt-out. On trials where the animal opts-out (uses a 'metacognitive' response) we are unable to measure performance, as no task is completed. On trials where the animal does not opt-out, performance measures are all we have. Thus, measures of metacognitive accuracy cannot be computed based on pairwise correlations between the two response types (see also [54]).

11.2 Psychological Determinants of Metacognitive Accuracy

In healthy individuals, metacognitive judgements are usually predictive of subsequent or past task performance [55]. What, then, underlies this ability to know that we know? On a direct-access view, metamemorial judgements are based upon a survey of memory contents, and thus draw upon the same information as a subsequent recognition or recall phase [28]. In contrast, inferential accounts suggest that JOL, FOK and confidence judgements draw upon various mnemonic cues that may only be partially related to the target [56] (see [57] for a review). Such cues include the fluency or ease with which information is processed [58, 59], the accessibility or relatedness of cue information to the target [60], and, for retrospective confidence judgements, the speed of a previous decision [16, 61]. Because available cues may only be indirectly related to the target, inferential accounts naturally accommodated dissociations between memory performance and metacognitive accuracy; in contrast, direct-access accounts predict a tight relationship between subjective and objective indices of knowledge.

A complementary perspective on the antecedents of metacognitive reports is provided by Type 2 SDT. Consider a perceptual decision task where post-decision wagers are elicited to tap knowledge of task performance. Optimal wagering behaviour requires computing the conditional probability of being correct given a previous choice [P(correct|choice)] to decide whether to wager high or low. There are various proposals as to how this might be achieved (e.g. [43, 62]). In an echo of direct-access accounts of metamemory discussed above, most involve tracking the strength of the underlying evidence entering into the choice process. Galvin and colleagues showed that the conditional probability of being correct or incorrect for a given decision signal is a linear transformation of Type 1 probability distributions [41]. Similarly, in a dynamic situation, Vickers [31] proposed that decision confidence could be derived from the absolute distance between the winning and losing integrators in an evidence accumulation framework (see also [52]). Confidence, therefore, is directly equated with the difficulty of the decision in these approaches [63, 64].

Two corollaries arise from this 'direct translation hypothesis' [65]. First, given that confidence is equated with choice probability (as derived from information governing choice), direct translation approaches cannot accommodate dissociations between object and meta levels. Second, if both performance and metacognitive judgements draw upon the same information, metacognitive accuracy, or the ability to discriminate correct from incorrect decisions, always increases as task performance itself increases. Importantly, both these hypotheses have been empirically falsified: for the same level of task performance, judgement confidence may differ considerably between conditions [66–68], and, when performance is held constant using a staircase procedure, metacognitive accuracy varies across individuals [20], and can be dissociated from performance through pharmacological [69], neural [19] and task-based [70] manipulations (Fig. 11.3).

Empirical dissociations between first-order and second-order components of decision-making have prompted a search for models that can accommodate such findings (see [71]). Recent models have been couched in an 'evidence accumulation' framework, in which samples of data are accumulated over time in order to model the temporal evolution of a decision [18, 72, 73]. Del Cul et al. [18] proposed a dual-route evidence accumulation framework in which evidence for behaviour (a forced-choice report of stimulus identity) and evidence for subjective

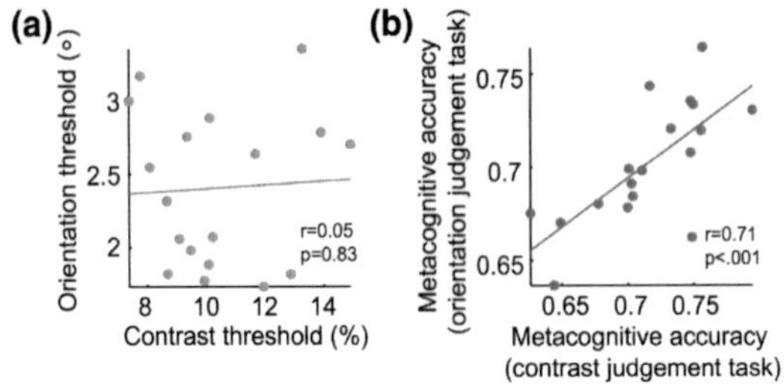

Fig. 11.3 Data from a visual decision task demonstrating a dissociation of metacognitive accuracy from task performance. Subjects made a visual decision (either an orientation or contrast judgement) and then provided a retrospective confidence rating. A measure of metacognitive accuracy was derived from these ratings by calculating the area under the Type 2 ROC function. Performance on the orientation judgement task did not predict task performance on the contrast judgement task (**a**). However, metacognitive accuracy was strongly correlated between tasks (**b**), suggesting that it is both independent of task performance and stable within individuals. Reproduced with permission from [70]

report (visibility) were accumulated separately. The fit of this model could account for the observed decoupling of subjective reports from performance in patients with damage to the prefrontal cortex (see [74] for an alternative account). In a related approach, Pleskac and Busemeyer [72] devised an evidence accumulation scheme that could account for a wide range of empirical regularities governing the relationship between choice and confidence ratings. The solution here was to allow accumulation to continue beyond the time at which the first-order decision is made. In other words the same noisy accumulator is then accessed to form the confidence judgement at a later timepoint. Interestingly, this model makes strong predictions about post-decision neural activity in the parietal and frontal cortices previously associated with pre-decision evidence accumulation [75], and recent developments of PDW methods in non-human primates may allow this and related hypotheses to be tested [76].

Despite being dissociable, it turns out that metacognitive accuracy generally scales with task performance [33, 77–80]. Note this regularity differs conceptually from the fact that trial-by-trial judgements of confidence tend to correlate with performance; such scaling is, after all, what measures of metacognitive accuracy attempt to capture. Instead, both within and across individuals, metacognitive accuracy itself covaries with performance on the task (Fig. 11.2b). A tied relationship between performance and metacognition presents a particular problem for studies of the neural correlates of metacognitive ability: how are we to disentangle brain systems involved in metacognition from those involved in performing the task itself (cf. [81])? In the following section, we keep this confound of performance in mind, and consider the extent to which it has been, and can be, addressed by studies of the neural basis of metacognitive accuracy.

11.3 Neural Basis of Metacognitive Accuracy

11.3.1 Studies of Metamemory

The first evidence regarding the neural basis of metacognition was obtained from neuropsychological cases [82]. Hirst and colleagues [83] suggested that metamemory might be impaired in patients with Korsakoff's syndrome, a neurological disorder characterised by severe anterograde amnesia that occurs as a result of chronic alcohol abuse and nutritional deficiency. Structural brain changes in Korsakoff's include increases in cerebrospinal fluid and severe volume loss in regions that include the orbitofrontal cortices and thalamus [84]. Shimamura and Squire [85] found that Korsakoff's patients have a selective impairment in the accuracy of FOK judgements compared to an amnesic control group, despite being equated on recognition memory performance. These findings suggested that metamemory impairment is due to damage in brain regions other than medial temporal lobe and diencephalic midline structures classically associated with amnesia. In line with this hypothesis, subsequent studies found that non-amnesic patients with frontal lobe damage also exhibit poor metamemory accuracy (e.g. [86]; see [87] for a review).

Although implicating frontal lobe structures in metacognitive accuracy, these early studies lacked anatomical specificity. Using lesion overlap measurements, Schnyer and colleagues found that damage to the right ventromedial prefrontal cortex (VMPFC) was associated with decreased FOK accuracy but intact confidence judgements, suggesting a possible dissociation between brain systems supporting different classes of metamemorial judgements [88] (Table 11.1). Patients in Schnyer et al.'s study also showed deficits in memory performance, but impairment in FOK accuracy could not be explained by these changes in performance alone. In support of a selective role for medial PFC in FOK judgements, patients with lesion overlap in the dorsal anterior cingulate cortex (ACC) who were matched in recognition performance to a control group showed a selective FOK deficit, despite intact confidence judgements [79]. The reverse dissociation was reported by Pannu et al. [89] who found that deficits in retrospective confidence judgements were predominantly associated with lateral frontal lesions. As we discuss below, together this evidence suggests prospective judgements are supported by medial PFC function, whereas retrospective judgements depend on lateral PFC.

Complementary functional brain imaging studies have shown that regions in the medial and lateral prefrontal cortex are active during metamemorial judgements, with activity in PFC modulated by both prospective and retrospective confidence judgements [90–94]. VMPFC (peak Montreal Neurological Institute (MNI) coordinate: −3, 30, −18) showed greater activity during accurate FOK judgements, and increased connectivity with medial temporal lobe memory structures in the FOK condition compared to a low-level control task [95]. Complementing this work, individual differences in metacognitive accuracy for prospective JOLs

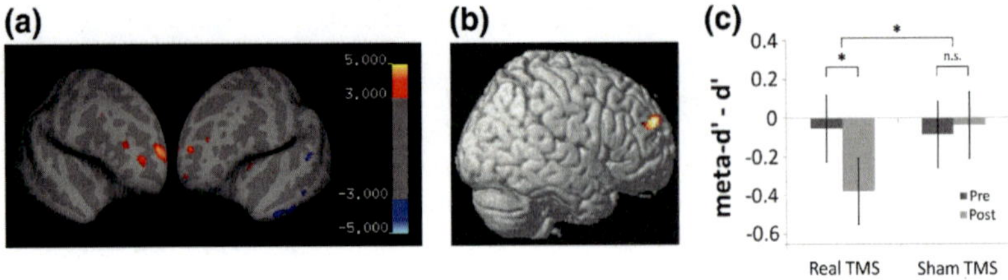

Fig. 11.4 Convergent evidence for a mediating role of rostrolateral PFC in metacognitive accuracy. **a** Across individuals, grey matter volume in rlPFC was found to positively correlate (*hot colours*) with metacognitive accuracy (Type 2 ROC area) after controlling for differences in task performance [20]. **b** In a complementary study, BOLD signal in right posterior-lateral BA10 was positively correlated with metacognitive accuracy (gamma) but not differences in task performance [98]. **c** The necessity of lateral PFC for metacognitive accuracy was confirmed by combining TMS with signal detection theory: following repetitive TMS to bilateral dlPFC, subjects exhibited reduced meta-*d'* (the Type 2 *d'* expected from a given level of Type 1 sensitivity) despite intact task performance [19]. Panels reproduced with permission

correlated with VMPFC activity (peak: −11, 42, −26) on accurate, but not inaccurate, prediction trials [78]; these differences were not explained by individual differences in memory performance.

11.3.2 Retrospective Confidence Judgements in Psychophysics

Other studies have begun to harness the methods of psychophysics to tightly clamp or adjust for differences in performance while simultaneously studying metacognition and its neural substrates (Fig. 11.4). As an example of this approach, Lau and Passingham [67] matched performance between two visual masking conditions, but found differences in threshold for metacognitive commentaries about the stimulus ('seen' responses) that were associated with activity in left dorsolateral PFC (dlPFC; peak: −46, 48, 14). Confirming a causal role for PFC in subjective report threshold, patients with lesions to rostrolateral prefrontal cortex (rlPFC, BA10) have an increased threshold for producing metacognitive commentaries about a stimulus compared to controls, despite objective performance being matched between groups [18]. The peak correlation between lesion and decrease in subjective report threshold was seen in left BA10 (peak: −32, 54, −6).

Taking an individual differences approach, Fleming, Weil et al. [20] constrained perceptual decision performance to be near-threshold (71 %) through use of a staircase procedure, while eliciting retrospective confidence ratings. Considerable variation in metacognitive accuracy (using Type 2 SDT analysis) was found despite task performance remaining constant across individuals. Through use of structural brain imaging, this variance in metacognitive accuracy was shown to positively correlate with grey matter volume in right rlPFC (BA10; peak: 24, 65, 18;

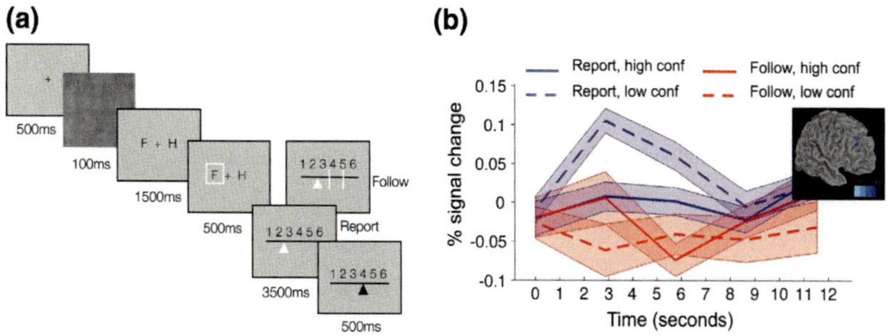

Fig. 11.5 **a** Schematic of a perceptual decision task used to examine metacognition-related brain activity. On each trial, participants were asked to categorise a noisy image as either a face or house by pressing one of two buttons held in their right hand. In the 'Report' condition, post-decision confidence was indicated using a sliding scale. In the 'Follow' condition, participants were instructed to slide the cursor into the zone indicated by the two vertical lines rather than make a confidence judgement. After 3.5 s, the cursor changed colour to indicate the participant's selected rating. **b** The timecourse of activity in the right rostrolateral PFC (peak: 27, 53, 25) shows a dual signature of activity during metacognitive judgements. Activity increases during the confidence judgement relative to the control condition, and this increase is greater for low compared to high confidence ratings. Adapted from Ref. [109]

Fig. 11.4a), and greater metacognitive accuracy was associated with increased white-matter integrity (fractional anisotropy) in a region of the corpus callosum known to project to the rlPFC [96]. Such findings are consistent with individual differences in localised brain structure affecting a region's functional properties [97]. In a study taking a similar approach using functional MRI, subjects performed a visual working memory test and provided retrospective confidence ratings. Metacognitive accuracy correlated with the level of activity in right posterior-lateral BA10 [98] (peak: 16, 56, 28), despite being uncorrelated with task performance (Fig. 11.4b).

While studies of individual differences can implicate particular brain regions in metacognitive efficiency, they are unable to provide information on their functional contribution to metacognition. To address this question, we asked subjects to carry out a near-threshold perceptual judgement task in the fMRI scanner [109]. On two-thirds of trials, subjects were asked to rate their confidence in their previous decision using a sliding cursor. On one-third of trials, they were asked not to reflect on their confidence, but simply to move the cursor into a region of the scale randomly sampled from one of their previous confidence judgements (Fig. 11.5a). This provided a control condition under which the perceptual decision and motor requirements were matched to that of the metacognitive task, allowing us to investigate activity specific to the confidence rating itself. As expected, subjects' confidence judgements tracked fluctuations in task performance, showing good metacognitive ability. Turning to the fMRI data, we found that bilateral rlPFC, dorsal ACC and right posterior parietal cortex showed a dual signature of activity during metacognitive trials (Fig. 11.5b). First, activity increased during metacognitive compared to control judgements, consistent with these regions being specifically involved in the appraisal of a previous decision. Second, the activity increase was greater for low compared to high confidence judgements, showing sensitivity to reported confidence. Finally, we found that the strength of the confidence signal in right rlPFC (peak: 27, 53, 25) predicted individual differences in metacognition across individuals.

While correlational analyses can reveal candidate brain regions mediating metacognitive accuracy, confirmation of their necessity ultimately requires intervention studies. By applying repetitive TMS to temporarily inactivate bilateral dlPFC, Rounis et al. [19] selectively decreased metacognitive accuracy while leaving performance on a perceptual task unaffected. Further, by explicitly modelling the link between Type 1 and Type 2 responses [44], they were able to show that dlPFC TMS decreased metacognitive accuracy below that expected from a direct translation account alone (Fig. 11.4c). Thus, there is a convergence of evidence in support of the idea that rostrolateral sectors of prefrontal cortex (BA10/46) play a mediating role in the accuracy of metacognitive judgements.

A role for rlPFC in metacognition is consistent with its anatomical position at the top of the cognitive hierarchy, receiving information from other prefrontal cortical regions, cingulate and anterior temporal cortex [99]. Further, compared to non-human primates, rlPFC has a sparser spatial organisation that may support greater interconnectivity [100]. One contribution of rlPFC to metacognitive commentary might be to represent task uncertainty in a format suitable for communication to others, consistent with activation here being associated with evaluating self-generated information [101, 102], and attention to internal representations [103]. Such a conclusion is supported by recent evidence from structural brain imaging that 'reality monitoring' and metacognitive accuracy share a common neural substrate in anterior PFC [104]. In contrast, dlPFC may maintain information about a previous decision, consistent with its role in working memory [105, 106]. However, in comparison to, for example, parietal cortex [107], reliable cytoarchitectonic boundaries are not yet established for human rlPFC [108]. Indeed, activations ascribed to either lateral rlPFC or dlPFC in this review cluster around a transition zone between BA10 and BA46 [98, 109], thus it is unclear whether they arise from a single functional region, or multiple subregions subserving different functions. Single-subject analyses (e.g. [110]) may aid in solving this puzzle.

11.3.3 Nature of Individual Differences

Harnessing individual differences can provide leverage on the neural correlates of metacognitive accuracy [20, 78, 98]. Such studies implicitly assume intrapersonal stability of metacognitive capacity. However, in the metamemory literature, evidence for a stable metacognitive ability is surprisingly weak [111, 112]. Given the interdependence of metacognition and performance discussed above, one explanation for this null result might be methodological in nature, as a performance-confidence relationship is naturally harder to quantify than performance itself. A similar line of thought led Keleman et al. to speculate that 'stable metacognitive performance might be detected using very large numbers of trials'. In support of this view, Fleming et al. [20] showed good split-half reliability ($r = 0.69$) in a perceptual decision task with hundreds of trials, and metacognitive accuracy has been shown to be stable across two perceptual tasks ($r = 0.71$), despite performance itself being uncorrelated ($r = 0.05$; Fig. 11.3) [70].

An important unanswered question is whether metacognitive accuracy is stable across distinct domains (e.g. memory and decision-making), as might be predicted by their overlapping neural substrates [113]. One recent study that addressed this question examined the relationship between regional brain volume and metacognitive accuracy in both memory and perceptual tasks [114]. Using a measure of metacognition that adjusts for performance confounds [44], metacognition scores were positively correlated across domains. However, structural brain imaging analysis revealed potentially separable neural substrates. Replicating the findings of Fleming et al. [20], perceptual metacognition was positively correlated with regional grey matter volume in the rlPFC, whereas metacognitive ability on the memory task was associated with precuneus volume. In addition, a formal model comparison indicated that these pathways were relatively independent. The authors went on to suggest that the behavioural correlation between metacognitive abilities across domains, while prima facie evidence for a general mechanism for metacognition, might be driven by the covariation of two separate systems in the healthy brain. It is clear that further work, for example using functional brain imaging techniques, is required to understand the precise contribution of different brain regions to domain-specific metacognition.

11.3.4 Summary

There is now considerable evidence that damage to the prefrontal cortex selectively impacts on the accuracy of metacognitive reports while leaving task performance relatively intact. Intriguingly, there is also evidence for a lateral–medial separation between neural systems supporting retrospective confidence judgements and prospective judgements of performance, respectively. The role of ventromedial PFC in prospective judgements of performance may be explained by its strong connections with both medial temporal lobe memory structures and the precuneus, and its role in use of mnemonic information to imagine the future [115, 116]. In contrast, the role of anterior and dorsolateral PFC in retrospective judgements of confidence may be more closely aligned to that of a performance monitor, integrating and maintaining information pertaining to the immediately preceding decision to facilitate accurate metacognitive commentary. In the next section, we focus in greater detail on these performance-monitoring functions to illustrate connections between metacognition and a separate but substantial literature on the neuroscience of cognitive control.

11.4 Relationship Between Metacognition and Cognitive Control

An influential suggestion is that decision-making systems should be sensitive to the current level of conflict between possible responses to mobilise additional 'cognitive control' resources in an adaptive fashion [117]. Activity in ACC and anterior insula is increased during heightened response conflict (see [118] and [119] for reviews), whereas lateral PFC activity correlates with behavioural adjustments, such as increased caution, following high-conflict trials [120, 121]. Further, the ACC is suggested to recruit lateral PFC to increase levels of control when conflict occurs [118]. This proposal for a cognitive control loop shares obvious similarities with concepts of monitoring and control in metacognition research (Fig. 11.1); indeed, a previous review proposed metacognition might be commensurate with cognitive control [122].

However, philosophers have discussed and debated two 'levels' of metacognition [123]: one involving declarative (conscious) meta-representation [124]; the other low-level, based on non-verbal epistemic feelings of uncertainty [125, 126]. For present purposes, we focus on monitoring processes that are consciously reportable, and thus available for deployment outside of a 'closed-loop' optimisation of the task at hand (see also [127]). Such reports can be empirically dissociated from monitoring and control, for example, skilled typists show subtle post-error adjustments in the absence of awareness, and yet accept blame for errors that are surreptitiously inserted by the experimenters on the screen [128]. Interestingly, subjective effects of heightened decision conflict may themselves be reportable in the absence of awareness of antecedents of this conflict [129], and thus it is not always simple to decide whether performance monitoring involves meta-representation.

What might govern the accessibility of performance monitoring information to awareness? We suggest that rostrolateral PFC is particularly important for the representation of information pertaining to a previous decision in a globally accessible frame of reference. In a direct comparison of confidence judgements following mnemonic and perceptual decisions, both ACC and right dlPFC activity increased with decreasing confidence [113]; however, only right dlPFC encoded confidence independent of changes in reaction time, leading the authors to suggest that while ACC responds to online decision conflict, dlPFC activity underlies metacognitive judgements. Furthermore, a recent study found that activity in rostrolateral PFC both increases during metacognitive reports and correlates with reported confidence [109]. Thus, accuracy of metacognitive commentaries, as dissociated from adjustments in performance, might be governed by the fidelity with which rlPFC integrates and maintains information from regions such as cingulate and insula involved in online adjustments in task performance, consistent with reciprocal anatomical connections between these regions [130].

If only a subset of nodes in this network is present, one might find effective performance monitoring in the absence of awareness. This pattern of results was observed in a patient with a large left prefrontal cortical lesion, who displayed intact performance adjustments in the Stroop task, without being able to report changes in the subjective sense of effort while performing the task [131]. As the patient displayed intact conflict-related N2 ERP responses during the Stroop task, the authors suggested that (implicit) monitoring and control is maintained by an intact right ACC, whilst a subjective feeling of effort would normally be mediated by the damaged lateral PFC. Such a conclusion is supported by recent evidence that lateral PFC activity is higher in subjects with a strong tendency to avoid cognitively demanding decisions [132]. Importantly for our hypothesis, if lateral PFC receives input from non-conscious monitoring loops, the reverse dissociation would not be predicted: we might be able to control objects we cannot report, but should not be able to report upon objects we cannot (cognitively) control.

The respective roles of distinct nodes in this network remain to be determined, but there is initial evidence for some form of division of labour. TMS to dlPFC impairs metacognition following correct but not incorrect decisions, suggesting a role in representing confidence rather than monitoring for response errors [19]. In contrast, reporting of response errors has been linked to the error-related positivity [133] with a possible source in insula cortex [119]. Indeed, accurate metacognitive commentaries about *performance* require access to information about both beliefs and responses. For example, after hitting a shot in tennis, you might have high confidence (low uncertainty) that the spot you chose to aim at is out of reach of your opponent (your belief), but low confidence in correctly executing the shot (your response). Thus, for commentaries to integrate information both about a belief and response, the 'frame of reference' in which information is encoded is crucial. If information is maintained in segregated sensorimotor loops, performance adjustments could be made based on deviations from an expected trajectory without this information being more generally available for, say, verbal report. It remains an open question as to the extent to which decision-making relies on 'embodied' or domain-general circuitry [134], but a role for the PFC in the abstract encoding of decision-related information, independent of response modality, has been found using fMRI conjunction analyses [135, 136]. It will be of interest to test whether this same activity is involved in metacognition.

11.5 Conclusions

Cognitive psychologists have developed a rich theoretical framework and empirical tools for studying self-assessments of cognition. A crucial variable of interest is the accuracy of metacognitive report with respect to object-level targets; in other words, how well do we know our own minds? A detailed, and eventually mechanistic, account of metacognition at the neural level is a necessary first step to understanding the failures of metacognition that occur following brain damage [87] and psychiatric disorder [137]. In this chapter, we summarised a variety of behavioural approaches for measuring the accuracy of metacognitive assessments, and reviewed the possible neural substrates of metacognitive accuracy in humans. We conclude that there are potentially separable brain systems for prospective and retrospective judgements of performance, and our synthesis of recent neuropsychological and brain imaging findings implicate the rostrolateral prefrontal cortex as important in mediating retrospective judgements of cognition. In this model, the rostrolateral PFC receives input from interoceptive cortex involved in 'closed-loop' monitoring and control, generating a metacognitive representation of the state of the system that can be deployed or reported outside of the current task at hand.

Acknowledgements Preparation of this chapter was supported by Wellcome Trust Programme Grant 078865/Z/05/Z to RJD, and a Sir Henry Wellcome Fellowship to SMF. We thank Matt Dixon, Chris Frith, Tali Sharot and Jon Simons for helpful comments on a previous draft of this manuscript.

References

1. Carruthers P (2011) The opacity of mind: an integrative theory of self-knowledge. Oxford University Press, USA
2. Nisbett RE, Wilson TD (1977) Telling more than we can know: verbal reports on mental processes. Psych Rev 84:231
3. Wilson TD, Dunn EW (2004) Self-knowledge: its limits, value, and potential for improvement. Ann Rev Psych 55:493–518
4. Johansson P, Hall L, Sikström S, Olsson A (2005) Failure to detect mismatches between intention and outcome in a simple decision task. Science 310:116–119
5. Hall L, Johansson P, Tärning B, Sikström S, Deutgen T (2010) Magic at the marketplace: choice blindness for the taste of jam and the smell of tea. Cognition 117:54–61
6. Pronin E (2007) Perception and misperception of bias in human judgment. Trends Cog Sci 11:37–43
7. Boring E (1953) A history of introspection. Psych Bull 50:169–189
8. James W (1950) The principles of psychology, vol 1. Dover Publications, New York
9. Friston K (2005) A theory of cortical responses. Phil Trans R Soc B 360:815–836
10. Koechlin E, Hyafil A (2007) Anterior prefrontal function and the limits of human decision-making. Science 318:594–598
11. Schmitz TW, Rowley HA, Kawahara TN, Johnson SC (2006) Neural correlates of self-evaluative accuracy after traumatic brain injury. Neuropsychologia 44:762–773
12. Metcalfe J, van Snellenberg J, DeRosse P, Balsam P Malhotra A (2014) Judgments of agency in schizophrenia: an impairment in autonoetic metacognition. In: Fleming SM, Frith C (eds) The cognitive neuroscience of metacognition. Springer, Berlin
13. Flavell J (1979) Metacognition and cognitive monitoring: a new area of cognitive–developmental inquiry. Am Psychol 34:906–911
14. Ericsson K, Simon H (1980) Verbal reports as data. Psych Rev 87:215–251
15. Nelson T (1984) A comparison of current measures of the accuracy of feeling-of-knowing predictions. Psych Bull 95:109–133
16. Nelson TO, Narens L (1990) Metamemory: a theoretical framework and new findings. Psychol Learn Motiv: Adv Res Theory 26:125–173
17. Shallice T, Burgess P (1996) The domain of supervisory processes and temporal organization of behaviour. Phil Trans R Soc B 351:1405–1411 discussion 1411–2

18. Del Cul A, Dehaene S, Reyes P, Bravo E, Slachevsky A (2009) Causal role of prefrontal cortex in the threshold for access to consciousness. Brain 132:2531
19. Rounis E, Maniscalco B, Rothwell J, Passingham R, Lau H (2010) Theta-burst transcranial magnetic stimulation to the prefrontal cortex impairs metacognitive visual awareness. Cogn Neurosci 1:165–175
20. Fleming SM, Weil RS, Nagy Z, Dolan RJ, Rees G (2010) Relating introspective accuracy to individual differences in brain structure. Science 329:1541–1543
21. Marti S, Sackur J, Sigman M, Dehaene S (2010) Mapping introspection's blind spot: reconstruction of dual-task phenomenology using quantified introspection. Cognition 115:303–313
22. Morsella E, Wilson LE, Berger CC, Honhongva M, Gazzaley A, Bargh JA (2009) Subjective aspects of cognitive control at different stages of processing. Atten Percept Psychophys 71:1807–1824
23. Dienes Z (2008) Subjective measures of unconscious knowledge. Prog Brain Res 168:49–64
24. Schooler JW (2002) Re-representing consciousness: dissociations between experience and meta-consciousness. Trends Cog Sci 6:339–344
25. Schwartz B, Bacon E (2008) Metacognitive neuroscience. In: Dunlosky J, Bjork R (eds) Handbook of memory and metamemory: essays in honor of Thomas O. Nelson. Psychology Press, New York, pp 355–371
26. Shimamura AP (2008) A neurocognitive approach to metacognitive monitoring and control. In: Dunlosky J, Bjork R (eds) Handbook of memory and metamemory: essays in honor of Thomas O. Nelson. Psychology Press, New York, pp 373–390
27. Arbuckle T (1969) Discrimination of item strength at time of presentation. J Exp Psych 8:126–131
28. Hart J (1965) Memory and the feeling-of-knowing experience. J Educ Psychol 56:208–216
29. Brown AS (1991) A review of the tip-of-the-tongue experience. Psych Bull 109:204–223
30. Peirce CS, Jastrow J (1885) On small differences in sensation. Mem Natl Acad Sci 3:73–83
31. Vickers D (1979) Decision processes in visual perception. Academic Press, New York
32. Eriksen CW (1960) Discrimination and learning without awareness: a methodological survey and evaluation. Psych Rev 67:279–300
33. Kunimoto C (2001) Confidence and accuracy of near-threshold discrimination responses. Conscious Cogn 10:294–340
34. Persaud N, McLeod P, Cowey A (2007) Post-decision wagering objectively measures awareness. Nat Neurosci 10:257–261
35. Hollard G, Massoni S, Vergnaud JC (2010) Subjective belief formation and elicitation rules: experimental evidence. Working paper
36. Becker GM, DeGroot MH, Marschak J (1964) Measuring utility by a single-response sequential method. Behav Sci 9:226–232
37. Dienes Z, Altmann G, Kwan L (1995) Unconscious knowledge of artificial grammars is applied strategically. J Exp Psychol Learn Mem Cogn 21:1322–1338
38. Macmillan N, Creelman C (2005) Detection theory: a user's guide. Lawrence Erlbaum, New York
39. Green D, Swets J (1966) Signal detection theory and psychophysics. Wiley, New York
40. Clarke F, Birdsall T, Tanner W (1959) Two types of ROC curves and definition of parameters. J Acoust Soc Am 31:629–630
41. Galvin SJ, Podd JV, Drga V, Whitmore J (2003) Type 2 tasks in the theory of signal detectability: discrimination between correct and incorrect decisions. Psych Bull Rev 10:843–876
42. Evans S, Azzopardi P (2007) Evaluation of a "bias-free" measure of awareness. Spat Vis 20:61–77
43. Fleming SM, Dolan RJ (2010) Effects of loss aversion on post-decision wagering: implications for measures of awareness. Conscious Cogn 19:352–363
44. Maniscalco B, Lau H (2012) A signal detection theoretic approach for estimating metacognitive sensitivity from confidence ratings. Conscious Cogn 21:422–430

45. Kersten D, Mamassian P, Yuille A (2004) Object perception as Bayesian inference. Ann Rev Psych 55:271–304
46. Kording K (2007) Decision theory: what 'should' the nervous system do? Science 318:606–610
47. Barthelmé S, Mamassian P (2009) Evaluation of objective uncertainty in the visual system. PLoS Comp Biol 5:1124–1131
48. Landy M, Goutcher R, Trommershäuser J, Mamassian P (2007) Visual estimation under risk. J Vis 7:4
49. Whiteley L, Sahani M (2008) Implicit knowledge of visual uncertainty guides decisions with asymmetric outcomes. J Vis 8:2
50. Trommershauser J, Maloney L, Landy M (2003) Statistical decision theory and trade-offs in the control of motor response. Spat Vis 16(3):255–275
51. Smith J, Couchman J, Beran M (2014) The highs and lows of theoretical interpretation in animal-metacognition research. In: Fleming SM, Frith C (eds) The cognitive neuroscience of metacognition. Springer, Berlin
52. Kepecs A, Uchida N, Zariwala HA, Mainen ZF (2008) Neural correlates, computation and behavioural impact of decision confidence. Nature 455:227–231
53. Kiani R, Shadlen M (2009) Representation of confidence associated with a decision by neurons in the parietal cortex. Science 324:759
54. Kepecs A, Mainen Z (2014) A computational framework for the study of confidence in humans and animals. In: Fleming SM, Frith C (eds) The cognitive neuroscience of metacognition. Springer, Berlin
55. Schwartz B, Metcalfe J (1996) Methodological problems and pitfalls in the study of human metacognition. In: Metcalfe J, Shimamura A (eds) Metacognition: knowing about knowing. MIT Press, Cambridge
56. Koriat A (1997) Monitoring one's own knowledge during study: a cue-utilization approach to judgments of learning. J Exp Psych Gen 126:349–370
57. Koriat A (2007) Metacognition and consciousness. In: Zelazo PD, Moscovitch M, Thompson E (eds) The Cambridge handbook of consciousness. Cambridge University Press, Cambridge, pp 289–325
58. Alter AL, Oppenheimer DM (2009) Uniting the tribes of fluency to form a metacognitive nation. Pers Soc Psychol Rev 13:219–235
59. Busey TA, Tunnicliff J, Loftus GR, Loftus EF (2000) Accounts of the confidence-accuracy relation in recognition memory. Psych Bull Rev 7:26–48
60. Koriat A (1993) How do we know that we know? The accessibility model of the feeling of knowing. Psych Rev 100:609–639
61. Baranski JV, Petrusic WM (1998) Probing the locus of confidence judgments: experiments on the time to determine confidence. J Exp Psych Hum Percept Perform 24:929–945
62. Clifford C, Arabzadeh E, Harris J (2008) Getting technical about awareness. Trends Cogn Sci 12:54–58
63. Insabato A, Pannunzi M, Rolls ET, Deco G (2010) Confidence-related decision making. J Neurophys 104:539–547
64. Rolls ET, Grabenhorst F, Deco G (2010) Choice, difficulty, and confidence in the brain. NeuroImage 53:694–706
65. Higham PA, Perfect TJ, Bruno D (2009) Investigating strength and frequency effects in recognition memory using type-2 signal detection theory. J Exp Psychol Learn Mem Cogn 35:57–80
66. Busey TA, Arici A (2009) On the role of individual items in recognition memory and metacognition: challenges for signal detection theory. J Exp Psychol Learn Mem Cogn 35:1123–1136
67. Lau HC, Passingham RE (2006) Relative blindsight in normal observers and the neural correlate of visual consciousness. Proc Natl Acad Sci USA 103:18763
68. Wilimzig C, Tsuchiya N, Fahle M, Einhäuser W, Koch C (2008) Spatial attention increases performance but not subjective confidence in a discrimination task. J Vis 8(7):1–10

69. Izaute M, Bacon E (2005) Specific effects of an amnesic drug: effect of lorazepam on study time allocation and on judgment of learning. Neuropsychopharmacology 30:196–204
70. Song C, Kanai R, Fleming SM, Weil RS, Schwarzkopf DS, Rees G (2011) Relating inter-individual differences in metacognitive performance on different perceptual tasks. Conscious Cogn 20:1787–1792
71. Yeung N, Summerfield C (2014) Metacognition in human decision making: confidence and error monitoring. In: Fleming SM, Frith C (eds) The cognitive neuroscience of metacognition. Springer, Berlin
72. Pleskac TJ, Busemeyer JR (2010) Two-stage dynamic signal detection: a theory of choice, decision time, and confidence. Psych Rev 117:864–901
73. Ratcliff R, Starns JJ (2009) Modeling confidence and response time in recognition memory. Psych Rev 116:59–83
74. Maniscalco B, Lau H (2009) Evaluating signal detection models of perceptual decision confidence. In: Cosyne Abstracts, Salt Lake City, USA
75. Gold J, Shadlen M (2007) The neural basis of decision making. Ann Rev Neurosci 30:535–574
76. Middlebrooks PG, Abzug ZM, Sommer MA (2014) Studying metacognitive processes at the single-neuron level. In: Fleming SM, Frith C (eds) The cognitive neuroscience of metacognition. Springer, Berlin
77. Kruger J, Dunning D (1999) Unskilled and unaware of it: how difficulties in recognizing one's own incompetence lead to inflated self-assessments. JPSP 77:1121–1134
78. Kao YC, Davis ES, Gabrieli JDE (2005) Neural correlates of actual and predicted memory formation. Nat Neurosci 8:1776–1783
79. Modirrousta M, Fellows LK (2008) Medial prefrontal cortex plays a critical and selective role in "feeling of knowing" meta-memory judgments. Neuropsychologia 46:2958–2965
80. Morgan M, Mason A (1997) Blindsight in normal subjects? Nature 385:401–402
81. Lau H (2010) Are we studying consciousness yet? In: Davies M, Weiskrantz L (eds) Frontiers of consciousness: chichele lectures. Oxford University Press, Oxford
82. Shimamura AP (2000) Toward a cognitive neuroscience of metacognition. Conscious Cogn 9:313–323
83. Hirst W (1982) The amnesic syndrome: descriptions and explanations. Psych Bull 91:435–460
84. Zahr N, Kaufman K (2011) Clinical and pathological features of alcohol-related brain damage. Nat Rev Neurol 7:284–294
85. Shimamura AP, Squire LR (1986) Memory and metamemory: a study of the feeling-of-knowing phenomenon in amnesic patients. J Exp Psychol Learn Mem Cogn 12:452–460
86. Janowsky JS, Shimamura AP, Kritchevsky M, Squire LR (1989) Cognitive impairment following frontal lobe damage and its relevance to human amnesia. Behav Neurosci 103:548
87. Pannu J, Kaszniak A (2005) Metamemory experiments in neurological populations: a review. Neuropsychol Rev 15:105–130
88. Schnyer DM, Verfaellie M, Alexander MP, LaFleche G, Nicholls L, Kaszniak AW (2004) A role for right medial prefontal cortex in accurate feeling-of-knowing judgements: evidence from patients with lesions to frontal cortex. Neuropsychologia 42:957–966
89. Pannu J, Kaszniak A, Rapcsak S (2005) Metamemory for faces following frontal lobe damage. J Int Neuropsychol Soc 11:668–676
90. Kikyo H, Ohki K, Miyashita Y (2002) Neural correlates for feeling-of-knowing. Neuron 36:177–186
91. Chua EF, Schacter DL, Rand-Giovannetti E, Sperling RA (2006) Understanding metamemory: neural correlates of the cognitive process and subjective level of confidence in recognition memory. NeuroImage 29:1150–1160
92. Chua EF, Schacter DL, Sperling RA (2009) Neural correlates of metamemory: a comparison of feeling-of-knowing and retrospective confidence judgments. J Cogn Neurosci 21:1751–1765

93. Kim H, Cabeza R (2007) Trusting our memories: dissociating the neural correlates of confidence in veridical versus illusory memories. J Neurosci 27:12190
94. Moritz S, Gläscher J, Sommer T, Büchel C, Braus DF (2006) Neural correlates of memory confidence. NeuroImage 33:1188–1193
95. Schnyer DM, Nicholls L, Verfaellie M (2005) The role of VMPC in metamemorial judgments of content retrievability. J Cogn Neurosci 17:832–846
96. Park HJ, Kim JJ, Lee SK, Seok JH, Chun J, Kim DI, Lee JD (2008) Corpus callosal connection mapping using cortical gray matter parcellation and DT-MRI. Hum Brain Mapp 29:503–516
97. Kanai R, Rees G (2011) The structural basis of inter-individual differences in human behaviour and cognition. Nat Rev Neurosci 12:231
98. Yokoyama O et al (2010) Right frontopolar cortex activity correlates with reliability of retrospective rating of confidence in short-term recognition memory performance. Neurosci Res 68:199–206
99. Ramnani N, Owen AM (2004) Anterior prefrontal cortex: insights into function from anatomy and neuroimaging. Nat Rev Neurosci 5:184
100. Semendeferi K, Teffer K, Buxhoeveden DP, Park MS, Bludau S, Amunts K, Travis K, Buckwalter J (2011) Spatial organization of neurons in the frontal pole sets humans apart from great apes. Cereb Cortex 21:1485–1497
101. Simons JS, Henson RNA, Gilbert SJ, Fletcher PC (2008) Separable forms of reality monitoring supported by anterior prefrontal cortex. J Cogn Neurosci 20:447–457
102. Yoshida W, Ishii S (2006) Resolution of uncertainty in prefrontal cortex. Neuron 50:781–789
103. Gilbert SJ, Spengler S, Simons JS, Frith CD, Burgess PW (2006) Differential functions of lateral and medial rostral prefrontal cortex (area 10) revealed by brain-behavior associations. Cereb Cortex 16:1783–1789
104. Buda M, Fornito A, Bergström ZM, Simons JS (2011) A specific brain structural basis for individual differences in reality monitoring. J Neurosci 31:14308–14313
105. Curtis C, D'Esposito M (2003) Persistent activity in the prefrontal cortex during working memory. Trends Cogn Sci 7:415–423
106. Sakai K, Rowe JB, Passingham RE (2002) Active maintenance in prefrontal area 46 creates distractor-resistant memory. Nat Neurosci 5:479–484
107. Scheperjans F, Hermann K, Eickhoff SB, Amunts K, Schleicher A, Zilles K (2008) Observer-independent cytoarchitectonic mapping of the human superior parietal cortex. Cereb Cortex 18:846–867
108. John JP, Yashavantha BS, Gado M, Veena R, Jain S, Ravishankar S, Csernansky JG (2007) A proposal for MRI-based parcellation of the frontal pole. Brain Str. Funct. 212:245–253
109. Fleming SM, Huijgen J, Dolan RJ (2012) Prefrontal contributions to metacognition in perceptual decision-making. J Neurosci 32:6117–6125
110. Smith R, Keramatian K, Christoff K (2007) Localizing the rostrolateral prefrontal cortex at the individual level. NeuroImage 36:1387–1396
111. Thompson WB, Mason SE (1996) Instability of individual differences in the association between confidence judgments and memory performance. Mem Cogn 24:226–234
112. Kelemen WL, Frost PJ, Weaver CA (2000) Individual differences in metacognition: evidence against a general metacognitive ability. Mem Cogn 28:92–107
113. Fleck MS, Daselaar SM, Dobbins IG, Cabeza R (2006) Role of prefrontal and anterior cingulate regions in decision-making processes shared by memory and nonmemory tasks. Cereb Cortex 16:1623–1630
114. McCurdy LY, Maniscalco B, Metcalfe J, Liu KY, de Lange FP, Lau H (2013) Anatomical coupling between distinct metacognitive systems for memory and visual perception. J Neurosci 33:1897–1906
115. Sharot T, Riccardi A, Raio C, Phelps EA (2007) Neural mechanisms mediating optimism bias. Nature 450:102–105

116. Hassabis D, Maguire E (2007) Deconstructing episodic memory with construction. Trends Cogn Sci 11:299–306
117. Botvinick MM, Braver TS, Barch DM, Carter CS, Cohen JD (2001) Conflict monitoring and cognitive control. Psych Rev 108:624–652
118. Ridderinkhof KR, Ullsperger M, Crone EA, Nieuwenhuis S (2004) The role of the medial frontal cortex in cognitive control. Science 306:443–447
119. Ullsperger M, Harsay HA, Wessel JR, Ridderinkhof KR (2010) Conscious perception of errors and its relation to the anterior insula. Brain Str. Funct. 214:629–643
120. MacDonald AW, Cohen JD, Stenger VA, Carter CS (2000) Dissociating the role of the dorsolateral prefrontal and anterior cingulate cortex in cognitive control. Science 288:1835–1838
121. Kerns JG, Cohen JD, MacDonald AW, Cho RY, Stenger VA, Carter CS (2004) Anterior cingulate conflict monitoring and adjustments in control. Science 303:1023–1026
122. Fernandez-Duque D, Baird JA, Posner MI (2000) Executive attention and metacognitive regulation. Conscious Cogn 9:288–307
123. Arango-Muñoz S (2010) Two levels of metacognition. Philosophia 39:71–82
124. Carruthers P (2009) How we know our own minds: the relationship between mindreading and metacognition. Behav Brain Sci 32:121–138
125. Proust J (2007) Metacognition and metarepresentation: is a self-directed theory of mind a precondition for metacognition? Synthese 159:271–295
126. Evans J (2008) Dual-processing accounts of reasoning, judgment, and social cognition. Annu Rev Psychol 59:255–278
127. Shea N, Heyes C (2010) Metamemory as evidence of animal consciousness: the type that does the trick. Biol Philos 25:95–110
128. Logan GD, Crump MJC (2010) Cognitive illusions of authorship reveal hierarchical error detection in skilled typists. Science 330:683–686
129. Wenke D, Fleming SM, Haggard P (2010) Subliminal priming of actions influences sense of control over effects of action. Cognition 115:26–38
130. Medalla M, Barbas H (2010) Anterior cingulate synapses in prefrontal areas 10 and 46 suggest differential influence in cognitive control. J Neurosci 30:16068–16081
131. Naccache L, Dehaene S, Cohen L, Habert M-O, Guichart-Gomez E, Galanaud D, Willer J-C (2005) Effortless control: executive attention and conscious feeling of mental effort are dissociable. Neuropsychologia 43:1318–1328
132. McGuire JT, Botvinick MM (2010) Prefrontal cortex, cognitive control, and the registration of decision costs. Proc Natl Acad Sci USA 107:7922–7926
133. Nieuwenhuis S, Ridderinkhof KR, Blom J, Band GPH, Kok A (2001) Error-related brain potentials are differentially related to awareness of response errors: evidence from an antisaccade task. Psychophysiology 38:752–760
134. Freedman DJ, Assad JA (2011) A proposed common neural mechanism for categorization and perceptual decisions. Nat Neurosci 14:143–146
135. Heekeren H, Marrett S, Ruff D, Bandettini P, Ungerleider L (2006) Involvement of human left dorsolateral prefrontal cortex in perceptual decision making is independent of response modality. Proc Natl Acad Sci USA 103:10023–10028
136. Ho TC, Brown S, Serences JT (2009) Domain general mechanisms of perceptual decision making in human cortex. J Neurosci 29:8675–8687
137. David AS, Bedford N, Wiffen B, Gilleen J (2014) Failures of metacognition and lack of insight in neuropsychiatric disorders. In: Fleming SM, Frith C (eds) The cognitive neuroscience of metacognition. Springer, Berlin

Chapter 12
The Cognitive Neuroscience of Metamemory Monitoring: Understanding Metamemory Processes, Subjective Levels Expressed, and Metacognitive Accuracy

Elizabeth F. Chua, Denise Pergolizzi and R. Rachel Weintraub

Abstract Metamemory has been broadly defined as knowledge of one's own memory. Based on a theoretical framework developed by Nelson and Narens (Psychol Learn Motiv 26:125–141, 1990), there has been a wealth of cognitive research that provides insight into how we make judgments about our memory. More recently, there has been a growing interest in understanding the neural mechanisms supporting metamemory monitoring judgments. In this chapter, we propose that a fuller understanding of the neural basis of metamemory monitoring involves examining which brain regions: (1) are involved in the process of engaging in a metamemory monitoring task, (2) modulate based on the subjective level of the metamemory judgment expressed, and (3) are sensitive to the accuracy of the metamemory judgment (i.e., when the subjective judgment is congruent with objective memory performance). Lastly, it is critical to understand how brain activation changes when metamemory judgments are based on different sources of information. Our review of the literature shows that, although we have begun to address the brain mechanisms supporting metamemory judgments, there are still many unanswered questions. The area with the most growth, however, is in understanding how patterns of activation are changed when metamemory judgments are based on different kinds of information.

E. F. Chua (✉) · D. Pergolizzi · R. R. Weintraub
Department of Psychology, Brooklyn College of the City University of New York, 2900 Bedford Ave, Brooklyn, NY 11210, USA
e-mail: echua@brooklyn.cuny.edu

D. Pergolizzi
e-mail: dpergolizzi@brooklyn.cuny.edu

R. R. Weintraub
e-mail: rweintraub@brooklyn.cuny.edu

E. F. Chua · D. Pergolizzi · R. R. Weintraub
Department of Psychology, The Graduate Center of the City University of New York, 365 5th Ave, New York, NY 10016, USA

S. M. Fleming and C. D. Frith (eds.), *The Cognitive Neuroscience of Metacognition*,
DOI: 10.1007/978-3-642-45190-4_12,

12.1 Introduction

Metamemory can be broadly defined as knowledge of one's own memory [58]. Research on metamemory has a long history in cognitive psychology, but in the past decade there has been a growing interest in understanding the neural mechanisms associated with metamemory (e.g., [16–19, 34, 45, 49, 51]). It has everyday relevance for patients with neurological and psychiatric disorders, and even the healthy aged, who have deficits in metamemory (e.g., [6, 60, 86]), for educators who want to promote learning (e.g., [48, 89]), for basic researchers interested in the fundamental computations carried out by specific brain areas and how they interact [16–19, 34, 50], and for people who swear they left their keys by the door only to find them in the kitchen. The goal of this chapter is to review the current literature on the cognitive neuroscience of metamemory monitoring, and to provide guidelines for future neuroimaging studies investigating metamemory.

12.1.1 Theoretical Framework of Metamemory: A Brief Overview

The guiding framework for studying metamemory is the Nelson and Narens [58] model that defined metamemory as the combination of *monitoring* and *control* processes (Fig. 12.1). Monitoring involves judging the success and/or progress of memory processing, and can be studied by asking for subjective introspections. Different kinds of monitoring occur during different stages of memory (e.g., encoding and retrieval), and thus there are several different tasks that ask for subjective reports at these different stages. The control component is the action-oriented component, which allows individuals to direct their behavior, typically by selecting information, choosing strategies, or ending processes. Extensive behavioral research has examined both monitoring and control processes, but the majority of research on the cognitive neuroscience of metamemory has focused on monitoring processes, which will be our main focus.

12.1.2 Metamemory Monitoring Tasks

Many tasks have been devised to probe metamemory monitoring at different stages during mnemonic processing ([58]; Fig.12.1). This can be done at a *global* level by asking people to make overall judgments about their memory performance (e.g., how many questions do you think you will get correct on the memory test that you will take shortly?), or at a trial-by-trial level by asking people to make a judgment after each trial. Judgments of learning (JOLs) are predictive judgments that are taken during, or shortly after, the study phase, and ask participants to judge how

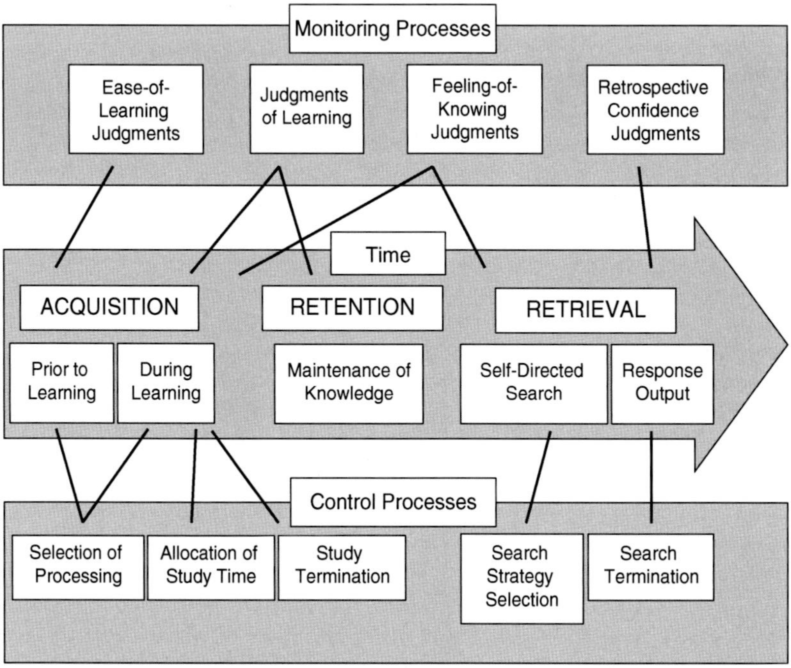

Fig. 12.1 The theoretical framework of monitoring and control processes in metamemory. Adapted with permission from Nelson and Narens [58]

likely they will later remember information. Feeling-of-knowing (FOK) judgments are taken during the retrieval phase and occur after a failed attempt at recall, and require participants to indicate how likely they are to recognize the information later. A related phenomenon is the Tip-of-the-Tongue (TOT) state, which has been distinguished based on its subjective feeling of imminent retrieval and accompanying frustration. On the postdictive side are retrospective confidence judgments (RCJs), which are made after recall or recognition and ask participants to indicate how certain they are that their responses are correct.

12.1.3 Measuring Metamemory Accuracy

Metamemory tasks require individuals to introspect about the contents of their mind and then make a subjective report (for review, see [3]). To get a measure of metamemory accuracy the discrepancy between the subjective reports and the objective results of the memory tests are compared (for review, see [7]). Broadly speaking, there are two main types of measurement, *absolute* and *relative* measures. Absolute measures, such as calibration curves and the Hamann index, are tied to the subjective rating scale used in the metamemory task and examine how well the subjective measures reflect performance on the memory task. However, people may use the scale differently, and may not be well calibrated, yet their performance does get better as their ratings increase, which shows a degree of metamemory accuracy in that there is some correlation between subjective ratings and objective performance. This is where *relative* measures of metamemory are important, and the most commonly used measure is the Goodman–Kruskal Gamma, a measures of association based on the difference between concordant and discordant pairs. More extensive reviews of measures of metamemory accuracy are reviewed elsewhere ([7, 60, 87]; Fleming and Dolan, this volume), and are beyond the scope of this chapter.

12.1.4 Tackling the Cognitive Neuroscience of Metamemory

Neuroimaging allows us to compare brain activity in many ways. We believe that this can bring us to a better understanding of the many facets of metamemory, and that it is necessary to understand: (1) brain regions that are involved in engaging *metamemory processes*, (2) brain regions that modulate based on the subjective *level* of the judgment expressed (e.g., high compared to low confidence), and (3) regions that correlate with *metamemory accuracy*, in which the subjective judgment and memory outcome are congruent (Table 12.1). Understanding the way the brain contributes to these three aspects of making metamemory judgments is fundamental in our attempts to understand metamemory and the brain. There is still much work to be done in understanding these three aspects across the different metamemory tasks. Once these basics are understood, the critical next steps are then to understand how these activations may change when metamemory decisions are based on different kinds of information (e.g., episodic vs. semantic, experiential vs. inferential). Although neuroimaging research at that level of specificity is currently rare, there are a few studies that meet this goal (e.g., [74, 85]). Finally, we will also examine common and distinct aspects of metamemory across tasks. Neuroimaging is correlational, so it is also necessary to examine metamemory in patients with brain lesions to determine if these brain regions are necessary for metamemory (Table 12.2), or to use brain stimulation techniques such as transcranial magnetic stimulation (TMS) or transcranial direct current stimulation (tDCS). Similar to their temporal order, we will begin with JOLs and end with RCJs.

12.2 Judgments of Learning

JOLs occur during or after learning, and are defined as predictions made about the ability to later retrieve information that is currently recallable [58]. In most JOL paradigms, participants study information and then give a subjective rating on how

Table 12.1 Brain regions associated with different aspects of metamemory using fMRI

Study	Judgment	Brain regions
Regions engaged in the processes of metamemory monitoring		
Do Lam et al. [45]	JOL	mPFC; OFC
Chua et al. [19]	FOK	vmPFC; DLPFC; PCC; lateral PPC; MTL
Chua et al. [17]	RCJ	DLPFC; precuneus; ventral PPC
Chua et al. [19]	RCJ	vmPFC; DLPFC; VLPFC; PCC; lateral PPC
Regions modulating by the subjective level of the judgment expressed		
Kao et al. [34]	JOL	VLPFC; vmPFC; amygdala; precuneus; lateral temporal; occipital cortex
Do Lam et al. [45]	JOL	mPFC; OFC; ACC
Chua et al. [19]	FOK	MTL; PCC; superior temporal; fusiform
Elman et al. [22]	FOK	VLPFC; DLPFC; mPFC; ventral PPC; PCC
Jing et al. [32]	FOK	DLPFC
Kikyo et al. [37]	FOK	VLPFC; DLPFC
Kikyo and Miyashita [36]	FOK	mPFC; aPFC; VLPFC; DLPFC; anterior temporal; lateral PPC
Maril et al. [49]	FOK	DLPFC; ACC; dorsal PPC
Maril et al. [50]	FOK	Medial and lateral parietal cortex
Schnyer et al. [74]	FOK	Medial and lateral PFC; PCC; MTL
Kikyo et al. [38]	TOT	DLPFC; ACC
Maril et al. [51]	TOT	mPFC; VLPFC; ACC; lateral temporal
Maril et al. [50]	TOT	aPFC; DLPFC; VLPFC; ACC
Chua et al. [17]	RCJ	mPFC; insula; PCC; MTL
Chua et al. [18]	RCJ	Lateral PFC; mPFC; ACC; MTL; PCC; lateral PPC
Hayes et al. [26]	RCJ	VLPFC; MTL; dorsal PPC
Henson et al. [28]	RCJ	DLPFC
Kim and Cabeza [39]	RCJ	Lateral PFC; PCC; Lateral PPC
Moritz et al. [56]	RCJ	ACC; PCC; MTL; dorsal PPC
Regions modulating by metamemory accuracy		
Kao et al. [34]	JOL	vmPFC
Schnyer et al. [74]	FOK	VLPFC; vmPFC; ACC; MTL
Yokoyama et al. [94]	RCJ	Fronto-polar cortex

Note JOL judgments of learning, *FOK* feeling-of-knowing, *RCJ* retrospective confidence judgments, *mPFC* medial prefrontal cortex, *OFC* orbitofrontal cortex, *DLPFC* dorsolateral prefrontal cortex, *VLPFC* ventrolateral prefrontal cortex, *vmPFC* ventromedial prefrontal cortex, *ACC* anterior cingulate cortex, *PPC* posterior parietal cortex, *PCC* posterior cingulated cortex, *aPFC* anterior prefrontal cortex

likely they think they are to recall the information later. These subjective ratings may be given immediately or after a delay. Typically, delayed JOLs are more accurate than immediate JOLs, which indicates that they may be based on different sources of information [57]. Immediate JOLs are thought to be based on monitoring working memory, whereas delayed JOLs are thought to be based on monitoring long-term memory [58].

Table 12.2 Neuropsychology of metamemory

Study	Lesion group	Comparison group	Spared	Impaired
Vilkki et al. [90]	R frontal	Controls		Word list recall; global JOLs
	R frontal	Posterior lesion	Word list recall	Global JOLs
	L frontal	Control		Word list recall; global JOLs
	L frontal	Posterior lesion	Global JOLs	Word list recall
Vilkki et al. [91]	R frontal	Controls	Face-location learning	Global JOL
	R frontal	Posterior lesions	Face-location learning	Global JOL
	L frontal	Controls	Face-location learning; global JOL	
	L Frontal	Posterior lesions	Face-location learning; global JOL	
Andres et al. [1]	TLE	Controls	Item JOL	Recall; recognition
Howard et al. [30]	TLE	Controls	Item JOL	Recall; recognition
Howard et al. [29]	TLE	Controls	Global and item JOLs	Recall; recognition
Bastin et al. [6]	FTD	Controls		Recall; recognition; FOK
Modirrousta and Fellows [55]	mPFC	Controls	JOL; RCJ	FOK
Pannu et al. [61]	Frontal	Controls	Recognition; FOK	RCJ
Perrotin et al. [62]	MCI	Controls	Commission errors (recall)	Omission errors (recall); FOK
Schmitter-Edgecombe and Anderson [73]	CHI	Controls	RCJ	Recall; recognition; FOK
Schnyer et al. [75]	Frontal	Controls	RCJ	Recall; recognition; FOK
Souchay et al. [86]	Older adults	Young controls	Semantic FOK	Recall; episodic FOK
Davidson et al. [21]	Parietal	Controls	Recognition	Recollection; RCJ
Simons et al. [83]	Parietal	Controls	Recognition	Recollection; RCJ

Note *JOL* judgments of learning, *FOK* feeling-of-knowing, *RCJ* retrospective confidence judgments, *TLE* temporal lobe epilepsy, *FTD* fronto-temporal dementia, *mPFC* medial prefrontal cortex, *MCI* mild cognitive impairment, *CHI* closed head injury

12.2.1 JOLs are Dissociable from Memory

A central question in metamemory research has been whether or not metamemory exists as something outside of memory. Neuropsychological studies have demonstrated dissociations between memory and JOLs, demonstrating that the two processes are at least partially independent [29, 90, 91] (Table 12.2). Converging evidence has also been found using event-related potentials (ERPs) [84] and functional magnetic resonance imaging (fMRI) [34, 45] (Table 12.1).

Patients with lesions to the frontal cortex tend to show impaired JOLs, either with intact memory [91], or impaired memory [90]. In one study, right frontal patients showed impaired global JOLs and intact memory for locations of faces, whereas patients with right posterior lesions showed intact global JOLs and impaired memory for locations of faces [91]. This double dissociation nicely demonstrates that memory and metamemory are separable, at least in the case of global JOLs (Cosentino, this volume). Similar work has been shown in word list learning paradigms with right frontal patients showing impairments in global JOLs compared to patients with posterior regions [90]. To further illustrate the double dissociation, patients with temporal lobe epilepsy (TLE) have impaired memory and intact JOLs [1, 29, 30]. Taken together, these findings suggest that the frontal cortex, and not the temporal lobes, is critical for JOLs.

Functional MRI and ERP studies also show that JOLs are separable from memory processes. Do Lam et al. [45] scanned participants during encoding and immediate JOLs, as separate tasks. There was a greater activity in the medial prefrontal cortex (mPFC) and orbitofrontal cortex (OFC) when participants made JOLs, even when encoding-related activity was masked out. Other regions showed similar levels of activity during JOLs and encoding. Interestingly, ERPs, which have excellent temporal resolution, have been able to narrow the similarities between JOLs and ERPs to early time windows, and differences between JOLs and encoding to late time windows [84]. Taken together, there is considerable evidence that JOLs cannot be reduced to memory processes alone.

In summary, the majority of evidence from neuroimaging and neuropsychology suggests that JOLs are separable from metamemory, and that the frontal cortex is important in JOLs and memory, whereas the temporal lobes are mainly important in memory processes and are not critical for JOLs. The nature of the processes carried out by the frontal cortex, and the specific locations within the frontal cortex, are informed by further data examining brain regions that modulate based on (1) the subjective level of JOL expressed, (2) the accuracy of the JOLs, and (3) the cognitive bases for JOLs.

12.2.2 The Rating Scale: Higher Versus Lower JOL

Critical to understanding the cognitive neuroscience of JOLs is examining the subjective rating expressed. TLE patients were able to use the JOL rating scale similarly to controls [29, 30], suggesting that other brain regions are responsible for signaling the subjective rating. Neuroimaging data are particularly valuable for this kind of understanding because they allow data analyses on a trial-by-trial basis, with the ability to directly compare trials with higher and lower ratings (Table 12.1). Kao et al. [34] used fMRI and examined which regions showed greater activity for predicted memory formation (i.e., High JOL vs. Low JOL) compared to actual memory formation (i.e., Remembered vs. Forgotten), and showed that the dorsal mPFC, anterior cingulate cortex (ACC), and left lateral prefrontal cortex (PFC) showed differences in activity for JOLs beyond actual memory performance. Similarly, in an associative memory task that isolated the JOL trial, there was greater activity in mPFC, OFC, and ACC for high versus low JOLs [45]. It is worth noting that the ACC activation reported by Kao et al. [34] is much more anterior (MNI coordinates: 4, 42, −4) than that reported by Do Lam et al. [45] (MNI coordinates: −4, 8, 30). Although fMRI work points most consistently to mPFC being modulated by JOL, topographic analyses of ERP data show a more centro-posterior component distinguishing high and low JOLs [84]. However, given the spatial resolution of ERPs and its poor ability to detect medial sources, it is unclear how these data map onto the fMRI data. Nevertheless, the fMRI literature suggests that activity in the mPFC modulates by the subjective JOL expressed most consistently, and other regions may vary based on stimuli and/or task demands.

As shown above, the most consistent region that modulated based on the subjective level of the JOL rating was the mPFC. Other non-metamemory fMRI studies have consistently shown greater activity in this region when engaged in self-referential processing (e.g., [27, 31, 35]), and also when reasoning about the mental states of others (e.g., [54, 71]). JOLs require introspecting on one's own memory, and it is likely that the activity in the mPFC reflects monitoring one's own mental state. Knowing which regions distinguish higher and lower JOLs is useful, but a fuller understanding of the cognitive neuroscience of JOLs is gained by examining the interactions between memory and JOLs.

12.2.3 Accurate JOL

One of the goals in research on metamemory is to determine the antecedents of accurate JOLs (i.e., when memory prediction is congruent with memory accuracy). The most common metamemory accuracy metric used is the Goodman–Kruskal Gamma coefficient. The ventral mPFC (vmPFC) was significantly correlated with the Gamma coefficient during accurate JOLs, but not inaccurate JOLs [34]. Findings from patients with frontal lesions have shown similar findings, with patients showing less accurate global JOLs compared to controls and patients with posterior lesions [90, 91]. However, patients with dysexecutive syndrome, presumably a prefrontal disorder, showed similar JOL accuracy, as measured by the Gamma coefficient, to controls [63]. Although there is some mixed evidence for the role of the frontal cortex in accurate JOLs, most evidence from fMRI (Table 12.1) and lesion studies (Table 12.2) suggest that the vMPFC is the most likely candidate for making accurate JOLs, but given the paucity of studies, further research is needed.

The vmPFC showed sensitivity to the subjective level of JOL rating expressed and JOL accuracy [34]. In other domains, the vmPFC has been implicated in encoding a value signal which then gets used for goal-directed behavior (e.g., [24]). Logically, it seems plausible that the vmPFC in JOLs reflects weighing of mnemonic evidence, analogous to its value, and deciding whether the information was learned well enough to warrant a particular judgment.

12.2.4 Basis of JOL

Critical to our understanding of JOLs is knowing what forms the basis of the judgment. Because JOLs are made about how well information is encoded, the most obvious basis for a JOL is the encoding phase. Kao et al. [34] used a masking procedure to examine which brain regions were activated for both actual encoding success and predicted encoding success, and showed that the lateral PFC was involved in both encoding and JOLs. This suggests that the same processes that contribute to successful encoding also contribute to JOLs. One possibility is that increased lateral PFC activity signals increased effort at encoding, which influences both the memory outcome and the JOL. Another hypothesized explanation for the lateral PFC activity was that it reflected partial retrieval of the target in working memory. Logically, this would mean that JOLs could be based, in part, on retrieval mechanisms.

In order to test the idea that retrieval operations influence JOLs, Do Lam et al. [45], used an inclusive masking approach to examine regions that showed increased activation associated with both memory predictions and successful recall. The mPFC was associated with both recall and JOLs. More generally, the mPFC has been implicated in performance monitoring (e.g., [69]), suggesting that in JOLs the vmPFC may be monitoring memory performance with respect to the recalled candidate information. Correspondingly, ERPs have also shown similar components for JOLs and recognition using a face task [85]. Thus, although JOLs are made in reference to encoding, it appears that individuals engage in retrieval processes to test how well they have learned information.

In addition to factors related to encoding and retrieval processes, more inferential factors may also influence JOLs. One study directly addressing this issue examined the role of distinctiveness in facial recognition and JOLs [85]. Two groups of subjects participated in a facial recognition task, with one group giving JOLs and the other giving distinctiveness ratings while ERPs were recorded. Behaviorally, individuals in the JOL group reported using distinctiveness to make their judgments, and distinctiveness ratings were just as predictive of recognition performance as JOLs. There were similar ERPs for distinctiveness ratings and

JOLs, and, assuming that similar patterns of brain activity reflect similar cognitive processes, JOLs in this paradigm were likely to have been based on distinctiveness. The use of distinctiveness may depend on the specific task, so further work using other tasks and other relevant factors is needed. Nevertheless, initial evidence shows the brain imaging can be an useful tool for investigating the bases of JOLs, and that, in addition to encoding and retrieval operations, inferential factors also influence JOLs.

12.3 Feeling-of-Knowing

Closely related to delayed JOLs is the feeling-of-knowing (FOK), but FOK differs from the delayed JOL in that it pertains only to non-recallable information. Similar to JOLs, FOK is a prospective metamemory judgment, and is made about future memory performance. The typical FOK paradigm uses a recall-judgment-recognition (RJR) task design [25]. In an RJR task, participants are asked to recall some target information. If they are unable to recall the information, participants make judgments of how likely they are to remember it at a later time, which constitutes the FOK judgment. Following the FOK judgment, they are presented with a recognition test and asked to choose the correct answer among a set of alternatives.

12.3.1 FOK is Dissociable from Memory

FOK has been shown to be a reasonable indicator of memory by demonstrating that having a high FOK for a non-recallable item results in a greater probability of successful subsequent recognition [25]. Because FOK has been related to memory accuracy, a critical question is whether FOK is merely an intermediate retrieval state between recognition and recall, or whether FOK judgments are dissociable from memory. One way to examine whether FOK is distinct from memory is to examine metamemory and memory performance in neuropsychological populations (e.g., [6, 61, 62, 73]) (Table 12.2). Several neuropsychological studies have highlighted the importance of the prefrontal cortex in FOK [6, 75]. Schnyer et al. [75] showed that frontal patients were impaired on FOK judgments compared to controls. However, these patients also showed worse memory performance, making it harder to interpret those data because there needs to be some minimum level of mnemonic information to monitor for metamemory to be accurate [41, 55]. Indeed, Schnyer et al. [75] performed covariate analyses and showed that memory did contribute to metamemory, but was not the only factor. Despite issues with memory accuracy, patients with the greatest impairments in FOK had more medial lesions, and they did not show the lowest performance. Another study was able to more directly test the role of the mPFC in FOK by examining instances when memory was matched in patients and controls, and the mPFC patients still showed significantly worse FOK accuracy compared to controls [55]. Although there are some ambiguities in the literature when memory performance differs between patients and controls, the majority of evidence suggests that memory and metamemory processes are dissociable and the mPFC is critical for accurate FOK judgments.

A second approach to examining whether memory and metamemory are separable is to use fMRI to compare metamemory and memory tasks [17–19] (Table 12.1). Compared to recognition, making an FOK judgment showed greater activity in the vmPFC, bilateral superior frontal, mid and posterior cingulate, and large lateral parietal/temporal area, including the inferior parietal lobule, the tempo-parietal junction (TPJ), and the superior temporal gyrus [19]. Collectively, these regions have been thought of as the "default" network (e.g., [11, 64, 65]), which has been implicated in internally directed thought (e.g., [52, 64]), mental simulation (e.g., [12]), and the self (e.g., [23]), all of which are relevant to FOK judgments. The vmPFC finding is consistent with lesion work, but it remains an open question whether a lesion in any part of this network would disrupt FOK judgments. Regardless, consistent converging evidence from lesion and fMRI studies indicate that FOK and memory are separable.

12.3.2 Levels of FOK

Another important aspect to understanding how the brain gives rise to metamemory is knowing which brain regions modulate based on the subjective level of FOK expressed (Table 12.1). Recent evidence has shown graded activation for FOK judgments, with greater activity for higher than lower levels of FOK [36, 37, 49, 74]. Earlier studies showed FOK as an intermediate level of activity between successful recall and failed recall in multiple prefrontal regions [37], or between successful recall and "don't know" response in left PFC, left posterior parietal cortex (PPC), and the ACC [49]. Similarly, comparisons of High versus Low FOK ratings showed greater activity in the ventrolateral PFC (VLPFC) and dorsolateral PFC (DLPFC) [19] and ventral PPC [22]. Few studies have reported greater activity for Low FOK compared to High FOK in any regions, but Elman et al. [22] showed this in the dorsal PPC. Similar patterns of ascending activity for higher compared to lower FOK ratings have been shown using a finer scale that rated FOK on a scale of 1–6 [36]. Several regions showed significant linear relationships to the FOK ratings including: the VLPFC, DLPFC, anterior PFC (aPFC), mPFC, cingulate cortex, as well as temporal and parietal regions. Altogether, these findings show the most consistent modulation of brain activity by subjective level occurs in the frontal cortices, with growing evidence that the PPC also modulates by subjective level.

12.3.3 Tip-of-the-Tongue: More than FOK?

Related to different levels of FOK is the tip-of-the-tongue (TOT) phenomenon, which is a subjective state that involves unsuccessful recall but a strong feeling that retrieval is imminent [10, 51, 77]; Diaz and Schwartz, this volume). Some TOT research focuses on linguistic aspects [80, 81], but here we will focus on metamemorial aspects. Kikyo et al. [38] made inferences about TOTs and suggested that TOTs elicited activation in the left DLPFC and the ACC. Maril et al. [51] used explicit behavioral responses to sort TOT trials and showed greater activation in the right middle frontal gyrus and ACC during TOT states compared to "Know" and "Don't Know" responses. Considering that the ACC has been implicated in conflict monitoring and effortful tasks (e.g., [9]), it is not surprising that it is active during TOT states, which are often described as frustrating and require effortful searching.

There is controversy over whether TOT and FOK are the same process, with TOT being a strong FOK (e.g., [3]), or whether the processes for FOK and TOT are distinct [50, 76]. One way to gain leverage on this controversy is to compare the neural correlates of TOT and FOK [38, 50]. When directly comparing FOK and TOT states, Maril et al. [50] found that TOT elicited greater activation in the ACC, right DLPC, right inferior PFC, and bilateral aPFC. However, both TOT and FOK elicited similar activation in the posterior medial parietal cortex and bilateral superior PFC. Taken together these findings suggest that, although there is some overlap with FOK, there are distinct brain regions responsible for the TOT state, most likely related to the feelings of frustration, conflict, imminent retrieval, or the decision to continue attempted retrieval.

12.3.4 Accurate FOK Judgments

In addition to studying the levels of FOK, researchers have also investigated accurate versus inaccurate FOK judgments. An accurate FOK judgment is consistent with performance at recognition, whereas an inaccurate FOK judgment is inconsistent with performance at recognition. Converging evidence from lesion [75] (Table 12.2) and fMRI studies [74] (Table 12.1) have implicated the mPFC in accurate FOK judgments. Specifically, frontal patients showed lower FOK accuracy, as measured by the Gamma correlation, a measure of relative accuracy, and Hamann index, a measure of absolute accuracy, compared to controls [75]. In a subsequent fMRI study, Schnyer et al. [74] compared accurate to inaccurate FOK judgments, and showed activation in a left hemisphere network of frontal and temporal cortical regions, including the medial and lateral frontal cortex, the hippocampus and parahippocampal gyrus, and the middle temporal gyrus. However, some areas in the right frontal cortex were also active, specifically the inferior frontal gyrus and the ACC.

Although Schnyer et al. [74] showed significant differences in activation for accurate and inaccurate FOK, several other studies that have tried to examine the full range of accurate and inaccurate FOK responses have not shown any significant differences [19, 32]. Indeed, when comparing High FOKs followed by correct recognition (an accurate FOK) to Low FOKs followed by incorrect recognition (an accurate feeling-of-not-knowing), there was greater activity in the left middle frontal gyrus [32]. This could explain why many studies that group different types of accurate FOKs (High FOKs followed by correct recognition and Low FOKs followed by inaccurate recognition) together and group different types of inaccurate FOKs together (High FOKs followed by incorrect recognition and Low FOK followed by correct recognition) often fail to find differences for accurate and inaccurate FOK.

12.3.5 Basis of FOK

A critical question that then arises is: on what are people basing these feelings-of-knowing or feelings-of-not-knowing? The leading hypotheses about how people make FOK judgments are: (1) cue familiarity (e.g., [53, 67, 78], (2) partial access to the sought-after information (e.g., [41, 42], or (3) a combination of cue familiarity and accessibility (e.g., [43]). Cognitive neuroscience research has only recently started to address these questions, and current evidence suggests that the mPFC plays a role in assessing accessibility of the retrieved information [74, 75]. Patients with mPFC damage who show impaired FOK, are able to make familiarity-based judgments, thus eliminating cue familiarity as a basis for their deficit in FOK [75]. More direct evidence comes from an fMRI study that used structural equation modeling to show that vMPFC activity was related to monitoring the outputs of retrieval, or content accessibility [74].

Different types of tasks, such as episodic versus semantic memory, appear to lead to FOK judgments based on different factors. Some participants show deficits in episodic FOK, but not semantic FOK [86]. Furthermore, evidence from fMRI shows that although some brain areas modulate based on level of FOK regardless of whether the information is episodic or semantic, other regions are task-specific [22, 68]. As one might expect, semantic FOKs activated the anterior temporal cortex, which has been associated with semantic knowledge, whereas strong episodic FOKs activated the ventral PPC, which is known to be involved in episodic retrieval [22], suggesting that the basis of FOK is task-sensitive. Broadly, this highlights the need to consider the bases of the FOK judgment across different studies.

12.4 Retrospective Confidence Judgments

Retrospective confidence judgments (RCJs) differ from prospective JOLs and FOKs, in that they require assessing one's confidence after recall or recognition. In neuroimaging, recognition tasks are more commonly used because of challenges in collecting verbal responses, so we will mainly focus on RCJs associated with recognition. Experimental tasks measure confidence by asking the participant to judge how confident they are in their recognition judgments, and either simultaneously with the recognition judgment (e.g., "sure old", [39, 40, 56]), or in a two-step process by asking the participant to rate their confidence immediately following a retrieval task [17–19]. Confidence judgments have been used outside of metamemory research, and have often been used in memory studies to assess the strength of the memory trace (e.g., [88]), or to separate recollection and familiarity (e.g., [66, 95, 97]). Although those studies are informative, their focus tends to make them difficult to interpret in terms of metamemory. Therefore, we have confined our review to studies that have a specific focus on confidence in recognition memory.

12.4.1 RCJs are Separable from Memory

Behavioral research shows that confidence and accuracy may be based on partially overlapping information (for review, see Busey et al. [14]), and are often positively correlated [46, 88, 95], raising the question of whether confidence and accuracy are separable. However, in several circumstances people report high confidence in memories that have never happened [47, 59, 70, 72]. Therefore, memory confidence and memory accuracy must rely, at least partly, on different information, and have different neural substrates.

The most direct evidence that recognition and confidence judgments are different processes comes from fMRI studies comparing these two tasks (Table 12.1). Compared to recognition, there was greater activity in bilateral PPC, insula, bilateral PFC, posterior cingulate cortex (PCC), and the right OFC during confidence judgments [17, 19]. These are similar to the "default network" [64] regions that were also involved in making FOK judgments.

Given that the fMRI studies highlight the parietal cortex in RCJs, a critical question is whether patients with lesions to the parietal cortex exhibit a dissociation between memory and metamemory. There is some anecdotal [21] and experimental [83] evidence that patients with parietal lesions may have impairments in retrospective confidence despite little or no impairment in memory tasks [8, 21, 82]. Experimentally, one parietal lesion patient, SM, showed lower conscious recollection rates compared to controls, using the "remember/know" paradigm [21]. From this finding, it is unclear whether this is a deficit in recollection or in the subjective experience associated with remembering. Anecdotal evidence from this patient suggests it is related to her subjective experience because (1) SM complained that she did not feel like she knew where her memories came from, (2) SM could not assess her confidence for the memories she retrieved, and (3) SM second-guessed many accurate recognition judgments, often asking for feedback on whether she was right or wrong.

In a study designed to tease apart recollection and subjective confidence, Simons et al. [83] showed a dissociation in memory accuracy and memory confidence in patients with parietal lesions. Across three experiments, patients and controls completed (1) old/new item recognition tasks with confidence judgments, and (2) source recollection tasks with confidence judgments. Patients with parietal lesions had significantly decreased confidence ratings compared to controls in the source recollection task, yet patients and controls had similar accuracy in their source judgments. Consistent with the idea that the parietal cortex is critical in subjective aspects of memory, such as confidence, TMS to the inferior parietal cortex showed a greater effect on subjective than objective memory performance [79]. Further brain stimulation studies that investigate the role of the frontal and parietal cortices in subjective confidence provide a promising avenue to investigate some of these issues further.

12.4.2 Regions that Modulate Based on the Subjective Confidence Level Expressed

In addition to understanding which brain regions are involved in the process of confidence assessment, it is also critical to know which brain regions modulate based on the subjective *level* of confidence expressed (Table 12.1). This includes understanding which regions show greater activity for high compared to low confidence responses, and which ones show greater activity for low compared to high confidence responses.

Similar to other metamemory judgments, regions in the PFC have been shown to modulate by the level of confidence expressed. However, unlike the other judgments, there is typically more prefrontal activity with lower levels of confidence [18, 26, 28]. Early work, using a single step design, implicated the lateral PFC with increased monitoring because there was greater activity in the right DLPFC for low confidence correct compared to high confidence correct item recognition judgments, regardless of whether the item was judged old or new [28]. Consistent with this, source memory paradigms have also shown greater activation in the VLPFC for low compared to high confidence recognition [26], and there were also greater evoked potentials (FN400) during low confidence than high confidence RCJs over the lateral PFC [93]. The previous studies were limited in that they examined only correct responses and used simultaneous confidence and recognition tasks, but Chua et al. [18] showed greater activity in the DLPFC, VLPFC, and ACC for low compared to high confidence responses for both correct and incorrect recognition in a two step judgment. Thus, these studies largely relate the PFC to low confidence recognition, a condition which is thought to require greater monitoring.

The PPC also plays an important role in signaling the subjective level of confidence expressed, and fMRI studies have indicated that superior and inferior parietal regions may play different roles in the way they signal subjective memory confidence (e.g., [18, 39]). Greater activity for high compared to low confidence in the *inferior* parietal cortex has been shown when individuals make separate confidence judgments [17], for true and false recognition [39], for hits and correct rejections [40], and for item and source memory [26]. In contrast, many of these studies have also shown greater activity in the *superior* parietal cortex for low compared to high confidence. Similarly, this holds true for studies of true recognition [39], item and source memory [26], for hits, misses, false alarms, and correct rejections [56], and hits and correct rejections [40]. Although there have been many consistencies, it is worth noting that the opposite patterns have been shown [18]. One interpretation of the typical inferior/superior distinction is that the parietal cortex is sensitive to the strength of memories. Differences that stray from these findings may reflect cases when individuals are using factors other than memory strength to make their confidence judgments. Future studies that manipulate the basis of the confidence judgment could shed light on some of these issues.

12.4.3 Basis of Recognition Confidence Judgments

Few neuroimaging studies have directly investigated confidence judgments based on different sources of information (e.g., memory strength vs. inferential processing; episodic vs. semantic; true vs. false recognition). Although there are many similarities in the regions that modulate by subjective confidence, the differences may reflect different bases for the confidence judgments. Kim and Cabeza [39] compared high and low confidence in memory for situations that were presumably based on different information: true and false recognition. False recognition in this paradigm relied on gist representations, thus allowing examination of high confidence false recognition when specific details of the remembered item are not present, which can then be compared to true recognition. For true recognition, frontal and parietal regions were significantly more activated for low confidence than high confidence responses, while medial temporal regions were significantly more activated for high confidence than low confidence responses. Conversely, for false recognition, medial temporal regions were significantly more activated for low confidence than high confidence, whereas frontal and parietal regions were significantly more activated during high confidence than low confidence [39]. Thus, patterns of activation associated with high and low confidence are related to the basis for those judgments.

12.4.4 Accuracy of the Confidence judgments

An outstanding issue is whether there are brain regions that contribute to accurate RCJs (i.e., confidence judgments that are congruent with memory accuracy). Because individuals vary in how well their confidence predicts their accuracy, we can correlate brain activity with the ability to make accurate confidence judgments (Table 12.1). Yokoyama et al. [94] first compared which regions activated more for RCJs compared to a control task (i.e., brightness discrimination). Second, they examined which voxels within these regions correlated with metamemory accuracy, as measured by the Gamma coefficient. The only region whose activity correlated with accurate metamemory performance was the right frontopolar cortex, suggesting this specific region is important in accurate self-monitoring. Although the frontopolar activity correlates across individuals, one necessary analysis is to compare accurate and inaccurate RCJs at the trial level within individuals. Thus far, many of these analyses have shown no significant differences in accurate and inaccurate trials [17–19]. Thus, it may be that variation in frontal function is related to the ability to make accurate RCJs, rather than frontopolar cortex being a signal for an accurate judgment.

To summarize, converging evidence from lesion, neuroimaging and electrophysiology consistently implicate the PFC and PPC during RCJs. These regions have various roles in making a confidence judgment, signaling the subjective level of confidence expressed, and leading to accurate confidence judgments. The prospective memory tasks discussed earlier mainly centered on prefrontal regions, and RCJs implicated the parietal cortex as also having an important role in metamemory. Explicitly comparing the neural substrates of the different tasks may help us elucidate the common neural mechanisms supporting the general demands involved in metamemory, and distinct mechanisms related to specific metamemory tasks.

12.5 Common and Distinct Metamemory Mechanisms

Thus far, we have examined three aspects of the brain mechanisms subserving three major metamemory tasks [19] by reviewing brain regions that: (1) are involved in metamemory *processes*, (2) modulate based on the subjective *level* of the judgment, and (3) correlate with *metamemory accuracy*. Next, we compare tasks across these three aspects (Tables 12.1 and 12.2).

12.5.1 Metamemory Processes

One hypothesis is that the process of metamemory monitoring, during which an individual turns his or her focus inward to the contents of memory, is consistent and reflects a universal component of metamemory. In contrast, it could be that the metamemory monitoring mechanisms differ depending on the task and the type of information being monitored.

There is good evidence that metamemory tasks share common processes (Table 12.1), most directly from comparisons of FOK and RCJ to non-metamemory tasks [19]. Both FOK and RCJ showed greater activity in the left and right TPJ, left and right superior temporal gyrus, vmPFC, and PPC, compared to recognition and attractiveness judgments. There was also consistently less activity in occipital, lateral PFC, and dmPFC during metamemory tasks compared to non-metamemory tasks. This indicates that there are common mechanisms supporting FOK and RCJ. It is more difficult to compare JOLs to these findings because JOLs are often compared to encoding, rather than retrieval. However, JOLs have been shown to activate the mPFC [45]. This provides indirect evidence that JOLs may also engage some of the same brain regions as other metamemory tasks, but further work is clearly needed.

The pattern of relative activations and deactivations seen when comparing metamemory monitoring and memory tasks suggests that the common processes of metamemory consist of shifting toward internal thoughts and away from external stimuli [19]; (see also Fox and Christoff, this volume). The vmPFC, lateral PPC, and PCC regions that showed greater activity for metamemory compared to non-metamemory tasks have previously been characterized as being part of the "default network" [11, 13, 23, 64]. Further characterization of the functions of the default network has implicated self-related processing, directing attention to internal processing, and mental simulation [12, 23, 64]. Metamemory is thought to engage all of these in the sense that it involves self-reflection, directing attention to internal thoughts and memory representations, and simulating the contents of memory. Furthermore, the regions that showed relative deactivations for metamemory compared to non-metamemory—less activation in the occipital cortex and the lateral prefrontal cortex—are consistent with less attention to the external environment [20].

In addition to shared mechanisms between metamemory tasks, fMRI has shown differential activation when directly comparing FOK and RCJ [19]. These likely reflect that FOK and RCJ are based on different sources of information. For example, there was greater activity in the left aPFC for RCJ than FOK, and greater activity in the hippocampus for FOK compared to RCJ [19], likely reflecting increased memory demands during the FOK task, which has been shown to rely, at least in part, on partial access to the to-be-retrieved information (e.g., [41]). Additionally, there was greater activity in the fusiform gyrus, which has been shown to be active during face processing (e.g., [33]), for FOK compared to RCJ. This likely reflects that in this paradigm, the cue was a face, and FOK may rely on cue familiarity (e.g., [53]).

12.5.2 Subjective Levels of the Metamemory Judgment

The next question is whether there are brain regions that modulate based on the subjective level of the metamemory judgment, regardless of the task. One possibility is that there are regions that signal overall certainty or doubt in one's memory, or may reflect that different amounts of monitoring occur under such conditions of certainty or doubt. However, it is also likely that the subjective judgment expressed is related to the specific bases of the different metamemory judgments, and would, therefore, differ across tasks (Table 12.1).

Direct comparisons have shown that different brain regions modulate based on the level of FOK or RCJ expressed. Chua et al. [19] showed greater activity in aVLPFC and aDLPFC for high compared to low FOK judgments. However, there were no differences based on the level of RCJ in these regions. These differences likely relate the fact that FOK and RCJ are based on different information.

However, comparisons across studies have suggested that some metamemory tasks may signal subjective level similarly. The lateral PPC may modulate based on both the level of FOK (e.g., [36, 49]) and RCJ (e.g., [26, 40, 56]). These effects may be more apparent in single task metamemory studies, and not in Chua et al. [19] because of increased power due to increased trial number. Both FOK and RCJs are given at retrieval and are thought to be based on either partial or full access to the sought after information. There are currently many theories being investigated about the role of the PPC in retrieval, some of which are very relevant to metamemory, including attention to memory, accumulation of mnemonic evidence, and decision making in relation to retrieval (for review, see [92]). Similarly, comparisons across studies suggest the subjective level expressed for FOK and JOLs may share common neural correlates in the PFC [19, 34, 36, 45, 49]. For JOL and FOKs, cue familiarity has been suggested as a shared monitoring process in JOLs (e.g., [44]) and FOK judgments (e.g., [78]), and the increasing PFC activity may relate to increasing familiarity (e.g., [96]).

On the whole, extant evidence suggests that there are no brain areas that signal certainty or doubt across different metamemory tasks. Instead, there are some commonalities across prospective tasks and across retrieval-based tasks, suggesting that brain regions that signal the subjective level of the judgment are specific to the basis of the judgment.

12.5.3 Metamemory Accuracy Across Types of Judgments

Anterior frontal regions have been implicated in metamemory accuracy, but these tend to be medial for JOLs [34] and FOK [55, 74, 75], and more lateral for RCJs [94] (Table 12.1). Consistent with the medial-lateral distinction, Schnyer et al. [75] showed that patients with more medial lesions were impaired on FOK, but intact on RCJ. In contrast, patients with more lateral lesions were impaired at RCJ, but not FOK [61]. In a study looking at patients with more circumscribed lesions on multiple metamemory tasks, Modirrousta and Fellows [55] showed that when patients with mPFC lesions were equated for memory performance with controls, patients were impaired on FOK, but not on RCJ or global JOLs. Given that the neuroimaging literature has suggested that trial-by-trial JOLs activate the mPFC, further work in patients with mPFC lesions on trial-by-trial JOLs is necessary for determining whether the mPFC is critical for JOLs, and prospective metamemory in general. Nevertheless, current evidence suggests that the mPFC plays a critical role in accurate trial-by-trial FOK judgments, and not RCJ, whereas its role in trial-by-trial JOLs remains unclear.

Broadly speaking, the aPFC seems to be a prime candidate for accurate metamemory. Current evidence suggests that the aPFC sits at the top of a hierarchy, coordinating signals from the DLPFC and VLPFC, making the aPFC ideal for monitoring and manipulating internally generated information [2, 13, 15], which is a key aspect of metamemory monitoring. The distinctions between lateral PFC for RCJ accuracy, and mPFC for JOL and FOK accuracy, most likely stems from the predictive nature of JOLs and FOKs. Indeed, recent research has suggested that the mPFC is important in generating predictions [4, 5]. Further research is clearly needed given the relatively few studies on the cognitive neuroscience of metamemory. However, the existing evidence suggests that more medial regions of the anterior prefrontal cortex may play a role in accurate predictions, whereas more lateral regions of the anterior prefrontal cortex play a role in accurate postdictions (see also, Fleming and Dolan, this volume).

12.6 Conclusions

The Nelson and Narens [58] model has been extremely useful in providing a structure for excellent experimental work in cognitive psychology. Now, we are beginning to be able to understand how metamemory is represented in the brain. In this chapter, we laid out what we believe to be the critical pieces in understanding how the brain gives rise to metamemory. First, we must understand what brain regions are used during the act of engaging in the *process* of metamemory monitoring. Second, we need to understand which brain regions modulate based on the subjective *level* of the metamemory judgment. Third, we need to understand whether there are brain regions that signal an *accurate* metamemory judgment. Bringing these three together, we then need to know how the brain regions involved in process, level, and accuracy change when judgments are based on different sources of information. Comparing different metamemory judgments gives us some leverage on this, but we also need to explain why there may be different bases for the final judgment within a particular judgment class. We encourage researchers to expand on our current understanding of metamemory monitoring processes in the brain at these different levels, as well as aspects of metamemory related to strategies and other control processes.

References

1. Andrés P, Mazzoni G, Howard CE (2010) Preserved monitoring and control processes in temporal lobe epilepsy. Neuropsychology 24:775–786. doi:10.1037/a0020281
2. Badre D (2008) Cognitive control, hierarchy, and the rostro-caudal organization of the frontal lobes. Trends Cogn Sci 12:193–200. doi:10.1016/j.tics.2008.02.004
3. Bahrick HP (2008) Thomas O. Nelson: his life and comments on implications of his functional view of metacognitive memory monitoring. In: Dunlosky J, Bjork RA (eds) Handbook of memory and metamemory. Psychology Press, New York, pp 1–10

4. Bar M (2007) The proactive brain: using analogies and associations to generate predictions. Trends Cogn Sci 11:280–289. doi:10.1016/j.tics.2007.05.005
5. Bar M (2009) The proactive brain: memory for predictions. Philos Trans R Soc Lond B Biol Sci 364:1235–1243. doi:10.1098/rstb 2008.0310
6. Bastin C, Feyers D, Souchay C et al (2012) Frontal and posterior cingulate metabolic impairment in the behavioral variant of frontotemporal dementia with impaired autonoetic consciousness. Hum Brain Mapp 33:1268–1278. doi:10.1002/hbm.21282
7. Benjamin AS, Diaz M (2008) Measurement of relative metamnemoic accuracy. In: Dunlosky J, Bjork RA (eds) Handbook of memory and metamemory. Psychology Press, New York, pp 73–94
8. Berryhill ME (2012) Insights from neuropsychology: pinpointing the role of the posterior parietal cortex in episodic and working memory. Front Integr Neurosci 6:31. doi:10.3389/fnint.2012.00031
9. Botvinick MM, Cohen JD, Carter CS (2004) Conflict monitoring and anterior cingulate cortex: an update. Trends Cogn Sci 8:539–546. doi:10.1016/j.tics.2004.10.003
10. Brown R, Mcneill D (1966) The "tip of the tongue" phenomenon. J Verbal Learn Verbal Behav 7:325–337
11. Buckner RL, Andrews-Hanna JR, Schacter DL (2008) The brain's default network: anatomy, function, and relevance to disease. Ann N Y Acad Sci 1124:1–38. doi:10.1196/annals.1440.011
12. Buckner RL, Carroll DC (2007) Self-projection and the brain. Trends Cogn Sci 11:49–57. doi:10.1016/j.tics.2006.11.004
13. Burgess P, Dumontheil I, Gilbert S (2007) The gateway hypothesis of rostral prefrontal cortex (area 10) function. Trends Cogn Sci 11:217–248
14. Busey TA, Tunnicliff J, Loftus GR, Loftus EF (2000) Accounts of the confidence-accuracy relation in recognition memory. Psychon Bull Rev 7:26–48
15. Christoff K, Gabrieli JDE (2000) The frontopolar cortex and human cognition: evidence for a rostrocaudal hierarchical organization within the human prefrontal cortex. Psychobiol Austin 28:168–186
16. Chua EF, Rand-Giovannetti E, Schacter DL et al (2004) Dissociating confidence and accuracy: functional magnetic resonance imaging shows origins of the subjective memory experience. J Cogn Neurosci 16:1131–1142. doi:10.1162/0898929041920568
17. Chua EF, Schacter DL, Rand-Giovannetti E, Sperling RA (2006) Understanding metamemory: neural correlates of the cognitive process and subjective level of confidence in recognition memory. NeuroImage 29:1150–1160. doi:10.1016/j.neuroimage.2005.09.058
18. Chua EF, Schacter DL, Sperling RA (2009) Neural basis for recognition confidence in younger and older adults. Psychol Aging 24:139–153. doi:10.1037/a0014029
19. Chua EF, Schacter DL, Sperling RA (2009) Neural correlates of metamemory: a comparison of feeling-of-knowing and retrospective confidence judgments. J Cogn Neurosci 21:1751–1765. doi:10.1162/jocn.2009.21123
20. Corbetta M, Shulman GL (2002) Control of goal-directed and stimulus-driven attention in the brain. Nat Rev Neurosci 3:201–215. doi:10.1038/nrn755
21. Davidson PSR, Anaki D, Ciaramelli E et al (2008) Does lateral parietal cortex support episodic memory? Evidence from focal lesion patients. Neuropsychologia 46:1743–1755. doi:10.1016/j.neuropsychologia.2008.01.011
22. Elman JA, Klostermann EC, Marian DE et al (2012) Neural correlates of metacognitive monitoring during episodic and semantic retrieval. Cogn Affect Behav Neurosci 12:599–609. doi:10.3758/s13415-012-0096-8
23. Gusnard DA, Akbudak E, Shulman GL, Raichle ME (2001) Medial prefrontal cortex and self-referential mental activity: relation to a default mode of brain function. Proc Natl Acad Sci USA 98:4259–4264. doi:10.1073/pnas.071043098
24. Hare TA, Camerer CF, Rangel A (2009) Self-control in decision-making involves modulation of the vmPFC valuation system. Science 324:646–648. doi:10.1126/science.1168450
25. Hart JT (1965) Memory and the feeling-of-knowing. J Edu Psychol 56:208–216

26. Hayes SM, Buchler N, Stokes J et al (2011) Neural correlates of confidence during item recognition and source memory retrieval: evidence for both dual-process and strength memory theories. J Cogn Neurosci 23:3959–3971. doi:10.1162/jocn_a_00086
27. Heatherton TF, Wyland CL, Macrae CN et al (2006) Medial prefrontal activity differentiates self from close others. Social Cogn Affect Neurosci 1:18–25. doi:10.1093/scan/nsl001
28. Henson RN, Rugg MD, Shallice T, Dolan RJ (2000) Confidence in recognition memory for words: dissociating right prefrontal roles in episodic retrieval. J Cogn Neurosci 12:913–923
29. Howard CE, Andrés P, Broks P et al (2010) Memory, metamemory and their dissociation in temporal lobe epilepsy. Neuropsychologia 48:921–932. doi:10.1016/j.neuropsychologia.2009.11.011
30. Howard CE, Andrés P, Mazzoni G (2013) Metamemory in temporal lobe epilepsy: a study of sensitivity to repetition at encoding. J Int Neuropsychol Soc (JINS) 19:1–10. doi:10.1017/S1355617712001646
31. Jenkins AC, Mitchell JP (2011) Medial prefrontal cortex subserves diverse forms of self-reflection. Soc Neurosci 6:211–218. doi:10.1080/17470919.2010.507948
32. Jing L, Niki K, Xiaoping Y, Yue L (2004) Knowing that you know and knowing that you don't know: a fMRI Study on feeling of knowing (FOK). Acta Psychol 36:426–433
33. Kanwisher N, McDermott J, Chun MM (1997) The fusiform face area: a module in human extrastriate cortex specialized for face perception. J Neurosci 17:4302–4311
34. Kao Y-C, Davis ES, Gabrieli JDE (2005) Neural correlates of actual and predicted memory formation. Nat Neurosci 8:1776–1783. doi:10.1038/nn1595
35. Kelley WM, Macrae CN, Wyland CL et al (2002) Finding the self? An event-related fMRI study. J Cogn Neurosci 14:785–794. doi:10.1162/08989290260138672
36. Kikyo H, Miyashita Y (2004) Temporal lobe activations of "feeling-of-knowing" induced by face-name associations. NeuroImage 23:1348–1357. doi:10.1016/j.neuroimage.2004.08.013
37. Kikyo H, Ohki K, Miyashita Y (2002) Neural correlates for feeling-of-knowing: an fMRI parametric analysis. Neuron 36:177–186
38. Kikyo H, Ohki K, Sekihara K (2001) Temporal characterization of memory retrieval processes: an fMRI study of the 'tip of the tongue' phenomenon. Eur J Neurosci 14:887–892
39. Kim H, Cabeza R (2007) Trusting our memories: dissociating the neural correlates of confidence in veridical versus illusory memories. J Neurosci 27:12190–12197. doi:10.1523/JNEUROSCI.3408-07.2007
40. Kim H, Cabeza R (2009) Common and specific brain regions in high-versus low-confidence recognition memory. Brain Res 1282:103–113. doi:10.1016/j.brainres.2009.05.080.Common
41. Koriat A (1993) How do we know that we know? The accessibility model of the feeling of knowing. Psychol Rev 100:609–639
42. Koriat A (1995) Dissociating knowing and the feeling of knowing: further evidence for the accessibility model. J Exp Psychol Gen 124:311–333
43. Koriat A, Levy-Sadot R (2001) The combined contributions of the cue-familiarity and accessibility heuristics to feelings of knowing. J Exp Psychol Learn Mem Cogn 27:34–53. doi:10.1037//0278-7393.27.1.34
44. Koriat A, Ma'ayan H (2005) The effects of encoding fluency and retrieval fluency on judgments of learning. J Mem Lang 52:478–492. doi:10.1016/j.jml.2005.01.001
45. Do Lam ATA, Axmacher N, Fell J et al (2012) Monitoring the mind: the neurocognitive correlates of metamemory. PloS One 7:e30009. doi:10.1371/journal.pone.0030009
46. Lindsay DS, Read JD, Sharma K (1998) Accuracy and confidence in person identification: the relationship is strong when witnessing conditions vary widely. Psychol Sci 9:215–218. doi:10.1111/1467-9280.00041
47. Loftus EF, Pickrell JE (1995) The formation of false memories. Psychiatr Ann 25:720–725
48. Maki RH, Willmon C, Pietan A (2009) Basis of metamemory judgments for text with multiple-choice, essay and recall tests. Appl Cogn Psychol 23:204–222. doi:10.1002/acp.1440
49. Maril A, Simons JS, Mitchell JP et al (2003) Feeling-of-knowing in episodic memory: an event-related fMRI study. NeuroImage. doi:10.1016/S1063-8119(03)00014-4

50. Maril A, Simons JS, Weaver JJ, Schacter DL (2005) Graded recall success: an event-related fMRI comparison of tip of the tongue and feeling of knowing. NeuroImage 24:1130–1138. doi:10.1016/j.neuroimage.2004.10.024
51. Maril A, Wagner AD, Schacter DL (2001) On the tip of the tongue: an event-related fMRI study of semantic retrieval failure and cognitive conflict. Neuron 31:653–660
52. Mason MF, Norton MI, Van Horn JD et al (2007) Wandering minds: the default network and stimulus-independent thought. Science 315:393–395. doi:10.1126/science.1131295
53. Metcalfe J, Schwartz BL, Joaquim SG (1993) The cue-familiarity heuristic in metacognition. J Exp Psychol Learn Mem Cogn 19:851–861
54. Mitchell JP, Macrae CN, Banaji MR (2006) Dissociable medial prefrontal contributions to judgments of similar and dissimilar others. Neuron 50:655–663. doi:10.1016/j.neuron.2006.03.040
55. Modirrousta M, Fellows LK (2008) Medial prefrontal cortex plays a critical and selective role in "feeling of knowing" meta-memory judgments. Neuropsychologia 46:2958–2965. doi:10.1016/j.neuropsychologia.2008.06.011
56. Moritz S, Gläscher J, Sommer T et al (2006) Neural correlates of memory confidence. NeuroImage 33:1188–1193. doi:10.1016/j.neuroimage.2006.08.003
57. Nelson TO, Dunlosky J (1991) When people's judgments of learning (JOLs) are extremely accurate at predicting subsequent recall: the "delayed-JOL effect". Psychol Sci 2:267–270. doi:10.1111/j.1467-9280.1991.tb00147.x
58. Nelson TO, Narens L (1990) Metamemory: a theoretical framework and new findings. Psychol Learn Motiv 26:125–141
59. Norman KA, Schacter DL (1997) False recognition in younger and older adults: exploring the characteristics of illusory memories. Mem Cogn 25:838–848
60. Pannu JK, Kaszniak AW (2005) Metamemory experiments in neurological populations: a review. Neuropsychol Rev 15:105–130. doi:10.1007/s11065-005-7091-6
61. Pannu JK, Kaszniak AW, Rapcsak SZ (2005) Metamemory for faces following frontal lobe damage. J Int Neuropsychol Soc (JINS) 11:668–676. doi:10.1017/S1355617705050873
62. Perrotin A, Belleville S, Isingrini M (2007) Metamemory monitoring in mild cognitive impairment: evidence of a less accurate episodic feeling-of-knowing. Neuropsychologia 45:2811–2826. doi:10.1016/j.neuropsychologia.2007.05.003
63. Pinon K, Allain P, Kefi MZ et al (2005) Monitoring processes and metamemory experience in patients with dysexecutive syndrome. Brain Cogn 57:185–188. doi:10.1016/j.bandc.2004.08.042
64. Raichle ME, MacLeod aM, Snyder aZ et al (2001) A default mode of brain function. Proc Natl Acad Sci USA 98:676–682. doi:10.1073/pnas.98.2.676
65. Raichle ME, Snyder AZ (2007) A default mode of brain function: a brief history of an evolving idea. NeuroImage 37:1083–1090; discussion 1097–1099. doi:10.1016/j.neuroimage.2007.02.041
66. Ranganath C, Yonelinas AP, Cohen MX et al (2004) Dissociable correlates of recollection and familiarity within the medial temporal lobes. Neuropsychologia 42:2–13. doi:10.1016/j.neuropsychologia.2003.07.006
67. Reder L, Ritter F (1992) What determines initial feeling of knowing? Familiarity with question terms, not with the answer. J Exp Psychol 18:435
68. Reggev N, Zuckerman M, Maril A (2011) Are all judgments created equal? An fMRI study of semantic and episodic metamemory predictions. Neuropsychologia 49:1332–1342. doi:10.1016/j.neuropsychologia.2011.01.013
69. Ridderinkhof KR, van den Wildenberg WPM, Segalowitz SJ, Carter CS (2004) Neurocognitive mechanisms of cognitive control: the role of prefrontal cortex in action selection, response inhibition, performance monitoring, and reward-based learning. Brain Cogn 56:129–140. doi:10.1016/j.bandc.2004.09.016
70. Roediger H, McDermott K (1995) Creating false memories: remembering words not presented in lists. J Exp Psychol: Learn Mem Cogn 21:803–814

71. Saxe R, Moran JM, Scholz J, Gabrieli J (2006) Overlapping and non-overlapping brain regions for theory of mind and self reflection in individual subjects. Soc Cogn Affect Neurosci 1:229–234. doi:10.1093/scan/nsl034
72. Schacter DL, Dodson CS (2001) Misattribution, false recognition and the sins of memory. Philos Trans R Soc Lond B Biol Sci 356:1385–1393
73. Schmitter-Edgecombe M, Anderson JW (2008) Feeling-of-knowing in episodic memory following moderate-to-severe closed-head injury. Neuropsychologia 21:224–234
74. Schnyer DM, Nicholls L, Verfaellie M (2005) The role of VMPC in metamemorial judgments of content retrievability. J Cogn Neurosci 17:832–846
75. Schnyer DM, Verfaellie M, Alexander MP et al (2004) A role for right medial prefontal cortex in accurate feeling-of-knowing judgements: evidence from patients with lesions to frontal cortex. Neuropsychologia 42:957–966. doi:10.1016/j.neuropsychologia.2003.11.020
76. Schwartz BL (2008) Working memory load differentially affects tip-of-the-tongue states and feeling-of-knowing judgments. Mem Cogn 36:9–19. doi:10.3758/MC.36.1.9
77. Schwartz BL, Metcalfe J (2011) Tip-of-the-tongue (TOT) states: retrieval, behavior, and experience. Mem Cogn 39:737–749. doi:10.3758/s13421-010-0066-8
78. Schwartz BL, Metcalfe J (1992) Cue familiarity but not target retrievability enhances feeling-of-knowing judgments. J Exp Psychol Learn Mem Cogn 18:1074–1083
79. Sestieri C, Capotosto P, Tosoni A et al (2013) Interference with episodic memory retrieval following transcranial stimulation of the inferior but not the superior parietal lobule. Neuropsychologia 51:900–906
80. Shafto MA, Burke DM, Stamatakis EA et al (2008) On the tip-of-the-tongue: neural correlates of increased word-finding failures in normal aging. J Cogn Neurosci 19:2060–2070
81. Shafto MA, Stamatakis EA, Tam PP, Tyler LK (2009) Word retrieval failures in old age: the relationship between structure and function. J Cogn Neurosci 22:1530–1540. doi:10.1162/jocn.2009.21321
82. Simons JS, Mayes AR (2008) What is the parietal lobe contribution to human memory? Neuropsychologia 46:1739–1742. doi:10.1016/j.neuropsychologia.2008.05.001
83. Simons JS, Peers PV, Mazuz YS et al (2010) Dissociation between memory accuracy and memory confidence following bilateral parietal lesions. Cereb Cortex 20:479–485. doi:10.1093/cercor/bhp116
84. Skavhaug I-M, Wilding EL, Donaldson DI (2010) Judgments of learning do not reduce to memory encoding operations: event-related potential evidence for distinct metacognitive processes. Brain Res 1318:87–95. doi:10.1016/j.brainres.2009.11.047
85. Sommer W, Heinz A, Leuthold H (1995) Metamemory, distinctiveness, and event-related potentials in recognition memory for faces. Mem Cogn 23:1–11
86. Souchay C, Moulin CJA, Clarys D et al (2007) Diminished episodic memory awareness in older adults: evidence from feeling-of-knowing and recollection. Conscious Cogn 16:769–784. doi:10.1016/j.concog.2006.11.002
87. Spellman BA, Bloomfield A, Bjork RA (2008) Measuring memory and metamemory. In: Dunlosky J, Bjork RA (eds) Handbook of memory and metamemory. Psychology Press, New York, pp 95–116
88. Stretch V, Wixted JT (1998) Decision rules for recognition memory confidence judgments. J Exp Psychol Learn Mem Cogn 24:1397–1410
89. Thiede KW, Anderson MCM, Therriault D (2003) Accuracy of metacognitive monitoring affects learning of texts. J Educ Psychol 95:66–73. doi:10.1037/0022-0663.95.1.66
90. Vilkki J, Servo A, Surma-aho O (1998) Word list learning and prediction of recall after frontal lobe lesions. Neuropsychology 12:268–277
91. Vilkki J, Surma-aho O, Servo A (1999) Inaccurate prediction of retrieval in a face matrix learning task after right frontal lobe lesions. Neuropsychology 13:298–305
92. Wagner AD, Shannon BJ, Kahn I, Buckner RL (2005) Parietal lobe contributions to episodic memory retrieval. Trends Cogn Sci 9:445–453. doi:10.1016/j.tics.2005.07.001
93. Woroch B, Gonsalves BD (2010) Event-related potential correlates of item and source memory strength. Brain Res 1317:180–191. doi:10.1016/j.brainres.2009.12.074

94. Yokoyama O, Miura N, Watanabe J et al (2010) Right frontopolar cortex activity correlates with reliability of retrospective rating of confidence in short-term recognition memory performance. Neurosci Res 68:199–206. doi:10.1016/j.neures.2010.07.2041
95. Yonelinas AP (2001) Consciousness, control, and confidence: the 3 Cs of recognition memory. J Exp Psychol Gen 130:361–379
96. Yonelinas AP, Otten LJ, Shaw KN, Rugg MD (2005) Separating the brain regions involved in recollection and familiarity in recognition memory. J Neurosci 25:3002–3008. doi:10.1523/JNEUROSCI.5295-04.2005
97. Yonelinas AP, Parks CM (2007) Receiver operating characteristics (ROCs) in recognition memory: a review. Psychol Bull 133:800–832

Chapter 13
Metacognitive Facilitation of Spontaneous Thought Processes: When Metacognition Helps the Wandering Mind Find Its Way

Kieran C. R. Fox and Kalina Christoff

Abstract Mind wandering (MW) and metacognition may give the impression of lying at the opposite poles of the spectrum of human cognition. MW involves undirected, spontaneous thought processes that often occur without our volition and sometimes despite our intentions. Metacognition, by contrast, involves the conscious, often intentional monitoring and evaluation of our own mental processes and behaviors. The neural correlates of MW and metacognition may also appear strictly distinct at first, considering the almost exclusive focus on default network regions' involvement in MW, in contrast to the emphasis on higher order prefrontal regions' role in metacognitive processing. In this chapter, we will argue that despite the apparent gulf between MW and metacognition, some of the most intriguing mental phenomena we humans are capable of experiencing involve an intimate, dynamic interplay between MW and metacognition. According to the standard view of their interaction, metacognition serves to correct the wandering mind, suppressing spontaneous thoughts and bringing attention back to more "worthwhile" tasks. In this chapter, we argue that this "negative" or suppressant view of their interactions represents only a part of the whole picture. Instead, we outline and discuss three examples of positive, facilitative interactions: creative thinking, mindfulness meditation, and lucid dreaming (being aware that one is dreaming while dreaming). We argue that at both the cognitive and neural levels, these phenomena appear to involve an intricate balance whereby spontaneous thought is allowed to arise naturally while at the same time accompanied by metacognitive monitoring of one's mental content and state of awareness. In ideal cases, this symbiotic relationship results in metacognition facilitating or optimizing spontaneous thought processes, so that they become more creative, less intrusive, and more likely to lead to novel conclusion and realizations.

K. C. R. Fox (✉) · K. Christoff
Department of Psychology, University of British Columbia, 2136 West Mall, Vancouver, BC V6T 1Z4, Canada
e-mail: kfox@psych.ubc.ca

S. M. Fleming and C. D. Frith (eds.), *The Cognitive Neuroscience of Metacognition*,
DOI: 10.1007/978-3-642-45190-4_13,

> Sound serious thoughts on worthy subjects [...] cannot be conjured up arbitrarily and at any time. All we can do is to keep the path clear for them [...] We need only keep the field open to sound ideas and they will come. Therefore whenever we have a free moment with nothing to do, we should not forthwith seize a book, but should for once let our mind become tranquil, and then in it something good may easily arise.
>
> *Arthur Schopenhauer* [123], p. 54

13.1 Introduction

Mind wandering (MW) and metacognition may appear to lie at opposite poles of the spectrum of human cognition. The former calls forth notions of daydreaming, spontaneous thoughts, perhaps even Freud's seething unconscious—a stream of undirected ruminations. In contrast, metacognition, the ability to reflect on and evaluate our own thoughts and behaviors, is often viewed as a high-level, deliberate process, the pinnacle of human thinking and a distinguishing hallmark of our species.

But could there be any overlap and interplay between the seemingly primitive flow of spontaneous and undirected musings, and the lofty self-reflective evaluations of metacognition? One standard view is that the brain networks involved in task-related cognition and in MW operate in an anticorrelated, almost mutually exclusive fashion [50, 51], but the view expressed by Schopenhauer [123] in the epigraph above suggests at least one potential overlap: the process of insight, or creativity. It suggests not only that thoughts and insights arise spontaneously, but that some (and only some) of these thoughts are sound and good—implying that self-generated content must subsequently be subjected to critical metacognitive evaluation.

In this chapter, we will argue that, despite the apparent gulf between MW and metacognition, some of the most intriguing mental processes human beings are capable of experiencing involve an intimate, dynamic interplay between "low-level" spontaneous mental processes and "high-level" metacognitive monitoring. What's more, recent evidence suggests that even MW itself, in the absence of metacognitive awareness, may share neural resources with brain regions traditionally viewed as metacognitive and executive ([23, 24]; Fig. 13.1b).

We begin with a brief overview of behavioral and cognitive neuroscience research that has explored these two cognitive processes independently of one another. We then review the standard view of their interaction, wherein metacognitive monitoring serves to correct the wandering mind, suppressing spontaneous thoughts and bringing attention back to more "worthwhile" tasks. We argue that this "negative" (i.e., suppressant) view of their interactions, although important, represents only a part of the whole picture. We go on to discuss three examples of positive, facilitative interactions: creative thinking, mindfulness meditation, and lucid dreaming (being aware that one is dreaming while dreaming).

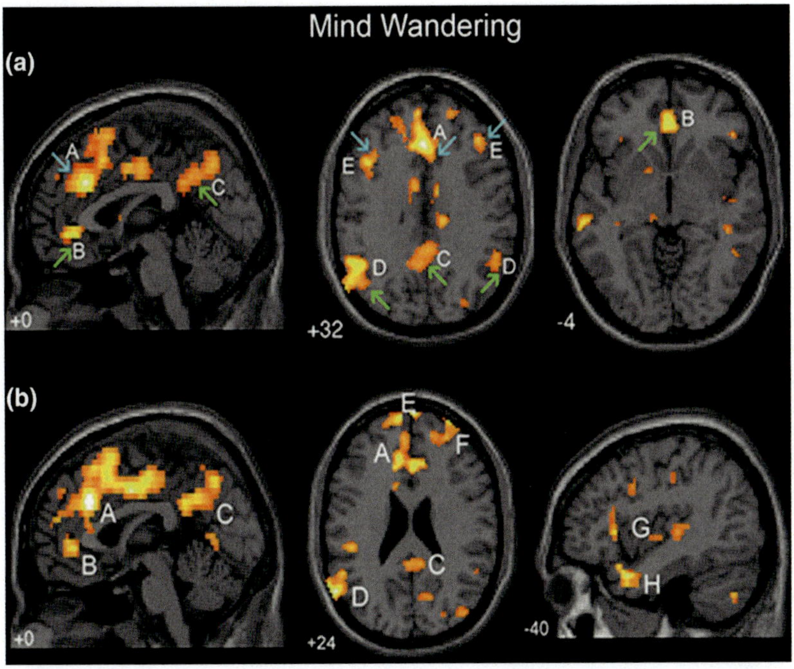

Fig. 13.1 Brain recruitment during mind wandering. **a** Mind wandering simultaneously recruits dorsolateral prefrontal cortex (*E*), anterior cingulate cortex (*A*), medial prefrontal cortex (*B*), inferior parietal lobule (*D*), and posterior cingulate cortex (*C*). **b** Mind wandering without meta-awareness, compared to mind wandering with meta-awareness, recruits a number of traditional metacognitive regions, including RLPFC (*F*) and RMPFC (*E*). Numbers indicate stereotactic coordinates in Montreal Neurological Institute (*MNI*) space. Reproduced with permission from Christoff et al. [23, 24]

The limited scope of this chapter necessitates broadly defined terms. We therefore use MW in a general sense to refer not only to thoughts that involve deviation from a particular task, but to all forms of undirected or spontaneous thought, such as daydreaming or "zoning out" [19]. On the other hand, by "metacognition" or "metacognitive monitoring" we mean the general "ability to reflect upon, comment about, and report a variety of mental states... [i.e.,] cognition about cognition" [43]. We use these terms not only in the literal sense of "thinking about thinking," but more broadly, to encompass meta-awareness, meta-attention, and metacognitive judgments about perception and performance.

Table 13.1 Core cortical components of the default mode network

Region	Approximate brain areas (BA)
Ventromedial prefrontal cortex	24, 10 m/10 r/10 p, 32 ac
Dorsal medial prefrontal cortex	24, 32 ac, 10 p, 9
Posterior cingulate/retrosplenial cortex	29/30, 23/31
Inferior parietal lobule	39, 40
Lateral temporal cortex	21
Hippocampus	–
Parahippocampus	35, 36
Entorhinal cortex	28, 34

Key cortical brain structures contributing to human default mode network activity, and potentially to the subjective state of mind wandering/spontaneous thought. Adapted from Buckner et al. [14]. *BA* Brodmann area

13.2 The Cognitive Neuroscience of Spontaneous Thought Processes

Extensive first-person reports of spontaneous thought and MW go back nearly a century (e.g., [149]), but it was in the 1960s, 1970s, and 1980s that thorough explorations of the subjective content of spontaneous thought (typically referred to then as "daydreaming") began revealing its complex nature (for reviews and seminal papers, see [6, 25, 47, 80, 132, 133, 131]). Based on these studies of content, MW mentation was shown to contain elements of fantasy [78, 79, 81], to be largely audiovisual in terms of sensory content [81], and to be largely based on memories and pre-existing behavioral repertoires [80, 81]. Studies suggest that spontaneous thought occupies a large proportion of our mental lives—anywhere from 30 to 50 % of our waking hours [75, 77, 81].

A number of studies have now examined brain activity during "rest" with intriguing results. For example, in an early report, Andreasen et al. [3] found that, compared to a nonmemory task, both autobiographical memory recall and "rest" revealed similar brain activations in numerous regions later found to be part of the "default mode network" (see Table 13.1). When asked what had been going through their minds at "rest," subjects regularly reported recollection of memories, planning for the future, and other thoughts [3]. The study of this "default mode" of brain function [113], and its relation to MW, was refined over time: early studies compared blocked periods of "rest" with blocked task periods (e.g., [3, 26, 129, 112]); later work made similar comparisons in a trial-by-trial, event-related fashion (e.g., [23, 24, 148, 127]); and the most recent studies have examined functional connectivity (temporally correlated activation and deactivation) across numerous default mode network hubs (e.g., [19, 62]).

Collating data from these three methods has allowed a tentative delineation of core cortical default mode network regions (Table 13.1; [14]). Researchers have hypothesized that activation of the posterior cingulate cortex (PCC) and the anterior medial PFC may reflect the affective, self-relevant nature of spontaneous thoughts

[5]. Medial PFC recruitment may also reflect acts of spontaneous mentalizing, i.e., imagining the thoughts and intentions of other individuals [138]. The temporopolar cortex may also contribute to spontaneous mentalizing [138]. By virtue of its anatomical connectivity with medial temporal lobe (MTL) structures and its role in autobiographical memory [61], the temporopolar cortex may also participate in experiencing spontaneously arising memories [26], especially those memories rich in sensory–perceptual detail [27].

With default mode network regions relatively well-defined, subsequent studies found that both retrospective [99] and online, trial-by-trial [23, 24] self-reported MW predicted increased activity in default mode network hubs (as well as other regions, however—a point to which we will return). Recent work has also found that self-reported intensity of engagement in internally directed thought predicted higher activation in default mode network hubs [148], and that self-reported frequency of thoughts about the past and future predicted the strength of functional connectivity between default mode network regions in MTL memory structures and in other default mode network parietal regions [5]. Taken together, first-person reports have provided a wealth of information about the subjective content of spontaneous thoughts and have tied spontaneous thought to activation of, and functional connectivity within, default mode network regions.

We stress, however, that default mode network activity and spontaneous thought are not merely the objective and subjective aspects (respectively) of a single phenomenon (see also [20]). Though we agree that there is now fairly strong evidence linking MW to recruitment of key default mode network regions (reviewed in [19]), several caveats are in order. Numerous studies noted above have used an a priori region of interest approach, which presupposes a link between the default mode network and MW, and often precludes looking at regions outside the default mode network; others have found activation of numerous regions beyond the default mode network during MW, including traditionally "metacognitive" regions like RLPFC and DLPFC ([23, 24, 99]; Fig. 13.1). Furthermore, multiple forms and definitions of spontaneous thought can be delineated [19]. Thus, although we use default mode network regions (Table 13.1) as a neuromarker for MW-related processes throughout the remainder of this chapter, we do so not out of certainty about the exclusivity of this relationship, but rather out of uncertainty about MW's true neural correlates. It should be emphasized that present evidence suggests [23, 24, 26, 99], and we suspect future work to confirm, that many brain regions outside the default mode network are also key neural substrates of spontaneous thought processes.

13.3 The Cognitive Neuroscience of Metacognition

Metacognition comes in many forms, but all tend to share the notion of a second, "meta" level of cognitive processing or awareness that is to some degree dissociable from a primary (or "object") level involving perception, decision making,

Table 13.2 Core cortical regions implicated in metacognition

Region	Approximate brain areas (BA)
Anterior prefrontal cortex (RLPFC/RMPFC)	10
Dorsolateral prefrontal cortex (DLPFC)	9/46
Anterior cingulate cortex (ACC)	32/24
Anterior insula	13

BA Brodmann area; *RLPFC* rostrolateral prefrontal cortex; *RMPFC* rostromedial prefrontal cortex

or attention [43]. This meta-level can relate, for example, to one's sense of the accuracy of one's own perceptions; certainty about the accuracy of one's decisions or performance; metacognitive evaluation of one's own ideas and theories; or meta-awareness of the quality of one's attention (e.g., focused vs. distracted).

A preliminary understanding of the neural underpinnings of metacognition has implicated rostrolateral, rostromedial, and dorsolateral prefrontal cortices (RLPFC, RMPFC, and DLPFC, respectively) in various metacognitive abilities [21, 22, 42, 44, 57, 58, 100, 114, 117, 120]. There also seem to be some finer distinctions between the metacognitive functions carried out by RLPFC, DLPFC and RMPFC [57, 58]. Metacognitive evaluation in the context of "cognitive" tasks, such as working memory, episodic memory retrieval, and abstract thought [11, 24, 118, 151] appear to involve the RLPFC rather than RMPFC. On the other hand, reflecting upon one's own emotions activates primarily the RMPFC, rather than RLPFC [87, 108, 109]. An alternative, but not mutually exclusive, subdivision between medial and lateral PFC contributions to metacognitive processing takes into account the temporal focus of metacognitive judgments: on this view, prospective judgments selectively recruit RMPFC, whereas retrospective judgments preferentially recruit RLPFC and DLPFC [42].

A more extended account of metacognition should also involve the anterior insula as an important center subserving conscious meta-awareness of emotions and the state of the body [29, 30, 32], and potentially as a key node relaying such information to higher PFC areas [42]. For example, Farb et al. [38] found a significant correlation between activation in the insula and lateral prefrontal cortex, including RLPFC, in subjects trained in mindfulness meditation that were asked to become aware of their thoughts, feelings, and body states (see Sect. 13.5, below). Consistent with these results, our group found improved self-regulation of anterior insula activity during a training paradigm that involved meta-awareness of one's own mental states, in parallel with improved RLPFC self-regulation based on real-time fMRI feedback from this region [100].

As with spontaneous thought, we use several regions (Table 13.2) as putative neuromarkers of the involvement of metacognitive processes, with the caveat that these areas are of course only a preliminary estimate of the neural structures central to metacognition, and a necessary simplification for the purposes of this brief chapter. Throughout, we focus specifically on RLPFC/RMPFC and DLPFC due to their basically unequivocal involvement in metacognition, but other regions too, including anterior cingulate cortex (ACC) and anterior insula, are discussed.

13.4 Mind Wandering as Illness, Metacognition as Cure

One kind of interaction between metacognition and MW has a corrective function. This is the case with the primarily suppressive, regulative role metacognition sometimes plays during goal-directed thought and behavior: it can note MW in the form of distractions (e.g., thoughts about competing external stimuli) and can redirect attention to the task at hand [122]. On this view, MW is conceptualized as an unwelcome detriment to the performance of more worthwhile tasks, and metacognition as the sentinel guarding against such costly, occasionally even dangerous, lapses (e.g., [135]).

This "negative" view, which highlights the role of metacognition in the suppression and disengagement from MW, has motivated the majority of research so far. It has led to a substantial number of studies focusing on the detrimental effects of MW on performance during a variety of traditional experimental tasks, such as memory encoding and reading comprehension (for reviews, see [122, 136]). The tendency to mind wander "too much," or too much about "negative" subject matter, has even been linked to clinical pathologies such as depression (reviewed in [135]) and attention-deficit/hyperactivity disorder (e.g., [128]). Such a negative view of MW was recently epitomized in a high-profile study whose title simply declared, "A wandering mind is an unhappy mind"[1] [77].

This focus has been unfortunate, but understandable given our cultural bias toward viewing MW as something negative, even pathological. In contrast to the more desirable pursuit of "rational" thought, MW is often portrayed as undesirable—a wasteful mental diversion and potentially dangerous distraction, a "mere whimsy without body and without subject" [102]—causing motorists to crash their cars [147], students to disregard their studies [154], and readers to skim over whole paragraphs before realizing they have absorbed none of the material on the page in front of them [121].

Overall, our culture values control and effort, and devalues spontaneity and leisure. Since metacognition is often associated with the former and MW with the latter, it is no wonder that research has so far been heavily influenced by this implicit mind-wandering-as-illness, metacognition-as-cure approach. Unfortunately, however, this has left us relatively ignorant of the more positive kinds of interactions through which metacognition may facilitate and even enhance the arising of spontaneous thought, thus enabling beneficial outcomes that would not otherwise be obtained.

[1] The empirical evidence presented by this paper in support of its title's claim is much more controversial than the title suggests. For example, far more spontaneous thoughts were rated as emotionally positive (42.5 %) than negative (26.5 %) [77].

Table 13.3 Three examples of mental phenomena during which metacognition may interact with mind wandering in a positive, facilitative fashion

State	Aspects of mind wandering	Aspects of metacognition
Creative thinking	Spontaneous generation of ideas, imagery, verse, music, solutions, insights, etc.	Evaluation of the novelty, quality, utility, and value of self-generated ideas; monitoring of the effectiveness of the creative process
Mindfulness ("insight") meditation	Arising of spontaneous thoughts; spontaneous "chaining" (elaboration) of thoughts; spontaneous emotional reactions	Monitoring the focus and quality of attention; maintaining a detached, nonelaborative mental stance
Lucid dreaming	Spontaneous generation of visual and auditory imagery, and often a fully immersive dream world resembling physical space; spontaneous construction of narratives, characters with personalities and motives, and theory of mind-like judgments	Recognition that the physical self is actually asleep in bed, and that the perceived "physical" environment is actually a mental representation; directing of the course of the dream and its imagery (rarely)

13.5 When Metacognition Helps the Wandering Mind Find Its Way

Though the "suppressant" MW-metacognition interactions are undoubtedly part of everyday life, in this chapter we aim to make a step toward redressing the imbalance of research focus by concentrating, albeit in a preliminary and speculative fashion, on three phenomena—creative thinking, mindfulness meditation, and lucid dreaming—that we believe represent examples of positive, facilitative interactions between MW and metacognition (Table 13.3).

13.5.1 Creative Thinking: Metacognitive Evaluation of Spontaneous Ideation

Creative thinking is a unique mental ability that relies on the skilled engagement of both deliberate, and spontaneous thought [25]. Often defined in terms of its product, creativity is the ability to produce ideas that are both novel (original and unique) and useful (appropriate and meaningful) [13, 54, 140]. In following with this two-fold definition of the creative product, emphasizing both its novelty and utility, psychological findings have suggested that creative thought involves two main components: the generation of new ideas, on the one hand, and the evaluation of any generated ideas as to their utility and originality, on the other [8, 16, 41, 69, 156]. This dichotomy is also present in subjective accounts by artists of their own creative process, which they often describe as alternating between rough sketching and critiquing [33, 49].

Table 13.4 Metacognitive and default mode network regions known to be involved in creative thinking

Metacognitive brain regions	Default mode network regions
DLPFC	Medial PFC
Dorsal ACC	PCC/retrosplenial cortex
RLPFC	IPL/lateral temporal cortex
Anterior insula	Medial temporal lobe (hippocampus, parahippocampus)

ACC anterior cingulate cortex; *DLPFC* dorsolateral prefrontal cortex; *IPL* inferior parietal lobule; *PCC* posterior cingulate cortex; *PFC* prefrontal cortex; *RLPFC* rostrolateral prefrontal cortex

Although somewhat over-simplifying matters, creative evaluation can be seen as heavily relying on metacognition, while creative generation likely relies on spontaneous thought processes. With the recognition that, when engaged simultaneously, metacognition might inhibit spontaneous generation, the optimal creative process is often considered to employ metacognitive evaluation and creative generation *sequentially*. Although these two components of the creative process certainly can and do occur in parallel, creating a temporal separation between the two is known to increase the creativity of outputs [8, 110]—a principle applied in the practice of "brainstorming." This iterative generation-evaluation process parallels the sequential nature of metacognitive judgments of perceptual decision making, for example confidence judgments about performance on a perceptual task (see [159]).

The facilitating effects of metacognition on creative generation are not, however, limited to simply preventing metacognition from occurring simultaneously with generation. Metacognitive evaluations can also be used to guide future creative generation efforts in directions that have been identified as novel and useful during previous evaluation phases [49]. In this way, metacognition can play a positive, facilitative role in the spontaneous generation of thoughts and ideas during the creative process.

Traditional metacognitive brain regions, as well as default mode network regions, are known to be involved in the creative process (for a review, see [20]; also Table 13.4). The DLPFC and dorsal ACC are known to be activated during a variety of creative tasks, including piano improvisation, creative story generation, word association, divergent thinking, fluid analogy formation, insight problem solving and visual art design [9, 18, 55, 83, 126]. Similarly, enhanced activations in the area of the inferior parietal lobule (IPL) and lateral temporal cortex (LTC), medial PFC, and PCC/retrosplenial cortex—three key hubs of the default mode network—have been observed during divergent thinking tasks, creative story generation, hypothesis generation, fluid analogy formation, remote associates insight problems, and jazz improvisation [55, 68, 72, 83, 90]. Recruitment of MTL regions such as the hippocampus and the parahippocampus are also observed [37, 40, 84].

What are the neural correlates of creative evaluation versus creative generation, and how do they interact at the neural level? A recent study from our group addressed these questions directly [37]. It revealed, on the one hand, simultaneous recruitment

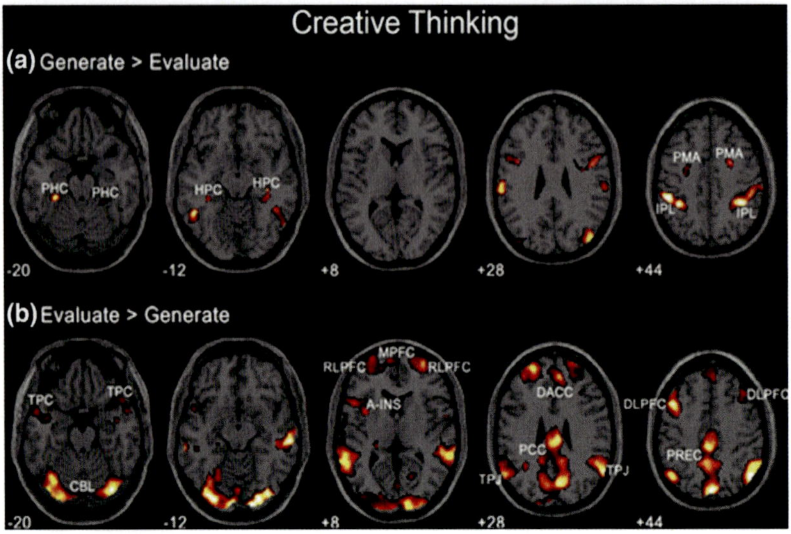

Fig. 13.2 Brain recruitment during the generation and evaluation phases of artistic creativity. Creative thinking recruits hippocampus, parahippocampus, and IPL during the generation of ideas (**a**), and subsequently involves activation of DLPFC, RLPFC, MPFC, and PCC during noetic metacognitive evaluation of one's own thoughts (**b**). Numbers indicate stereotactic coordinates in Montreal Neurological Institute (*MNI*) space. *A-INS* anterior insula; *CBL* cerebellum; *DACC* dorsal anterior cingulate cortex; *DLPFC* dorsolateral prefrontal cortex; *HPC* hippocampus; *IPL* inferior parietal lobule; *MPFC* medial prefrontal cortex; *PCC* posterior cingulate cortex; *PHC* parahippocampus; *PMA* premotor area; *PREC* precuneus; *RLPFC* rostrolateral prefrontal cortex; *TPC* temporopolar cortex; *TPJ* temporoparietal junction. Reproduced with permission from Ellamil et al. [37]

of metacognitive brain regions and default mode network regions during the process of creative evaluation (Fig. 13.2b). Three metacognitive regions—RLPFC, RMPFC, and the anterior insula—were specifically identified as being part of metacognitive creative evaluation, even though they have not been emphasized in terms of their contribution to the creative process in the literature so far.

On the other hand, the results revealed that the process of creative generation is preferentially linked to recruitment of the IPL, as well as the hippocampus and parahippocampus—the two MTL regions that have also been implicated in default mode network functioning (Fig. 13.2a; see also [14]). The parahippocampus may form new, or access old, associations that are then recombined by the hippocampus with other information to construct episodic simulations [119]. Previous studies have also indirectly linked the MTL to the spontaneous generation of thoughts and memories, spontaneous re-activation of memories in humans [56], spontaneous mental processing during rest [10, 26, 139] and including replay of memories during rest [45, 141]. The associative and spontaneous nature of MTL function suggests that it may be important for creative thought by facilitating the generation of novel ideas and associations, as well as the recombination of old ones.

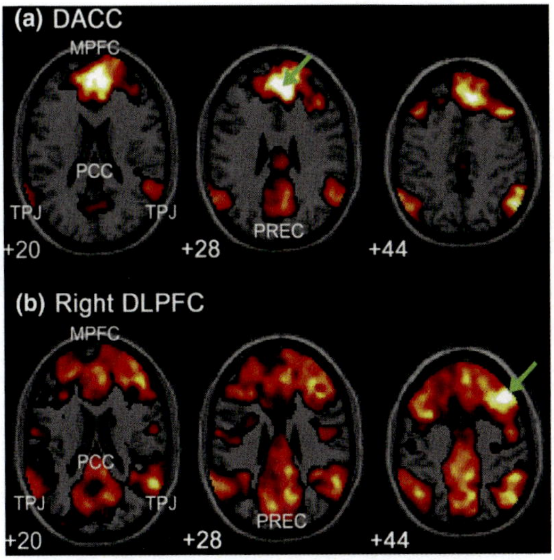

Fig. 13.3 Functional connectivity between metacognitive and default mode network regions during the evaluation phase of creative thinking. Functional connectivity analyses using seed regions in (**a**) dorsal ACC and (**b**) right DLPFC (indicated by green arrows) reveal strong positive temporal correlations of activity between default mode network and metacognitive brain regions. *DACC* dorsal anterior cingulate cortex; *DLPFC* dorsolateral prefrontal cortex; *MPFC* medial prefrontal cortex; *PCC* posterior cingulate cortex; *PREC* precuneus; *TPJ* temporoparietal junction. Reproduced with permission from Ellamil et al. [37]

In addition to being co-activated during creative evaluation, metacognitive and default mode network regions also exhibited positive functional connectivity during the creative process (Fig. 13.3). This finding provides specific neural evidence for the existence of temporally coupled, possibly facilitative interactions between these two networks in the process of creative evaluation.

How might metacognition facilitate spontaneous thought during the creative process? First, low levels of metacognitive control during the generation phase may enable an associative mode of information processing that facilitates and ensures the generation of novel ideas [67]. This may allow access to more diverse, non-obvious pieces of information to combine and use as building blocks for novel ideas, or more comprehensive and unusual connections [150]. Second, metacognitive evaluation of already-generated ideas during the evaluation phase may assign positive cognitive and emotional associations to those ideas or directions of creative thought. These positive associations may then be used during subsequent generation phases in order to guide the further generation of novel ideas. Significantly, the metacognitive regions involved in creative evaluation are not limited to strictly cognitive metacognition regions, but also include self- and emotional evaluative regions such the medial PFC and the anterior insula, suggesting the potential importance of affective and visceroceptive forms of evaluative processing during creative thought.

In summary, during creative thinking metacognition appears to facilitate spontaneous thought by first being selectively attenuated during generation phases in order to "make way" for spontaneous thoughts to emerge, and second, by being used during the evaluation phase to identify fruitful directions toward which the generation of spontaneous thought can be directed in subsequent generation phases. One positive outcome of these facilitative interactions may be the arrival at novel conclusions, solutions, and insights that may not otherwise be reached by MW alone, without the positive evaluation and facilitation from meta-awareness.

13.5.2 Mindfulness Meditation as Meta-Awareness of Mind Wandering

Meditation can be thought of as a broad set of mental techniques for focusing and training attention, regulating emotion, enhancing awareness of the body, and various other processes [96, 134]. A crucial component of meditation is a persistent metacognitive monitoring of one's progress in, and execution of, the practices. At the same time, the arising of spontaneous thoughts is a virtually universal experience among practitioners of meditation [60, 145]. In contrast to creative thinking, where spontaneous thought generation and metacognitive evaluation are ideally separated in time, during meditation the two processes ideally occur simultaneously, so that metacognition is present in parallel with any spontaneously arising thoughts.

Two broad strategies can be delineated in response to MW during meditation practice, both of which involve metacognitive monitoring. One common technique involves the simple focusing of attention on the sensations associated with respiration—typically, to the exclusion of all else. The practitioner must also monitor the effectiveness with which they are maintaining attentional stability: laxity (e.g., drowsiness or lack of focus) and outright lapses (e.g., MW) are to be not only noticed, but usually corrected for as well [91]. That is, not only should attention be sustained on a single object, but meta-attention must also be continuously employed [2, 145] during such a "focused attention" meditation [96]. In focused attention meditation, the role of metacognition is in noticing lapses of attention, and then redirecting focus to a chosen object. As such, it strongly resembles the negative, "suppressant" MW-metacognition interaction discussed above.

A second strategy releases the meditator from the need for a single object of focus during practice. Instead, the practitioner maintains an open attentional stance: they neither give preference to, nor attempt suppression of, any stimulus that arises, be it incoming sensation or internal thoughts and emotions. Commonly referred to as "mindfulness" [73], "open monitoring" [96], or "Insight" meditation [88], this practice involves a nonreactive, nonjudgmental, nonelaborative mental stance, during which any object of attention is acceptable so long as metacognitive monitoring of one's stream of thought and emotional reactions is

Table 13.5 Brain regions activated during mindfulness meditation

Metacognitive brain regions	Default mode network brain regions
RLPFC	Posterior cingulate cortex
DLPFC	Inferior parietal lobule
Insula (anterior)	Hippocampal formation

DLPFC dorsolateral prefrontal cortex; *RLPFC* rostrolateral prefrontal cortex

continuously maintained. In contrast to focused attention meditation, during mindfulness meditation the role of metacognition is to maintain detachment from, or restrain elaboration of, thoughts and sensory input, and further to regulate arousal so that one does not become over-involved emotionally [73, 91, 96].

Neuroimaging studies of mindfulness meditation have often shown greater activations in both default mode network and metacognitive brain regions (Table 13.5). The former include greater recruitment during mindfulness of PCC [70], IPL [38] and the hippocampal formation [92]. Activations in metacognitive regions include results in RLPFC [98, 115] and DLPFC [38, 98]. There are exceptions to this trend, however, with some studies showing default mode network or metacognitive region deactivation during mindfulness meditation (e.g., [38, 70]). As noted above (Sect. 13.3), the insula has been hypothesized to play a role in metacognition [42], and so significant insular cortex activations during mindfulness meditation are also of interest [38, 53, 95, 98]. Again, there are exceptions to this observation, too (e.g., [70]).

If meditation practitioners are indeed consistently engaged in metacognitive monitoring, it is possible that this skill may be trained by its persistent engagement [97]. Though the evidence to date remains tentative, work by our own group [48] and others [106, 142] suggests that metacognitive abilities might be enhanced in long-term meditation practitioners. A persistent engagement of metacognitive skills alongside attention to spontaneous thoughts is not only consistent with the functional neuroimaging results discussed above, but would also likely entail a corresponding reorganization of brain structure. Speaking to this possibility, numerous studies have now examined brain structure differences in both long-term meditation practitioners (with thousands of hours of experience) and novices undergoing short-term training. The subjects come from a wide variety of contemplative backgrounds, but essentially all have training in some form of meditation that could be classified as either focused attention or mindfulness. Among many other intriguing differences in both gray and white matter, across cortical and subcortical regions (reviewed in [46]), structural heterogeneities in several default mode network (Table 13.1) and metacognitive (Table 13.2) regions are salient. In 21 structural neuroimaging studies of meditation to date contrasting meditators versus controls, several have found structural enhancement of RLPFC (BA 10) [76, 88, 152], DLPFC [76, 88], and the insula [64, 76, 88, 143]. Default mode network regions are also consistently altered in meditation practitioners, including differences in hippocampus [64, 65, 93, 94] and parahippocampus [76, 89], as well as PCC [65, 66].

We recently conducted a review and meta-analysis of all structural neuroimaging studies of meditation. We found meta-analytic clusters of cross-study structural enhancement in RLPFC (BA 10), ACC, anterior insula, and hippocampus (among other regions), suggesting that the structure of metacognitive and default mode network areas is consistently and significantly altered in relation to meditation practice [46].

What might be the benefits of such an open, nonjudgmental metacognitive stance toward spontaneous thought processes? A primary contention in classic Buddhist thought is that mindfulness meditation leads to a gradual lessening of one's identification with passing thoughts and emotions, and thereby to improved well-being (e.g., [2, 145]). This could prove beneficial in the context of negative, depressive thoughts, for instance—such mental phenomena could come to be seen as merely ephemeral experiences, rather than traits that define one's identity. Indeed, such metacognitive detachment from self-identification with negative rumination has been proposed to be a key mechanism underlying the beneficial effects of mindfulness meditation for clinical disorders such as depression and anxiety [28, 144].

A related possibility is that of decreased automaticity in the associations among spontaneous thoughts: although the incidence of spontaneous thoughts per se might not decrease with mindfulness practice, an open, nonjudgmental metacognitive stance might reduce the "chaining" or elaboration of the thoughts that do arise. Reduced elaboration of habitual cognitive and emotional associations might then allow for greater cognitive-emotional flexibility and novel, more adaptive, behavioral responses (e.g., [103]). Furthermore, some spontaneous thoughts—especially those previously judged to be of negative or of a personally "unacceptable" nature—may be suppressed before they reach awareness through a habitual elaborative process that may over time become automatic. The emotional sequelae of those "unconscious" thoughts may affect mood negatively and without the person's awareness. By maintaining an open, nonjudgmental metacognitive mindset, meta-awareness during mindfulness meditation may therefore enable such habitually suppressed thoughts and their emotional consequences to come more fully into conscious awareness, allowing increased insight into the functioning of one's own mind and a greater flexibility in directing mental activity toward personally beneficial goals.

In summary, mindfulness meditation is a unique phenomenon during which brain regions associated with both MW and metacognition appear to be activated, and during which metacognition may occur simultaneously with MW, facilitating the emergence of spontaneous thoughts that may otherwise not reach awareness. This process may enable the meditator to reach new realizations and conclusions and may allow for improved behavioral and mental flexibility.

13.5.3 Lucid Dreaming: Meta-Awareness of the Dream State

Lucid dreaming is perhaps the least researched and most elusive of our examples of potential facilitative interactions between metacognition and spontaneous thought. This seemingly paradoxical phenomenon, wherein one is aware that one is dreaming while in the dream state (and can in some cases direct the dream's course and content), has fascinated humanity for millennia. Ancient written records from both the East and West have elaborated on the notion of lucid dreaming: the Indian scriptures known as the Upanishads [111], for instance, discuss the possibility of maintaining conscious awareness throughout the sleep cycle; Aristotle in his writings on sleep and dreaming [52] noted that, "Often when one is asleep, there is something in consciousness which declares that what then presents itself is but a dream;" and archaic Tibetan Buddhist meditation practice manuals [59, 157] discuss methods of attaining, and beneficial effects of, dream lucidity at length.

As lucid dreaming involves meta-awareness of the true state of the physical self (asleep in bed), as well as recognition that the apparent dreamworld is in fact a projection of the self, it can be considered a form of autonoetic (i.e., self- as opposed to perception-focused) metacognition [74, 101]. But is regular (nonlucid) dreaming a form of spontaneous thought? In a recent review and meta-analysis of the subjective content and neural basis of dreaming, we argue that it likely is [47]. First, the subjective reports from daytime MW and nighttime dreams overlap considerably in terms of sensory content, bizarreness, emotionality, and so on. Second, brain activations during dreaming (compared to waking) show a pattern highly similar to that of the resting state/default mode network [47]. The combined neurophysiological and experiential evidence has led us to propose that nighttime dreaming can be considered as a more intense and immersive version of waking MW or daydreaming [47]. Interestingly, compared to waking rest, nonlucid dreaming typically involves the deactivation of prefrontal cortical regions involved in executive control and metacognitive monitoring, including DLPFC [47, 63, 105], which may explain the lack of meta-awareness during regular dreaming.

If dreaming is an even more immersive form of MW, can the light of metacognitive awareness still penetrate to such depths? Paralleling the ancient accounts mentioned above, some contemporary researchers argue that indeed it can (e.g., [12, 159]), but lucid dreaming continues to meet with considerable skepticism. As the voluntary musculature of the body is paralyzed during rapid eye movement (REM) sleep, when lucid dreaming has been assumed to take place, communicating one's meta-awareness in a verifiable way to outside observers had seemed impossible. It was eventually noted, however, that voluntary control of the muscles of the eyes appeared intact, and that observable eye movements during REM seemed to correlate with direction of gaze in the subjective dream experience [116]. In the early 1980s, a team at Stanford University published the first objective evidence of lucid dreaming by using complex, pre-arranged patterns of eye movements to signal meta-awareness from within verified REM sleep [86].

Further work found other correspondences between subjective reports of lucid dreaming activity and various physiological measures, including increased respiration during dreamed speech and greater electromyographic (EMG) activity during dreamed muscle flexion [39]. Recent work has now complemented these early results by taking advantage of sophisticated methods combining simultaneous electroencephalography (EEG) and fMRI [34].

The latest work has begun to reveal features that distinguish lucid from regular dreaming at the neural level. A recent study employing EEG found that, compared to nonlucid dreaming, lucid dreaming showed greater overall coherence levels across the entire EEG frequency spectrum, as well as greater 40 Hz (γ-band) power localized to frontal and frontolateral regions of the brain [153]. The finding of high gamma activity is of particular interest, since γ-band ($\sim$30–70 Hz) synchrony has been argued to be a key neural correlate of conscious awareness, with the ensuing capacity for self-reflection (e.g., [31]).

Localization of EEG signals to particular cortical areas is contentious, however, and the gold standard for studying lucid dreaming has long been considered fMRI, due to this method's high temporal and spatial resolution. To date only a single case study of lucid dreaming measured with combined EEG/fMRI has been reported [35]. The results, though highly tentative, are suggestive: lucid REM sleep dreaming, as compared to regular REM dreaming, showed higher activation in numerous cortical regions [35]. Most relevant to the present discussion were activity increases in right DLPFC as well as bilateral RLPFC, both of which have been strongly linked to metacognitive awareness (see Sect. 13.3). Their increased activity was therefore argued to be the basis of the heightened self-reflective awareness present during lucid dreaming [35] (Fig. 13.4).

But to what end does one engage metacognition during dreaming? The reasons are many and varied. Ancient Tibetan Buddhist texts, for example, view lucid dreaming as a chance to practice deep meditation, and as an aid to understanding the impermanent, partially mind-constructed nature of the waking, physical world [59, 158]. Professional athletes have attempted to use lucid dreaming as an opportunity to rehearse demanding or possibly dangerous physical activities [85]—in line with fairly ample evidence that mental practice, including dreaming of recently learned skills [157], improves actual performance (reviewed in [36]).

Others view lucid dreaming as a potential adjunct to psychotherapy [146]. Many regular (nonlucid) dreams are characterized by negative emotion [107, 124], and intriguingly, the attainment of lucidity is frequently triggered by nightmares [125]. Metacognitive awareness in dreams, then, may also serve to attenuate the high levels of fear and negative emotion in dreams or nightmares [125], while at the same time facilitating the continuation of the spontaneous dream mentation that would otherwise abruptly end if intense negative emotion led to sudden awakening.

Though the cognitive neuroscience of lucid dreaming remains in its infancy, the preliminary work outlined above suggests an intriguing cognitive state that demands rigorous and extensive research. Much work will be required to further understand how immersion in a spontaneously generated, immersive dream world

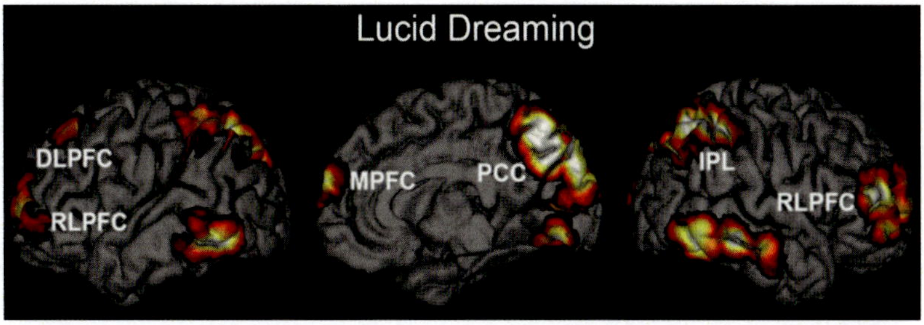

Fig. 13.4 Brain recruitment during lucid dreaming. Lucid dreaming involves simultaneous recruitment of default mode network and metacognitive regions, including rostrolateral (*RLPFC*) and dorsolateral prefrontal cortex (*DLPFC*), as well as medial prefrontal cortex (*MPFC*), inferior parietal lobule (*IPL*), and posterior cingulate cortex (*PCC*). Modified and reproduced with permission from Dresler et al. [35]

can be simultaneously accompanied by metacognitive awareness of the illusory, self-generated nature of one's perceptions and experiences. Just as important will be research into the putative benefits of lucid dreaming, including the potential for mental training, and cultivation of positive emotions and experiences.

13.6 Conclusions and Some Remaining Questions

In this chapter, we focused on the contrast between suppressive and facilitative interactions between metacognition and MW in order to bring more attention to the usually overlooked positive effects of metacognition during MW. But a number of questions still remain: Could the suppressive and facilitative interactions be simply flip sides of the same coin—that of selective pressures exerted by metacognition on spontaneously generated mental contents? Is continuous metacognition, occurring in parallel with the stream of consciousness, possible—and indeed desirable? And are there any other examples of human cognition, in addition to the three we have outlined here, during which there may be positive interactions between metacognition and MW?

13.6.1 Survival of the Fittest in the Cortical Ecosystem?

Nature's profligacy is notorious: a single tree may throw millions of seeds to the wind on the off chance that but one will find fertile ground. So long as slight variations characterize individual units, however, the high cost of such extravagance may conceivably be justified by the immense reward of a single success perpetuating the individual, and possibly the species.

Could the human brain function in a similar fashion, generating an unending array of ideas, plans, and solutions, in order that a single triumph might justify hundreds, perhaps thousands, of failures and mere fantasies? Could metacognition serve to decide among these innumerable ideas and thoughts, and judge their value or utility? This framework was most famously applied by Donald T. Campbell to scientific and artistic creativity, as well as problem solving generally [16, 17, 130]. Campbell's "selectionist" theory of creativity retains enormous influence today. He considered spontaneous thoughts as quasi-random variation of pre-existing ideas and patterns of behavior; metacognitive evaluation as selective pressure; and long-term memory as the substrate allowing for "heritability" or persistence of selected variants.

Such "selectionist" accounts are consistent with the kind of facilitative MW-metacognition interactions we have discussed throughout this chapter, and are certainly worthy of further investigation (cf. [130]). It is worth noting, however, that the analogy with evolutionary selection, albeit useful to some degree, may also obscure other possible facilitative long-term effects that metacognition may have on the spontaneous generation of thoughts. For example, it is possible that by positively evaluating certain spontaneously generated ideas, metacognition makes related ideas more likely to spontaneously arise in the future (as in the case of creative thought). This kind of interaction may be missed if our understanding is framed solely in selectionist terms, which emphasize competition between entities and the "survival of the fittest." In contrast, when it comes to spontaneously generated thoughts and ideas, metacognition may enable an active prospective biasing of certain semantic domains and therefore types of ideas at the neural level, which may then make it more likely for these types of ideas to be spontaneously generated in the future. This prospective biasing would need to be examined and explained in neural rather than evolutionary terms, because of the obvious differences in the way biological species and mental ideas are produced.

13.6.2 Is Continuous Metacognition Possible?

A large body of research suggests that "self-regulation"—the ability to control oneself, delay gratification, and maintain vigilance—is a limited resource ([104]; but see also [71]). It seems plausible that a related higher-order skill like metacognitive monitoring is also subject to "depletion" with continued use, although to our knowledge this remains an unexplored question. As we discuss above, however, it has been suggested that repeated use during, for example, meditation, might not just temporarily deplete metacognitive resources, but may also ameliorate metacognitive skills such as introspection—at least over the long term [48, 97]. Relatedly, advanced meditation practitioners have claimed that with a certain amount of training a qualitative change occurs, after which metacognitive monitoring is effortless and virtually perpetual—attention can be directed to any object, for any length of time, without distraction [155]. As noted above with respect to creativity, metacognition might be a double-edged sword that, if over-applied, can interfere with certain processes, such as creative generation. Whether continuous metacognition is indeed an enviable skill or state remains to us an open question, then. But the plausibility, and indeed desirability, of a continuous state of metacognitive monitoring (not only during MW and meditation, but all thoughts and actions whatsoever) is salient in even the earliest Buddhist writings [2]. Although such claims remain highly speculative from a scientific standpoint, we consider them intriguing questions that could be addressed by future work.

13.6.3 Other Constructive Interactions Between Spontaneous Thought and Metacognition

Above we have outlined three processes suggestive of a "positive" or facilitative interaction between metacognition and spontaneous thought processes, but there may of course be others as well [4]. Related to creativity, for example, is the phenomenon of sudden insights or "Aha" moments, during which one is sometimes unaware of the MW process until a "correct" and/or fully formed solution presents itself spontaneously (e.g., [33, 82]). Such sudden presentations of apparently pre-evaluated ideation raise the intriguing possibility that high-level metacognitive evaluation of some kind could also take place semi-unconsciously (for further discussion see, e.g., [7]). Trial-and-error problem solving presents another related case, in which a somewhat more focused, albeit still creative and spontaneous, approach is brought to bear on a particular issue. Here, spontaneous thought processes might be more closely monitored and guided by metacognition (than during, say, artistic creativity) in order to avoid immaterial distractions and ensure a swift solution. Spontaneous musical improvisation (e.g., [90]) seems to be a related case, wherein the two stages of creative thinking are condensed into one, and metacognitive evaluation accompanies spontaneous ideation quasi-simultaneously. Imagining detailed future situations also appears to recruit a combination of default mode network and PFC metacognitive areas (e.g., [1]), suggesting that prospection (thinking about the future) too may involve the spontaneous generation of scenarios with a simultaneous metacognitive valuation of their likelihood or utility (see [15], for a review).

13.6.4 Conclusions

Aside from the everyday interaction whereby metacognition quells or helps us disengage from MW, we have argued here that there are also a number of mental states during which metacognitive evaluation functions instead to facilitate or guide spontaneous thought processes toward personally relevant, higher-order goals. These may be goals such as artistic or scientific creativity, improved understanding of a complex problem, insight into the operation of one's own mind, or greater flexibility and adaptability of emotional and behavioral responses. We believe that this "positive" interplay is indicative of some of the most intriguing mental states we as humans are capable of experiencing. We reviewed evidence that neuroscientific measures of these states support the notion of interplay between spontaneous thought and metacognitive judgment or awareness, including both simultaneous and sequential recruitment of midline default mode network and metacognitive brain regions, as well as evidence for positive functional connectivity between the two during processes such as creative thinking. We also elaborated on some of the possible cognitive mechanisms whereby metacognition may positively interact with MW, facilitating spontaneous mentation and opportunities for arriving at conclusions and realizations that may not otherwise be reached by spontaneous thought processes alone.

Donald Campbell once remarked, "Mental meandering, mind wandering… is an essential process. If you are allowing that mentation to be driven by the radio or the television or other people's conversations, you are just cutting down on… your intellectual exploratory time" (quoted in [33]). Perhaps it is only with the assistance of metacognition that we can make the best use of our mental meanderings and help our wandering mind find its way during those highly valuable, and possibly uniquely human, intellectual explorations.

Acknowledgments We thank Dr. Michael Czisch for kind permission to reproduce portions of figures from an original publication [35] in this chapter's Fig. 13.4. Work on this chapter was supported by grants from the Canadian Institute of Health Research (CIHR) and the Natural Sciences and Engineering Research Council (NSERC) of Canada awarded to K.C., as well as by a University of British Columbia Four-Year Fellowship and NSERC Vanier Canada Graduate Scholarship awarded to K.C.R.F.

References

1. Addis DR, Wong AT, Schacter DL (2007) Remembering the past and imagining the future: common and distinct neural substrates during event construction and elaboration. Neuropsychologia 45(7):1363–1377
2. Analayo (2003) Satipatthana: the direct path to realization. Windhorse Publications, Ltd, Cambridge, p 336
3. Andreasen NC, O'Leary DS, Cizadlo T, Arndt S, Rezai K et al (1995) Remembering the past: two facets of episodic memory explored with positron emission tomography. Am J Psychiatry 152:1576–1585
4. Andrews-Hanna JR (2012) The brain's default network and its adaptive role in internal mentation. Neuroscientist 18(3):251–270
5. Andrews-Hanna JR, Reidler JS, Huang C, Buckner RL (2010) Evidence for the default network's role in spontaneous cognition. J Neurophysiol 104:322–335
6. Antrobus J, Singer JL, Greenberg S (1966) Studies in the stream of consciousness: experimental enhancement and suppression of spontaneous cognitive processes. Percept Mot Skills 23:399–417

7. Baars BJ (2010) Spontaneous repetitive thoughts can be adaptive: postscript on "mind wandering". Psychol Bull 136(2):208–210
8. Basadur M, Graen GB, Green SG (1982) Training in creative problem solving: effects on ideation and problem finding and solving in an industrial research organization. Organ Behav Human Perform 30:41–70
9. Berkowitz AL, Ansari D (2008) Generation of novel motor sequences: the neural correlates of musical improvisation. NeuroImage 41:535–543
10. Binder JR, Frost JA, Hammeke TA, Bellgowan PSF, Rao SM, Cox RW (1999) Conceptual processing during the conscious resting state: a functional mri study. J Cogn Neurosci 11:80–93
11. Braver TS, Bongiolatti SR (2002) The role of frontopolar cortex in subgoal processing during working memory. Neuroimage 15(3):523–536
12. Brooks JE, Vogelsong J (2000) The conscious exploration of dreaming. 1st Book Library, Bloomington
13. Bruner JS (1962) The conditions of creativity. In: Gruber H, Terrell G, Wertheimer M (eds) Contemporary approaches to creative thinking. Atherton, New York, pp 1–30
14. Buckner RL, Andrews-Hanna JR, Schacter DL (2008) The brain's default network: anatomy, function, and relevance to disease. Ann NY Acad Sci 1124:1–38
15. Burgess PW, Gonen-Yaacovi G, Volle E (2011) Functional neuroimaging studies of prospective memory: what have we learnt so far? Neurospsychologia 49:2246–2257
16. Campbell DT (1960) Blind variation and selective retention in creative thought as in other knowledge processes. Psychol Rev 67:380–400
17. Campbell DT (1974) Unjustified variation and selective retention in scientific discovery. In: Ayala FJ, Dobzhansky TG (eds) Studies in the philosophy of biology. Macmillan, London, pp 139–169
18. Carlsson I, Wendt PE, Risberg J (2000) On the neurobiology of creativity. Differences in frontal activity between high and low creative subjects. Neuropsychologia 38:873–885
19. Christoff K (2012) Undirected thought: neural determinants and correlates. Brain Res 1428:51–59
20. Christoff K (2013) Thinking. In Ochsner K, Kosslyn SM (eds) The Oxford Handbook of Cognitive Neuroscience, The Cutting Edges, Vol 2. Oxford, Oxford University Press, pp. 318–333
21. Christoff K, Gabrieli JDE (2000) The frontopolar cortex and human cognition: evidence for a rostrocaudal hierarchical organization within the human prefrontal cortex. Psychobiology 28:168–186
22. Christoff K, Ream JM, Geddes LPT, Gabrieli JDE (2003) Evaluating self-generated information: anterior prefrontal contributions to human cognition. Behav Neurosci 117:1161–1168
23. Christoff K, Gordon AM, Smallwood J, Smith R, Schooler JW (2009) Experience sampling during fMRI reveals default network and executive system contributions to mind wandering. Proc Natl Acad Sci USA 106(21):8719–8724
24. Christoff K, Keramatian K, Gordon AM, Smith R, Mädler B (2009) Prefrontal organization of cognitive control according to levels of abstraction. Brain Res 1286:94–105
25. Christoff K, Gordon AM, Smith R (2011) The role of spontaneous thought in human cognition. In: Vartanian O, Mandel DR (eds) Neuroscience of decision making. Psychology Press, New York, pp 259–284
26. Christoff K, Ream JM, Gabrieli JDE (2004) Neural basis of spontaneous thought processes. Cortex 40:623–630
27. Conway MA (2001) Sensory-perceptual episodic memory and its context: autobiographical memory. Phil Trans R Soc B 356:1375–1384
28. Corcoran KM, Farb NAS, Anderson A, Segal ZV (2009) Mindfulness and emotion regulation: outcomes and possible mediating mechanisms. In: Kring AM, Sloan DM (eds) Emotion regulation and psychopathology: a transdiagnostic approach to etiology and treatment. The Guilford Press, New York, pp 339–358

29. Craig AD (2004) Human feelings: why are some more aware than others? Trends Cog. Sci. 8(6):239–241
30. Craig AD (2009) How do you feel—now? The anterior insula and human awareness. Nat Rev Neurosci 10:59–70
31. Crick F, Koch C (1990) Towards a neurobiological theory of consciousness. Neurosciences 2:263–275
32. Critchley HD, Wiens S, Rotshtein P, Öhman A, Dolan RJ (2004) Neural systems supporting interoceptive awareness. Nat Neurosci 7:189–195
33. Csikszentmihalyi M (1996) Creativity. HarperPerennial, New York
34. Dresler M, Koch SP, Wehrle R, Spoormaker VI, Holsboer F, Steiger A, Samann PG, Obrig H, Czisch M (2011) Dreamed movement elicits activation in the sensorimotor cortex. Curr Biol 21(21):1833–1837
35. Dresler M, Wehrle R, Spoormaker VI, Koch SP, Holsboer F, Steiger A, Obrig H, Samann PG, Czisch M (2012) Neural correlates of dream lucidity obtained from contrasting lucid versus non-lucid REM Sleep: a combined EEG/fMRI case study. Sleep 35(7):1017–1020
36. Driskell JE, Copper C, Moran A (1994) Does mental practice enhance performance? J App Psychol 79(4):481–492
37. Ellamil M, Dobson C, Beeman M, Christoff K (2012) Evaluative and generative modes of thought during the creative process. NeuroImage 59(2):1783–1794
38. Farb NAS, Segal ZV, Mayberg H, Bean J, McKeon D, Fatima Z, Anderson AK (2007) Attending to the present: mindfulness meditation reveals distinct neural modes of self-reference. Soc Cogn Affect Neurosci 2:313–322
39. Fenwick P, Schatzman M, Worsley A, Adams J, Stone S, Baker A (1984) Lucid dreaming: correspondence between dreamed and actual events in one subject during rem sleep. Biol Psychol 18(4):243–252
40. Fink A, Grabner RH, Benedek M, Reishofer G, Hauswirth V, Fally M et al (2009) The creative brain: investigation of brain activity during creative problem solving by means of eeg and fmri. Hum Brain Mapp 30:734–748
41. Finke RA, Ward TB, Smith SM (1992) Creative cognition: theory, research and applications. MIT Press, Cambridge
42. Fleming SM, Dolan RJ (2012) The neural basis of metacognitive ability. Phil Trans R Soc B 367:1338–1349
43. Fleming SM, Dolan RJ, Frith CD (2012) Metacognition: computation, biology, and function. Phil Trans R Soc B 367:1280–1286
44. Fleming SM, Weil RS, Nagy Z, Dolan RJ, Rees G (2010) Relating introspective accuracy to individual differences in brain structure. Science 329(5998):1541–1543
45. Foster DJ, Wilson MA (2006) Reverse replay of behavioural sequences in hippocampal place cells during the awake state. Nature 440:680–683
46. Fox KCR, Sedlmeier P, Nijeboer S, Dixon ML, Floman JL, Ellamil M, Rumak S, Christoff K (submitted) Does meditation alter brain structure? A review and meta-analysis of morphometric neuroimaging in meditation practitioners. Psychol. Bull
47. Fox KCR, Nijeboer S, Solomonova E, Domhoff GW, Christoff K (2013) Dreaming as mind wandering: evidence from functional neuroimaging and first-person content reports. Front Hum Neurosci 7(412):1–18
48. Fox KCR, Zakarauskas P, Dixon ML, Ellamil M, Thompson E, Christoff K (2012) Meditation experience predicts introspective accuracy. PLoS ONE 7(9):e45370
49. Fox M (1997) Red herring. Critique 6:56–57
50. Fox MD, Zhang D, Snyder AZ, Raichel ME (2009) The global signal and observed anticorrelated resting state brain networks. J Neurophysiol 101(6):3270–3283
51. Fox MD, Snyder AZ, Vincent JL, Corbetta M, Van Essen DC, Raichle ME (2005) The human brain is intrinsically organized into dynamic, anticorrelated functional networks. Proc Natl Acad Sci USA 102(27):9673–9678
52. Gallop D (1991) Aristotle on sleep and dreams. Cambridge University Press, Cambridge

53. Gard T, Hölzel BK, Sack AT, Hempel H, Lazar SW, Vaitl D, Ott U (2012) Pain attenuation through mindfulness is associated with decreased cognitive control and increased sensory processing in the brain. Cereb Cortex 22(11):2692–2702
54. Gardner H (1989) To open minds. Basic, New York
55. Geake JG, Hansen PC (2005) Neural correlates of intelligence as revealed by fMRI of fluid analogies. NeuroImage 26:555–564
56. Gelbard-Sagiv H, Mukamel R, Harel M, Malach R, Fried I (2008) Internally generated reactivation of single neurons in human hippocampus during free recall. Science 322:96–101
57. Gilbert SJ, Spengler S, Simons JS, Frith CD, Burgess PW (2006) Differential functions of lateral and medial rostral prefrontal cortex (area 10) revealed by brain-behavior associations. Cereb Cortex 16:1783–1789
58. Gilbert SJ, Spengler S, Simons JS, Steele JD, Lawrie SM, Frith CD, Burgess PW (2006) Functional specialization within rostral prefrontal cortex (Area 10): a meta-analysis. J Cogn Neurosci 18(6):932–948
59. Gillespie G (1988) Lucid dreams in Tibetan Buddhism. In: Gackenbach J, LaBerge SP (eds) Conscious mind, sleeping brain. Plenum Press, New York, pp 27–66
60. Goenka SN (2000) The discourse summaries. Vipassana Research Publications, Onalaska, p 144
61. Graham KS, Lee AC, Brett M, Patterson K (2003) The neural basis of autobiographical and semantic memory: new evidence from three PET studies. Cogn Affect Behav Neurosci 3(3):234–254
62. Greicius MD, Krasnow B, Reiss AL, Menon V (2003) Functional connectivity in the resting brain: a network analysis of the default mode hypothesis. Proc Natl Acad Sci USA 100:253–258
63. Hobson JA, Pace-Schott EF, Stickgold R (2000) Dreaming and the brain: toward a cognitive neuroscience of conscious states. Behav Brain Sci 23:793–842
64. Hölzel BK, Ott U, Gard T, Hempel H, Weygandt M, Morgen K, Vaitl D (2008) Investigation of mindfulness meditation practitioners with voxel-based morphometry. Soc Cog Affect Neurosci 3:55–61
65. Hölzel BK, Carmody J, Vangel M, Congleton C, Terramsetti SM, Gard T, Lazar SW (2011) Mindfulness practice leads to increase in regional brain gray matter density. Psych Res Neuroimaging 191:36–43
66. Hölzel BK, Lazar SW, Gard T, Schuman-Olivier Z, Vago DR, Ott U (2011) How does mindfulness meditation work? Proposing mechanisms of action from a conceptual and neural perspective. Perspect Psycholog Sci 6(6):537–559
67. Howard-Jones PA, Murray S (2003) Ideational productivity, focus of attention, and context. Creativity Res J 15:153–166
68. Howard-Jones PA, Blakemore S-J, Samuel EA, Summers IR, Claxton G (2005) Semantic divergence and creative story generation: an fMRI investigation. Cogn Brain Res 25(1):240–250
69. Israeli N (1962) Creative processes in painting. J Gen Psychol 67:251–263
70. Ives-Deliperi VL, Solms M, Meintjes EM (2011) The neural substrates of mindfulness: an fMRI investigation. Soc Neurosci 6(3):231–242
71. Job V, Dweck CS, Walton GM (2010) Ego depletion–Is it all in your head? Implicit theories about willpower affect self-regulation. Psychol Sci 21:1686–1693
72. Jung-Beeman M, Bowden EM, Haberman J, Frymiare JL, Arambel-Liu S, Greenblatt R et al (2004) Neural activity when people solve verbal problems with insight. PLoS Biol 2:500–510
73. Kabat-Zinn J (1994) Mindfulness meditation for everyday life. Hyperion, New York, p 304
74. Kahan TL, LaBerge SP (1994) Lucid dreaming as metacognition: implications for cognitive science. Conscious Cogn 3:246–264

75. Kane MJ, Brown LH, McVay JC, Silvia PJ, Myin-Germeys I, Kwapil TR (2007) For whom the mind wanders, and when: an experience-sampling study of working memory and executive control in daily life. Psychol Sci 18(7):614–621
76. Kang D, Jo HJ, Jung WH, Kim SH, Jung Y, Choi C, Lee US, An SC, Hang JH, Kwon JS (2013) The effect of meditation on brain structure: cortical thickness mapping and diffusion tensor imaging. Soc Cogn Affect Neurosci 8(1):27–33
77. Killingsworth MA, Gilbert DT (2010) A wandering mind is an unhappy mind. Science 330:932
78. Klinger E (1977) Meaning and void: inner experience and the incentives in people's lives. University of Minnesota Press, Minneapolis
79. Klinger E (1990) Daydreaming. Tarcher, Los Angeles
80. Klinger E (2008) Daydreaming and fantasizing: thought flow and motivation. In: Markman KD, Klein WMP, Suhr JA (eds) Handbook of imagination and mental simulation. Psychology Press, New York, pp 225–239
81. Klinger E, Cox WM (1987) Dimensions of thought flow in everyday life. Imagin Cogn Pers 7:105–128
82. Koestler A (1990) The act of creation. Arkana, London
83. Kounios J, Fleck JI, Green DL, Payne L, Stevenson JL, Bowden EM et al (2008) The origins of insight in resting-state brain activity. Neuropsychologia 46:281–291
84. Kowatari Y, Lee SH, Yamamura H, Nagamori Y, Levy P, Yamane S et al (2009) Neural networks involved in artistic creativity. Hum Brain Mapp 30:1678–1690
85. LaBerge SP, Rheingold H (1990) Exploring the world of lucid dreaming. Ballantine Books, New York
86. LaBerge SP, Nagel LE, Dement WC, Zarcone VP (1981) Lucid dreaming verified by volitional communication during REM sleep. Percept Motor Skills 52:727–732
87. Lane RD, Fink GR, Chau PM-L, Dolan RJ (1997) Neural activation during selective attention to subjective emotional responses. NeuroReport 8(18):3969–3972
88. Lazar SW, Kerr CE, Wasserman RH, Gray JR, Greve DN, Treadway MT, McGarvey M, Quinn BT, Dusek JA, Benson H, Rauch SL, Moore CI, Fischl B (2005) Meditation experience is associated with increased cortical thickness. NeuroReport 16:1893–1897
89. Leung M, Chan CCH, Yin J, Lee C, So K, Lee TMC (2013) Increased gray matter volume in the right angular and posterior parahippocampal gyri in loving-kindness meditators. Soc Cogn Affect Neurosci 8(1):34–39
90. Limb CJ, Braun AR (2008) Neural substrates of spontaneous musical performance: an fMRI study of jazz improvisation. PLoS ONE 3(2):e1679
91. Lodrö G, Hopkins J (1998) Calm abiding and special insight. Snow Lion Publications, Ithaca
92. Lou HC, Kjaer TW, Friberg L, Wildschiodtz G, Holm S, Nowak M (1999) A 15O-H2O PET study of meditation and the resting state of normal consciousness. Hum Brain Mapp 7:98–105
93. Luders E, Toga AW, Lepore N, Gaser C (2009) The underlying anatomical correlates of long-term meditation: larger hippocampal and frontal volumes of gray matter. NeuroImage 45:672–678
94. Luders E, Thompson PM, Kurth F, Hong J-Y, Phillips OR, Wang Y, Gutman BA, Chou Y-Y, Narr KL, Toga AW (2012) Global and regional alterations of hippocampal anatomy in long-term meditation practitioners. Hum Brain Mapp. doi:10.1002/hbm.22153
95. Lutz A, McFarlin DR, Perlman DM, Salomons TV, Davidson RJ (2013) Altered anterior insula activation during anticipation and experience of painful stimuli in expert meditators. NeuroImage 64:538–546
96. Lutz A, Slagter HA, Dunne JD, Davidson RJ (2008) Attention regulation and monitoring in meditation. Trends Cog Sci 12(4):163–169
97. Lutz A, Thompson E (2003) Neurophenomenology: Integrating subjective experience and brain dynamics in the neuroscience of consciousness. J Consciousn Stud 10:31–52

98. Manna A, Raffone A, Perrucci MG, Nardo D, Ferretti A, Tartaro A, Londei A, Del Gratta C, Belardinelli MO, Romani GL (2010) Neural correlates of focused attention and cognitive monitoring in meditation. Brain Res Bull 82:46–56
99. Mason M, Norton MI, Van Horn JD, Wegner DM, Grafton ST, Macrae CN (2007) Wandering minds: the default mode network and stimulus-independent thought. Science 315:393–395
100. McCaig RG, Dixon ML, Keramatian K, Liu I, Christoff K (2011) Improved modulation of rostrolateral prefrontal cortex using real-time fMRI training and meta-cognitive awareness. NeuroImage 55:1298–1305
101. Metcalfe J, Son LK (2012) Anoetic, noetic, and autonoetic metacognition. In: Beran M, Brandl JR, Perner J, Proust J (eds) Foundations of metacognition. Oxford University Press, Oxford
102. Montaigne M (1580/1910) Of diversion, essays of montaigne. Edwin C. Hill, New York
103. Moore A, Malinowski P (2009) Meditation, mindfulness, and cognitive flexibility. Conscious Cogn 18(1):176–186
104. Muraven M, Baumeister RF (2000) Self-regulation and depletion of limited resources: does self-control resemble a muscle? Psychol Bull 126(2):247–259
105. Muzur A, Pace-Schott EF, Hobson JA (2002) The prefrontal cortex in sleep. Trends Cog Sci 6(11):475–481
106. Nielsen L, Kaszniak AW (2006) Awareness of subtle emotional feelings: a comparison of long-term meditators and nonmeditators. Emotion 6(3):392–405
107. Nielsen TA, Deslauriers D, Baylor GW (1991) Emotions in dream and waking event reports. Dreaming 1:287–300
108. Ochsner KN, Gross JJ (2005) The cognitive control of emotion. Trends Cogn Sci 9(5):242–249
109. Ochsner KN, Ray RD, Cooper JC, Robertson ER, Chopra S, Gabrieli JDE, Gross JJ (2004) For better or for worse: neural systems supporting the cognitive down- and up-regulation of negative emotion. NeuroImage 23(2):483–499
110. Parnes SJ, Meadow A (1959) Effects of brainstorming instructions on creative problem-solving by trained and untrained subjects. J Educ Psychol 50(4):171–176
111. Prabhavananda S, Manchester F (2002) Breath of the eternal: the Upanishads. Signet Classics, New York
112. Raichle ME, MacLeod AM, Snyder AZ, Powers WJ, Gusnard DA & Shulman GL (2001) A default mode of brain function. Proc Nat Acad Sci U.S.A. 98(2): 678–682
113. Raichle ME (2011) The restless brain. Brain Connect 1:3–12
114. Ramnani N, Owen AM (2004) Anterior prefrontal cortex: insights into function from anatomy and neuroimaging. Nature Rev Neurosci 5:184–194
115. Ritskes R, Ritskes-Hoitinga M, Stodkilde-Jorgensen H, Baerentsen K, Hartman T (2003) MRI scanning during Zen meditation. Constructiv Human Sci 8(1):85–89
116. Roffwarg HP, Dement WC, Muzio JN, Fisher C (1962) Dream imagery: Relationship to rapid eye movements of sleep. Arch Gen Psychiatry 7:235–258
117. Rounis E, Maniscalco B, Rothwell JC, Passingham RE, Lau H (2010) Theta-burst transcranial magnetic stimulation to the prefrontal cortex impairs metacognitive visual awareness. Cogn Neurosci 1(3):165–175
118. Rugg MD, Wilding EL (2000) Retrieval processing and episodic memory. Trends Cogn Sci 4(3):108–115
119. Schacter DL, Addis DR (2009) On the nature of medial temporal lobe contributions to the constructive simulation of future events. Philos Trans R Soc B Biol Sci 364:1245–1253
120. Schmitz TW, Kawahara-Baccus TN, Johnson SC (2004) Metacognitive evaluation, self-relevance, and the right prefrontal cortex. NeuroImage 22:941–947
121. Schooler JW, Reichle ED, Halpern DV (2004) Zoning out while reading: evidence for dissociations between experience and metaconsciousness. In: Levitin DT (ed) Thinking and seeing: visual metacognition in adults and children. MIT Press, Cambridge, pp 204–226

122. Schooler JW, Smallwood J, Christoff K, Handy TC, Reichle ED, Sayette MA (2011) Meta-awareness, perceptual decoupling and the wandering mind. Trends Cog Sci 15(7):319–326
123. Schopenhauer A (2010) Parerga and paralipomena, vol 2. Oxford University Press, Oxford
124. Schredl M (2010) Characteristics and content of dreams. Int Rev Neurobiol 92:135–154
125. Schredl M, Erlacher D (2004) Lucid dreaming frequency and personality. Personal Individ Diff 37:1463–1473
126. Seger CA, Desmond JE, Glover GH, Gabrieli JDE (2000) Functional magnetic resonance imaging evidence for right-hemisphere involvement in processing unusual semantic relationships. Neuropsychology 14:361–369
127. Shannon BJ (2006) Functional anatomic studies of memory retrieval and the default mode. Washington University in St. Louis. St. Louis, MO
128. Shaw GA, Giambra LM (1993) Task-unrelated thoughts of college students diagnosed as hyperactive in childhood. Devel Neuropsychol 9:17–30
129. Shulman GL, Fiez JA, Corbetta M, Buckner RL, Miezin FM et al (1997) Common blood flow changes across visual tasks: II.: decreases in cerebral cortex. J Cogn Neurosci 9:648–663
130. Simonton DK (1999) Creativity as blind variation and selective retention: is the creative process Darwinian? Psychol Inq 10(4):309–328
131. Singer JL, McCraven V (1961) Some characteristics of adult daydreaming. J Psychol 51: 151–164
132. Singer JL (1966) Daydreaming: An introduction to the experimental study of inner experience. New York: Crown Publishing Group/Random House
133. Singer JL & Antrobus JS (1972) Daydreaming, imaginal processes, and personality: A normative study. The function and nature of imagery 175–202
134. Slagter HA, Davidson RJ, Lutz A (2011) Mental training as a tool in the neuroscientific study of brain and cognitive plasticity. Frontiers Human Neurosci 5:17
135. Smallwood J, Fishman DJ, Schooler JW (2007) Counting the cost of an absent mind: mind wandering as an underrecognized influence on educational performance. Psychonom Bull Rev 14(2):230–236
136. Smallwood J, Schooler JW (2006) The restless mind. Psychol Bull 132(6):946–958
137. Smith R, Keramatian K, Christoff K (2007) Localizing the rostrolateral prefrontal cortex at the individual level. NeuroImage 36:1387–1396
138. Spiers HJ, Maguire EA (2006) Spontaneous mentalizing during an interactive real world task: an fMRI study. Neuropsychologia 44(10):1674–1682
139. Stark CEL, Squire LR (2001) When zero is not zero: the problem of ambiguous baseline conditions in fMRI. Proc Nat Acad Sci USA 98:12760–12766
140. Sternberg RJ (1985) Implicit theories of intelligence, creativity, and wisdom. J Pers Soc Psychol 49:607–627
141. Sutherland GR, McNaughton B (2000) Memory trace reactivation in hippocampal and neocortical neuronal ensembles. Curr Opin Neurobiol 10:180–186
142. Sze JA, Gyurak A, Yuan JW, Levenson RW (2010) coherence between emotional experience and physiology: does body awareness training have an impact? Emotion 10(6):803–814
143. Tang YY, Lu Q, Geng X, Stein EA, Yang Y, Posner MI (2010) Short-term meditation induces white matter changes in the anterior cingulate. Proc Nat Acad Sci USA 107(35):15649–15652. doi:10.1073/pnas.1011043107
144. Teasdale JD (1999) Metacognition, mindfulness and the modification of mood disorders. Clin Psychol Psychother 6(2):146–155
145. Thera N (1954) The heart of buddhist meditation. Buddhist Publication Society, Kandy
146. Tholey P (1988) A model for lucidity training as a means of self-healing and psychological growth. In: Gackenbach J, LaBerge SP (eds) Conscious mind, sleeping brain. Plenum Press, New York, pp 263–287
147. Trick LM, Enns JT, Mills J, Vavrik J (2004) Paying attention behind the wheel: a framework for studying the role of attention in driving. Theor Issues Ergon Sci 5:385–424

148. Vanhaudenhuyse A, Demertzi A, Schabus M, Noirhomme Q, Bredart S, Boly M, Phillips C, Soddu A, Luxen A, Moonen G, Laureys S (2010) Two distinct neuronal networks mediate the awareness of environment and of self. J Cogn Neurosci 23(3):570–578
149. Varendonck J (1921) The psychology of daydreams. Macmillan, New York
150. Vartanian O, Martindale C, Kwiatkowski J (2007) Creative potential, attention, and speed of information processing. Personality Individ Differ 43:1470–1480
151. Velanova K, Jacoby LL, Wheeler ME, McAvoy MP, Petersen SE, Buckner RL (2003) Functional-anatomic correlates of sustained and transient processing components engaged during controlled retrieval. J Neurosci 23(24):8460–8470
152. Vestergaard-Poulsen P, van Beek M, Skewes J, Bjarkam CR, Stubberup M, Bertelsen J, Roepstorff A (2009) Long-term meditation is associated with increased grey matter density in the brain stem. NeuroReport 20:170–174
153. Voss U, Holzmann R, Tuin I & Hobson JA (2009) Lucid dreaming: A state of consciousness with features of both waking and non-lucid dreaming. Sleep, 32(9): 1191–1200
154. Wagman M (1968) University achievement and daydreaming behavior. J Couns Psychol 15:196–198
155. Wallace BA (2006) The attention revolution. Wisdom Publications Inc, Somerville
156. Wallas G (1926) The art of thought. Cape, London
157. Wamsley EJ, Tucker M, Payne JD, Benavides JA, Stickgold R (2010) Dreaming of a learning task is associated with enhanced sleep-dependent memory consolidation. Curr Biol 20:850–855
158. Wangyal Tenzin (1998) The Tibetan yogas of dream and sleep. Snow Lion Publications, Boston, p 217
159. Yeung N, Summerfield C (2014) Shared mechanisms for confidence judgments and error detection in decision making. In: Fleming S, Frith C (eds) The cognitive neuroscience of metacognition. Springer, Berlin
160. Yuschak T (2006) Advanced lucid dreaming. Lulu Enterprises, Raleigh

Chapter 14
What is the Human Sense of Agency, and is it Metacognitive?

Valerian Chambon, Elisa Filevich and Patrick Haggard

Abstract Agency refers to an individual's capacity to initiate and perform actions, and thus to bring about change, both in their own state, and in the state of the outside world. The importance of agency in human life cannot be understated. Social responsibility is built on the principle that there are "facts" of agency, on which individuals can generally agree. At the individual level, the experience of agency is considered a crucial part of normal mental life. Abnormal sense of agency (SoA)—such as in the well-documented "delusion of control"—is recognised as one of the key symptoms of mental disorders. Yet, beyond abnormalities of control that pertain to psychiatric conditions, normal SoA can be easily fooled. Errors in agency attribution and agency experience have received much attention in recent experimental literature. In everyday life, coincidental conjunctions between our actions and external events commonly occur. The fact that the SoA can be over or underestimated, or that judgements of agency can be wrong, testifies to a significant gap between what individuals think or believe their control capabilities are, and what these capabilities really are. The ability to experience these computations as the causes driving and shaping our actions may account for our ability to correct our behaviours when, precisely, they seem to escape our control. In this sense, any reliable theory about human agency must explain how we can sometimes be deluded about our own agency, but also must account for why we are not deluded all the time. In this chapter, we first identify which signals may contribute to an SoA, and how they might be integrated. We will ask whether human cognition of agency is best analysed as an *experience* or as an *inference*. We evaluate the existing data in relation to two contrasting accounts

V. Chambon
INSERM, ENS, Paris, France

E. Filevich
Max-Planck Institute for Human Development, Berlin, Germany

P. Haggard (✉)
ICN-UCL, London, UK
e-mail: p.haggard@ucl.ac.uk

S. M. Fleming and C. D. Frith (eds.), *The Cognitive Neuroscience of Metacognition*, 321
DOI: 10.1007/978-3-642-45190-4_14, © Springer-Verlag Berlin Heidelberg 2014

for agency, namely prospective versus purely retrospective approaches. We draw on two major classes of data throughout: psychological data that aims to capture the experience of agency, and physiological data that aims to identify the neural basis of this experience. Finally, we will consider whether the human SoA should really be called 'metacognitive'. In particular, we directly compare key features of metacognition of agency with perceptual metacognition.

14.1 What is Agency?

Agency refers to an individual's capacity to initiate and perform actions, and thus to bring about change, both in their own state, and in the state of the outside world. The importance of agency in human life cannot be understated. Societies depend on the idea that there are "facts" of agency, on which individuals can generally agree. This allows societies to hold individuals responsible for their own actions and for their consequences, thus rewarding or punishing the individual for what they do. Legal responsibility, and payment for labour provide two pervasive examples.

There are at least two aspects of agency. First, agency is an objective fact, demonstrated by individuals' behaviours and the consequences of those behaviours. But agency has a first-person component as well: it involves distinct cognitive processes and subjective experience unique to the agent. The experience of agency is considered a crucial part of normal mental life. Abnormal sense of agency (SoA)—as in the well-documented "delusion of control"—is recognised as one of the key symptoms of mental disorders. Further, links between SoA and health and well-being in the general population have been clearly established [7]. Nevertheless, the basis of the SoA is poorly understood.

Here we investigate what aspects of agency, if any, are metacognitive. We do this by analysing agency into a number of components, and by investigating how each component is computed in the human brain. We use two major distinctions to investigate the basis of SoA. First, we distinguish the types of *signals* contributing to SoA. This allows us to distinguish prospective SoA based on predictive signals linked to action intentions, from retrospective SoA based on action outcomes. Second, we distinguish the types of cognitive *processes* operating on those signals, to ask whether agency is best analysed as an *experience* or as an *inference*. In each case, we ask whether the particular component of agency can be considered metacognitive or not, and why. Finally, we compare the metacognition of agency with the features of the more widely studied perceptual metacognition.

We will draw on two major classes of data throughout: psychological data that aims to capture the experience of agency, and neural data that aims to identify the neural basis of this experience. The ability to *experience* these computations as the causes driving and shaping our actions, may account for the ability to correct our actions when action control is suboptimal. In this sense, any reliable theory about human agency must also explain how we can sometimes be deluded about our own agency, but also must account for why we are not deluded *all the time*.

To investigate whether SoA is or is not metacognitive, we need a clear definition of metacognition. Under a wide definition, metacognition is the general ability to monitor mental states and processes. This monitoring may be explicit or not. Explicit monitoring leads to meta-representations that allow reflecting upon, commenting about and reporting on the metal processes. Experience monitoring, however, may be implicit and may not allow explicit judgements based on first-order processes. Rather, the operation of the first-order processes is experienced. The concepts of first and second-order processing are central in current work on metacognition. Two main approaches have been taken, towards the study of metacognition. First, in psychophysical tests, experimenters may ask human volunteers to make simple (typically visual) perceptual judgements, and to also report their confidence on each of their responses [32]. Confidence judgements are thought to depend on purely internal aspects of the processing of first-order perceptual input signals. A second important body of work has investigated the relation between knowledge, and "knowing that you know" [42, 43]. In both cases, second-order processing within the brain itself generates an experience that can play a functional role in the organism's mental life and behaviour. In both cases, the distinguishing feature of metacognition is the presence of an internal, first-order signal as the content of a second-order representation or process.

Based on this view, we can now consider (a) which signals contribute to agency, (b) whether SoA is metacognitive in virtue of the nature of those signals, and why, (c) whether the SoA is similar, or essentially different, from other metacognitions, given that its 0th-order contents (i.e. actions and outcomes) differ from the 0th-order contents studied in other well-established areas of metacognition, such as perception (e.g. visual input) and knowledge (e.g. facts about the world). Addressing these questions must inevitably begin with a clear, analytical understanding of SoA.

14.2 Experiences of Agency

Agency can be defined from the point of view of an external observer, as it is related to the objective fact that individuals can make actions, and change their environment. However, agency also involves distinct cognitions and experiences on the part of the agent. Following Synofzik et al. [69] we use the term 'sense of agency' (SoA) to refer to the feeling or experience that individuals may have in relation to their own actions, and to the consequences of their actions, when *they* control those actions. We use the term 'judgement of agency' (JoA) to refer to an explicit judgement made by an individual regarding whether they, or another individual, brought about the action, or the external event. Note that both SoA and JoA are cognitive constructs rather than external physical facts. To this extent they are both subjective, rather than objective, and both can be wrong. For example, an individual can have an illusion of agency when they in fact are not the agent, as we will see later. Note also that JoA and SoA are normally related. In particular, an individual may judge that they are the agent of an event because they have an SoA with respect to that event. Likewise, and individual may judge they are not the agent, because they lack an SoA. We return to the relation between SoA and JoA later in this chapter.

The relationship between SoA and JoA is also important for the organization of societies. All known human societies depend on attribution of blame, and thus on individual responsibility for action. That is, societies depend on the idea that there are facts of agency, on which individuals can generally agree. Third-person judgements of agency must then have clear and objective truth conditions. However, agreement about judgements of agency and responsibility is only possible if individuals' brains support a subjective experience of agency. Only then will individuals feel and understand their actions and responsibilities, and accept society's third-person judgements of agency. To be useful, judgements of agency must align both with facts of agency, and with SoA, in most cases, though not necessarily in all. Therefore, the experience of agency is considered a crucial part of normal mental life. Abnormal SoA is recognised as one of the key symptoms of mental disorders, and links between SoA and health and well-being have been clearly established [7].

Despite this importance, SoA has only recently been addressed within cognitive science, perhaps because appropriate methods of measurement have been lacking. In particular, the SoA, as in general the experience of voluntary action, has been described as 'thin' and 'elusive'. Few psychophysical studies have sought to identify the factors that influence SoA and JoA.

14.3 Analytic Structure of Agency

14.3.1 Agency Impressionism

As a first step in an experimental analysis of SoA, we should characterise the experience of agency itself, and consider how it can be measured. On one view, agency is an atomic experience, and without any internal analytic structure. It is an impression that individuals directly and authoritatively have in cases where they are in fact responsible for an external sensory event. We call this view *agency impressionism*, by analogy with Michotte's concept of a causal impression [53]. Indeed, it was classically suggested that a direct impression of one's own motoric agency formed the basis for cognition of general causation in the external world [16]. A principal difficulty for agency impressionism is to explain why, if agency is directly perceived, illusions and misperceptions of agency may nevertheless occur [76].

14.3.2 Relational View

An alternative view is based on the *relational* aspect of agency. The facts of agency depend on a particular relation between an individual and an event, expressed by the proposition "I did that". On the relational view, these two components "I" and "that" remain present in the experience of agency itself. SoA is not, therefore, an immediate perceptual experience in the same way that a sound or a smell may be, because it involves a second-level relation between two primary elements, the agent and the event. Relational theories would view SoA as more than an atomic percept. Following Hume's view of causation [39], relations cannot be perceived directly. Rather, the relational view suggests that the mind supplies the relation between agent and event, based on the conjunction of the percepts of the cause (one's own intentional action), and the effect (the action outcome).

The relational view has two strong merits. First, it explains how SoA can be generalised, substituted or extrapolated from one case to another. An individual's SoA when they switch on a light has much in common with their SoA when they switch on a radio, or cause some similar event in the outside world. The "that" in "I did that" can be substituted. The agent and action remain the same, though the outcomes differ, and the feeling of being in control also remains broadly the same. By the same token, the SoA may be similar when an individual uses their hand or their head to switch on the light [34]. Similarly, the "I" in "I did that" can be substituted. One individual's SoA is assumed to be much like another's, so that one can understand another's SoA by observing the relation between their actions and subsequent outcomes [25]. The idea of agency as a relation implies that the key aspects of SoA should remain constant even when the basic content of action and outcome vary.

Recent experiments broadly confirm the relational view. The effects of time delays between action and outcome have been particularly extensively studied. In particular the intentional binding effect [35] reliably shows that actions are perceived as shifted in time towards the outcomes that they cause, while outcomes are perceived as shifted back in time towards the actions that cause them. This temporal attraction emphasises the temporal contiguity and conjunction between action and effect [39]. The effect is reduced or absent in cases of involuntary or passive movement [35]. Equally, when participants make numerical judgements about the interval between action and effect, their judgements show a perceptual compression, relative to intervals that begin with equivalent passive movements [20]. These data provide strong evidence that the relation between action and outcome is indeed a core component of SoA. They are also consistent with a broadly Humean associationist account of SoA. There may be no direct experience of agency over and above the experiences of the action and outcome itself, yet the mind may associate experiences of actions and outcomes so that they stand in a characteristic relation to each other. In the next section, we consider the signals that are related, and how the relation might be computed.

14.3.3 Signals for Agency

A person who grasps the relation "I did that" must be sensitive to two different signals, corresponding to the "I" and the "that". This suggests that SoA presupposes a relation between two quite distinct components, which we call *attribution* and *instrumentality*. Attribution concerns the "I" component. From a signal-processing point of view, someone who grasps that "I did that" must be capable of discriminating between themselves and other agents. That is, they must be able to attribute the outcome to "I", rather than "you", or any other cause. This requires a signal sensitive to *one's own* agency, i.e. some neural event that is present when one is the agent, and only when one is the agent. In philosophy, the direct, first-person access to one's own intentional states provides this signal [61]. In contrast, in neuroscience, awareness of one's own intentions remains controversial [29, 45], and the idea of immunity from error through self-identification has been questioned. Nevertheless, experimental studies show that self-recognition through active movement is superior to that with passive movement [71]. Efferent signals—i.e. signals that are sent from the brain's motor centres via the spinal cord to the muscles—therefore play an important role in discriminating "I" from other agents, and may provide the basis for attribution aspects of agency. Importantly, errors in agency attribution should then occur when efferent signals provide little discriminative information about agency, for example in situations where several people act at once.

Instrumentality refers to the "that" component of "I did that". To have an SoA, an individual must discriminate between events that she did cause, and events that she did not. Again, signals regarding one's intentional actions may play a key role. To know that "I did p", but "I did not do q", it may be sufficient to have access to an efferent signal that correlates well with p, and correlates poorly with q. Several studies have investigated the role of motor identity (i.e. response—stimulus associations), temporal relations and statistical contingency in the representation of agency [73]. In some cases, one can cause something to happen despite not intending to do so. One may even retrospectively acquire an SoA in such cases, by coming to believe that one *had* intended to do so. However, these are cases where SoA is decoupled from facts of agency. The primary task of a metacognitive account of agency is to deal with how our factual agency is experienced [60].

Interestingly, most previous studies of "agency", focus *either* on attribution ("I"), *or* on instrumentality ("that"), but do not clearly distinguish between the two aspects. This has lead to considerable confusion: often studies of "agency" meet with the reaction "that's not what we mean by agency". We believe that, in many cases, this critique should really be translated as "What does your account of instrumentality imply for attribution?", or "What does your account of attribution imply for instrumentality?".

The crucial link between attribution and instrumentality is that both depend on action signals. But a signal-processing approach clearly shows that these signals can provide information of two different kinds. In computing attribution, efferent signals are used to discriminate between agents, and can support explicit judgements of agency. In computing instrumentality, efferent signals are used to discriminate between outcomes. Efferent signals provide an SoA relating to some outcomes, but not others. The implications of this distinction for metacognition are discussed in Sect. 14.7.

14.4 Computational Models of Agency

14.4.1 Comparator Models

We previously took "I did that" as the cardinal expression of agency. This shows that agency also implies a specific process of action control (corresponding to "did") that relates these terms. Specifically, agency implies a control mechanism that has goals, and that controls actions to achieve them. This concept was successfully formalised as a comparator model [52, 78]. These models translate intentions into outcomes, by continually monitoring whether action consequences occur, or do not occur, as predicted. Though originally formulated as models of motor control, comparator models have also been increasingly used to explain the subjective SoA.

Because of their importance in the agency literature, we will present these models in some detail (see Fig. 14.1). A typical framework comprises two internal models: an inverse model and a forward model (see [30]). The desired goal is first fed to the inverse model which selects appropriate motor commands for achieving the goal. These commands are then executed, by sending them from the brain to the musculature. At the same time, a copy of the motor commands ("efference copy") is fed to the forward model, which predicts their effects. Thus, the motor control system can predict the current state of the body in advance of delayed sensory feedback about the effects of a motor command. This predictive information can then be used in two critical comparisons.

First, it can be compared to the desired goal state, to assess whether further motor commands are required, or whether the action has achieved its goal. Second, it can be compared to sensory information about actual effects of action. This second comparison assesses whether sensory information is or is not a predicted consequence of the current motor command. Crucially, this second comparison can distinguish self-caused sensory events (reafferences) from external events (exafferences). Thus, it functions as an agency-detector: no error means "I did that", while any error signal means "I didn't do that". On this view, agency can be attributed by low-level, pre-reflective mechanisms that learn to predict consequences of motor commands [9, 69].

Several studies confirm a role for motor prediction in agency judgement (see [14] for a review). Introducing a temporal or a spatial transformation between an action and its visual consequences reduces participants' sense of control in proportion to the mismatch induced. In one task, participants received distorted visual

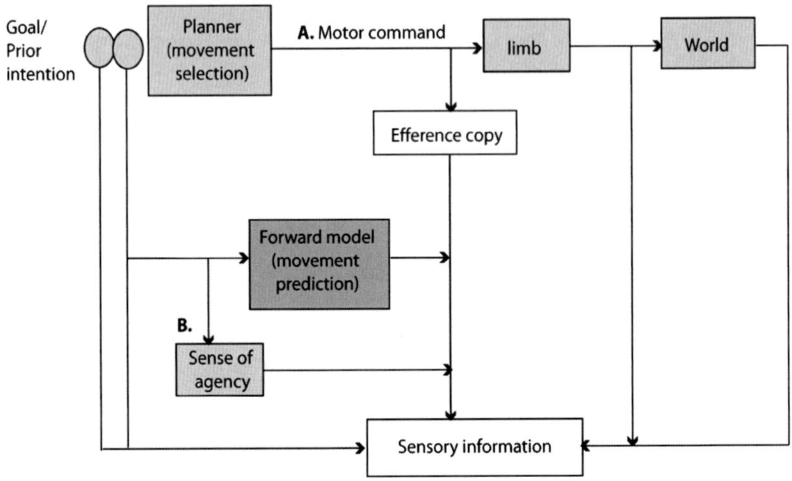

Fig. 14.1 A computational framework for action. Point *A* marks a point after movement selection where conscious awareness of intention might arise. Point *B* marks an integration of efference, predicted feedback and sensory information, which might lead to the sense of agency (adapted from Haggard 2005)

feedback of their hand moving a joystick. When the movement of the virtual hand did not correspond to the subjects' movement [23], or when an angular bias was introduced between the subject's and the virtual hand's movement, participants more readily attributed it to another agent [13, 21, 27, 68]. Note that manipulating temporal relations between actions and outcomes had similar effects [13, 15, 22, 28, 44, 49].

On comparator accounts, a positive SoA is the default operation when no error occurs. It is the experiential output of subpersonal processes that mostly run outside consciousness [69]. Crucially, although SoA relies on real-time motor signals, it can only be *computed* after those signals are compared with reafferent (visual, motor, or proprioceptive) feedback. Thus, a reliable, explicit SoA may only be formed when reafferent signals become available for matching with intentions. Thus, one cannot feel agency over any event until that event has been registered and processed in the brain. Although agency is informed by online signals about motor guidance and control, it can only be *retrospectively* attributed [9].

14.4.2 "Belief-like" Models

An alternative model treats agency not as a result of sensorimotor computations, but as an inference about authorship. Prior thought about an event, and general predictability of the event boost the experience of agency [3, 48, 63, 75]. This series of findings strongly suggests that agency does not simply depend on predictive motor signals. Instead, agency may be based on a general mechanism for estimating event likelihoods. When prior conscious thought about doing X co-occurs with X itself, a causal relationship is retrospectively *assumed* [39]—between the self and an external event, so that the event is inferred as having been caused through one's own will or action [74]. On this view, the experience of action would be necessarily *reconstructed* as an output of this secondary, belief-fixation mechanism. Thus, both belief and comparator models are reconstructive. Agency attribution, as a way of rationalising our actions and experiences, could thus primarily depend on conceptual, reflective processes or states—such as ad hoc theorising about oneself [72] or personal background beliefs [31]—, and not only on a signals within comparator. Importantly, belief-based models of agency allow that SoA is a consequence of JoA, rather than a cause.

14.5 Neural Bases of Agency

Reduced SoA following spatial and temporal mismatches between anticipated and actual action consequences is associated with increased activation in the angular gyrus (AG, [21–23]. Activation of AG should code for feelings of non-agency under ambiguous experience, rather than for positive self-agency experience [56]. The cerebellum may also signal discrepancies between predicted and actual sensory consequences of movements [5, 6, 59]. Other candidates for the comparator role have also been suggested, including premotor cortex [18, 19]. Interestingly, the opposite pattern of activation has been observed in the insula. Insula activation is positively correlated with control felt by subjects over visual consequences of their action [21]. However, this activation has also been interpreted as related to sense of body ownership, rather than agency [70].

By contrast, the belief-like account of agency might recruit higher cortical centres such as the prefrontal cortex (PFC), which provide conscious monitoring [67] rather than sensorimotor integration. Specifically, the dorsal lateral part of the PFC has been implicated in conflict monitoring and detection such as between intention and sensory outcome (e.g. [24, 66]). The supplementary and pre-supplementary motor areas might also be recruited when motor intention matches with a sensory feedback, to give rise to the intentional binding, mentioned before [56]. Finally, the interplay between these medial frontal areas and the PFC may be crucial for SoA. On one view, mismatches in cases of non-agency detected by AG are transmitted to PFC where alternative accounts of agency would be computed retrospectively (see [59]).

We may ask whether comparator models and belief models are truly metacognitive. That is, are judgements of agency generated by second-order processes that process purely internal signals? In the case of comparator models, the answer is a clear 'yes': the model is based on an efference copy and an internal predictor that operates in advance of action itself. If a signal that roughly corresponds to an internal intention contributes to SoA or JoA, then these states are, at least partly, metacognitive. However, it is much harder to prove that any particular JoA crucially depends on these internal signals. In particular, the internal signals are highly correlated with signals provided by the sensorimotor action itself. Thus, alternative, non-metacognitive accounts based on reconstructive inference from non-internal signals are always available. For example, if I judge that I switched on the light, the comparator model would view the judgement as driven by an efferent signal corresponding to the intention to switch on the light. However, the same judgement could also be an inference or assumption that one had switched on the light, driven by one's knowledge that it was dark, one's first-level experience that one's hand is touching the light switch, and one's first-level experience that the lights have come on [74].

Belief-like models need not be metacognitive. As the above example of the light switch shows, the belief model might begin with a "prior conscious thought" that it is dark. Somatosensory feedback from the hand on the switch, and visual feedback from the lights coming on are then sufficient to infer agency. This inference then leads to reconstruction of a first-person, explicit, judgement of agency. This judgement need not be related to first-order internal processes. Thus, previous studies linked to comparator and belief models provide only modest support for the view of agency as a form of metacognition.

In our view, only one class of evidence conclusively demonstrates that SoA is based on the internal efferent signals of the comparator model. Patients with anosognosia for hemiplegia [4] report an SoA over actions which they are, in fact, unable to make because of paralysis. This experience appears to be driven the internal signal corresponding to the intention to act [33]. Because of deficient feedback-based monitoring due to the lesion, this signal is sufficient to generate an SoA in the patients [26].

In addition to evidence from patient populations, we consider an alternative, more recent class of evidence from studies in healthy volunteers in the Sect. 14.6.

14.6 Beyond Comparators: Experiential Metacognition of Agency

14.6.1 Action Selection Contributes to Feelings of Control

Previous studies have shown that judgements of agency tend to be related to how participants think that they perform in a task [51]. Similarly, errors in task performance may lead to a *feeling* of something dysfluent during the task, without any explicit awareness of an error, and without ability to explicitly report the error (see [51]). The term 'epistemic feeling' has been coined to describe this subjective, online, experience of an error [12, 58]. Importantly, these epistemic feelings strongly influence the SoA, as shown by recent subliminal priming studies.

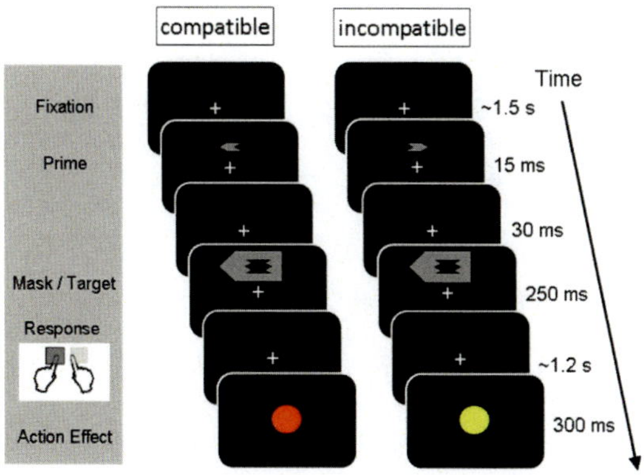

Fig. 14.2 Schematic of trial procedure and stimuli (cued-choice conditions only). Example trials from the two possible combinations of the prime-action compatibility (compatible: *left panel*; incompatible: *right panel*). The appearance of the effect was randomly jittered 150, 300 or 450 ms after the keypress to avoid ceiling effects in perceived control. Adapted from Wenke et al. [77]

We have recently identified a situation where an avowedly internal signal appears to contribute to the SoA [77]. We showed that the SoA could be modulated by using subliminal priming to affect the *fluency* of action selection processes. Interestingly, this procedure allowed us to manipulate the subjective sense of control, without manipulating the *predictability* of action outcomes. We interpret this as an implicit, non-conceptual form of metacognition [9, 58].

In this experiment, participants pressed left or right keys in response to left- or right-pointing arrow targets. Prior to the target, subliminal left or right arrow primes were presented, unbeknownst to the subject. Prime arrow directions were either identical (compatible condition) or opposite (incompatible condition) to the subsequent target (Fig. 14.2). Responding to the target caused the appearance of a colour after a jittered delay. The colour patch can thus be considered as the action outcome. The specific colour shown depended on whether the participant's action was compatible or incompatible with the preceding subliminal prime, but did not depend on the prime identity or the chosen action alternative alone. Unlike previous studies, therefore, the primes did not predict action effects, nor could any specific colour be predicted on the basis of the action chosen. Participants rated how much control they experienced over the different colours at the end of each block [77].

Analyses of reaction times showed that compatible primes facilitated responding whereas incompatible primes interfered with response selection. More importantly, priming also modulated the sense of control over action effects: participants experienced more control over colours that followed actions compatible with the preceding primes than over colours that followed prime-incompatible actions. Thus, subliminal priming made action selection processes more or less *fluent*, and this modulation of fluency affected the sense of control over action outcomes.

These results have several important cognitive implications. First, they suggest that the SoA depends strongly on processes of action selection that necessarily occur *before* action itself. Second, strong SoA may be associated with fluent, uncontested action selection. In contrast, conflict between alternative possible actions, such as that caused by incompatible subliminal priming, may reduce the feeling of control over action outcomes. Third, this prospective contribution of action selection processes to SoA is distinct from predicting the outcomes of action, since action outcomes were equally (un-) predictable for compatible and incompatible primes. That is, these primes did not prime effects of action as in previous studies (e.g. [1, 46, 62, 76]). Therefore, participants could not retrospectively base their control judgements on match between primes and effects alone. Rather, their stronger experience of control when primes were compatible could only be explained by the *fluency* of action selection—i.e. by a signal experienced *before* the action was made, and the effect was displayed.

Finally, participants did not consciously perceive the subliminal primes. Therefore, participants' sense of control could not be based on (conscious) beliefs about the primes. Instead, action priming itself presumably directly influenced the subjective sense of control. Pacherie [61] (see also [69]) has suggested that action selection conflict need not necessarily be conscious [57]. Such conflict may elicit the feeling "that something is wrong", without (necessarily) leading to knowledge about *what* is wrong. Wenke et al.'s study shows that subjects can rely on this first-person, implicit feeling to make judgements about their own control over action effects.

14.6.2 Dissociating Fluency of Action Selection from Performance Monitoring

Monitoring fluency signals generated *during* action selection could therefore be an important marker for the experience of agency. If so, agency would clearly have a metacognitive component, because these signals are generated internally by the process of action selection. However, it is also possible that participants might have estimated agency based on implicit monitoring of their own performance, such as their reaction times (RTs). Since RTs are lower on compatibly primed trials [17, 64, 65], participants would therefore feel more control on compatible trials, because they respond more rapidly. On this second view, agency would depend on *retrospective* monitoring of action execution performance [50], not on *prospective* monitoring of premotor fluency signals. Importantly, SoA would have a metacognitive aspect according to the latter view, but not the former.

To distinguish between these two accounts of sense of control, Chambon and Haggard [10] used an experimental procedure that dissociated fluency of action selection from performance monitoring. Specifically, they increased the interval between mask and target to take advantage of a Negative Compatibility Effect

(NCE) in priming. Longer mask-target latencies *increase* RTs following compatible primes, relative to incompatible primes [65]. By combining this factor with Wenke et al.'s design for assessing sense of control, it was possible to directly compare the contrasting retrospective (performance monitoring) and prospective (action selection) accounts. Specifically, if sense of agency depends on intentional fluency, it should be greater when actions are compatibly versus incompatibly primed, irrespective of whether priming benefits or impairs performance. Alternatively, if SoA depends only on performance monitoring, it should be stronger for rapid versus slower responding, irrespective of whether priming is compatible or incompatible with the action executed.

Crucially, reversing the normal relationship between prime-target compatibility and RTs did not alter subjective sense of control. Thus, in compatible NCE trials, participants experienced *stronger* control despite *slower* response times and higher error rates, compared to incompatible NCE trials. These results suggest that the feeling of control normally experienced by subjects on compatible trials does not depend on retrospectively monitoring performance, thereby strengthening the evidence for a prospective contribution of action selection fluency to SoA.

In both Wenke et al. [77] and Chambon and Haggard [10] experiments, priming did not influence the actual objective level of control that participants had over the colours presented after their actions. Indeed, the contingency between action and colour effect was similar for compatibly primed and incompatibly primed trials. So the prospective sense of control identified in these experiments is in fact an illusion of control, since it is not based on differences in the actual statistical relation between action and effect. In other words, action selection is irrelevant to actual action/effect contingency, and thus to the agent's actual ability to drive external events. However, this prospective sense of control may nevertheless be a convenient proxy for actual control, because we often just know what to do and what will happen next. In that sense, *fluent* action selection is generally a good advance predictor of actual statistical control over the external environment [37]. If prospective agency is a particular conscious experience generated by action programming, which we learn to use as a convenient marker of our own factual agency, it might indeed qualify as a metacognition.

14.6.3 Prospective Agency: Neural Underpinnings

Taken together, these findings suggest that neural activity in action preparation circuits *prospectively* informs agency, independent of outcome predictability of the outcome, and actual performance. Tracking dysfluency in action selection networks [54, 59] could be the basis for this prospective SoA. Recently, Chambon and collaborators [11] adapted the prospective agency paradigm for functional neuroimaging (fMRI). They studied whether the angular gyrus (AG), which has been shown to compute *retrospective* agency by monitoring mismatches between actions and subsequent outcomes [21, 22], may also code for a *prospective* sense

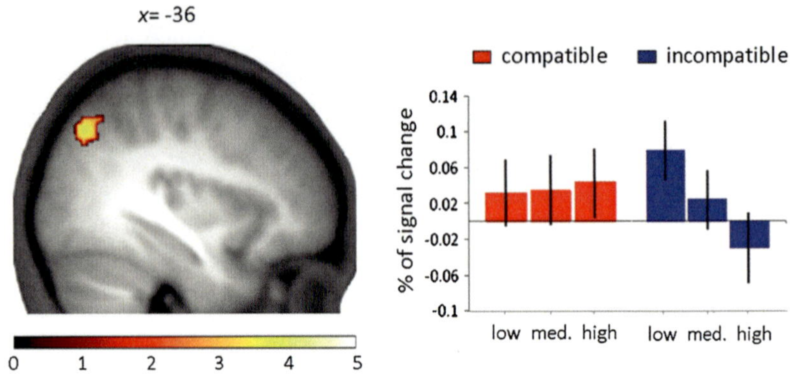

Fig. 14.3 Parametric interaction of control and compatibility in the angular gyrus (AG). Left AG is differentially modulated by participants' control ratings depending on how fluent action selection is; scale shows *t*-value. Adapted from Chambon et al. [11]

of control, by monitoring action selection processes in advance of the action itself, and independently of action outcomes. This would inform one *whether one's actions are appropriately following through one's original intentions*. If a dysfluency, or causal break between intention and action occurs, SoA over outcomes would be reduced.

Again, participants experienced greater control over action effects when the action was compatibly versus incompatibly primed. More importantly, this prospective contribution of action selection processes to SoA was accounted for by exchange of signals between specific frontal action selection areas and the parietal cortex. First, Chambon et al. found that activity in the angular gyrus was sensitive to mismatches, but not matches, between prime arrow and actual response to the target arrow. Moreover, this activity due to the prime-target mismatch predicted the magnitude of subsequent sense of control: for incompatible trials only, activity in the AG decreased as sense of control over outcomes increased (Fig. 14.3a). Importantly, this neural coding of non-agency occurred at the time of action selection *only*, as in Wenke et al.'s original experiment.

Second, activity in the AG (signalling non-agency) in incompatible trials was negatively correlated with activity in the dorso-lateral prefrontal area (DLPFC) (Fig. 14.3b). This pattern of fronto-parietal interaction would reflect contribution of action selection brain areas to sense of control. Indeed, DLPFC has long been associated with top-down cognitive control and selection of appropriate responses according to current instructions or task demands [41, 55]. In particular, a key control function of DLPFC is to resolve conflicts by allowing responses with weaker activation levels to gain priority over stronger ones under appropriate circumstances. In incompatible trials, DLPFC may therefore provide conflict resolution between action alternatives (i.e. left or right key press), through reducing activations for incompatibly-primed responses. Since AG activation negatively correlates with the subjective sense of control in incompatible trials only, strong executive contribution of DLPFC to resolve conflict in these trials would produce a weaker activation of AG, corresponding to a greater sense of control. Overall, this suggests that AG may monitor signals of conflict resolution generated during action selection within DLPFC, to prospectively inform subjective judgements of control over action outcomes.

14.7 So is Agency Metacognitive?

Metacognition is a relatively broad term that encompasses a variety of different processes. These processes all have in common that they are second-order representations of first-order mental states. The first-order mental states that are meta-represented can range from simple forms of visual perception, in cases of perceptual confidence in detection judgements, to knowledge [42], to (perhaps) agency. The first-order mental processes and states of visual perception are clearly very different from those of action control, computationally, neutrally and phenomenally. Could there then be a single second-order process that monitors them, or does each type of first-level process require its own content-specific metacognitive monitoring circuit? A common, metacognitive monitoring circuit for all first-order processes would imply a strongly hierarchical, quasi-homuncular, cognitive organization. In contrast, independent metacognitive systems would imply a highly distributed mechanism.

To answer the question of domain-generality versus domain-specificity, we examine each alleged metacognitive domain in turn, and draw comparisons between them.

In the case of agency, Miele et al. [54] argue that JoA are metacognitive and meta-representational for two reasons. First, judgements of agency are conscious, in contrast to action monitoring and action correction, which may be unconscious [8]. Second, according to Miele et al., JoA are meta-representational. By this, it is meant that judgements of agency take first-order action representations as their content. However, this latter point seems problematic. If judgements of agency are judgements about "my actions", then Miele et al.'s point stands. However, this alternative, retrospective, inferential view would not require judgements of agency to be metacognitive. If judgements of agency were simply narrative explanations of somatosensory input ("why my body moved"), then the content is not a first-level representation of action, but, ultimately, a basic-level somatosensory signal. The critical distinction seems to be whether internal, efferent signals tag body movements as being specifically "my action". If they do, then agency is metacognitive. But, if they do not, then agency may not be based on any first-order mental states, and should therefore not be considered as a metacognitive process.

It has long been argued by ideomotor theorists that retrospective SoA is only possible because (1) the computations underlying motor control are largely unconscious, for reasons of cognitive economy, and (2) the consciously available information regarding action is largely a representation of action *effects* [38].

However, we have shown that even "vague" signals such as selection fluency, which do not produce conscious phenomenology in the conventional sense, can nevertheless participate in distinctive, first-person experiences, such as SoA. The results described above show that one can experience the cause of a conflict arising during action planning/selection even though the cause behind this conflict cannot be fully represented (or defined, or named, or even accurately identified). In other words, such conflict may elicit the feeling "that something is wrong", without (necessarily) leading to awareness or knowledge about *what* is wrong (see page 331 of this chapter). Similarly, fluent action selection appears to contribute to the "buzz" of agency, because the agent *just knows* what to do. The vagueness of these feelings is interesting: because one cannot perfectly, and consciously, represent what causes this experience of fluency or conflict, the feeling is easily mistaken for something else. In our case, selection fluency is interpreted or experienced as real agency, and selection conflict is experienced as reduced agency.

This aspect of agency experience is clearly metacognitive in the sense that it is driven by the first-level motor signals associated with planning and selecting an action. However, the prospective SoA that we have described refers to these signals in a purely experiential, rather than in a representational, format. Put differently, SoA is second-order (in the sense that it directed to identifiable first-level signals), but it is not a second-order *representation*, because the specific first-level content does not form part of the second-level phenomenology. In the terms of Muñoz [58], action selection fluency contributions to prospective agency would fit better with a "control" theory of metacognition than with a "meta-representational" theory.

14.7.1 Metacognition of Agency Versus Perceptual Metacognition

Because metacognition can be so clearly defined in the domain of perception, it is useful to compare perceptual metacognition and agency metacognition. Muñoz [58] argued that a second-order (or metacognitive) representation requires "the self attribution of a mental concept together with a first-order representation". This is easily operationalised, in the case of perceptual metacognition, by the example "I know that I see that it rains". In this example, seeing that it rains is a first-order representation of an event in the external world. "I know that I see" is, in turn, the second-order representation. However, this definition of metacognition will not do for agency, because of some specific peculiarities of action signals.

First, agency judgements do not directly represent action fluency signals, at least not in the same way that "I know that I see" represents visual signals. In particular, action fluency signals lack strong phenomenology [47], and do not form a clear first-order representation. People commonly act without being fully aware

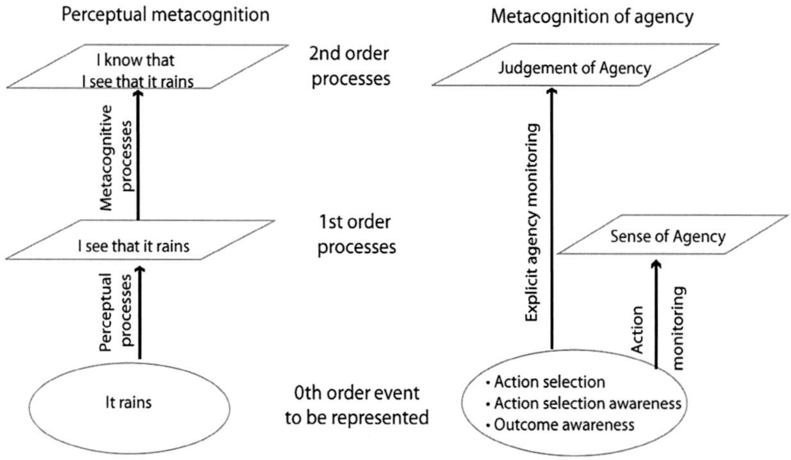

Fig. 14.4 Comparison between perceptual metacognition (*left panel*) and metacognition of agency (*right panel*). In metacognition of perception, the first-order processes are thought to be accessible to hierarchical second-order metarepresentational processes. The key feature of agency, on the other hand, is the parallel existence of two monitoring processes: A first-order process of action monitoring, and a second-order process of explicit agency monitoring both depend on the same zero-th order events, but via dissociable paths. The two paths may contribute to SoA and to JoA, respectively (see text for full details)

of their actions, leading to the textbook observation that action control is often automatic [12]. In contrast, pace the special case of blindsight, vision often provides strong phenomenology with clear first-order representational content that can be monitored, evaluated and described. To summarise, we suggest that basic-level sensorimotor signals may be processed in either or both of two dissociable ways (see schematic Fig. 14.4). They may be used by first-level processing for action monitoring, or by second-level processing for agency judgement. However, these two routes are independent, and dissociable. Unconscious adjustment of actions demonstrates the possibility of first-level processing without metacognitive SoA [8, 40]. Anosognosia for hemiplegia provides an unusual case of SoA for actions where first-level processing is effectively absent because of primary sensorimotor damage.

There is therefore a dissociable parallelism between SoA (the 'first-order' signal) and JoA (the 'second-order' signal). This implies that the two cognitive processes, namely unconscious movement monitoring and conscious agency evaluation depend on the same underlying signal, but not on each other. This is strikingly opposite to what happens in cases of perceptual metacognition, in which the hierarchically organised second-order process is formed by directly accessing the first-order process.

14.8 Implications

The nature of intentional action and agency are hotly disputed. This may be because the societal importance of these concepts is so widely recognised. Thus, individuals are responsible for their own actions, and outcomes of these actions, before society and before the law. Further, the legal concept of *mens rea* implies that individuals consciously intend particular outcomes, and that their actions realise those intentions. That is, legal responsibility depends on an SoA, which is present as part of the intentional generation of action, and at the time of controlling one's actions. However, this view has recently come under attack from two quite distinct forms of determinism. First, neurobiological determinism holds that conscious intentions are not directly controlled by persons, but are rather conscious consequences of unconscious neural events in the brain that prepare actions. Holding people responsible for unconscious, neurobiological events seems at odds with traditional ideas of 'free will' on which legal responsibility is based [36]. A second, rather different version of determinism is equally problematic for traditional ideas of legal responsibility. Social-psychological determinism suggests that people's 'voluntary' behaviour is in fact caused by subtle, often social influences of which they may be quite unaware [2].

The questions of free will, determinism and agency have been debated many times. Here, we have identified a prospective aspect of SoA, based on experimental analyses. Therefore, we simply ask, what implications does a prospective SoA have for the ideas of voluntary action and legal responsibility. First, our work shows that frontal executive processes for planning action are involved in the SoA. Our work therefore supports the idea that people are aware of actions and action outcomes (just) before they act. The neurobiological machinery that underlies planning and volition can also process action outcomes. On the other hand, our work clearly shows that this system can be driven by subliminal primes. In that sense, it is not strictly voluntary, in the sense that intentional action selection is not truly endogenous, but driven by an *external* prime.

14.9 Conclusions

To summarise, experimental analyses of responsibility suggest that the sense of being in control of one's own actions, and through them the external world, can be studied experimentally. We have distinguished between attributional and instrumental aspects of SoA. Most research to date has focused on neural mechanisms that match the predicted and actual consequences of action, and these mechanisms can be used for computing either instrumentality or attribution. These mechanisms are necessarily reconstructive, since they rely on delayed action consequences. We have argued that SoA also depends on a prospective aspect, in which fluent selection between alternative actions in the frontal cortex is monitored by parietal mechanisms at the time of action selection itself. A component of agency is therefore computed in advance of action execution, based on a purely internal signal. This aspect of agency must, in our view, be metacognitive, but it is an experiential rather than a judgemental form of agency. Finally, we have suggested a peculiarity of agency judgement, lacking in perceptual judgement. Specifically, the 'automatic' nature of action processing means that first-level processing of action signals can occur without second-level, metacognitive, explicit self-attributive JoA. Thus, voluntary actions are generally accompanied by prospective SoA, which may be termed metacognitive. People may also make retrospective judgements of agency, which may or may not be metacognitive, depending on whether they are simply inferences about action events, or depend on internal action signals. The implications of prospective agency for voluntary control of action and legal responsibility require future research.

Acknowledgments PH was supported by an ESRC Professorial Fellowship, by EU FP7 project VERE (WP8), and by ERC Advanced Grant HUMVOL.

VC was supported by a postdoctoral bursary from the Fyssen foundation, and by EU FP7 project VERE (WP8).

References

1. Aarts H, Custers R, Wegner DM (2005) On the inference of personal authorship: enhancing experienced agency by priming effect information. Conscious Cogn 14:439–458
2. Ackerman JM, Nocera CC, Bargh JA (2010) Incidental haptic sensations influence social judgments and decisions. Science 328:1712–1715
3. Banks WP, Isham EA (2009) We infer rather than perceive the moment we decided to act. Psychol Sci 20:17–21
4. Berti A, Bottini G, Gandola M et al (2005) Shared cortical anatomy for motor awareness and motor control. Science 309:488–491
5. Blakemore S-J, Frith CD, Wolpert DM (2001) The cerebellum is involved in predicting the sensory consequences of action. NeuroReport 12:1879–1884
6. Blakemore S-J, Sirigu A (2003) Action prediction in the cerebellum and in the parietal lobe. Exp Brain Res 153:239–245
7. Bobak M, Pikhart H, Rose R et al (2000) Socioeconomic factors, material inequalities, and perceived control in self-rated health: cross-sectional data from seven post-communist countries. Soc Sci Med 51:1343–1350
8. Castiello U, Paulignan Y, Jeannerod M (1991) Temporal dissociation of motor responses and subjective awareness a study in normal subjects. Brain 114:2639–2655
9. Chambon V, Haggard P (2013) 14 Premotor or Ideomotor: how does the experience of action come about? Action Sci Found Emerg Discipl 359
10. Chambon V, Haggard P (2012) Sense of control depends on fluency of action selection, not motor performance. Cognition
11. Chambon V, Wenke D, Fleming SM et al (2013) An online neural substrate for sense of agency. Cereb Cortex 23:1031–1037
12. Charles L, van Opstal F, Marti S, Dehaene S (2013) Distinct brain mechanisms for conscious versus subliminal error detection. NeuroImage
13. David N, Cohen MX, Newen A et al (2007) The extrastriate cortex distinguishes between the consequences of one's own and others' behavior. Neuroimage 36:1004–1014

14. David N, Newen A, Vogeley K (2008) The "sense of agency" and its underlying cognitive and neural mechanisms. Conscious Cogn 17:523–534
15. David N, Stenzel A, Schneider TR, Engel AK (2011) The feeling of agency: empirical indicators for a pre-reflective level of action awareness. Front, Psychol 2
16. De Biran M (1841) Oeuvres philosophiques. Ladrange
17. Dehaene S, Naccache L, Le Clec'H G et al (1998) Imaging unconscious semantic priming. Nature 395:597–600
18. Desmurget M, Reilly KT, Richard N et al (2009) Movement intention after parietal cortex stimulation in humans. Science 324:811
19. Desmurget M, Sirigu A (2009) A parietal-premotor network for movement intention and motor awareness. Trends Cogn Sci 13:411–419. doi:10.1016/j.tics.2009.08.001
20. Engbert K, Wohlschläger A, Haggard P (2008) Who is causing what? The sense of agency is relational and efferent-triggered. Cognition 107:693–704
21. Farrer C, Franck N, Georgieff N et al (2003) Modulating the experience of agency: a positron emission tomography study. Neuroimage 18:324–333
22. Farrer C, Frey SH, Van Horn JD et al (2008) The angular gyrus computes action awareness representations. Cereb Cortex 18:254–261
23. Farrer C, Frith CD (2002) Experiencing oneself vs another person as being the cause of an action: the neural correlates of the experience of agency. Neuroimage 15:596–603
24. Fink GR, Marshall JC, Halligan PW et al (1999) The neural consequences of conflict between intention and the senses. Brain 122:497–512
25. Fogassi L, Ferrari PF, Gesierich B et al (2005) Parietal lobe: from action organization to intention understanding. Science 308:662–667
26. Fotopoulou A, Tsakiris M, Haggard P et al (2008) The role of motor intention in motor awareness: an experimental study on anosognosia for hemiplegia. Brain 131(12):3432–3442
27. Fourneret P, Jeannerod M (1998) Limited conscious monitoring of motor performance in normal subjects. Neuropsychologia 36:1133–1140
28. Franck N, Farrer C, Georgieff N et al (2001) Defective recognition of one's own actions in patients with schizophrenia. Am J Psychiatry 158:454–459
29. Fried I, Mukamel R, Kreiman G (2011) Internally generated preactivation of single neurons in human medial frontal cortex predicts volition. Neuron 69:548–562
30. Frith CD, Blakemore SJ, Wolpert DM (2000) Abnormalities in the awareness and control of action. Philos Trans R Soc Lond B Biol Sci 355:1771–1788
31. Gallagher S (2004) Neurocognitive models of schizophrenia: a neurophenomenological critique. Psychopathology 37:8–19
32. Galvin SJ, Podd JV, Drga V, Whitmore J (2003) Type 2 tasks in the theory of signal detectability: discrimination between correct and incorrect decisions. Psychon Bull Rev 10:843–876
33. Garbarini F, Rabuffetti M, Piedimonte A et al (2012) "Moving" a paralysed hand: bimanual coupling effect in patients with anosognosia for hemiplegia. Brain 135:1486–1497. doi:10.1093/brain/aws015
34. Gergely G, Bekkering H, Király I (2002) Developmental psychology: rational imitation in preverbal infants. Nature 415:755
35. Haggard P, Clark S, Kalogeras J (2002) Voluntary action and conscious awareness. Nat Neurosci 5:382–385
36. Haggard P, Libet B (2001) Conscious intention and brain activity. J Conscious Stud 8:47–64
37. Haggard P, Chambon V (2012) Sense of agency. Curr Biol 22:390–392
38. Hommel B, Musseler J, Aschersleben G, Prinz W (2001) The theory of event coding (TEC): a framework for perception and action planning. Behav Brain Sci 24:849–877
39. Hume D (1978) A treatise of human nature [1739]. Br Moralists 1650–1800
40. Johnson H, van Beers RJ, Haggard P (2002) Action and awareness in pointing tasks. Exp Brain Res 146:451–459
41. Koechlin E, Ody C, Kouneiher F (2003) The architecture of cognitive control in the human prefrontal cortex. Science 302:1181–1185. doi:10.1126/science.1088545

42. Koriat A (1993) How do we know that we know? The accessibility model of the feeling of knowing. Psychol Rev 100:609
43. Koriat A (2012) The subjective confidence in one's knowledge and judgements: some. Found Metacognition 213
44. Leube DT, Knoblich G, Erb M et al (2003) The neural correlates of perceiving one's own movements. Neuroimage 20:2084–2090
45. Libet B, Gleason CA, Wright EW, Pearl DK (1983) Time of conscious intention to act in relation to onset of cerebral activity (readiness-potential). The unconscious initiation of a freely voluntary act. Brain. J Neurol 106(Pt 3):623–642
46. Linser K, Goschke T (2007) Unconscious modulation of the conscious experience of voluntary control. Cognition 104:459–475
47. Logan GD, Crump MJC (2010) Cognitive illusions of authorship reveal hierarchical error detection in skilled typists. Science 330:683–686
48. Lynn MT, Berger CC, Riddle TA, Morsella E (2010) Mind control? Creating illusory intentions through a phony brain–computer interface. Conscious Cogn 19:1007–1012
49. MacDonald PA, Paus T (2003) The role of parietal cortex in awareness of self-generated movements: a transcranial magnetic stimulation study. Cereb Cortex 13:962–967
50. Marti S, Sackur J, Sigman M, Dehaene S (2010) Mapping introspection's blind spot: reconstruction of dual-task phenomenology using quantified introspection. Cognition 115:303–313. doi:10.1016/j.cognition.2010.01.003
51. Metcalfe J, Greene MJ (2007) Metacognition of agency. J Exp Psychol Gen 136:184–199. doi:10.1037/0096-3445.136.2.184
52. Miall RC, Wolpert DM (1996) Forward models for physiological motor control. Neural Netw 9:1265–1279
53. Michotte A (1963) The perception of causality
54. Miele DB, Wager TD, Mitchell JP, Metcalfe J (2011) Dissociating neural correlates of action monitoring and metacognition of agency. J Cogn Neurosci 23:3620–3636
55. Miller EK, Cohen JD (2001) An integrative theory of prefrontal cortex function. Annu Rev Neurosci 24:167–202. doi:10.1146/annurev.neuro.24.1.167
56. Moore JW, Ruge D, Wenke D et al (2010) Disrupting the experience of control in the human brain: pre-supplementary motor area contributes to the sense of agency. Proc R Soc B Biol Sci 277:2503–2509
57. Morsella E, Wilson LE, Berger CC et al (2009) Subjective aspects of cognitive control at different stages of processing. Atten Percept Psychophys 71:1807–1824. doi:10.3758/APP.71.8.1807
58. Muñoz SA (2010) Metarepresentational versus control theories of metacognition. Work. Twenty-Fourth AAAI Conference of Artificial Intelligence
59. Nahab FB, Kundu P, Gallea C et al (2011) The neural processes underlying self-agency. Cereb Cortex 21:48–55
60. Pacherie E (2007) The sense of control and the sense of agency. Psyche (Stuttg) 13:1–30
61. Pacherie E (2008) The phenomenology of action: a conceptual framework. Cognition 107:179–217
62. Sato A (2009) Both motor prediction and conceptual congruency between preview and action-effect contribute to explicit judgment of agency. Cognition 110:74–83
63. Sato A, Yasuda A (2005) Illusion of sense of self-agency: discrepancy between the predicted and actual sensory consequences of actions modulates the sense of self-agency, but not the sense of self-ownership. Cognition 94:241–255
64. Schlaghecken F, Eimer M (2000) A central-peripheral asymmetry in masked priming. Percept Psychophys 62:1367–1382
65. Schlaghecken F, Rowley L, Sembi S et al (2007) The negative compatibility effect: a case for self-inhibition. Adv Cogn Psychol 3:227
66. Schnell K, Heekeren K, Schnitker R et al (2007) An fMRI approach to particularize the frontoparietal network for visuomotor action monitoring: detection of incongruence between test subjects' actions and resulting perceptions. Neuroimage 34:332–341

67. Slachevsky A, Pillon B, Fourneret P et al (2001) Preserved adjustment but impaired awareness in a sensory-motor conflict following prefrontal lesions. J Cogn Neurosci 13:332–340
68. Synofzik M, Thier P, Lindner A (2006) Internalizing agency of self-action: perception of one's own hand movements depends on an adaptable prediction about the sensory action outcome. J Neurophysiol 96:1592–1601
69. Synofzik M, Vosgerau G, Newen A (2008) Beyond the comparator model: a multifactorial two-step account of agency. Conscious Cogn 17:219–239. doi:10.1016/j.concog.2007.03.010
70. Tsakiris M, Costantini M, Haggard P (2008) The role of the right temporo-parietal junction in maintaining a coherent sense of one's body. Neuropsychologia 46:3014–3018
71. Tsakiris M, Haggard P, Franck N et al (2005) A specific role for efferent information in self-recognition. Cognition 96:215–231
72. Vosgerau G, Newen A (2007) Thoughts, motor actions, and the self. Mind Lang 22:22–43
73. Waszak F, Cardoso-Leite P, Hughes G (2012) Action effect anticipation: neurophysiological basis and functional consequences. Neurosci Biobehav Rev 36:943–959
74. Wegner D (2002) The illusion of conscious will. MA, Cambridge
75. Wegner D, Sparrow B, Winerman L (2004) Vicarious agency: experiencing control over the movements of others. J Pers Soc Psychol 86:838
76. Wegner D, Wheatley T (1999) Apparent mental causation: sources of the experience of will. Am Psychol 54:480–492
77. Wenke D, Fleming SM, Haggard P (2010) Subliminal priming of actions influences sense of control over effects of action. Cognition 115:26–38
78. Wolpert DM, Ghahramani Z, Jordan MI (1995) An internal model for sensorimotor integration. Sci-NEW YORK THEN Wash- 1880–1880

Part IV
Neuropsychiatric Disorders of Metacognition

Chapter 15
Failures of Metacognition and Lack of Insight in Neuropsychiatric Disorders

Anthony S. David, Nicholas Bedford, Ben Wiffen and James Gilleen

Abstract Lack of insight or unawareness of illness is the hallmark of many psychiatric disorders, especially schizophrenia (SCZ) and other psychoses, and could be conceived of as a failure in metacognition. Research in this area in the mental health field has burgeoned with the development and widespread use of standard assessment instruments and the mapping out of the clinical and neuropsychological correlates of insight and its loss. There has been a growing appreciation of the multifaceted nature of the concept and of the different *objects* of insight such as the general awareness that one is ill, to more specific metacognitive awareness of individual symptoms, impairments and performance. This in turn has led to the notion that insight may show modularity and may fractionate across different domains and disorders, supported by work which directly compares metacognition of memory deficits and illness awareness in patients with SCZ, Alzheimer's disease (AD) and brain injury (BI). The focus of this chapter will be on the varieties of metacognitive failure in psychiatry, particularly the psychoses. We explore cognitive models based on self-reflectiveness and their possible social and neurological bases including data from structural and functional MRI. The medial frontal cortex appears to play an important role in self-appraisal in health and disease.

Keywords Insight · Awareness · Schizophrenia · Psychosis · Self-reflectiveness · Neuroimaging

---

This chapter is adapted from: David AS, Bedford N, Wiffen B, Gilleen J (2012) Failures of metacognition and lack of insight in neuropsychiatric disorders. Phil Trans R Soc B 367:1379–1390

A. S. David (✉) · N. Bedford · B. Wiffen · J. Gilleen
Institute of Psychiatry, King's College London, London, UK
e-mail: Anthony.david@kcl.ac.uk

N. Bedford
e-mail: Nicholas.bedford@kcl.ac.uk

J. Gilleen
e-mail: james.1.gilleen@kcl.ac.uk

S. M. Fleming and C. D. Frith (eds.), *The Cognitive Neuroscience of Metacognition*, DOI: 10.1007/978-3-642-45190-4_15,

Everyone complains of his lack of memory, but nobody of his want of judgment.

de Rochefoucauld (1613–1680)

15.1 Introduction

The topic of metacognition has had a huge stimulating effect on what might be termed cognitive neuropsychiatry—that is, the field which, 'seeks to promote the study of cognitive processes underlying psychological and behavioural abnormalities'. In particular the sorts of abnormalities which have been illuminated when viewed in the light of metacognition include most notably, autistic deficits (not considered here) but also psychotic symptoms. In a volume dedicated to metacognition and severe adult mental disorder, Saxe and Offen [1] described two meanings for the term in this context. The first they called "attributive metacognition" which concerns the ability to attribute beliefs and desires to oneself and was seen as a variety of self-knowledge. The second meaning they termed "strategic metacognition", which denotes the ability to monitor and control ongoing mental activities. The definition continues:

> Attributive and strategic metacognition differ from one another both in the objects of thoughts (beliefs and desires versus mental activities and plans) and actions taken (attribution in the service of explanation versus monitoring in the service of control) p. 14.

The following chapter concentrates on the strategic type of metacognition but a certain blurring of the boundaries occurs when the outcome of the latter leads to a revision in the former. For example, monitoring of a cognitive operation may reveal deficits and impairments which then require a revision in self-knowledge, specifically the knowledge that one is impaired or ill or in need of help. Similarly, we may wish to expand the range of 'objects' under scrutiny to include not just mental operations and day-to-day beliefs and desires, but a particular set of beliefs about the self or personality which presumably change (if at all) at a slower pace.

The terminology used in clinical circles to capture these notions also requires some comment. In neuropsychiatry, the terms *anosognosia*, and *lack of awareness* are often used synonymously to describe a collection of attitudes and behaviours directed at one's illness. Anosognosia may be used to convey lack of awareness of specific functions seen after brain injury (BI), leading to for example hemiplegia [2]. In contrast, *insight* (and occasionally, somewhat colloquially, denial or *being in denial*) is typically used to describe the phenomenon in psychiatric disorders, such as schizophrenia (SCZ), addictions, bipolar disorder and even personality disorders, and in neurological conditions, such as AD and BI. Here the expressed awareness in question refers to that of being ill in general and more specifically, the capacity to judge impairment of, say, memory or social behaviour. It is also applied to judgements of the content of experiences or symptoms, such as delusions and hallucinations, as not being real [3, 4]. Rather than lack of insight being a symptom in its own right it is more usefully thought of as a relational concept: insight into something [5]. Broadly, this may encompass stated awareness of an objective, obvious (physical) deficit such as hemiplegia, or behaviour such as excessive alcohol intake, through to objects that are verifiable but invisible. One may or may not have insight into a mental failure or deficit—true metacognition—such as amnesia. This is different again from *objects* which are purely subjective experiences such as hallucinations, where the insight concerned may take the form: did I just hear the voice of my dead mother or was my mind playing tricks on me?

Insight research, particularly in relation to SCZ and the psychoses, has burgeoned over the last 20 years with the development of operationalised definitions and easy to administer rating scales. It was probably the realisation that insight was not a unitary phenomenon but rather a multidimensional construct that revived interest in the field. This may have been because such a formulation rang true with clinicians. Furthermore, the separation of awareness of deficits or symptoms from their attribution chimed with advances in other areas of cognitive and behavioural science, such as social psychology (attribution theory) and developmental psychology ('theory of mind').

Lack of insight was once held to be the *sine qua non* of 'psychosis' [3] but this has given way to a more nuanced and quantitative view. David [3, 4] proposed three insight dimensions: recognition of having a mental illness, compliance with treatment and the ability to label unusual mental events (e.g. hallucinations) as pathological. Amador et al. [6] split insight into five components: four relate to (un)awareness of having a mental disorder, of the effects of medication, of consequences of illness and of specific symptoms, and the final component is the attribution of symptoms to illness. Popular measures of insight include the Schedule for the assessment of insight—Expanded version (SAI-E) [7, 8]; and the semi-structured interview: Scale to assess unawareness of mental disorder (SUMD) [6]. Many authors, particularly in the dementia and BI fields, have made use of patient–carer discrepancy questionnaires [2, 9, 10].

Dividing insight into sub-components clarifies one aspect, namely detecting and labelling unusual mental events as pathological (as per schemes advanced by David and Amador and colleagues) which are metacognitive in the strategic sense defined above. Indeed, within schizophrenic psychopathology, the types of events or objects of contemplation extend to delusions, negative symptoms and thought disorder at the very least and it appears that insight into one aspect does not necessarily predict insight into another (see Fig. 15.1b, [11]).

In this chapter we will summarise work specifically addressing the fractionation of insight in different disorders across domains of cognitive functions. Next, we will highlight some of the challenges in extending studies of metacognition from 'cold' information processing to 'hotter' areas of self-concept and the presentation of self and some preliminary findings. We will then describe some of the key findings on insight-related metacognition in psychiatric and neuropsychiatric disorders, including neuroimaging research.

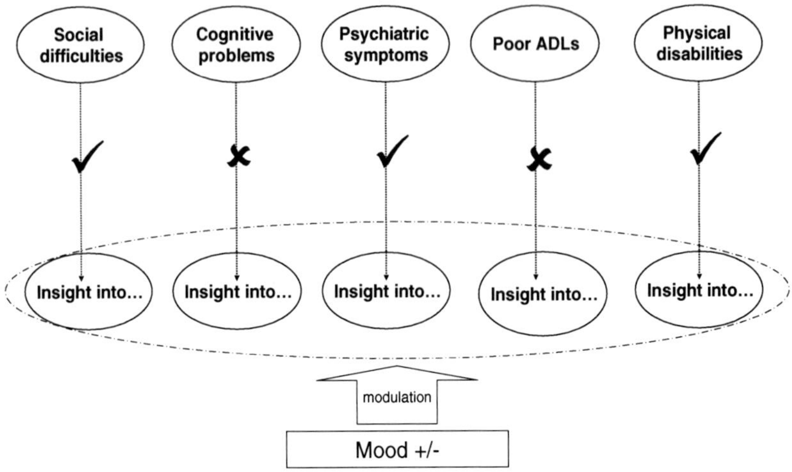

Fig. 15.1 Illustrating a multiple modality specific awareness systems (modularity) but with general modulation by factors such as mood. In this theoretical example insight is preserved into 'objects' such as social difficulties, psychiatric symptoms and physical disability but not cognitive problems and poor activities of daily living (*ADLs*). Adapted from Gilleen et al. [10]

15.2 Metacognition Across Diagnostic Groups

Most clinical studies of metacognition consider a single patient population, but there is a case for comparing different patient groups on the same measures. Just as patients with SCZ have, in addition to their core symptoms, cognitive impairments and behavioural and social deficits, so patients with AD and BI may have a range of psychopathologies which they may or may not be able to monitor and appraise and of which they have varying degrees of awareness. We can ask whether the same factors associated with metacognition are consistent across the groups [see e.g. 12–14]. Modularity of 'awarenesses' is perhaps the most likely pattern from the neurological literature [15]. This is illustrated in Fig. 15.1. In a large pan-European BI study [16] considerable within-diagnosis heterogeneity of awareness was found. Similarly in the dementia field, different levels of awareness have been noted by contrasting behaviour with cognition [17–19] (see Fig. 15.1).

We recently compared aspects of insight and metacognition in three different neuropsychiatric populations: SCZ, BI and probable AD [9, 10, 20 in preparation] (Table 15.1). The former were mostly sub-acute, chronic and treated out-patient plus some inpatients at the Maudsley Hospital, London, while the Alzheimer group were locally dwelling subjects identified as part of a larger cohort study. The BI patients were a heterogeneous group with a mixture of traumatic, hypoxic and vascular aetiologies and with behavioural problems. Naturally, the patients were not matched on factors such as age and length of illness.

Table 15.1 Demographic and insight data on clinical groups

Variable mean (SD)	Schizophrenia $N = 31$	Brain injury $N = 26$	Alzheimer's $N = 27$
Age, years	38.3 (10.4)	40.0 (12.1)	82.4 (4.3)
Sex, m/f	16/15	22/4	14/13
Premorbid IQ, (NART)	102.3 (12.8)	102.2 (13.8)	109.1 (12.8)
SAI-E	11.2 (7.15)	15.4 (5.7)	7.0 (6.4)
SUMD awareness of mental illness	3.37 (1.6)	1.92 (1.43)	4.04 (1.4)
DEX discrepancy scores[a] (mean and range)	2.48 (−33 to 31)	−14.76 (−55 to 15)	−25.96 (−62 to 13)

NART National adult reading test (estimate of premorbid IQ)
SAI-E Schedule for the Assessment of Insight—Expanded
SUMD Scale to Assess Unawareness of Mental Disorder
[a] Self-rating of difficulties minus informant rating of difficulties yielding a negative score. The more negative, the greater the discrepancy (greater patient unawareness)

The groups were compared on measures of estimated premorbid IQ (National Adult Reading Test; NART) and clinician rated and patient–carer rated awareness scales. All were rated on the SAI-E, and the SUMD and the Dysexecutive (DEX) Questionnaire (Table 15.1).

The DEX Questionnaire from the Behavioural Assessment of the DEX Syndrome (BADS) [21] is a 20-item measure of functioning which addresses problems such as impulsivity, apathy, distractibility and unconcern for social rules and difficulties with abstract thinking. Informants rate patients' functioning and the patient rates him/herself on the scales and the difference between patient and informant scores creates a discrepancy score, the greater and more negative the discrepancy between the scores, the greater the unawareness of the patient. Items are scored on a 5-point scale from 0-never to 4-very often. Hence, DEX discrepancy scores can range from −80 to +80. So if a patient is rated by a carer as, "losing his temper at the slightest thing", 'very often = 4', while the patient himself says that this occurs 'occasionally = 1' then there is a discrepancy of −3 on that one item. A score of zero indicates perfect awareness in that the patient agrees with the level of impairment scored by their respective informant.

The validity of the discrepancy index to measure insight and awareness or metacognitive failure, may be questioned since it assumes that the informant is the 'gold standard' [see 17, 18]. Nevertheless, the methodology has been found to be valuable and consistently shows underestimation of deficits by patients in relation to informal caregivers.

Across the groups, patient- and informant-ratings of behavioural problems, as measured by the total DEX score, were highly discrepant in the BI and AD disease groups (see Table 15.1), representing low awareness of behavioural impairments but this was much less so in the SCZ group, and suggests that patients exhibit different levels of unawareness of behavioural deficits. This is despite the fact that the SCZ patients as a group were rated by clinicians on the SAI-E as having rather poor insight into their psychiatric disorder and symptoms (11.2 ± 7.15), worse than the BI group (15.4 ± 5.7 out of 28), a few of whom denied that they had any impairments or deficits. Pooling all patients, DEX-discrepancy and SAI-E scores correlated moderately and significantly at $r = 0.32$; $r = 0.46$ in the BI group alone, but $r < 0.2$ (NS) in the SCZ and AD groups. In other words, not only do patients with different neuropsychiatric disorders show very different levels of awareness into DEX problems, there is little relationship between a patient's ability to view their DEX behaviours as problematic and their overall insight according to a clinician rating, perhaps because they are governed by different processes.

15.3 Meta-Memory

Metacognitive awareness can also be calculated as a discrepancy score between actual performance and ratings by the patient or other persons. Clare et al. [22] have developed the Memory Awareness Rating Scale (MARS) which provides a variety of awareness of memory measures based on discrepancy scores between predicted and actual functioning on the Rivermead behavioural memory test (RBMT; [23]) and pre-test, post-test self-ratings and informant ratings. Importantly, the MARS offers an 'isomorphic' and ecologically valid measure of awareness in as far as the questions are easily understood in terms of everyday memory tasks, and the questions that form the ratings are analogous to the RBMT sub-tests thus allowing direct comparison between ratings and functioning. Thus a patient may be asked to predict how many words they think they will remember from a list, pre-test (and the same question is put to the informant about the patient). The test is then given and an objective score is obtained. Next the patient is asked how they think they did.

With the same sample already described, the BI and SCZ patients scored comparably on the RBMT scale—in the 'moderately impaired' range. The AD patients scored in the very impaired range and scored significantly lower than the SCZ ($p < 0.001$) and the BI groups ($p < 0.001$) (see Table 15.2). Of interest is that all groups reported similar levels of functioning on the pre-test rating scale despite clear memory functioning differences.

BI and AD patients ($t(1, 26) = -13.78$, $p < 0.001$), but not SCZ patients ($t = (1,30)$, $p = 0.72$) significantly overestimated their memory functioning before completing the memory test. Following test completion, BI ($p = 0.01$) and AD patients ($p < 0.001$), but not SCZ patients still significantly overestimate their memory functioning. In terms of informant ratings, informants for the AD ($p < 0.001$) but not the BI or the SCZ group also significantly overestimated the memory functioning of the respective patients.

Correlation analyses showed that in the SCZ sample, post-diction ($r = 0.58$, $p = 0.001$) but not prediction ($r = 0.18$, NS) scores correlated significant with memory scores, whereas informant scores were associated with memory scores at a trend level ($r = 0.38, p = 0.052$). In the BI group, informant ($r = 0.57, p < 0.005$) and post-diction ($r = 0.89, p < 0.001$) ratings correlated highly with actual memory scores, while prediction scores showed a strong trend to be associated with memory

Table 15.2 Mean (SD) RBMT memory scores, out of 48, and discrepancy scores betweenRBMT and pre-test self-ratings, post-test self-ratings and informant ratings

Group	RBMT	Discrepancy mean scores (sd)			
		Self pre-test	Self post-test	Informant	% Improve
Schizophrenia	30.65 (9.51)	0.71 (10.78)	−0.63 (7.76)	−1.40 (11.38)	[a]1.34
Brain injury,	27.31 (14.75)	*−7.96 (13.67)*	*−4.08 (7.92)*	−3.83 (12.89)	47.6
Alzheimer's disease	4.11 (4.42)	*−28.63 (10.80)*	*−13.22 (7.35)*	*−11.02 (7.79)*	53.8

Negative scores reflect over-estimation of functioning. Percent improvement from pre- to post-ratings are also shown. Italicised scores show significant difference with RBMT scores. [a] The average rating switched from under- to over-estimation of functioning (a total change of 1.34 points), and so improvement scores cannot be calculated, but in any case, represent a negligible change

scores ($r = 0.39, p = 0.051$). Lastly, in the AD group, neither informant nor postdiction scores correlated significantly with memory scores, whereas prediction scores showed a strong negative trend association ($r = -0.38$, $p = 0.053$)—the lower the RBMT score the greater the predicted memory performance.

The SCZ patients were poor at pre-rating their own memory performance ($r = 0.18$), the BI group somewhat more accurate, however, the AD group were particularly poor at predicting memory performance. Indeed, while the SCZ and BI groups showed positive correlations between prediction and actual performance—if they were better they generally said they were better—the AD group showed a *negative* correlation between the two scores indicating that as their actual memory scores became *worse* they thought they were *better*.

On average, AD and BI patients showed an approximately 50 % improvement in their estimation of functioning, after testing. This would conform to Agnew and Morris' notion of "mnemonic anosognosia" [24]—which describes an inability to update one's default appraisal of one's memory but with at least some intact ability to detect errors and monitor performance 'on-line'. In other words, this aspect of strategic metacognition was not failing entirely but the output of this process (presumably being provoked each time there was a memory lapse in the real world), failed to register. The SCZ patients' reasonable estimate of their memory appeared to be dissociable from their insight into the core aspects of their disorder.

A similar approach was taken by Williamson et al. [25] recently in a small study using items from the Neuropsychological Assessment Battery who compared 10 AD patients with 10 with fronto-temporal dementia—who are clinically characterised as severely lacking in 'insight' attributed to frontal lobe deficits. They found that, unlike controls, patients consistently overestimated their performance, the FTD group more so than the AD group (see also [26, 27]). Overall, neither showed much variation in estimations pre- versus post- test although, as in our study, the AD patients did overestimate their performance less on the memory subtest, post test.

15.4 Clinical Insight

As noted above 'lack of insight'—into the experiential and behavioural manifestations of major mental disorders has been regarded as a defining feature (see [4, 28] for reviews). In an effort to clarify the relationship of awareness and psychopathology, Mintz et al. [29] conducted a meta-analysis of 40 relevant studies and found small negative associations between awareness and global, positive and negative symptoms accounting for 7.2, 6.3, and 5.2 % of the variance in awareness respectively. This would strongly suggest that symptomatology plays a small part in, and is relatively independent from, the degree of awareness displayed by SCZ patients.

Several studies have shown that SCZ patients present with less insight than patients with other diagnoses such as bipolar disorder and major depressive disorder [30] and schizo-affective disorder and mood disorder with and without psychosis [31–33], or similar levels of insight as bipolar patients but less than patients with unipolar affective disorder [34]. However, others have found no significant differences between different patient groups [35, 36].

15.5 Mood

One of the more reliable findings in the literature is the positive correlation between metacognitive ability leading to awareness of illness, and low mood or depression (and between elevated mood and lack of awareness), which has been shown across different patient groups [4, 5, 17, 18]. Although findings are variable, many studies have reported that increased awareness in SCZ is associated with greater depressive symptoms [37–42] including a meta-analysis [29]. In this way, poor insight is often conceptualised as a form of denial, in order to maintain self-esteem, while good insight equally is regarded as an example of 'depressive realism'. The mechanism underlying this may be conceptualised in signal detection terms as a modulation of response bias. The exception that proves the rule is psychotic depression. Here, a point is reached wherein the usual relationship between low mood and better insight breaks down and the psychosis predominates (see [43]).

15.6 Insight and Neurocognition

Several studies have suggested a relationship between intelligence (IQ) and insight in SCZ [44]. The largest individual study to investigate this relationship in over 500 psychosis patients reported that lack of insight did reflect a generalised cognitive deficit rather than a specific relationship with a particular function [45]. Others claim a more specific association with executive functioning [46, 47], particularly as assessed using the Wisconsin Card Sorting Test (WCST). The WCST is generally thought to be a measure of set-shifting ability, where impairment has been hypothesised to be analogous to patients' inability to shift from an previously established 'set' (that of being well); to a more accurate, post-morbid 'set' (of being ill). Cooke et al. [48] examined 29 studies which included a measure of WCST performance and awareness. Of these there were nine studies where all WCST measures and nine where some measures correlated with awareness. All findings were in the anticipated direction, with lower awareness being associated with poorer WCST performance. The most comprehensive and quantitative systematic review and meta-analysis of work in this area [49] suggests that WCST performance has more in common with awareness than other measures such as IQ, or memory, with 13 studies creating a pooled effect size of $r = 0.23$.

The pattern of neuropsychological impairment associated with poor insight in first episode psychosis (FEP) is also unclear. Several studies have found general cognitive impairment to be related to insight [50–52], whilst others have shown insight to be linked to working memory [53], or verbal memory [45]. Koren et al. [54] were perhaps the first to consider an aspect of metacognition in relation to the clinical concept of insight, that is, the ability to accurately judge one's own performance and found a relationship with executive functioning.

In summary, clinical insight which includes the ability to re-label symptoms as pathological and to recognise that one is suffering from a disorder which merits treatment, seems to require a degree of executive functioning ability and could be considered to be an executive function in itself. However within the psychosis, where there have been the most studies, the magnitude of this effect is modest suggesting the importance of other contributory variables.

15.7 Cognitive Insight

More recently, a distinction has been proposed between 'cognitive' insight and clinical insight. This was introduced by Beck et al. [55], and separates a person's awareness and acceptance of illness (clinical insight), from their cognitive style or attributive metacognitive ability; specifically flexibility towards their beliefs, judgements and experiences. The Beck Cognitive Insight Scale (BCIS; [55, 56]) is a self-report questionnaire developed to measure cognitive insight. The scale has two theoretically driven and empirically derived factors: self-certainty and self-reflectiveness. Self-certainty refers to overconfidence in the judgments and attributions that one makes (e.g. "I know better than anyone else what my problems are"), while self-reflectiveness (e.g. "Some of the ideas I was certain were true turned out to be false") refers to recognition of one's own fallibility and acceptance of correction. A composite index can also be calculated by subtracting the self-certainty score from the self-reflectiveness score.

Research using the scale [57] has shown a correlation between increased severity of delusions and decreased cognitive insight on at least one BCIS subscale

[58–61]. However, findings have not been consistent (see [62, 63]). The self-reflectiveness subscale has been the less consistent of the two, with one study finding active delusions associated with higher (rather than the predicted lowed) self-reflectiveness [64].

The scale has also been used in non-clinical samples with the promise of providing a normative understanding to the insight construct. It has been shown [65] that amongst students, theoretical delusion proneness was significantly positively correlated with self-certainty, but not self-reflectiveness. Comparisons between patients and controls have had mixed results. Self-reflectiveness seems to be lower in deluded patients than for controls, and self-certainty seems to be higher [64, 66], as might be expected. Others, however, have found no differences between healthy controls and individuals with SCZ or bipolar disorder on either subscale [67]. There are reasons why the scale is problematic for use in controls. Specifically, several items refer to attitudes towards 'unusual experiences' (e.g. 'my unusual experiences may be due to my being extremely upset or stressed'), which may be interpreted inconsistently by healthy controls, who may not have had such experiences.

Another question for research is what is the relationship between 'cognitive' and 'clinical' insight? The expected positive correlation between clinical scale scores and cognitive insight scores is found in most [56, 61, 68], but not all [63, 69] studies. Much like with clinical insight, it is intuitive to suggest that failures in self-reflection and poor evaluation of one's own thinking may be, at least in part, caused by an inability to perform the complex metacognitive operations required and that high levels of self-certainty may be related to mental inflexibility. Investigation of the neuropsychological correlates of cognitive insight [70] has shown that the composite score was related to verbal learning and memory in a sample of 51 FEP patients. The composite index appears to be related to visual working memory, with self-certainty related to both verbal and visual memory as well as non-perseverative errors from the Wisconsin Card Sort Task [71], see also [68]. An investigation using the Metacognition Assessment Scale [72], found that the 'understanding one's own mind' subscale (which correlated with BCIS total score at $r = 0.43$) was significantly related to several measures of executive function.

The only published study to date which assessed the relationship between neuropsychological function and cognitive insight in healthy controls found that the index score was significantly correlated with perseverative errors on the WCST [73], which is the opposite direction to the results using patient samples.

We recently carried out a study in 107 patients experiencing their first episode of psychosis, and 72 healthy controls from South London as part of the National Institute of Health Research (NIHR) Biomedical Research Centre, Genetics and Psychosis study. Simple correlations showed a positive association between 'cognitive' and 'clinical' insight in patients (Pearson's r ranged from 0.34 to 0.48) for the subscales and composite respectively and the total SAI-E (Wiffen et al. [74]).

In terms of cognitive functioning, there were some moderate correlations between BCIS scores and a battery of neuropsychological tests, but these were exclusively confined to the patient group. A regression analysis showed that cognitive variables explained 11.9 % of the variance (R squared change $= 0.119$,

$p = 0.017$) along with psychotic symptoms scores and IQ. Immediate verbal memory was the only neuropsychological variable to contribute independently to the final model.

Whilst metacognition in the form of self-reflection and self-certainty may reflect some sort of cognitive style in healthy participants which may put them at risk of psychotic disorder, it may not necessarily reflect the same mechanism as in patients. Indeed, the present results suggest that the style is moderated by memory (and positive psychotic symptoms) in patients while there is little or no evidence of this in healthy controls.

There was a small trend towards a correlation between the composite score and depression score. Results in the literature on this have again been mixed [63, 75]. The relationship between depression and cognitive insight parallels the same finding for clinical insight [29]. Conceptually, it is more likely that the correlation with cognitive insight is driven by self-reflectiveness rather than certainty since the former can take on a ruminative quality typical of depressive thinking. However, the association found here is relatively small ($r = 0.21$), so should be interpreted with caution.

In sum, there is still much work to be done to establish whether there is a general thinking style involving self-reflectiveness and self-certainty captured by the BCIS analogous to insight into psychotic illness. One conclusion from the work reviewed is that despite suffering from delusions and other phenomena, psychotic patients show surprisingly little evidence for a gross disturbance in this thinking style or an overarching metacognitive failure. However, this may be due to inherent difficulties in measuring such concepts and possible confounds such as mood and intellectual functions (especially memory). There is also the uncomfortable fact that assessing cognitive insight with questionnaires like the BCIS threatens an infinite regress: you must be able to reflect accurately on your ability to reflect accurately.

15.8 Insight and Self-Reflection

One of the issues raised by the BCIS work is the difficulty in finding a normal equivalent of clinical insight. After all, asking a self-aware healthy person to perform the metacognitive task of saying whether they suffer from a mental illness or symptoms thereof should lead to an emphatic 'no' while the patient with SCZ who lacks all insight into their condition will give the same response, with the same certainty. However, self-reflection directed at personality traits is something everyone, in principle, should be able to perform meaningfully. And the extent to which this process is equivalent to illness awareness will be considered later.

Work by Nick Bedford for his Ph.D. [76–78] addressed this question using a variety of novel paradigms. A simple starting point was to examine the acceptance of trait adjectives by patients and controls, both positive and negative, some of which were related to mental illness but in a way that might be seen applicable to many people, with or without a clinical psychiatric diagnosis. He carefully

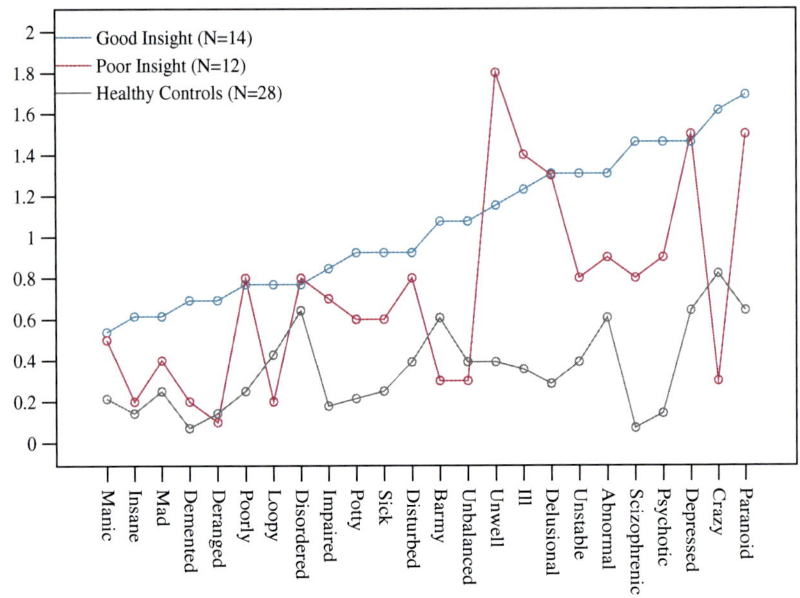

Fig. 15.2 Mean *Trait Ownership Scale* ratings for individual mental illness traits

constructed a list of 96 trait adjectives matched on important psycholinguistic parameters and carried out a number of memory and evaluative studies using them. Examples included: Mental illness related: unstable, crazy, disordered, psychotic; Negative: evil, cruel, hostile, dishonest; Positive: wonderful, great, special, clever; Physical illness-related terms were also included to act as a control for psychiatric illness-related terms: diabetic, cancerous, paralysed, etc. Note there is no attempt to relate judgement to a 'gold standard'.

However, one of the most striking results came from simply contrasting the extent to which demographically matched participants—healthy volunteers ($n = 28$), SCZ patients with good ($n = 14$) and poor insight ($n = 12$) according to standard scales—would admit to the trait applying to themselves, on a Likert Scale from 0 (not at all) to 4 (extremely). See Fig. 15.2. From inspection of the figure it appears that the patients who are repudiating any mental illness in themselves show variation according to the way the illness is described. So being 'unwell' or 'ill' seems to be acceptable even more so in the poor insight than the good insight patients, yet being 'crazy' seems to be abhorrent to poor insight patients while quite acceptable to good insight patients and even some healthy volunteers. The technical term 'schizophrenic' is, not surprisingly, acceptable to the good insight patients, not at all to the controls and yet is endorsed to an intermediate degree by the remaining patients.

Participants were also asked to rate the desirability of the traits on a similar 5-point scale. It would be reasonable to hypothesise that social and interpersonal factors affect trait ownership and this complicates (biases) the assessment of metacognitive 'accuracy' in determining self-reflective ability—e.g. self-serving and other biases [79]. Overall there was a tendency for ownership and desirability to co-vary, especially for negative traits—i.e. the less desirable a trait, the less strongly it was 'owned'. Mental illness trait desirability did not correlate with ownership at all in healthy subjects ($r = -0.02$) but correlated in the good insight ('accepting') patients at $r = 0.28$ (NS) and $r = 0.45$ (NS) in the poor insight ('denial') patients suggesting that perceived desirability may have been one but certainly not the most powerful driver for mental illness trait ownership [77, 78]. Correlations between desirability and positive trait ownership were, however, especially strong in poor insight patients ($r = 0.82$, $p < 0.05$) but not the other groups.

Nevertheless, there is still the concern that metacognitive processes aimed at personal illness evaluation inevitably end up requiring the individual to 'admit' to something bad about themselves. Being ill in some way is never a good thing. Hence there is a potential confound in all such self-reflection tasks. We attempted to tackle this by devising vignettes in which, paradoxically, the protagonist stood to benefit if they admitted that they had a mental illness or condition [77]. For example, Tom, a person with mental health problems gets a Council Tax bill. He fears that he will struggle to pay it. However, the bill is slightly lower for people who are unemployed and much lower for people who have psychiatric problems. The participant is then given a number of options on how they would proceed if they were Tom, ranging from admitting he has psychiatric problems and paying the lower amount to paying the amount in full. In short, *all* participants were influenced by the relative advantageousness of admitting to mental health problems, but the SCZ patients less so than controls. There was no clear difference between the low and high insight patients. To conclude, it appears that many patients with mental illness are reluctant to acknowledge their mental illness traits or history. This does not seem to be easily explained on the basis of perceived benefits/losses. Instead, the possibilities include a 'genuine' inability to reflect accurately one's life story or current mental contents (a failure of strategic metacognition), or, the activity of self-reflection may be performed but the appraisal of the material, the 'object' under scrutiny, is systematically biased or at fault (a failure of attributive metacognition).

15.9 Clinical Insight and Brain Structure

Imaging findings first suggested an association between poor insight and reduced total brain volume [50, 80, 81], reduced frontal lobe volume [82–85] reduced cingulate gyrus and temporal lobe grey matter volume [86, 87]. There are now several studies, but much study variation in the location of brain-insight correlates and in some instances a failure to identify any brain abnormalities associated with poor insight [12, 36, 89] see Table 15.3. One explanation for this inconsistency could be the use of different image analysis techniques such as region of interest

Table 15.3 Summary of neuroimaging studies in relation to insight in psychosis

First authors (year) [Reference]	Patients	Main findings (association with reduced insight and brain indices)	Insight measure
Antionus (2011) [92]	Sz (n = 36)	Fronto-temporal/temp-parietal white matter	SUMD
Berge (2010) [90]	Sz, FE (n = 21)	↓Medial frontal bilat; sup frontal, R inf temporal, inf frontal grey; VBM	SUMD
Buchy (2010) [91]	Sz, FE (n = 79)	L frontal, temp (and parietal) cortical thinning	SUMD
Morgan (2010) [93]	Psychosis, FE (n = 82)	↓Posterior cingulate and right precuneus/cuneus grey density	SAI-E
Cooke (2008) [88]	Sz/Sz Aff, OPs (n = 52)	↓L temporal and parietal; precuneus grey; VBM	SAI-E/BIS
Sapara (2007) [85]	Sz, chronic,OPs (n = 28)	↓Prefrontal grey	SAI-E
Shad (2006) [86]	Sz (n = 14)	↓R dorsolateral prefrontal and ↓awareness; ↑R orbitofrontal and abnormal attributions	SUMD
Bassitt (2006) [89]	Sz (n = 50)	No assoc. with prefrontal grey/white vols	SUMD
McEvoy (2006) [50]	Sz (n = 251)	↓Total grey/white/whole brain	ITAQ
Ha (2004) [87]	Sz OPs (n = 35)	↓Grey L post/R ant. cingulate and bilateral inf. temporal	PANSS
Rossell (2003) [12]	Sz (males) (n = 78)	No assoc. whole brain, white/grey vols	SAI-E
Laroi (2000) [83]	Sz (n = 20)	Frontal lobe atrophy (CT)	SUMD
Flashman (2001) [84]	Sz spectrum (n = 30)	↓Frontal lobe volume	SUMD
Flashman (2000) [80]	Sz spectrum (n = 30)	↓Whole brain volume	SUMD
David (1995) [35]	Mixed psychosis (n = 128)	No assoc. with ventricular vol. (CT)	PSE
Takai (1992) [82]	Sz, chronic (n = 22)	Ventricular enlargement	PANSS

Magnetic resonance imaging (MRI) unless otherwise stated
CT computed tomography; *ITAQ* Insight and Treatment Attitudes Questionnaire; *OPs* outpatient; *PANSS* Positive and Negative Symptoms of Schizophrenia Scale; *PSE* Present State Examination; *SAI-E* Schedule for the Assessment of Insight (expanded); *SUMD* Scale for the assessment of Unawareness of Mental Disorder; *Sz* schizophrenia patients; *FE* first episode; *BIS* Birchwood Insight Scale

measurements or voxel-based morphometry (VBM) methods of analysis [87–89]. In some studies a single insight assessment item has been used [35, 81, 86] while in others, insight schedules were employed. The most recent studies have started to employ novel imaging methods, for example, cortical thickness [90] and white matter integrity [91].

Nevertheless, frontal abnormalities predominate. In our study, Kevin Morgan et al. [92] used VBM methods in a large sample of FEP patients and found deficits, particularly with respect to attribution of symptoms in cingulate cortex, perhaps related to the midline cerebral system for self-processing [93, 94] as well as right posterior deficits—reminiscent of regions implicated in neurological cases of anosognosia of hemiplegia and neglect [95]. Damage to any of these putative systems could potentially account for impaired self-awareness. Research in other psychiatric disorders is needed before we can say whether or not these findings are disorder specific.

15.9.1 Functional Imaging

The functional correlates of self-reflection have been the topic of several imaging experiments. These have been carefully reviewed and summarised [93, 94] and found to reveal a fairly consistent picture, namely that there is a 'cortical midline system' which is reliably engaged in such tasks encompassing the medial frontal (ventro-medial, Brodmann Areas (BA) 10,11) and dorso-medial, BA 9) and cingulate cortex. Furthermore, Fleming et al. [96] showed that metacognitive ability on a perceptual task was related to grey mater volume in BA 10. A quantitative meta-analysis was performed by van der Meer et al. [97], which highlighted the anterior portion of the system as most often engaged when self-appraisal was contrasted to other-appraisal. This region overlaps with that noted in structural imaging studies in SCZ as differentiating low and high insight patients—a tantalising hint that there might be structural–functional convergence. This prompted further work in which 11 of the 12 patients studied by Bedford mentioned above also underwent functional MRI during a version of the self-reflection-attribution of traits task. We hypothesised that the cortical midline system would be less active in self- versus other appraisal when considering the self-relevance of trait adjectives (with each also contrasted to a letter monitoring baseline condition).

Results of this ongoing study [98] seem to suggest that the medial frontal cortex is indeed critical to abnormal self-appraisal in patients with SCZ since this region was the only one to reach statistical threshold for the interaction between patients and controls and self- versus other appraisal (taking all trait terms together). As shown in Fig. 15.3, activation in the left superior frontal gyrus (BA 9) close to the midline (Talairach coordinates −6, 53, 32) increased when controls considered themselves as opposed to a famous 'other' (in this case, Tony Blair), while patients failed to show this increase—which we interpret as demonstrating a failure to differentiate sufficiently self and other metacognitive processing. These findings are in line with recent fMRI work [99] which showed that patients with SCZ activated less medial prefrontal regions and relatively more mid and posterior cingulate cortex during a similar self-reflection task to the one used by Bedford et al., the difference being that in this task the patient's mother was the 'other person' comparator. Similar central midline brain regions are highlighted in an fMRI study of students scoring highly on a questionnaire recording psychotic experiences but seem to show increased rather than reduced activation [100].

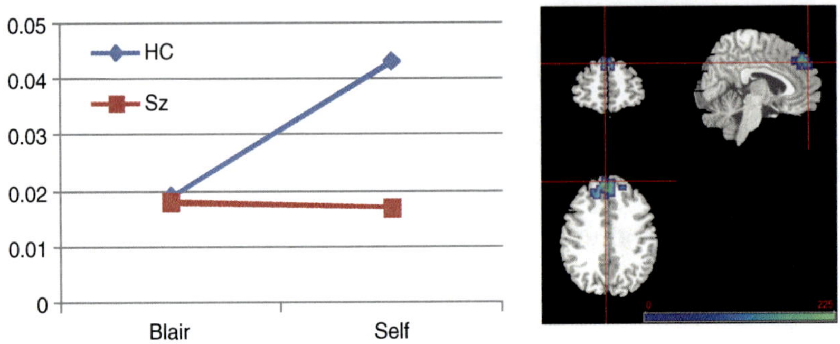

Fig. 15.3 Functional MRI study showing region of significant activation difference between schizophrenia patients (*red boxes*) and healthy controls (*blue trapezoid*) during self-reflection task for self versus Blair. *Left* superior frontal gyrus (BA9)—*x*, *y*, *z* coordinates: 6, 53, 32 [57 voxels]

15.10 Conclusions

It is possible to place the clinical, especially psychiatric, concept of insight into illness within a metacognitive framework. From this a number of fairly reliable associations emerge: worse symptoms tends to go with worse insight, with the exception of mood (lower mood, better insight); in contrast, better cognitive ability, general (IQ) and specific (executive function; memory) are associated with better insight. The effect of low mood may be mediated by a more conservative response bias, or as some clinicians would have it, may reflect the consequences of insight into a disorder. This is an area ripe for research and may exploit the emerging notion of 'cognitive insight' which seeks to provide a normative metacognitive framework relevant to psychiatry. Hence the effects of mood on metacognition can be studied naturalistically without reference to actual disorder. The association between cognitive impairment and poor insight may point to common information processing deficits underlying impaired metacognition and self-refection/appraisal.

Exploration of failures of metacognition across different neuropsychiatric conditions (SCZ, AD and BI) reveals that Alzheimer and brain injured patients demonstrate on-line awareness of memory problems, but we inferred a failure to use this to update knowledge about memory ability. We also showed that there could be marked variability across diagnostic groups both within and across domains (i.e. memory versus symptom awareness). This fractionation, even within the narrow sphere of psychotic symptoms, arguably raises questions about the modularity or content specificity of metacognition in general. That is to say, studying failures of metacognition may illuminate healthy metacognition. There are also parallels to be drawn between the notion of insight in psychopathology as a relational concept—there are potentially many objects of insight and good insight or accurate metacognition that may pertain to some but not all—accords well with the non-clinical literature on meta-memory [101].

This principle may be extended into the neurological instantiation of metacognition which again, may be revealed more strongly by inference from pathological systems. A clear cerebral localisation for a system supporting self-reflection centring on the medial frontal and posterior medial cortex is becoming accepted. Summarising structural neuroimaging findings in relation to insight in psychotic disorders, particularly SCZ, we find that a clear pattern has yet to emerge despite several studies using different imaging modalities. However, taking some of the findings along with preliminary functional imaging work of self-reflection in patients presented here, points to an important role for dorso-medial frontal cortex in mediating metacognition in relation to psychopathology.

References

1. Saxe R, Offen S (2010) Seeing ourselves: what vision can teach us about metacognition. In: Dimaggio G, Lysaker PH (eds) Metacognition and severe adult mental disorders. Routledge, Hove, East Sussex, pp 13–30
2. Prigatano GP (2009) The study of anosognosia. Oxford University Press, New York
3. David AS (1990) Insight and psychosis. Brit J Psychiat 156:798–808
4. Amador XF, David AS (eds) (2004) Insight and psychosis: awareness of illness in schizophrenia and related disorders, 2nd edn. Oxford University Press, Oxford
5. Markova IS (2005) Insight in psychiatry. Cambridge University Press, Cambridge
6. Amador XF, Strauss DH, Yale SA et al (1993) Assessment of insight in psychosis. Am J Psychiat 150:873–879
7. Wiffen BDR, Rabinowitz J, Lex A et al (2010) Correlates, change and 'state or trait' properties of insight in schizophrenia. Schizophr Res 122:94–103
8. Sanz M, Constable G, Lopez-Ibor I et al (1998) A comparative study of insight scales and their relationship to psychopathological and clinical variables. Psychol Med 28:437–446
9. Gilleen J, Greenwood K, David AS (2009) Anosognosia in schizophrenia and other neuropsychiatric disorders: similarities and differences. In: Prigatano GP (ed) The study of anosognosia. Oxford University Press, Oxford, pp 255–290
10. Gilleen J, Greenwood K, David AS (2010) Lack of insight and awareness in schizophrenia and neuropsychiatric disorders. In: Myoshi K, Morimura Y, Maeda K (eds) Neuropsychiatric disorders. Springer, Tokyo, pp 33–49
11. Rossell SL, Coakes J, Shapleske J et al (2003) Insight: its relationship with cognitive function, brain volume and symptoms in schizophrenia. Psychol Med 33:111–119
12. Kircher TTJ, Koch K, Stottmeister F et al (2007) Metacognition and reflexivity in patients with schizophrenia. Psychopathology 40:254–260
13. Medalia A, Thysen J (2008) Insight into neurocognitive dysfunction in schizophrenia. Schizophr Bull 34:1221–1230
14. Bayard S, Capdevielle D, Boulenger JP et al (2009) Dissociating self-reported cognitive complaint from clinical insight in schizophrenia. Eur Psychiatry 24:251–258
15. McGlynn SM, Schacter DL (1989) Unawareness of deficits in neuropsychological syndromes. J Clin Exp Neuropsychol 11:143–205
16. Teasdale TW, Christensen AL, Willmes K et al (1997) Subjective experience in brain injured patients and their close relatives: a European brain injury questionnaire study. Brain Inj 11:543–564

17. Clare L (2004) Awareness in early-stage Alzheimer's disease: a review of methods and evidence. Brit J Clin Psychol 43:177–196
18. Clare L (2004) The construction of awareness in early-stage Alzheimer's disease: a review of concepts and models. Brit J Clin Psychol 43:155–175
19. Vasterling JJ, Seltzer B, Foss JW et al (1995) Unawareness of deficit in Alzheimer's disease. Domain-specific differences and disease correlates. Neuropsych Neuropsych Behav Neurol 8:26–32
20. Gilleen J, Greenwood K et al (2011) Domains of awareness in schizophrenia. Schizophr Bull 37:61–72
21. Burgess PW, Alderman N, Emslie H et al (1996) The dysexecutive questionnaire. In: Wilson BA, Alderman N, Burgess PW, Emslie H, Evans JJ (eds) Behavioural assessment of the dysexecutive syndrome. Thames Valley Test Company, Bury St. Edmunds, UK
22. Clare L, Wilson BA, Carter G et al (2002) Assessing awareness in early-stage Alzheimer's disease: development and piloting of the Memory Awareness Rating Scale. Neuropsych Rehab 12:341–362
23. Wilson B, Cockburn J, Baddeley A (1985) The rivermead behavioural memory test (RBMT). Thames Valley Test Company, Reading
24. Agnew SK, Morris RG (1998) The heterogeneity of anosognosia for memory impairment in Alzheimer's disease: a review of the literature and a proposed model. Aging and Mental Health 2:7–19
25. Williamson C, Alcantar O, Rothlind J et al (2010) Standardised measurement of self-awareness deficits in FTD and AD. J Neurol Neurosurg Psychiat 81:140–145
26. Banks SJ, Weintraub S (2009) Generalized and symptom-specific insight in behavioral variant frontotemporal dementia and primary progressive aphasia. J Neuropsychiatr Clin Neurosci 21:299–306
27. O'Keeffe FM, Murray B, Coen RF et al (2007) Loss of insight in frontotemporal dementia, corticobasal degeneration and progressive supranuclear palsy. Brain 753:764
28. Osatuke K, Ciesla J, Kasckow JW et al (2008) Insight in schizophrenia: a review of etiological models and supporting research. Compr Psychiat 49:70–77
29. Mintz AR, Dobson KS, Romney DM (2003) Insight in schizophrenia: a meta-analysis. Schizophr Res 61:75–88
30. Michalakeas A, Skoutas C, Charalambous A et al (1994) Insight in schizophrenia and mood disorders and its relation to psychopathology. Acta Psychiat Scand 90:46–49
31. Ghaemi SN, Rosenquist KJ (2004) Is insight in mania state-dependent? A meta-analysis. J Nerv Ment Dis 192:771–775
32. Amador XF, Flaum M, Andreasen NC et al (1994) Awareness of illness in schizophrenia and schizoaffective and mood disorders. Arch Gen Psychiat 51:826–836
33. Pini S, Cassano GB, Dell'Osso L et al (2001) Insight into illness in schizophrenia, schizoaffective disorder, and mood disorders with psychotic features. Am J Psychiat 158:122–125
34. Varga M, Magnusson A, Flekkoy K et al (2007) Clinical and neuropsychological correlates of insight in schizophrenia and bipolar I disorder: does diagnosis matter? Compr Psychiat 28:583–591
35. David A, van Os J, Jones P et al (1995) Insight and psychotic illness. Cross-sectional and longitudinal associations. Brit J Psychiat 167:621–628
36. Cuesta MJ, Peralta V, Zarzuela A (2000) Reappraising insight in psychosis multi-scale longitudinal study. Brit. J. Psychiat 177:233–240
37. Carroll A, Fattah S, Clyde Z (1999) Correlations of insight and insight change in schizophrenia. Schizophr Res 35:247–253
38. Smith TE, Hull JW, Israel LM et al (2000) Insight, symptoms, and neurocognition in schizophrenia and schizoaffective disorder. Schizophr Bull 26:193–200
39. Pyne JM, Bean D, Sullivan G (2001) Characteristics of patients with schizophrenia who do not believe they are mentally ill. J Nerv Ment Dis 189:146–153

40. Moore O, Cassidy E, Carr A et al (1999) Unawareness of illness and its relationship with depression and self-deception in schizophrenia. Eur Psychiat 14:264–269
41. Iqbal Z, Birchwood M, Chadwick P et al (2000) Cognitive approach to depression and suicidal thinking in psychosis. 2. Testing the validity of a social ranking model. Brit J Psychiat 177:522–528
42. Schwartz RC (2001) Self-awareness in schizophrenia: its relationship to depressive symptomatology and broad psychiatric impairments. J Nerv Ment Dis 189:401–403
43. Owen GS, Richardson G, David AS et al (2009) Mental capacity, diagnosis, and insight in psychiatric inpatients: a cross sectional study. Psychol Med 39:1389–1398
44. David AS (1999) To see ourselves as others see us Aubrey Lewis's Insight. Brit J Psychiat 174:210–216
45. Keshavan MS, Rabinowitz J, DeSmedt G et al (2004) Correlates of insight in first episode psychosis. Schizophr Res 70:187–194
46. Young DA, Zakzanis KK, Bailey C et al (1998) Further parameters of insight and neuropsychological deficit in schizophrenia and other chronic mental disease. J Nerv Ment Dis 186:44–50
47. Young DA, Campbell Z, Zakzanis K et al (2003) A comparison between an interview and a self-report method of insight assessment in chronic schizophrenia. Schizophr Res 63:103–109
48. Cooke MA, Peters ER, Kuipers E et al (2005) Disease, deficit or denial? Models of poor insight in psychosis. Acta Psychiat Scand 112:4–17
49. Aleman A, Agrawal N, Morgan KD et al (2006) Insight in psychosis and neuropsychological function: meta-analysis. Brit J Psychiat 189:204–212
50. McEvoy JP, Johnson J, Perkins D et al (2006) Insight in first-episode psychosis. Psychol Med 36:1385–1393
51. Morgan KD, David AS (2004) Neuropsychological studies of insight in patients with psychotic disorders. In: Amador XF, David AS (eds) Insight and psychosis: awareness of illness in schizophrenia and related disorders. NY, Oxford University Press, New York
52. Parellada M, Boada L, Fraguas D et al (2011) Trait and state attributes of insight in first episodes of early-onset schizophrenia and other psychoses: a 2-year longitudinal study. Schizophr Bull 37:38–51. doi:10.1093/schbul/sbq109
53. Mutsatsa SH, Joyce EM, Hutton SB et al (2006) Relationship between insight, cognitive function, social function and symptomatology in schizophrenia: the West London first episode study. Eur Arch Psychiat Clin Neuros 256:356–363
54. Koren D, Seidman L, Poyurovsky J et al (2004) The neuropsychological basis of insight in first-episode schizophrenia: a pilot metacognitive study. Schizophr Res 70:195–202
55. Beck AT, Warman DM (2004) Cognitive insight: theory and assessment. In: Amador XF, David AS (eds) Insight and psychosis: Awareness of illness in schizophrenia and related disorders, 2nd edn. Oxford University Press, New York, NY, pp 79–87
56. Beck AT, Baruch E, Balter JM et al (2004) A new instrument for measuring insight: the Beck Cognitive Insight Scale. Schizophr Res 68:319–329
57. Riggs SE, Grant PM, Perivoliotis D et al (2012) Assessment of cognitive insight: a qualitative review. Schizophr Bull 38(2):338–350
58. Bora E, Erkan A, Kayahan B et al (2007) Cognitive insight and acute psychosis in schizophrenia. Psychiat Clin Neurosci 61:634–639
59. Buchy L, Mall A, Joober R et al (2009) Delusions are associated with low self-reflectiveness in first-episode psychosis. Schizophr Res 112:187–191
60. Engh JA, Friis S, Birkenaes AB et al (2009) Delusions are associated with poor cognitive insight in schizophrenia. Schizophr. Bull. doi:10.1093/schbul/sbn193
61. Pedrelli P, McQuaid JR, Granholm E et al (2004) Measuring cognitive insight in middle-aged and older patients with psychotic disorders. Schizophr Res 71:297–305
62. Favrod J, Zimmermann G, Raffard S et al (2008) The Beck Cognitive Insight Scale in outpatients with psychotic disorders: further evidence from a French-speaking sample. Can J Psychiat 53:783–787

63. Tranulis C, Lepage M, Malla A (2008) Insight in first episode psychosis: who is measuring what? Early Interv Psychiatry 2:34–41
64. Warman DM, Lysaker PH, Martin JM (2007) Cognitive insight and psychotic disorder: the impact of active delusions. Schizophr Res 90:325–333
65. Warman DM, Martin JM (2006) Cognitive insight and delusion proneness: an investigation using the Beck Cognitive Insight Scale. Schizophr Res 84:297–304
66. Martin JM, Warman DM, Lysaker PH (2010) Cognitive insight in non-psychiatric individuals and individuals with psychosis: an examination using the Beck Cognitive Insight Scale. Schizophr Res 121(1–3):39–45
67. Engh JA, Friis S, Birkenaes A et al (2007) Measuring cognitive insight in schizophrenia and bipolar disorder: a comparative study. BMC Psychiatry 7:71
68. Cooke MA, Peters ER, Fannon D et al (2010) Cognitive insight in psychosis: the relationship between self-certainty and self-reflection dimensions and neuropsychological measures. Psychiat Res 178:284–289
69. Greenberger C, Serper MR (2010) Examination of clinical and cognitive insight in acute schizophrenia patients. J Nerv Ment Dis 198:465–469
70. Lepage M, Buchy L, Bodnar M et al (2008) Cognitive insight and verbal memory in first episode of psychosis. European Psychiatry 23:368–374
71. Orfei MD, Caltagirone C, Cacciari C et al (2011) The neuropsychological correlates of cognitive insight in healthy participants. App. Cog Psychol 25(6):927–932
72. Lysaker PH, Warman DM, Dimaggio G (2008) Metacognition in schizophrenia: associations with multiple assessments of executive function. J Nerv Ment Dis 196:384–389
73. Orfei MD, Spoletini I, Banfi G et al (2010) Neuropsychological correlates of cognitive insight in schizophrenia. Psychiatry Res 178:51–56
74. Wiffen BDR, O'Connor JA, Russo M et al (submitted) The neuropsychological basis of cognitive insight in first episode psychosis and healthy controls
75. Colis M, Steer R, Beck AT (2006) Cognitive insight in inpatients with psychotic, bipolar, and major depressive disorders. J Psychopathol Behav Assess 28:242–249
76. Bedford NJ (2009) Denial of illness in schizophrenia as a disturbance of insight and self-awareness. Schizophrenia—academic dissertations [MESH]. Institute of Psychiatry Theses Ph.D System No 001296598, University of London, pp 361. http://library.kcl.ac.uk/
77. Bedford N, David A (2008) Impression management of deficiencies and denial of illness in schizophrenia: reluctance to expose mental illness unmoderated by its level of advantageousness. Schizophr Res 102(suppl 2):117
78. Bedford N, David A (2009) Poor insight in psychosis: cognitive deficit or bias? Schizophr Bull 35(suppl 1):281
79. Pronin E (2008) How we see ourselves and how we see others. Science 320:1177–1180
80. Flashman LA, McAllister TW, Andreasen NC et al (2000) Smaller brain size associated with unawareness of illness in patients with schizophrenia. Amer J Psychiat 157:1167–1169
81. Takai A, Uematsu M, Ueki H et al (1992) Insight and its related factors in chronic schizophrenic patients: a preliminary study. Eur J Psychiat 6:159–170
82. Laroi F, Fannemel M, Ronneberg U et al (2000) Unawareness of illness in chronic schizophrenia and its relationship to structural brain measures and neuropsychological tests. Psychiat Res 100:49–58
83. Flashman LA, McAllister TW, Johnson SC et al (2001) Specific frontal lobe subregions correlated with unawareness of illness in schizophrenia: a preliminary study. J Neuropsychiatr Clin Neurosci 13:255–257
84. Sapara A, Cooke M, Fannon D et al (2007) Prefrontal cortex and insight in schizophrenia: a volumetric MRI study. Schizophr Res 89:22–34
85. Shad MU, Muddasani S, Keshavan MS (2006) Prefrontal subregions and dimensions of insight in first-episode schizophrenia—a pilot study. Psychiat Res 146:35–42
86. Ha T, Youn T, H K et al (2004) Grey matter abnormalities in paranoid schizophrenia and their clinical correlations. Psychiat Res-Neuroimag 132:251–260

87. Cooke MA, Fannon D, Kuipers E et al (2008) Neurological basis of poor insight in psychosis: a voxel-based MRI study. Schizophr Res 103:40–51
88. Bassitt DP, Neto MRL, de Castro CC et al (2007) Insight and regional brain volumes in schizophrenia. Eur Arch Psychiat Clin Neurosci 257:58–62
89. Berge D, Carmona S, Rovira M et al (2010) Gray matter volume deficits and correlation with insight and negative symptoms in first-psychotic-episode subjects. Acta Psychiatr Scand 1–9. doi:0.1111/j.1600-0447.2010.01635.x
90. Buchy L, Ad-Dab'bagh Y, Malla A et al (2010) Cortical thickness is associated with poor insight in first-episode psychosis. J Psychiat Res. doi:10.1016/j.jpsychires.2010.10.016
91. Antonius D, Prudent V, Rebani Y et al (2011) White matter integrity and lack of insight in schizophrenia and schizoaffective disorder. Schizophr Res 128:76–82
92. Morgan KD, Dazzan P, Morgan C et al (2010) Insight, grey matter and cognitive function in first-onset psychosis. Brit J Psychiat 197:141–148
93. Schmitz TW, Johnson SC (2007) Relevance to self: a brief review and framework of neural systems underlying appraisal. Neurosci Biobehav Rev 31:585–596
94. Northoff G, Heinzel A, Bermpohl GM et al (2006) Self-referential processing in our brain—a meta-analysis of imaging studies on the self. Neuroimage 31:440–457
95. Berti A, Bottini G, Gandola M et al (2005) Shared cortical anatomy for motor awareness and motor control. Science 309:488–491
96. Fleming SM, Weil RS, Nagy Z et al (2010) Relating introspective accuracy to individual differences in brain structure. Science 329:1541–1543
97. Van der Meer L, Costafreda SC, Aleman A et al (2010) Self-reflection and the brain: a theoretical review and meta-analysis of neuroimaging studies with implications for schizophrenia. Neurosci Biobehav Rev 34:935–946
98. David AS, Bedford N, Gilleen J et al (2001) The etiology of lack of insight in schizophrenia. Schizophr Bull 37(suppl 1):14
99. Holt DJ, Cassidy BS, Andrews-Hanna JR et al (2011) An anterior-to-posterior shift in midline cortical activity in schizophrenia during self-reflection. Biol Psychiat 69:415–423
100. Modinos G, Renken R, Ormel J et al (2011) Self-reflection and the psychosis-prone brain: an fMRI study. 2011. Neuropsychology 25:295–305
101. Nelson TO, Narens L (1990) Metamemory: a theoretical framework and some new findings. In: Bower GH (ed) The psychology of learning and motivation, vol 26. Academic Press, San Diego, pp 125–173

Chapter 16
Judgments of Agency in Schizophrenia: An Impairment in Autonoetic Metacognition

Janet Metcalfe, Jared X. Van Snellenberg, Pamela DeRosse, Peter Balsam and Anil K. Malhotra

Abstract We investigated judgments of agency in participants with schizophrenia and healthy controls. Participants engaged in a computer game in which they attempted to touch downward falling X's and avoid touching O's. On some trials participants were objectively in perfect control. On other trials they were objectively not in complete control because the movement of the cursor on the screen was distorted with respect to the position of the mouse by random noise (turbulence), or it was lagged by 250 or 500 ms. Participants made metacognitive judgments of agency as well as judgments of performance. Control participants' judgments of agency were affected by the turbulence and lag variables—indicating that they knew they were objectively not in control in those conditions. They were influenced by their assessments of performance. The patients also used their assessments of performance but neither turbulence nor lag affected their judgments of agency. This indicated an impairment in agency monitoring. The patients,

---

This chapter is adapted from: Metcalfe J, Van Snellenberg JX, DeRosse P, Balsam P, Malhotra AK (2012) Judgments of agency in schizophrenia: An impairment in autonoetic metacognition. Phil Trans R Soc B 367:1391–1400

J. Metcalfe (✉)
Department of Psychology, Columbia University, New York, USA
e-mail: jm348@columbia.edu

J. X. Van Snellenberg · P. Balsam
Department of Psychiatry, New York State Psychiatric Institute, Division of Translational Imaging, Columbia University College of Physicians and Surgeons, New York, USA

P. DeRosse · A. K. Malhotra
Division of Psychiatry Research, The Zucker Hillside Hospital, New York, USA

P. DeRosse · A. K. Malhotra
Center for Psychiatric Neuroscience, Feinstein Institute for Medical Research, New York, USA

P. Balsam
Department of Psychology, Barnard College, New York, USA

S. M. Fleming and C. D. Frith (eds.), *The Cognitive Neuroscience of Metacognition*, DOI: 10.1007/978-3-642-45190-4_16, © Springer-Verlag Berlin Heidelberg 2014

unlike the healthy controls, used only publically available external cues about performance in making judgments of 'agency' and did not rely on any additional access to internal self-relevant cues that were diagnostic in indicating whether or not they were, in fact, in control.

Keywords Metacognition · Agency · Schizophrenia

The question of how an individual is able to determine whether it was the self or an alternative cause that was responsible for an action—metacognition of agency—is the concern of this chapter. This ability is crucial for learning and understanding one's own causal effect on the world, for all social interactions, and especially for coordination of individual and joint action (where the allocation of effort depends on knowing what one is doing and what the other is doing, and titrating one's own actions to accommodate those of others). This metacognitive capacity also underlies higher order social judgments such as those that are necessary for the assignment of credit and blame. Understanding of how people make judgments of agency and how other metacognitive judgments relate to these self-referential judgments is important in many domains. But, not all people make these judgments in the same way, and some have great difficulty in doing so accurately. In particular, the inability to keep the self straight—to know what is self-produced and what is externally produced—characterizes a large part of the core deficit in patients with schizophrenia. Investigation of the cues that are used to make these self-relevant judgments as well as specification of the cues that patients with schizophrenia are unable to recruit may increase our understanding both of schizophrenia and of the processes underlying how people know about their own agency.

Following Tulving [1], Metcalfe and Son [2] have argued that there are three levels of metacognitive judgments: anoetic judgments (which are judgments about objects or events currently present in the world), noetic judgments (which are judgments concerned with internal representations, but without self-relevance), and autonoetic judgments (which are self-knowing judgments in which reference to the individual's self is implicated). While many researchers have argued that a central reason for studying metacognition is that it is the hallmark of human self-reflective consciousness, this characteristic only applies to autonoetic metacognition, and not to the other kinds. Reflection upon the self is not involved in either noetic or anoetic metacognition. Anoetic metacognition involves a judgment about a stimulus that remains present at the time of judgment. It is sometimes thought that it is not even metacognition proper since no internal representation, or cognition, need be involved [3]. Most animals are capable of anoetic 'metacognition.' And although noetic metacognition, in which a judgment is made about an internal representation, is thought by virtually all researchers to really be metacognition (and some non-human animals have this capability, [4–6]) it does not necessarily implicate *self-reflective* consciousness. No *self* need be involved. Judgments of agency, though, are truly autonoetic metacognitive judgments, being both self-reflective and self-knowing [7]. They are a reflection on a cognition concerning the extent of one's own personal involvement and responsibility for an action [8]. Isolating the cues that people use to accurately make these particular self-relevant judgments concerning how they know they are the agent, is, then, of specific interest for understanding the nature of human self-reflective consciousness. Considerable recent research has been devoted to the problem of how the cues to agency are isolated and combined [9] as well as to their diagnosticity [10, 11]. People with schizophrenia frequently have difficulty with attributions of just this sort.

Jeannerod [12], Synofzik et al. [13] and Voss et al. [14] have noted that patients with the positive symptoms of schizophrenia, such as hearing voices and experiencing hallucinations and delusions, have difficulty in accurately reflecting upon their own agency. Such symptoms are also related to imaging findings showing hyperactivity in areas of the brain, in particular, the temporal parietal junction [15] that relate to the detection of a discrepancy between one's own intentions and the outcome that ensues [16, 17]. These brain activity differences almost certainly relate to impairments in action monitoring, and have been related to the mechanisms specified by the comparator model (see, Chambon et al. [18] this volume). There are also known deficits in such patients in frontal brain areas that are associated with self-relevant processing [19–23] which likely relate to deficits in metacognitive judgments [24, 25]. With such impairments, one might easily make the mistake either of thinking that one's own internal thoughts came from outside and were produced by someone else rather than by oneself, or of believing that one was controlling events that were externally caused. Whether the representation one perceives came about because of one's own thought—or image—generation processes or was externally produced, is, at base, an attribution of agency, that is, a judgment about who or what was causal in producing the percept. These and other kinds of thought processes and inferences associated with schizophrenia [26–28] might well result from impairments in a circuit that normally, accurately and efficiently, evaluates agency. Finally, there are some indications that feelings of being in control are linked to striatal reward systems and preSMA [29] suggesting a potential link to the dopamine hypothesis of the locus of impairment in schizophrenia.

Although healthy adults are usually able to make accurate judgments of agency [30], even they can sometimes be fooled about whether or not they were the agent [31, 32]. Furthermore, people at different stages of development make judgments of agency that are systematically sensitive to different parameters [33]. The findings of illusions of agency, and of systematic differences in these judgments even in healthy adult populations, substantiate the idea that there are a number of distinct cues that contribute to agency judgments. Both the cues and the judgment processes appear to be malleable.

The idea that metacognitive judgments of agency are based on cues, rather than direct knowledge [34, 35] is consistent with the widely held view that other metacognitive judgments are cue based. There are many cases, detailed in the voluminous metacognitive literature, in which it has been shown that certain judgments rely on different cues from one another (see, [36–42] for discussion and evidence concerning the cue-based nature of different metacognitive judgments).

Understanding which cues are used for making judgments of agency, as well as what neural circuitry underlies them, is important in ameliorating distortions seen in these judgments in people with schizophrenia. Studying the locus of the deficit in patients who have impairments in this particular metacognitive domain may allow more intensive scrutiny of the cues and mechanisms contributing to these central metacognitive judgments in healthy people. The investigation of metacognition of agency, though relatively new, points to four cues, or sources of information, that appear to contribute to these judgments, only one of which—the discrepancy detection cue—is valid [9, 11, 43, 44]. Interestingly, while the judgments themselves are concerned with whether the self was or was not responsible for an action, people often use cues to make these judgments that are neither internal nor self-referential.

16.1 Cues Contributing to Judgments of Agency

16.1.1 Judgments of Performance

Perhaps surprisingly, the single most important factor that has emerged as a predictor of people's agency judgments is another metacognitive judgment, namely, judgments of performance. While judgments of agency are autonoetic—being explicitly about the role of the self in an action—judgments of performance need not reference or even reflect the self. They are merely noetic (i.e., judgments about a representation, but without the necessary involvement of the self that would make them autonoetic). In the task that we will investigate (see [29]), people play a computer-based game of having a cursor touch X's and avoid touching descending O's by moving a computer mouse. At the end of each trial they are asked for a judgment of performance. The judgment of performance does not, itself, require that the individual participant be the agent. Such a judgment about the proportion of X's touched and O's avoided on the last trial could be made even if someone other than the subject had been controlling the mouse. In short, this judgment is noetic, not self-referential, and need not imply access to the participant's own role as the person controlling the mouse to touch the X's. Even though this assessment says nothing about who was responsible for the action, people's perception of performance is, nevertheless, an important cue used to make judgments of agency: when performance is perceived to be good agency is claimed; when performance is perceived to be poor, agency is denied.

Regression analyses directed at determining the sources of information that contribute to normal adults' metacognition of agency have revealed that people's perception of their level of success on the task on each trial is a strong contributor [10, 28]. The self-relevant autonoetic agency judgment, then, is based in large part on a non-self-referential noetic judgment concerning the goodness of performance. It is not necessary to be the agent at all, or to evaluate any internal or visceral cue to which one has privileged access, to make a judgment of performance. For instance, in one experiment [10] the target X's that the person was supposed to 'pop' by touching their representations with the computer cursor popped only 75 % of the time, even when the person had executed the planned action perfectly and had touched the X, that is, had exercised perfect control over everything that could be controlled by self-action. But whether the resultant outcome was the intended popped X or not was not up to the participant, but rather was due to external circumstances beyond the participant's control. This distal cue about success or failure at the task was, nevertheless, integrated into people's judgments of agency through their perception of performance. Whether the program was set to pop the X's 100 % of the time, 75 % of the time, or never was not up to the agent, but to the external world. And yet, people apparently use these judgments about performance outcomes to a large but, importantly, not exclusive extent to evaluate whether or not they were in control of an action.

While acknowledging that people do use their judgments of performance to make agency judgments, Metcalfe and Greene [30] and Metcalfe et al. [33] have argued that since these performance judgments do not indicate the *source* of the action, they should be factored out of the analyses, to allow investigation of whether people were sensitive, in a *veridical way,* to cues implying that the self was or was not the agent. They also noted that people's judgments of agency should be assessed relative to their *perception* of their performance, rather than their actual performance, since it is not how the person is doing objectively that counts, but rather how they think they are doing. The use of judgment of performance, to anchor people's judgments of agency, also provides some leverage on how individuals' use the rating scales. The question of interest in evaluating the accuracy of people's judgments of agency is whether—in the conditions in which they are not fully in control—they pick up on their lack of control, over and above their perception of their overt performance. Thus, to evaluate people's metacognition of agency in past experiments [30, 33] people's judgments of agency were compared to their judgments of performance. Because there may be scaling effects in how people ground the judgment of agency scale the Control condition—in which there were no distortions of their actual control—was used to anchor their use of the performance and the agency scales. In the analyses that follow in this article, we, too, will use the difference between judgments of performance and judgments of agency, and will use the Control condition as the baseline against which to evaluate differences in these two judgments that occur in the experimental conditions in which objective control was distorted. In addition, we will also use regression analyses to investigate the contribution of this cue.

Before leaving the topic of the role of judgments of performance on judgments of agency, it is notable that other researchers have also shown that noetic metacognitive and autonoetic judgments sometimes appear to be intertwined. Cosentino et al. [45] have shown that noetic judgments of learning are strongly related to autonoetic judgments of agency. Similarly, David et al. [46] (this volume) have discussed the relation of metacognition to anosagnosia, and Cosentino et al. [47] have shown that inaccuracy in metamemory judgments (again, noetic judgments) are associated with a lack of awareness of memory deficit (anosagnosia)—an awareness of one's own capabilities that would seem to involve self-knowing consciousness. And, finally, Fleming et al. [24] (and see [25, 48]) showed that the same brain area (BA10) that Miele et al. [29] isolated as being more strongly activated in making self-relevant judgments of agency as contrasted with judgments of performance, is also, itself, implicated in noetic metacognitive judgments. Indeed, as Fleming et al. [24] showed, individual differences in the accuracy of (noetic) metacognitive judgments was shown to be related to structural brain differences in this area. Thus, the two kinds of judgments—while conceptually distinct—may be functionally related at a deep level.

16.1.2 Discrepancy Monitoring of the Correspondence Between Plan and Outcome

Frith et al. [26] (and see [49–52]) have proposed a brain-based framework for motor control that relates in a natural way to people's metacognitions of agency. The 'comparator' model was originally devised to explain how people make fine-grained corrections of motor movements, and has been shown to be valuable in illuminating one source of information that could provide a focal cue in agency judgments. According to this scheme, when a person has a goal, it gives rise to an internal model of their intentions (inverse model) and expectations (forward model) about achieving the goal. This initiates a motor plan which provides the specifications about what needs to be done to achieve the goal. The plan or expectation runs off in real-time simultaneously with the person's motor actions. A comparator mechanism evaluates the correspondence of the actions and the plan. A match between the expectation and the outcome indicates that the person's intentions corresponded to what happened, and no motor adjustment need occur. A discrepancy provides a signal to the motor system indicating that the movement needs to be adjusted to achieve the goal. The discrepancy can also be used by the metacognitive system as a cue indicating that something or someone else was interfering with the intended action: the person was not in complete control. For example, if turbulence or noise were introduced into the instrument the person is controlling, then the plan for the motor actions would fail to match what happened because of the noise. In such a situation, the discrepancy may be a cue used in a judgment process that provides a reliable indicator that the person was not completely in control.

In schizophrenia, either the plan or the internal feedback from the person's own actions may be distorted, and this may give rise to misattributions of control [53] that are the result of such a discrepancy monitoring mechanism. The model even points to components of a brain network (the temporal parietal junction, with cerebellum involvement) where one might seek to find evidence of this discrepancy. Given that discrepancy detection is closely linked to motor control, one might expect to see altered motor control in patients whose impaired metacognition of agency is due to an impairment in the forward model. For example, Synofzik et al. [13] showed that patients with positive symptoms showed a higher threshold for detecting discrepancies in feedback rotations, indicating an impairment in the precision of sensory predictions. Additionally, many patients with schizophrenia exhibit motor impairments, as well as abnormalities indicating irregularities in this action monitoring system.

However, not all patients exhibit such motor impairments. Knoblich et al. [54] conducted an experiment in which the participants—both healthy controls and people with schizophrenia—attempted to keep their stylus on a circling moving dot on the screen. When the dot accelerated off course, such that the participants had to change their action pattern to allow it to continue following the correct path, participants were supposed to keep the dot on the circle and to indicate that they detected the distortion that was occurring. People with schizophrenia were able to do the motor task—altering their motor behavior—as well as the healthy control participants. However, they were much slower and less likely to consciously *notice* the distortion (i.e., to have metacognition about the change) than were healthy controls. Thus, it appears that the metacognitive assessment can sometimes be dissociated from the motor aspects of the task in schizophrenia, suggesting that metacognitive judgment processes themselves may be independent of the action monitoring guiding motor performance.

16.1.3 Reward

While it is logically possible that the feeling of being in control is just a lack of feeling out of control [55] it is also possible that positive feelings of agency are, themselves, neurally coded, and distinct from such a proposed default state of not being out of control. Feelings of being *in control* have been claimed by a number of 'positive' psychologists (e.g., [56]) who stress the role of self-determination, to be both intrinsically rewarding, and to be associated with learning. Consistent with this notion, in Miele et al. [29] fMRI study, it was found that trials in which participants reported a high level of feeling 'in control' were associated with increased activity in the presupplementary motor areas, the rostral cingulate zone and the dorsal striatum, regions that are linked to self-initiated action and reward. The activation of this intention and reward-related system, in conjunction with feelings of being in control, rather than, say, deactivation of the temporal parietal junction area (which would indicate a default state of *not* being *out of* control) lends some credibility to the idea that feeling 'in control' may, itself, be a separable state with consequences.

Kirkpatrick et al. [57] work also converges on the idea that the reward system is related to positive agency judgments. In their study, methamphetamine users, after receiving either the drug or a placebo, engaged in a motor agency task similar to the one used in the present article. Insofar as methamphetamine has its effects on the dopamine/reward system, effects of the drug itself on the judgments of agency might be thought to be mediated by the reward system. Interestingly, then, although there was no difference in performance on the task depending upon whether the participants were on methamphetamine or not, their judgments of agency were increased, under conditions of objectively perfect control, when they were on the drug rather than on placebo. The drug, evidently, made them feel more agentic, or more in control.

Finally, Tricomi et al. [58] have found that reward-related areas were activated during conditioned learning, but only when the participants were aware of the contingency between their button presses and the outcomes. Being aware of the contingency between one's actions and the outcome, of course, could be rephrased as knowing that one was in control. If we interpret the results in this way, they would suggest that learning, associated with activation in the striatum, is related to feelings of agency. These data, then, suggest that knowledge of agency may be necessary for reward-related learning.

The potential involvement of the reward system in metacognition of agency may be of importance in schizophrenia because of the involvement of dopamine in schizophrenia. It is possible, perhaps even likely, that people with schizophrenia have abnormal responses to reward that relate in a complex way to misperceptions of agency. The role of reward and its impact upon people's metacognitions of being in control may, therefore, have especial interest in this context.

16.1.4 Temporal Delay

A judgment of agency is a special case of a judgment of causality, in which the question is whether the *self* is the causal agent. It would, therefore, be expected that factors affecting people's perception of causality would also affect judgments of agency. Perhaps the most studied of these factors that affect judgments of causality is temporal contiguity. Michotte [59] (and see [60]) has shown that when one moving object makes contact with another, and then the second, without any delay, begins to move, this interaction is perceived as causal with the first object causing the movement of the second. Michotte called this phenomenon the 'launching' effect. The perception of causality is systematically diminished as a lag is interposed between the movement of A and the movement of B. It follows that feelings of agency should also be decreased if a delay is interposed between one's act and the result.

In keeping with this idea, Blakemore et al. [61] have shown that when there is no temporal delay between the act of attempted self-tickling and the resultant self-stimulation, healthy individuals are unable to tickle themselves. They argue that the reason an individual cannot tickle him/herself is that the concordance between plan and outcome results in a diminution in the perceived stimulation. No such diminution occurs with a mismatch, and the tickle sensation is, hence, perceived when another person is responsible for the tickling. However, when a delay is interposed between the act of tickling oneself and the resultant self-stimulation (by means of a mechanical device) healthy individuals can self-tickle, underlining the role of temporal delay. It is notable, in this context, that Blakemore et al. [62] found that, unlike healthy participants, patients with schizophrenia were able to tickle themselves even without a temporal delay being interposed.

The data of Schlottman and Shanks [63] show systematic decreases in causality ratings as delay is increased. Nevertheless, even at fairly large delays people still judged A and B to have a causal relation, consistent with the Kantian idea that causality is inferred as long as there is *any* rule that is seen to mediate between A and B. In the experiment below, we equated the amount of discrepancy between the position of the cursor and the position of the mouse in a pure 'turbulence' condition, in which no rule mediates, and in a time delayed condition where there was a mediating rule. Past research has indicated that healthy adult participants feel less out of control in the time delayed condition, which has a mediating rule, than in the turbulence condition which does not [33]. People with schizophrenia, though, may have difficulty picking up on such a subtle mediating rule, and hence may not use this cue.

16.1.5 The Judgment Process

Finally, while the cues used and the sensitivity to them may vary from person to person, and some or all of them may be impaired in people with schizophrenia, it is possible that these cues to agency could all be normal, and yet an impairment in metacognition of agency could still result. It is possible that the judgment process itself could be distorted. An fMRI study has shown that there is a difference in neural processing in anterior prefrontal cortex between making a judgment of agency as contrasted to making a judgment of performance [29]. In other research, this area has been shown to be associated with other kinds of self-referential processing [21, 64, 65] and metacognitive judgments [24, 25]. The self-referential metacognitive judgment appears to be distinctive. It is possible that patients could have either intact or impaired ability to make such self-referential judgments. However, if this judgment process were impaired, agency judgments would be expected to be impacted even in the presence of veridical cues.

16.2 Experiment

The task employed was the same as has been used in past experiments [33] in which metacognition of agency was compared between young adults, children, and elders. As mentioned above, participants played a computer game in which they moved the mouse to touch downward falling X's on the screen and, at the same time, to avoid touching O's. Objective control of the cursor by the mouse could be undistorted, in the control condition (i.e., the person was objectively in full control) or could be altered by means of a lag in the relation between the mouse position and the cursor position or by turbulence (random noise) intervening between the mouse position and the cursor position. At the end of each trial, the participant made a judgment of his/her own control, that is a judgment of agency, as well as a judgment of performance. This task allowed us to investigate whether manipulations that objectively altered the person's control were open to accurate metacognitive assessment.

16.2.1 Method

16.2.1.1 Participants

The schizophrenia patient group included 22 patients recruited from The Zucker Hillside Hospital (ZHH), a division of the North Shore-Long Island Jewish Health System (NSLIJHS), in Glen Oaks, NY. to a protocol designed to assess functional disability in stable outpatients. Inclusion criteria for patients included clinical stability as defined by no hospitalization in the last 6 months, between 18 and 59 years of age with a DSM-IV diagnosis of schizophrenia or schizoaffective disorder, and no substance abuse in the preceding 1 month. All patients were on antipsychotic medication at time of testing. The mean age of the patient sample was 42.3 years (SD = 11.1) and 40.9 % were female. Healthy comparison subjects were recruited from the general population through the ZHH Healthy Control Initiative. Potential controls were excluded if they had a DSM-IV axis I diagnosis or a first-degree relative with a known or suspected axis I disorder. The mean age of the control sample was 38.1 years (SD = 11.3) and 45.0 % were female. Patients and controls with a history of CNS trauma, neurological disorder (including seizures), mental retardation, or known genetic disorder were excluded. All subjects provided written informed consent to a protocol approved by the NSLIJHS Institutional Review Board.

16.2.1.2 Diagnostic Measures

Patients' diagnoses were established with the structured clinical interview for DSM-IV (SCID) [66] and confirmed by diagnostic consensus conference, which utilizes expert clinical opinion alongside SCID and corroborating medical record information. Brief psychiatric rating scale (BPRS) mean was 27.2 (5.8) and the scale for the assessment of negative symptoms (SANS) was 29.3 (12.2). Comparison subjects were assessed with the SCID–Non-Patient Edition to rule out axis I diagnoses.

16.2.1.3 Apparatus

All experiments were conducted on individual iMac computers, used with a mouse, and mouse pad. Participants were tested individually.

16.2.1.4 Procedure

The instructions were: "Throughout this experiment you are going to play a game in which you will use the computer mouse to move a box on a gray track. Your job is to touch all of the X's as they come into range and to avoid touching any of the O's. After each trial, you will be asked to assess your performance. If you felt you got all of the X's, and avoided all of the O's, you should click to the far right of the blue bar, indicating everything correct. If you felt you got none of the X's, and touched all of the O's, then you should click to the far left, indicating nothing correct. You may also click anywhere in between. You will also be asked to assess how in control you felt. If you felt you were in complete control, click to the far right of the red bar. If you felt that you had no control, click to the far left. You may also click anywhere in between."

In this experiment, the performance judgment was always made before the judgment of agency. The constant order was used to minimize possible confusion. Previous experiments that have used either only an agency judgment or only a performance judgment on each trial [29, 30] have produced comparable results to those that have used both judgments on every trial [33].

Participants practiced both playing the game and making judgments, under the supervision of the experimenter, who made sure that the participant understood how the task and how the rating scales worked by having the participant report what each judgment meant, following each practice trial. The practice trials were repeated as many times as was necessary. After the practice trial(s), the experimenter asked if there were any questions, and if there were, he or she answered them. At the end of the experiment, the participant was questioned about what they had done, and was paid and thanked for participating.

16.2.1.5 Design

The experiment included six within-participant conditions: a Control condition in which the participant had perfect control of the mouse, a short lag condition (Lag1) in which the cursor responsiveness lagged the mouse position by 250 ms., a long lag (Lag2) condition in which the cursor position lagged the mouse position by 500 ms., a small amount of turbulence (Turb1) condition, which was discrepancy-matched (as will be described shortly) to the Lag1 condition, a large amount of turbulence (Turb2) condition, which was discrepancy-matched to the Lag2 condition, and a Magic condition, which artificially inflated performance, and was important so that participants did not become discouraged. The magic condition showed no differences between the control and patient participants on any measure, and is, therefore, not included in the analyses or discussed further. There were four replications of all conditions.

The amount of noise in the Turbulence conditions was matched to the amount of discrepancy between the mouse position and the cursor position in the Lag conditions. This was done by measuring every 8 ms., on the first lag trial, the discrepancy between the mouse position and the cursor position, and then rerandomizing these signed difference scores and adding them to the cursor position at each 8 ms, interval in the appropriate turbulence condition. This added noise was smoothed to prevent sudden jerks. Because of this matching algorithm, the amount of discrepancy—where discrepancy is considered to be the difference between the mouse position and the cursor position at each sampled position over the entire 15 s trial—was the same in the Lag condition and in the matched Turbulence condition, and so Type of Discrepancy and Amount of Discrepancy could be treated as factors. The difference between the two type of discrepancy conditions was that the discrepancy between the mouse position and the cursor position in the Turbulence condition was random, while in the Lag condition it was lawfully mediated by a time lag rule: If one were to shift the cursor position function back by 250 or 500 ms., in the Lag conditions, it would match the mouse position function perfectly. The Lag and Turbulence conditions, with high and low levels of discrepancy, therefore, comprised a 2×2 design. In the Control condition, which is used as a baseline, there was no discrepancy between the cursor position and the mouse position.

To equate the discrepancy as outlined above, the Lag condition had to come first, which constrained the randomization of the order of conditions within block, though all conditions were well distributed over the entire session. The data were, therefore, analyzed both with and without the first lag trial. Because there was no difference depending on its inclusion, it was included in the analyses that are reported below. The data from the four trials in each condition for each participant were collapsed.

The two metacognitive dependent variables of central interest were people's Judgment of Performance (i.e., how well did they think they had done on touching the X's and avoiding the O's), and their Judgment of Agency (i.e., how in control did they think they were). Both were measured on an analog scale coded from 0 to 1.0. We also computed performance using hit rate (i.e., the proportion of times the person touched in range X's) and false alarm rate (i.e., the proportion of times the person, incorrectly, touched O's). In past experiments on this paradigm, as in the present experiment, hit rate and d' were highly correlated, and only the former has shown a strong relation to people's judgments of performance with false alarm rate having only a very small impact on their judgments [30]. We therefore report only hit rate here.

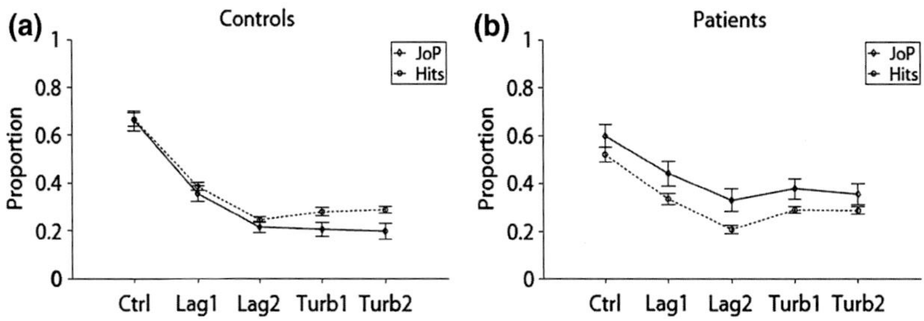

Fig. 16.1 Hit rates and judgments of performance in control participants, in (**a**), and patients with schizophrenia, in (**b**)

16.2.2 Results

In the results that follow, in cases where a participant did not finish all trials, their data are included as long as they completed at least two trials in each condition. A value of $p < 0.05$ was used to determine significance.

16.2.2.1 Performance

As is shown in Fig. 16.1, there was a main effect of condition on hit rate, $F(4, 160) = 136.04$; $p < 0.01$. There was also a main effect of group on hit rate, $F(1, 40) = 6.15$, $p < 0.02$ but this effect was qualified by significant interaction between condition and group, $F(4, 160) = 6.27$, $p < 0.01$. Post hoc tests showed that that the healthy controls performed significantly better than did the patients only in the control condition, $t(40) = 3.43$, $p < 0.01$.

16.2.2.2 Metacognition of Performance

Figure 16.1 also shows that judgments of performance closely tracked hit rate in both groups. There were strong correlations between hit rate and judgments of performance (collapsing across conditions, within participants, and using Fisher's r-to-Z transformation throughout to normalize the distributions). The mean correlation ($\pm$ SD) for control participants was 0.87 ± 0.48, which was significantly greater than zero [$t(19) = 11.52, p{<}0.01$]. For patients, the mean correlation was 0.64 ± 0.43, which was also significantly greater than zero [$t(21) = 7.81, p{<}0.01$]. Although the correlation for controls was significantly greater than that for patients [$t(40) = 3.81, p{<}0.01$], the correlations shown between performance and judgments of performance by the patients were still very high and comparable to the correlations found, in this same paradigm, with children, $r = 0.67$, and elders, $r = 0.81$ [33]—groups that showed very good metacognition of agency.

A measure of calibration for each participant in each condition was computed based on the difference between their hit rate and their judgment of performance. As can be seen from Fig. 16.1, judgments of performance were only slightly lower than hit rate for the control participants, whereas judgments of performance were higher than performance for the patients. Statistically, while there was neither a main effect of condition nor an interaction between condition and group, there was a significant calibration main effect of group [$F(1, 40) = 9.72$, $p<0.01$]. This difference in calibration between groups—showing that the healthy controls were slightly under confident while the patients were overconfident—may relate to a 'reward' related difference in perception between the two groups: the patients, but not the controls, thought they had done better than they had.

In summary, both the patients' actual performance and their noetic metacognition, as measured by the correlation between their judgments of performance and their performance, were well above chance, though not as good as those of the healthy control participants.

16.2.2.3 Metacognition of Agency

We next asked whether participants picked up on their lack of control, appropriately, over and above their perception of their overt performance. To evaluate whether people experienced a greater decrement in their feelings of agency in the turbulence and lag conditions, we computed summary 'agency' scores, namely, the contrast: (Judgment of Performance$_C$-Judgment of Agency$_C$)-(Judgment of Performance$_E$-Judgment of Agency$_E$) where the subscript C refers to the control condition and E refers to either Turb1, Turb2, Lag1 or Lag2.[1] This summary score tests for an interaction between the performance and agency ratings.

As can be seen from Fig. 16.2, the control participants were sensitive to both the turbulence and the lag conditions' effect on decreasing their control—they showed strongly negative contrast scores. In contrast, the patients were insensitive to these manipulations—showing contrast scores of zero. There was a significant main effect of group, $F(1, 40) = 4.55$, $p < 0.05$, but no other main effects or interactions. Furthermore, one sample t-tests revealed that control participants had significantly negative contrast scores in all four conditions (all p's < 0.01), while patients' contrast scores were not different from zero in any of the conditions (all p's > 0.66). These patient data reveal a lack of metacognition of agency unlike that seen in any group that we have studied to date. In contrast to the data presented here, all previous groups tested on this paradigm have shown significantly negative values on all four contrast scores.

[1] In the control condition the patients had agency judgments lower (58.60, SD = 21.63) than performance judgments (60.28, SD = 21.76), while the healthy controls' agency judgments were higher (69.79, SD = 22.22) than performance judgments (65.85, SD = 20.09). This latter pattern has been found in other studies with healthy participants.

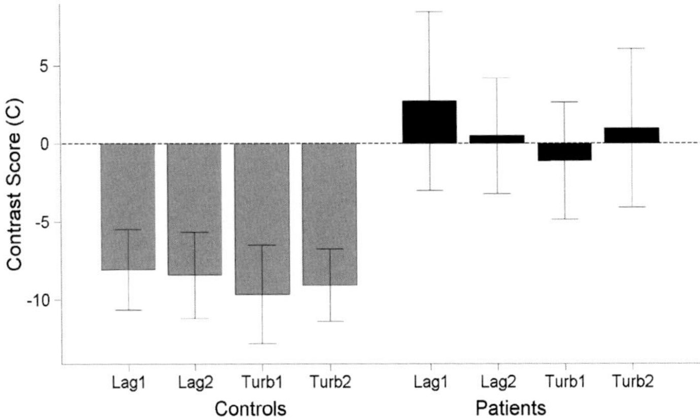

Fig. 16.2 Contrast scores in the turbulence and lag conditions for control participants, and for patients with schizophrenia. Scores provide a summary of the interaction between JoPs and JoAs in the turbulence and lag conditions, relative to the control condition. Negative scores in the turbulence and lag conditions in which participants were not fully in control, indicate that the person picked up on their lack of control, over and above their perception of their overt performance

Although numerically the controls showed slightly more negative scores in the turbulence conditions than in the lag conditions (as has been shown with young adults in past experiments, 21), the interaction was not significant.

We conducted correlation analyses to determine which, if any, symptoms in the patients' diagnostic profiles, predicted the above contrast scores. Insofar as negative contrast scores indicated sparing of metacognition of agency, we hypothesized that some symptoms, or some lack of symptoms, might predict such sparing. However, the results of these analyses failed to show significant results selective to any symptoms. We conducted a similar analysis with the raw Judgment of Agency scores, and, again found no correlation between particular symptoms and scores.

Finally, we conducted a regression analysis to investigate what information contributed to participants' judgments of agency. As can be seen from the (normalized) beta values given in Fig. 16.3, the control participants' judgments of agency were predicted by their judgments of performance as well as by the turbulence and lag conditions in which control was objectively impaired. In contrast, the patients' judgments of agency were predicted *only* by their judgments of performance. Control participants' judgments of agency were influenced by their judgments of performance [$t(19) = 11.37$, $p < 0.01$], as were those of patients [$t(21) = 11.03$, $p < 0.01$], and the two groups were not different ($t(40) = 1.02$, $p = 0.32$). However, with all other predictors, there was a difference between the controls and the patients. Control participants' judgments of agency were significantly influenced by the Lag1 condition [$t(19) = 3.87$, $p < 0.01$], by the Lag2 condition [$t(19) = 4.07$, $p < 0.01$], by the Turb1 condition [$t(19) = 4.02$, $p < 0.01$], and by the Turb2 condition [$t(19) = 4.31, p < 0.01$]. In contrast, as can

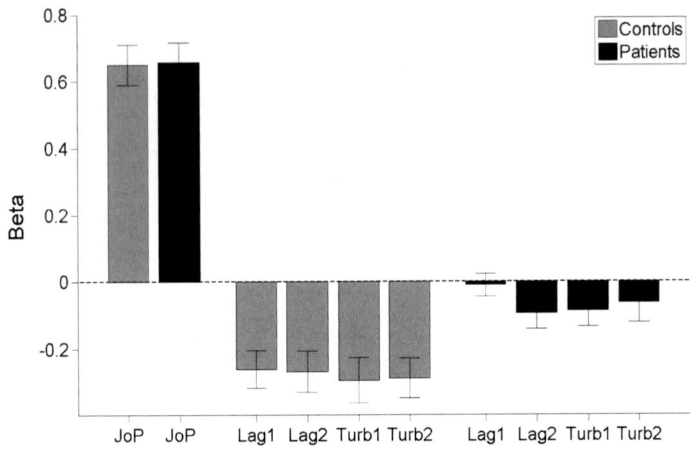

Fig. 16.3 Cues used by control participants and patients with schizophrenia to make judgments of agency. The figure shows mean betas from regression models calculated for each participant, with individual trials as the unit of analysis, and in which Judgment of Agency was the criterion variable. The potential predictors were Judgment of Performance and the experimental conditions—Lag1, Lag2, Turbulence 1 and Turbulence 2

be seen from the figure, the patients' judgments of agency were not significantly influenced by any of these conditions. As might be expected, in each of the four cases, the influence of the condition on judgment of agency was significantly greater for the control participants than for the patients (respectively for Lag1, Lag2, Turb1 and Turb2: $t(40) = 3.69$, $p < 0.01$; $t(40) = 3.04$, $p < 0.01$; $t(40) = 2.94$, $p < 0.01$; $t(40) = 2.98$, $p < 0.01$).

16.3 Discussion

The results presented here provide further indication that people's metacognition of agency is based on specific cues that are evaluated by a judgment process. The results also provide support for the separation of noetic and autonoetic metacognition. The healthy controls used both noetic (performance-related, that could be purely external) and autonoetic (internal) cues in making their agency judgments. These data indicate, however, that the patients used only the noetic cues, and did not recruit the autonoetic cues in making their judgments of agency.

The patients with schizophrenia performed very well on many aspects of the task. Moreover, their noetic metacognition, as given by the high correspondence between their judgments of performance and their actual performance, was good, though not quite as good as that of healthy controls. Thus, they did not show a profound deficit in all kinds of processing, or even in all kinds of metacognitive processing. However, they showed no sensitivity whatsoever to internal factors that objectively provide the kind of cues that healthy controls use to determine, accurately, when they are in control and when they are not. Unlike healthy control participants, the patients with schizophrenia appeared to be unaware of the presence of turbulence in the mouse controls, or by the fact that the response of the cursor was altered by a time lag of up to half a second. Healthy control participants know, very reliably, that they are 'out of control' under those circumstances.

The patients were not random in making their judgments of agency. The regression analyses showed that they did use one cue that is also used by healthy control subjects, namely the perceived goodness of performance. Furthermore, they appear to use this cue to about the same extent as do the healthy controls. But this was the *only* cue that the patients with schizophrenia used. Judgments of agency were, apparently, made without evaluation of any internal or visceral cues, or, indeed without any reference to the self, insofar as judgments of performance could be made purely externally and visually, by simply observing and remembering what happened in the trial. It was not necessary to know who was the agent (or, indeed, that there even was a human agent) to make such a judgment. Thus, the cues that provided the input for the purportedly self-referential judgments were, for the patients, lacking in any privileged or internal information. It is of some interest that Synofzik et al. [13] have observed a similar reliance on external visual cues in patients with schizophrenia. They propose that this reliance on external cues occurs because patients' internal cues are unreliable. The fact that basic noetic metacognitive processes—though not the self-referential ones—seem to be spared in the patient group, indicates, along with other evidence surveyed herein, that self-referential or autonoetic metacognition, while building on more basic metacognitive processes that are noetic in nature, may, nevertheless be both different and separable.

It is clear from this study that some aspects of the positive symptoms of schizophrenia could arise from a deficit in the perception of agency based on either a difficulty in perceiving the autonoetic cues or on using those cues to make judgments. It would be interesting to know if this difficulty encompasses all autonoetic cues or only those related to discriminating the relation between actions and outcomes. Further, it might be possible to explicitly train patients to discriminate autonoetic cues, perhaps producing a significant reduction in some positive symptoms such as delusions. Current pharmacologic treatments only temporarily meliorate positive symptoms. In contrast, interventions such as the one suggested above hold the potential for more permanent alterations by directly treating the underlying deficit.

Finally, the present results point to the interweaving of different kinds of metacognitive cues in the service of an externally posed task. This study asked for a judgment that directly focused on participants' own personal involvement as a causal agent. And yet, despite the task requirements, those judgments were made by using cues related to external outcomes that have no necessary connection to the role of the self in the action. Judgments about agency were, in healthy control participants, also based on internal cues indicating distorted objective control.

These internal cues provided reliable information about the individual's role as an agent. In contrast, the patients used only the performance cues, and did not access the internal cues that could allow accurate evaluation of the causal role of the self in action. The dissociation between the judgments of patients and healthy controls provides support for the importance of a distinction between noetic and autonoetic metacognition.

References

1. Tulving E (1985) Memory and consciousness. Can Psychol 26:1–12. doi:10.1037/h0080017
2. Metcalfe J, Son LK (2012) Anoetic, noetic, and autonoetic metacognition. In: Beran M, Brandl JL, Perner J, Proust J (eds) The foundations of metacognition. Oxford University Press, Oxford, UK, p 289–301
3. Metcalfe J, Kober H (2005) Self-reflective consciousness and the projectable self. In: Terrace HS, Metcalfe J (eds) The missing link in cognition: origins of self-reflective consciousness. Oxford University Press, Oxford, pp 57–83
4. Hampton RR (2001) Rhesus monkeys know when they remember. Proc Natl Acad Sci USA 98:5359–5362
5. Kornell N, Son LK, Terrace HS (2007) Transfer of metacognitive skills and hint seeking in monkeys. Psychol Sci 18:64–71
6. Smith JD (2009) The study of animal metacognition. Trends Cogn Sci 13:389–396
7. Frith CD, Frith U (2012) Mechanisms of social cognition. Annu Rev Psychol 63:287–313
8. Synofzik M, Vosgerau G, Newen A (2008) I move, therefore I am: a new theoretical framework to investigate agency and ownership. Conscious Cogn 17:411–424
9. Moore JW, Fletcher PC (2012) Sense of agency in health and disease: a review of cue integration approaches. Conscious Cogn 21:59–68
10. Metcalfe J, Eich TS, Miele DB (2013) Metacognition of agency: proximal action and distal outcome. Exp brain res, published online 29 Jan 2013
11. Metcalfe J (2013) 'Knowing' that the self is the agent. In: Metcalfe J, Terrace HS (eds) Agency and joint attention. Oxford University Press, Oxford, pp 238–255
12. Jeannerod M (2009) The sense of agency and its disturbances in schizophrenia: a reappraisal. Exp Brain Res 192:527–532. doi:10.1007/s00221-008-1533-3
13. Synofzik M, Thier P, Leube DT, Schlotterbeck P, Lindner A (2010) Misattributions of agency in schizophrenia are based on imprecise predictions about the sensory consequences of one's actions. Brain 133:262–271. doi:10.1093/brain/awp291
14. Voss M, Moore J, Hauser M, Gallinat J, Heinz A, Haggard P (2010) Altered awareness of action in schizophrenia: a specific deficit in predicting action consequences. Brain 133:3104–3112. doi:10.1093/brain/awq152
15. Decety J, Lamm C (2007) The role of the right temporo-parietal junction in social interaction: how low-level computational processes contribute to meta- cognition. Neuroscientist 13:580–593 PMID: 17911216
16. Farrer C, Franck N, Frith CD, Decety J, Georgieff N, d'Amato T et al (2004) Neural correlates of action attribution in schizophrenia. Psychiatry Res 131:31–44. doi:10.1016/j.pscychresns.2004.02.004
17. Frith C (2012) Explaining delusions of control: the comparator model 20 years on. Conscious Cogn 21:52–54
18. Chambon V, Filevich E, Haggard P (2014) What is the human sense of agency, and is it metacognitive? In: Fleming SM, Frith CD (eds) The Cognitive Neuroscience of Metacognition. Springer, Berlin

19. Carter CS, MacDonald AW, Ross LL, Stenger VA (2001) Anterior cingulate cortex activity and impaired self-monitoring of performance in patients with schizophrenia: an event-related fMRI study. Am J Psychiatry 158:1423–1428. doi:10.1176/appi.ajp.158.9.1423
20. Flashman LA, McAllister TW, Johnson SC, Rick JH, Green RL, Saykin AJ (2001) Specific frontal lobe subregions correlated with unawareness of illness in schizophrenia: a preliminary study. J Neuropsychiatry Clin Neurosci 13:255–257. doi:10.1176/appi.neuropsych.13.2.255
21. Gilbert SJ, Spengler S, Simons JS, Steele JD, Lawrie SM, Frith CD, Burgess PW (2006) Functional specialization within rostral prefrontal cortex (area 10): a meta-analysis. J Cogn Neurosci 18:932–948. doi:10.1162/jocn.2006.18.6.932
22. Juckel G, Schlagenhauf F, Koslowski M, Wüstenberg T, Villringer A, Knutson B, Wrase J, Heinz A (2006) Dysfunction of ventral striatal reward prediction in schizophrenia. Neuroimage 29:409–441. doi:10.1016/j.neuroimage.2005.07.051
23. Pomarol-Clotet E, Canales-Rodríguez EJ, Salvador R, Sarró S, Gomar JJ, Vila F, Ortiz-Gil J, Iturria-Medina Y, Capdevila A, McKenna PJ (2010) Medial prefrontal cortex pathology in schizophrenia as revealed by convergent findings from multimodal imaging. Mol Psychiatry 15:823–830. doi:10.1038/mp.2009.146
24. Fleming SM, Weil RS, Nagy Z, Dolan RJ, Rees G (2010) Relating introspective accuracy to individual differences in brain structure. Science 329:1541–1543. doi:10.1126/science.1191883
25. McCurdy LY, Maniscalco B, Metcalfe J, Liu KY, de Lange F, Lau H (2013) Anatomical coupling between distinct metacognitive systems for memory and visual perception. J Neurosci 33:1897–1906
26. Frith CD, Blakemore S, Wolpert DM (2000) Interactive report explaining the symptoms of schizophrenia: abnormalities in the awareness of action. Brain Res Rev 31:357–363
27. Jeannerod M (2009) The sense of agency and its disturbances in schizophrenia. Exp Brain Res 192:527–531. doi:10.1007/s00221-008-1533-3
28. Pacherie E, Green M, Bayne T (2006) Phenomenology and delusions: who put the ‘alien’ in alien control? Conscious Cogn 15:566–577
29. Miele DM, Wager TD, Mitchell JP, Metcalfe J (2011) Dissociating neural correlates of action monitoring and metacognition of agency. J Cogn Neurosci 23:3620–3636. doi:10.1162/jocn_a_00052
30. Metcalfe J, Greene MJ (2007) Metacognition of agency. J Exp Psychol Gen 136:184–199. doi:10.1037/0096-3445.136.2.184
31. Wegner DM, Wheatley T (1999) Apparent mental causation: sources of the experience of will. Am Psychol 54:480–492. doi:10.1037//0003-066X.54.7.480
32. Wegner DM, Sparrow B, Winerman L (2004) Vicarious agency: experiencing control over the movements of others. J Pers Soc Psychol 86:838–848. doi:10.1037/0022-3514.86.6.838
33. Metcalfe J, Eich TS, Castel A (2010) Metacognition of agency across the lifespan. Cognition 116:267–282. doi:10.1016/j.cognition.2010.05.009
34. Carruthers P (2011) The opacity of mind: an integrative theory of self-knowledge. Oxford University Press, Oxford
35. Ryle G (1949) The concept of mind London: Hutchinson Page references are to the 2000 republication. Penguin Books, London
36. Benjamin AS, Bjork RB, Schwartz BL (1998) The mismeasure of memory: when retrieval fluency is misleading as a metamnemonic index. J Exp Psychol Gen 127:55–68. doi:10.1037//0096-3445.127.1.55
37. Dunlosky J, Metcalfe J (2010) Metacognition. Sage, Thousand Oaks
38. Finn B, Metcalfe J (2008) Judgments of learning are influenced by memory for past test. J Mem Lang 58:19–34. doi:10.1016/j.jml.2007.03.006
39. Fleming SM, Dolan RJ (2012) The neural basis of metacognitive ability. Philos Trans R Soc B 367:1338–1349
40. Kelly CM, Jacoby LL (2000) Recollection and familiarity: process-dissociation. In: Tulving E, Craik FIM (eds) The Oxford handbook of memory. Oxford University Press, London, pp 215–222

41. Koriat A (2002) Metacognition research: an interim report. In: Perfect TJ, Schwartz BL (eds) Applied metacognition. Cambridge University Press, Cambridge, pp 261–286
42. Koriat A (2008) Easy comes, easy goes? The link between learning and remembering and its exploitation in metacognition. Mem Cogn 36:416–428. doi:10.3758/MC.36.2.416
43. van der Wel R, Knoblich G (2013) Cues to agency: time can tell. In: Metcalfe J, Terrace HS (eds) Agency and joint attention. Oxford University Press, Oxford, pp 256–267
44. Adams RA, Stephan KE, Brown HR, Frith CD, Friston KJ (2013) The computational anatomy of psychosis. Frontiers in Psychiatry 4:47, published online 2013 May 30
45. Cosentino S, Metcalfe J, Holmes B, Steffener J, Stern Y (2011) Finding the self in metacognitive evaluations: a study of metamemory and agency in non-demented elders. Neuropsychology 25:602–612. doi:10.1037/a0023972
46. David AS, Bedford N, Wiffen B, Gilleen J (2014) Failures of metacognition and lack of insight in psychiatric disorders. In: Fleming SM, Frith CD (eds) The Cognitive Neuroscience of Metacognition. Springer, Berlin
47. Cosentino SA, Metcalfe J, Butterfield B, Stern Y (2007) Objective metamemory testing captures awareness of deficit in Alzheimer's disease. Cortex 43:1004–1019. doi:10.1016/S0010-9452(08)70697-X
48. Yokoyama O, Miura N, Watanabe J, Takemoto A, Uchida S, Sugiura M, Horie K, Sato S, Kawashima R, Nakamura K (2010) Right frontopolar cortex activity correlates with reliability of retrospective rating of confidence in short-term recognition memory performance. Neurosci Res 68:199–206. doi:10.1016/j.neures.2010.07.2041
49. Blakemore S-J (2003) Deluding the motor system. Conscious Cogn 12:647–655. doi:10.1016/j.concog.2003.07.001
50. Haggard P (2008) Human volition: towards a neuroscience of will. Nat Rev Neurosci 9:934–946. doi:10.1038/nrn2497
51. Wolpert DM (1997) Computational approaches to motor control. Trends in Cognitive Sciences 1:209–216. doi:10.1016/S1364-6613(97)01070-X
52. Wolpert DM, Ghahramani Z, Jordan MI (1995) An internal model for sensorimotor integration. Science 269:1880–1882. doi:10.1126/science.7569931
53. Blakemore S-J, Wolpert DM, Frith C (2002) Abnormalities in the awareness of action. Trends Cogn Sci 6:237–242. doi:10.1016/S13646613(02)01907-1
54. Knoblich J, Stottmeister F, Kircher T (2004) Self-monitoring in patients with schizophrenia. Psychol Med 34:1561–1569. doi:10.1017/S0033291704002454
55. Synofzik M, Vosgerau G, Newen A (2008) Beyond the comparator model: a multifactorial two-step account of agency. Conscious Cogn 17:219–239. doi:10.1016/j.concog.2007.03.010
56. Bandura A (2001) Social cognitive theory: an agentic perspective. Annu Rev Psychol 52:1–26. doi:10.1111/1467-839X.00024
57. Kirkpatrick M, Metcalfe J, Greene M, Hart C (2008) Effects of intranasal methamphetamine on metacognition of agency. Psychopharmacology 197:137–144. doi:10.1007/s00213-007-1018-2
58. Tricomi EM, Delgado MR, Fiez JA (2004) Modulation of caudate activity by action contingency. Neuron 41:281–292. doi:10.1016/S0896-6273(03)00848-1
59. Michotte A (1963) The perception of causality. Basic Books, New York
60. Hubbard TL, Blessum JA, Ruppel SE (2001) Representational momentum and Michotte's "launching effect" paradigm (1946/1963). J Exp Psychol Learn Mem Cogn 27:294–301. doi:10.1037/0278-7393.27.1.294
61. Blakemore SJ, Frith CD, Wolpert DM (1999) Spatio-temporal prediction modulates the perception of self-produced stimuli. J Cogn Neurosci 11:551–559. doi:10.1162/089892999563607
62. Blakemore SJ, Wolpert DM, Frith CD (2000) Why can't you tickle yourself? Neuro Report 11:R11–R16. doi:10.1097/00001756-200008030-00002

63. Schlottman A, Shanks DR (1992) Evidence for a distinction between judge and perceived causality. Q J Exp Psychol 44:321–342. doi:10.1080/02724989243000055
64. Jenkins AC, Mitchell JP (2010) Medial prefrontal cortex subserves diverse forms of self-reflection. Soc Neurosci 6:211–218. doi:10.1080/17470919.2010.507948
65. Ochsner KN, Beer JS, Robertson ER, Cooper JC, Kihlstrom JF, D'Esposito M, Gabrieli JDE (2005) The neural correlates of direct and reflected self-knowledge. Neuroimage 28:797–814. doi:10.1016/j.neuroimage.2005.06.069
66. First MB, Spitzer RL, Gibbon M, Williams JBW (1995) Structured clinical interview for DSM-IV axis i disorders, patient edition (SCID-P), version 2. New York State Psychiatric Institute, Biometrics Research, New York

Chapter 17
Metacognition in Alzheimer's Disease

Stephanie Cosentino

Abstract It has long been recognized that a significant proportion of patients with Alzheimer's disease (AD) display some degree of unawareness for disease related memory loss. Historically, the majority of studies examining awareness in AD have implemented subjective assessment tools including clinician rating scales and informant based discrepancy measures. In the past two decades, there has been increasing focus on the objective assessment of metacognition in AD. These studies have made important strides in advancing our understanding of the nature of awareness deficits in AD and the mechanisms which may contribute to metacognitive variability in AD. However, there are several methodological issues that may complicate interpretation of existing data and that require consideration as this field moves forward. This chapter will: (1) review several commonly used subjective and objective approaches to measuring memory awareness in AD; (2) highlight important dissociations that characterize metacognitive functioning in AD; (3) evaluate specific models of metacognitive deficits (i.e., anosognosia) in AD and; (4) discuss future directions for metacognitive research in AD.

17.1 Introduction

Alzheimer's disease (AD), the most common cause of late life dementia, affects an estimated 5.2 million Americans and has a projected prevalence of 13.8 million by the year 2050 [107]. This neurodegenerative disease is traditionally diagnosed in the context of progressive change in memory and at least one other cognitive domain that is sufficient to interfere with everyday functioning [7, 65]. More

S. Cosentino (✉)
Division of Cognitive Neuroscience in the Department of Neurology, Gertrude H. Sergievsky Center and Taub Institute for Research on Alzheimer's Disease and the Aging Brain, Columbia University Medical Center, 630 West 168th Street, P&S Mailbox 16, New York 10032, USA
e-mail: sc2460@cumc.columbia.edu

S. M. Fleming and C. D. Frith (eds.), *The Cognitive Neuroscience of Metacognition*, DOI: 10.1007/978-3-642-45190-4_17,

recently, this diagnostic framework has been revised to accommodate non-amnestic presentations of AD whose earliest symptoms may relate to language, visuospatial abilities, executive functioning, or even social cognition [8, 66]. Indeed, while memory impairment is considered a classic and hallmark feature of AD, there is marked heterogeneity in the actual presentation of this disease with subsets of individuals demonstrating relatively greater impairment in other domains [30, 42, 68, 69].

The extent to which specific cognitive domains are impaired reflects the underlying distribution of neuropathology, a phenomenon which is perhaps most apparent in cases of posterior cortical atrophy [6] and frontal variant AD [49]. In less extreme examples, cognitive profiles can be fairly different across participants such that comparable memory deficits are accompanied by quite different degrees of spatial, semantic, or executive dysfunction, reflecting the relative involvement of parietal, temporal, and prefrontal networks, respectively. This heterogeneity in cognitive functioning is mirrored by heterogeneity in self-awareness, or metacognitive functioning, across individuals with AD [22]. It has long been recognized that a significant proportion of patients with AD display some degree of unawareness for disease related memory loss [54, 76, 88, 99]. Level of symptom awareness can vary dramatically among individuals with similar levels of global cognition and memory [24]. Estimates regarding the prevalence of impaired memory awareness in AD range widely, from as low as 25 % [33] to as high as 81 % [88], almost certainly reflecting differences in the criteria used to define awareness, the method of assessing awareness, and the disease severity of the sample.

Growing evidence suggests that level of awareness reflects the integrity of cognitive and neural networks that are critical for processes of self-evaluation [50, 52, 77, 90, 114]. However, the precise cognitive and neural components of such networks have long been debated and remain fairly ambiguous. Moreover, awareness in AD does not appear to map clearly onto disease severity [32, 70, 88, 99, 116], level of memory impairment [33, 70, 88, 104], depression [88, 99, 106], psychosis [58, 59], or executive dysfunction [31, 39, 58, 70, 88, 97, 105, 110].

In the past two decades, there has been increasing focus on the objective assessment of metacognition in AD, [20, 47, 74, 83, 84, 101]. These studies have made important strides in advancing our understanding of the nature of awareness deficits in AD and the mechanisms which may contribute to metacognitive variability in AD. However, there are several methodological issues that may complicate interpretation of existing data and that require consideration as this field moves forward. This chapter will: (1) review several commonly used subjective and objective approaches to measuring memory awareness in AD; (2) highlight important dissociations that characterize metacognitive functioning in AD; (3) evaluate specific models of metacognitive deficits (i.e., anosognosia) in AD; and (4) discuss future directions for metacognitive research in AD.

17.2 Approaches to Measuring Memory Awareness

Historically, the majority of studies examining awareness in AD have implemented subjective assessment tools including clinician rating scales [33, 63, 88, 116] and informant-based discrepancy measures [59, 99]. In the former, the clinician or examiner rates the participant on an ordinal scale, generally according to his or her answer to an open ended question regarding his or her memory. In the latter, participants and knowledgeable informants independently rate the participant across a range of abilities relating to cognition, behavior, and/or activities of daily living to determine the extent to which the participant's ratings differ from those of the informant. These tools are valuable in that they provide a clinically meaningful and quickly obtained snapshot of awareness level. They also characterize an individual's everyday perception of themselves, sometimes referred to as a global or off-line level of awareness [41, 85, 93]. Moreover, global awareness has been shown to relate to local or online levels of awareness, that is objectively measured metamemory in the context of a specific task [24, 40, 41, 93]. These associations are independent of disease severity measured with global cognitive performance or memory, suggesting that there is a specific self-referential quality that is captured by each type of assessment. Despite this association, however, the correlation between global (i.e., subjective/clinically based) and local (i.e., objective/task based) metamemory assessments is imperfect. Indeed, in a recent study, approximately 39 % of participants received discrepant ratings, with awareness appearing intact on global but not local measures in 9 %, and the reverse pattern in 30 % of individuals with AD [19]. There is also preliminary evidence that awareness metrics differ in their correlates, with global but not local ratings being associated with depression, for example [18]. Future work should explore these associations in more depth, with consideration of premorbid personality style and its effects on different levels of awareness.

Global measures of awareness have important advantages including their reflection of everyday or "real world" levels of awareness. In some respects, a global score may be better able to inform practically and clinically relevant issues including the extent to which participants appreciate the need to seek assistance or devise strategies for completing cognitively demanding activities. For example, everyday decision-making capacity related to managing medications in everyday life appears to relate primarily to global awareness, independent of cognition, and is unrelated to local awareness [25]. However, a disadvantage is that global awareness scores are limited in their ability to inform the nature of impaired awareness. That is, an overall subjectively derived score is less amenable than local metacognitive metrics to examination of its component properties and the manner in which they change under different task manipulations. In order to further advance the study of memory awareness in AD, and to shed light on the errors that give rise to this deficit, it is necessary to dissect memory awareness into clear and identifiable components that can be measured objectively [22, 54, 100].

A growing number of studies have applied objective tasks to examine memory awareness, or metamemory, in AD [9, 11, 12, 16, 20, 24, 34, 41, 57, 61, 64, 72, 73, 82, 83, 90, 101]. These studies have been critical in allowing us to examine the nature of metacognitive errors in individuals with AD and to then draw inferences about the basis of the clinical syndrome of impaired memory awareness. A variety of approaches have been used to collect information about the integrity of self-assessment including global predictions and retrospective ratings regarding the overall number of words that would be (or were) remembered, as well as item by item predictions or postdictions regarding the likelihood of accurately recognizing (or having recognized) a given stimuli at test. A number of objective metrics are used to synthesize information from predictions with information about accuracy, and to determine either the extent to which an individual is over or under confident (i.e., calibration/absolute accuracy) or the extent to which predictions for performance covary with accuracy (i.e., resolution/relative accuracy). While critical for fully appreciating the implications of findings from metacognitive studies in AD and potential discrepancies across studies, a detailed discussion and comparison of these metrics is beyond the scope of this chapter (see Fleming and Dolan, this volume; Schwartz and Diaz, this volume). When relevant, however, attention will be drawn to measurement issues and their implications for understanding disordered memory awareness in AD.

17.3 Dissociations in Metacognition in AD and Implications for Models of Anosognosia

Six years ago, Souchay performed a comprehensive review of metamemory studies in AD, concluding that there appeared to be a fractionation of metamemory, with preservation of specific abilities and degradation of others [100]. First, multiple sources of evidence converged on the idea that individuals with AD appreciate, or are sensitive to, characteristics of the stimuli or task at the time of encoding that influence their recallability [74]. Specifically, predictions in AD are appropriately reduced for delayed tasks as compared to immediate memory tasks [64], recall tasks as compared to recognition [73], and as a function of item difficulty [75] and distinctiveness [16]. Thus, knowledge regarding the factors that influence encoding is comparable to that seen in healthy older adults and therefore cannot account for the metamemory deficit seen in a subset of individuals with AD.

Moreover, it is not the case that objectively measured metamemory deficits simply reflect difficulty with the nature of metacognitive tasks, due to working memory or comprehension difficulties. This is clear when one considers the dissociation between semantic metamemory and episodic metamemory [11, 24, 57]. That is, individuals with AD tend to perform similarly to controls when asked whether or not they will recognize answers to general knowledge questions, but are frequently impaired in their predictions regarding the likelihood that they will recognize newly learned information [24, 83, 101]. This dissociation has been conceptualized to reflect the greater potential for reliance on cue familiarity rather than the retrieval of partial information regarding the target word in semantic tasks [100]. Both factors have been theorized to contribute to Feeling of Knowing (FOK) judgments, predictions regarding the likelihood that a specific piece of information will be recognized [55, 67]. The potentially greater role of cue familiarity in semantic FOK tasks derives from the fact that such tasks are generally probed in the form of a question whereas episodic FOKs are often probed by a single word, and therefore more reliant on the retrieval of partial information regarding the target word. Souchay [100] also suggests that the dissociation in semantic versus episodic FOK judgments in AD may reflect preserved noetic consciousness (knowing) related to semantic memory in the context of altered autonoetic consciousness (self-knowing, or remembering) as an extension of deficient episodic memory [109].

A second dissociation that has been reported relates to relatively impaired predictions for performance in the absence of task experience (prospective judgments) versus improved estimations of performance after experience with the task (retrospective judgments) [9, 73, 74]. This dissociation has the potential to inform the basis of impaired metamemory in AD, implicating an intact mechanism by which to evaluate performance retrospectively, yet a seemingly compromised ability to hold onto information about memory failures over the long term to enable accurate prospective ratings. This pattern of metamemory performance would be consistent with the mnemonic subtype of anosognosia as outlined by the Cognitive Awareness Model (CAM) [5, 46]. According to the CAM, information about a memory failure is first processed through a comparator mechanism; that is, the memory failure is compared with information in one's personal knowledge base to categorize it either as a regular or irregular occurrence. These occurrences are then stored in one's personal knowledge base. Mnemonic anosognosia is theorized to occur when one is unable to update the personal knowledge base by consolidating new memory failures over time, despite recognizing the memory failure as aberrant when it occurs [5, 46]. As a result, individuals appear to form expectations for performance based on an outdated, or "petrified" sense of self [71]. This form of anosognosia has been contrasted with an executive anosognosia in which failure is theorized to occur at the level of the comparator, that is, at the level of error monitoring or detection.

While the relative advantage of retrospective ratings of memory performance as compared to prospective ratings appears to support the mnemonic model of anosognosia in AD, closer consideration of this dissociation challenges the validity of applying this model. Unlike the presence of episodic memory loss, the presence of memory awareness deficits in AD (whether measured clinically or objectively) is not universal [24]. Almost all studies that have examined metamemory in AD combine individuals with heterogeneous levels of memory awareness, potentially clouding our ability to understand the exact nature of the metamemory impairment in only a subset of the participants. One potential problem with this approach relates to the fact that healthy young and old adults make more accurate retrospective ratings of memory than prospective ratings [21]. Thus, individuals with AD who are aware of their memory loss would also be expected to have better retrospective than prospective ratings, a phenomenon that could pull up the group mean and make it appear that those individuals with metamemory impairment improve significantly after exposure to the task.

This possibility is supported by the pattern of results obtained in a recent study of online versus offline memory awareness in AD [93]. 20 individuals with AD made predictions and post-task estimations of performance on the Rey Auditory Verbal Learning Test. Consistent with earlier studies, participants improved their estimations of performance after exposure to the task. Interestingly, however, the degree of improvement was correlated with the offline subjective measure of awareness (informant based discrepancy score), such that individuals whose estimations improved the least were those that had the least amount of reported awareness of memory loss on a day to day basis. This suggests that individuals who are the least aware of their memory loss in general (i.e., those who have anosognosia) do not benefit as fully from exposure to the task as those who have greater day to day awareness of their deficits. Depending on how one interprets these findings, there are different implications for a cognitive model of anosognosia in AD. In one respect, it could be said that individuals with the greatest degree of anosognosia have compromise not only to the personal knowledge base but to a comparator mechanism as well. Alternatively, one could argue that the correlation between everyday anosognosia and lack of adjusted estimations after task exposure suggests that the primary basis of anosognosia in AD is at the level of the comparator mechanism rather than the personal knowledge base. In other words, it is possible that the petrified self could be an "epiphenomenon" of a more central executive dysfunction [71].

Let us consider a different set of findings that may lend support to a mnemonic model of anosognosia in AD. Several studies have shown that memory accounts for differences in metamemory across healthy elders and individuals with AD [101, 102]. That is, group differences in metamemory disappear when memory is entered as a covariate in between group analyses. However, it is important to consider whether this is the same thing as memory being correlated with metamemory in individuals with AD. In fact, within AD participants (as opposed to AD and healthy elders combined), FOK was unrelated to a memory composite score [101]. As the authors suggest, this could reflect a lack of statistical power in the limited group of participants. However, it is also worth considering the question that if all participants with AD, by definition, have considerable impairment in storing new information, how is it that some patients are able to update the personal knowledge base, and others are not? In other words, how can some participants "remember that they forget" while others cannot? It has been suggested that heterogeneity in awareness may be due to individual differences in the preservation of personal semantics, or the presence of residual abilities to acquire new semantic knowledge [71]. It is also possible, however, that other non-memory based aspects of cognition related to error detection may play a key role in contributing to impaired memory awareness, and it has been suggested that this type of deficit is present in at least some individuals with AD [5, 46]. Moving forward, computational models of metacognitive efficiency that explicitly model influences of task performance on metacognition may help to clarify the extent to which deficits in memory itself contribute to disordered metamemory in AD [62], see also Maniscalco and Lau, this volume.

While not a problem-free solution, isolated examination of individuals who are clinically classified as unaware may shed light on the relationship between error detection and metamemory deficits in AD. As an aside, one potential problem with selection of this subsample is that it restricts analysis to individuals who have impaired awareness at a global rather than local level, and while related, these two constructs are not synonymous as discussed earlier. Another approach could be to select only those individuals who perform poorly on objective metamemory testing to determine whether specific types of errors contribute to lowered scores, and whether or not particular task manipulations improve metamemory scores. However, this approach could be limited by regression to the mean effects, in which task performance improves under different task conditions not because of the critical task manipulation, but simply because performance on the original task was required to be poor. Recent experience in our lab has reinforced this idea; specifically, when the unaware sample was defined on the basis of poor metamemory scores, metamemory improved under all task conditions and the opposite pattern was observed for the aware sample. That is, those individuals who were selected on the basis of intact metamemory scores demonstrated significantly lower metamemory scores under all other task conditions [27]. As this phenomenon obscures the ability to determine the factors which enhance test performance in unaware individuals, we return to the idea that identification of unaware individuals may be best accomplished by clinical ratings. While imperfect, this procedure has the benefit of informing the basis of the clinical syndrome of disordered awareness seen in a large proportion of individuals with AD.

Examination of metamemory errors in participants selected to be clinically unaware of their memory loss revealed a pattern of results that may be more consistent with an error detection problem, or an executive anosognosia, than a primary problem with storing information about memory failures [24]. Specifically, clinically unaware individuals failed to adjust their predictions for performance after experience with the task and became overconfident over the course of several metamemory trials, as compared to clinically aware individuals with AD and healthy controls who moved toward under-confidence. This dissociation in performance across aware and unaware participants argues that the latter either do not recognize when memory failures are occurring, or they do not use information about memory failures to form judgments for future performance. Qualitative examination of responses on metamemory testing suggests that a problem with error detection may be primary. Specifically, in instances in which participants did not recognize the correct answer during the test phase of a metamemory task, clinically aware individuals more often answered "I don't know" than did clinically unaware individuals who were more likely to endorse a recognition foil.

Lack of awareness for individual errors could reflect several different issues including a failure to systematically attend to one's performance, or to detect memory errors as they occur. Recent work from our lab has examined these two possibilities by comparing performance across three versions of a metamemory task [27]. In the standard condition of the task, individuals are asked to make predictions about their ability to recognize newly learned information. For details on the task, see [24]. In the query condition, participants are asked to make predictions as well as retrospective ratings of accuracy after each item. Two metamemory scores are calculated in this condition. In the first (the query score), metamemory scores are still calculated using predictions for performance; the critical manipulation is that by adding item by item retrospective ratings, individuals are forced to evaluate their performance after each trial. In other words, they are forced to attend to their performance and systematically evaluate themselves. The second score (the retrospective score) is calculated using the postdictions to determine if assessment of performance is intact immediately following a memory failure. Improved performance in this condition only, as compared to the standard and query conditions, would suggest that a deficit in error detection is not a primary source of metamemory impairment in AD. Interestingly, in individuals clinically rated as unaware, performance does not improve significantly in either the query or retrospective condition. The fact that cuing participants to systematically evaluate themselves after each trial does not improve predictions suggests that metamemory is not compromised secondary to a failure to attend to one's own performance. Rather, the fact that obtaining estimations immediately following the test item did not improve performance suggests that a failure to detect the memory failure as it occurs may be a primary basis of anosognosia in AD.

However, two recent studies examining explicit versus implicit levels of awareness in AD suggest that the basis of disordered awareness as it relates to error awareness is likely to be more nuanced than this [61, 72]. Specifically, while individuals with AD demonstrated an impaired ability to retrospectively judge level of performance with explicit ratings, measures of implicit awareness including speed of reading dementia-related words versus neutral words, and emotional reactivity on self-report and analysis of facial expressions after memory failures suggested that individuals with AD had some level of awareness about poor task performance. Interestingly, explicit and implicit metrics of awareness were unrelated in both AD participants and healthy elders, implicating separate pathways [72]. The integrity of implicit awareness informs the basis of awareness deficits in AD to the extent that it demonstrates some preservation of error detection, perhaps selectively shifting the metacognitive dysfunction away from the comparator mechanism to one of its output routes, namely the explicit route to the personal knowledge base, but not the implicit route which bypasses the personal database as outlined by the CAM [5].

A failure in communication between the comparator mechanism and the personal database, rather than the comparator mechanism per se, could potentially explain why metamemory impairment in AD is not consistently related to a central executive dysfunction. While executive functioning has been tied to memory monitoring in several studies of healthy older adults [26, 103] its association with memory monitoring in AD has been less clear. Several (but not all) studies using subjective measures of awareness have tied lack of awareness to executive dysfunction [31, 58, 70, 82, 88, 104]. There are certainly reasons to expect such an association, including the substantial literature in both clinical and healthy populations implicating the role of the PFC in supporting executive functioning and aspects of self-awareness, and the conceptual similarities between executive functioning and memory monitoring. However, findings from empirical examinations of the relationship between objectively measured metamemory and executive function in AD and MCI do not support this idea [84, 101, 102]. A recent study designed to disentangle the cognitive correlates of metamemory in AD suggests that a specific cognitive domain may be less influential for metamemory than the extent to which a task relies on a critical set of brain regions [95]. Specifically, FOK in a large sample of individuals with AD ($n = 68$) was preferentially related to nonverbal measures of both memory (figure learning) and executive functioning (design fluency). Given the relatively greater role for the right hemisphere in supporting these nonverbal tasks [43–45] and the long documented effect of right hemisphere damage on self-awareness in a variety of illnesses [1–4, 15, 38, 87, 94], these findings preliminarily suggest that damage to a critical fronto-temporal network, seemingly right greater than left, may contribute to metamemory impairment in AD.

Indeed, there is steadily growing evidence that this metacognitive disturbance may reflect compromise to a critical set of brain regions spanning frontal, midline, and temporal regions that are integral for processes of self-assessment. Specifically, reduced awareness of memory loss has been related to decreased functional connectivity between the medial prefrontal cortex (mPFC) and anterior cingulate cortex (ACC) [90], attenuated activation in the mPFC and posterior cingulate cortex (PCC) during a self-appraisal task [89], and prefrontal hypoperfusion [88]. The mPFC and PCC both appear to play an important role in accurate self monitoring and self-appraisal in healthy adults [50, 51], and the ACC plays an important role in error detection, responding to errors, and conflict monitoring [14, 17, 108]. These cortical midline regions emerged as a highly consistent set of structures in a meta-analysis examining a number of neuroimaging studies investigating the neural basis of self-referential processes [77]. The results of cluster and factor analysis suggested that within the larger set of cortical midline structures, there may be three clusters representing ventral (medial orbitofrontal cortex, ventromedial PFC, and subgenual and pregenual ACC), dorsal (dorsomedial PFC and supragenual ACC), and posterior (PCC, retrosplenial cortex, and medial parietal cortex) regions with functional specializations. These regions did not diverge across stimuli type (e.g., mnemonic vs. emotional), and were thus theorized to differ with regard to their specific processes in supporting aspects of self-assessment, with the ventral region critical for tagging stimuli as self-referential, the dorsal region for appraising self-related stimuli, and the posterior region for placing self-related stimuli in a temporal context and linking them with past self-related stimuli. In addition to these regions, the lateral PFC was prominently involved across a number of self-referential tasks; however this seemed to be the case primarily for those tasks that had a strong cognitive component related to linguistic or mnemonic processes, for example.

A recent study by Zamboni et al. [115] examining the neural basis of impaired self-awareness in AD supported the role of the mPFC and cingulate, while also implicating the bilateral anterior temporal lobes for the assessment of self-specific information related to cognitive and behavioral traits in individuals with AD. Specifically, while healthy elders and participants with MCI used mPFC and anterior temporal regions while making ratings of others as well as themselves, individuals with AD showed significantly decreased activation in both of these regions when conducting the self-rating as opposed to the other rating. Moreover, level of activation in the mPFC during the self- condition was associated with two informant based ratings of anosognosia, independent of disease severity, or memory loss. This study was one of the first to demonstrate a role for the anterior temporal lobe in supporting aspects of self-assessment in AD. The authors suggest that this may reflect the importance of the ATL in supporting knowledge about people in general, and learning facts about others [98, 113]. Interestingly, the role of the temporal lobes was highlighted in an earlier study by Salmon and colleagues who reported a negative correlation between anosognosia as measured by a caregiver based discrepancy score and metabolism on FDG-PET imaging in regions including the bilateral temperoparietal junctions and inferior temporal cortices, and the left superior frontal sulcus in a large sample ($n = 209$) of individuals with mild to moderate AD [92]. This relationship was independent of a variety of disease related factors including global cognitive decline measured with the Mini Mental State Examination and Clinical Dementia Rating Scale, as well as age and apathy as rated on the Neuropsychiatric Inventory. Finally, recent work has highlighted a potentially important role for the right insula in supporting memory awareness across both healthy elders and individuals with AD [23]. This is consistent with previous imaging work implicating the insula in aspects of self-awareness in healthy adults including recognizing one's own face and detecting performance errors [29], as well as a long standing literature pointing to a preferential role for the right hemisphere in subserving accurate self-assessment across diverse clinical populations.

The brain regions that support memory awareness in AD are coming into focus, and it is becoming increasingly clear that these regions comprise a broad anterior to posterior network with integral roles not only for cortical midline regions including the medial PFC, and anterior and posterior cingulate, but for multiple regions within the temporal lobes as well. The precise functions served by each of these hubs in the network, as well as other potential regions, remain a matter of debate as does the specific nature of metamemory errors in AD. It is quite possible that the basis of metamemory impairment in AD is heterogeneous, reflecting involvement of different neural regions and cognitive processes within the broader network across different patients. This possibility is underscored by clinical observations with patients who enter an evaluation at the family's request and who have no personal complaints regarding memory or other cognitive abilities. In some cases, such individuals leave the evaluation with an increased sense of cognitive difficulties based on experience with the tasks. In other cases, however, failure on the tasks appears to do little to increase deficit awareness and is instead greeted with perplexity, or superficial reasons for poor performance (e.g., "The lighting is not good" or "I don't remember these words because they aren't important to me"). Moving forward, the development of metacognitive tasks to understand the cognitive and neuroanatomic basis of awareness deficits in AD would be greatly informed by careful consideration of the clinical syndrome of disordered memory awareness in this population.

17.4 Future Directions

The examination of metacognition in AD has gained momentum over the past decade. Still, there is a great deal of work that needs to be done in the future. In addition to methodological issues raised throughout the chapter, there are several big picture issues for consideration in future research. First, the field would benefit from increased attention to assessment of metacognition at the earliest end of the disease spectrum. A growing literature has documented the presence of impaired self-awareness in individuals with Mild Cognitive Impairment (MCI) suggesting that in at least some individuals, disordered awareness is present from the onset of memory loss rather than as a symptom that emerges over the course of the disease [84, 89]. Prevalence rates of impaired memory awareness in MCI have been reported to be equal to those seen in AD [111, 112], although this is not consistent across studies [81, 91]. Despite the frequency and clinical relevance of impaired memory awareness in early AD and MCI, much is unanswered regarding its emergence and course in the context of memory loss, the specific cognitive errors and neural changes that underlie this symptom, or the scope of its impact on everyday life; thus little is known about how to manage it. Just as early examination of individuals is critical for an accurate differential diagnosis of dementia, examination of metamemory changes at their earliest stage will offer important information regarding the nature of such changes. The value of early assessment reflects the progressively global nature of AD, and the resulting involvement of multiple brain areas that can obscure the specific nature of a cognitive or metacognitive deficit. Moreover, the longitudinal evaluation of metamemory from the earliest stages of the disease spectrum will provide new and important information regarding its course and outcomes.

In addition to pushing forward our understanding of the nature and course of metacognitive deficits in AD, greater attention needs to be paid toward the examination of the practical effects of metacognitive deficits in AD, such as impaired capacity for everyday decision making. Decision-making capacity is fundamental to an individual's independence, and invariably deteriorates at some point along the dementia continuum, compromising autonomy in financial matters [60], medical care [35, 78], and informed consent [13]. In some patients with AD, decision making can be affected relatively early in the disease, and the extent to which patients perceive themselves to have impaired functioning is likely to affect the manner in which they make decisions. Indeed, previous work has demonstrated an important role for metacognitive factors including memory awareness and disease awareness in determining individuals' decisions about taking a hypothetical treatment for AD [53], discontinuing driving [28, 48], and managing medications [25]. These associations raise the question of how memory awareness influences everyday decisions about other cognitively demanding activities, and at what point along the disease spectrum impaired metamemory may begin to compromise such decisions. For example, poor awareness of memory loss may prevent an individual from implementing strategies to ensure accurate bill paying. Indeed, it has been shown that individuals with MCI and dementia are often unaware not only of cognitive decline but of functional limitations as well [32, 36, 60, 79, 80]. Although a myriad of instruments exist to document everyday functioning [37, 56, 86], these instruments do not assess an individual's decision to monitor or modify their participation in such activities based on concerns about their memory or other thinking abilities. Moreover, while there are several validated and comprehensive tools to assess decision-making capacity, such measures are lengthy and involve an in-depth investigation into a single decision [10, 25, 53]. Measures are needed to briefly survey individuals' decisions about a range of cognitively demanding activities which they may discontinue or approach differently in the context of significant memory impairment e.g., managing medications, managing finances, preparing meals, scheduling appointments, taking public transportation, driving, working, etc. Importantly, examining an individual's decision to cease or modify their engagement in these activities is different than simply assessing function. All individuals with declining cognition will eventually demonstrate impairment on a measure of everyday function which represents the extent to which subjects have difficulties with a given task. However, individuals with intact metamemory will likely put into place better systems for, and make better decisions about, navigating functional difficulties, and will therefore have fewer problems as a result of their functional decline, and preliminary data support this hypothesis [96]. In other words, while two individuals might have similar functional ratings on driving, awareness of one's cognitive limitations is likely to influence the extent to which a person engages in that activity, and thus the likelihood of undesirable consequences of the functional disability (e.g., a car accident).

This leads to a third direction for future research, which includes examination of the value, feasibility, and potential means of remediating metacognitive deficits in AD. If metamemory deficits stem from dysfunctional error detection, or a disconnection between the comparator mechanism and the personal knowledge base, it is possible that some sort of feedback could be critically important for improving metamemory. However, it remains to be determined if improving metamemory would be desirable. In one respect, memory awareness could lead to increased anxiety or depression. Alternatively, preserved memory awareness could extend an individual's autonomy through the preservation of everyday decision making, in which case remediation of metamemory deficits could be highly valuable for both the individual and the family. Cognitive studies are needed to examine whether or not feedback facilitates metamemory, and the manner in which it does. For example, feedback may differentially enhance awareness of memory failures, that is, information that was unsuccessfully remembered despite expectations that it would be remembered. Alternatively, feedback may differentially enhance awareness of memory successes, or information that was successfully remembered despite expectations that it would not be. Another important issue to be determined is whether improvements in metamemory in the context of feedback are restricted to specific items, or whether feedback on specific items generalizes to other items within the task. Finally, it would be critical to determine the extent to which any benefit of feedback generalizes to a task without feedback. At the least, investigation of the value of feedback will provide additional information regarding the nature of metamemory deficits, and inform the mechanisms by which such deficits might ultimately be remediated.

References

1. Adair JC, Gilmore RL, Fennell EB, Gold M, Heilman KM (1995) Anosognosia during intracarotid barbiturate anesthesia: unawareness or amnesia for weakness. Neurology 45(2):241–243 (Clinical trial comparative study research support, non-US Government research support, US Government, non-P.H.S)
2. Adair JC, Na DL, Schwartz RL, Fennell EM, Gilmore RL, Heilman KM (1995) Anosognosia for hemiplegia: test of the personal neglect hypothesis. Neurology 45(12):2195–2199 (Clinical trial randomized controlled trial research support, U.S. Government, non-P.H.S. research support, U.S. Government, P.H.S)
3. Adair JC, Schwartz RL, Na DL, Fennell E, Gilmore RL, Heilman KM (1997) Anosognosia: examining the disconnection hypothesis. J Neurol Neurosurg Psychiatry 63(6):798–800 (Research support, non-U.S. Government research support, U.S. Government, non-P.H.S. research support, U.S. Government, P.H.S)
4. Adair JC, Acothley R, Knoefel JF (2006) White matter abnormalities predict symptom awareness in mild cognitive impairment [abstract]. J Int Neuropsychol Soc 12(S1):253
5. Agnew SK, Morris RG (1998) The heterogeneity of anosognosia for memory impairment in Alzheimer's disease: a review of the literature and a proposed model. Aging Mental Health 2:9–15
6. Alladi S, Xuereb J, Bak T, Nestor P, Knibb J, Patterson K, Hodges JR (2007) Focal cortical presentations of Alzheimer's disease. Brain 130(10):2636–2645. doi: 10.1093/brain/awm213 (research support, non-U. S. Government)
7. American Psychiatric Association (1994) Diagnostic and statistical manual of mental disorders, 4th edn. American Psychiatric Press, Washington, DC
8. American Psychiatric Association (2013) Diagnostic and statistical manual of mental disorders, 5th edn. American Psychiatric Publishing, Arlington, VA
9. Ansell EL, Bucks RS (2006) Mnemonic anosognosia in Alzheimer's disease: a test of agnew and morris (1998). Neuropsychologia 44:1095–1102
10. Appelbaum PS, Grisso T (1988) Assessing patients' capacities to consent to treatment. N Engl J Med 319(25):1635–1638. doi:10.1056/NEJM198812223192504

11. Backman L, Lipinska B (1993) Monitoring of general knowledge: evidence for preservation in early Alzheimer's disease. Neuropsychologia 31(4):335–345
12. Barrett AM, Eslinger PJ, Ballentine NH, Heilman KM (2005) Unawareness of cognitive deficit (cognitive anosognosia) in probable AD and control subjects. Neurology 64(4):693–699
13. Black BS, Brandt J, Rabins PV, Samus QM, Steele CD, Lyketsos CG, Rosenblatt A (2008) Predictors of providing informed consent or assent for research participation in assisted living residents. Am J Geriatr Psychiatry 16(1):83–91. doi:10.1097/JGP.0b013e318157cabd (16/1/83)
14. Botvinick M, Nystrom LE, Fissell K, Carter CS, Cohen JD (1999) Conflict monitoring versus selection-for-action in anterior cingulate cortex. Nature 402(6758):179–181. doi:10.1038/46035 (Research support, non-U.S. Government research support, U.S. Government, non-P.H.S. research support, U.S. Government, P.H.S)
15. Breier JI, Adair JC, Gold M, Fennell EB, Gilmore RL, Heilman KM (1995) Dissociation of anosognosia for hemiplegia and aphasia during left-hemisphere anesthesia. Neurology 45(1):65–67 (Clinical trial comparative study controlled clinical trial research support, U.S. Government, non-P.H.S)
16. Budson AE, Dodson CS, Daffner KR, Schacter DL (2005) Metacognition and false recognition in Alzheimer's disease: further exploration of the distinctiveness heuristic. Neuropsychology 19(2):253–258
17. Carter CS, Braver TS, Barch DM, Botvinick MM, Noll D, Cohen JD (1998) Anterior cingulate cortex, error detection, and the online monitoring of performance. Science 280(5364):747–749
18. Cines S, Farrell M, Karlawish J, Sullo E, Huey E, Jimenez D, Cosentino S (2014) Disentangling the link between awareness and depression in Alzheimer's disease. Poster to be presented at the annual meeting of the International Neuropsychological Society, Seattle, WA (in press)
19. Cines, S., Sullo, E., Karlawish, J., Cosentino, S. (2012) A tale of two measures: when clinical ratings and objective scores of memory awareness conflict. J Int Neuropsychol Soc 18(1):262. Poster presented at 40th annual meeting of the International Neuropsychological Society, Montreal, QC
20. Clare L, Whitaker CJ, Nelis SM (2010) Appraisal of memory functioning and memory performance in healthy ageing and early-stage Alzheimer's disease. Neuropsychol Dev Cogn B Aging Neuropsychol Cogn 17(4):462–491. doi:10.1080/13825580903581558 (919993002 [pii])
21. Connor LT, Dunlosky J, Hertzog C (1997) Age-related differences in absolute but not relative metamemory accuracy. Psychol Aging 12(1):50–71
22. Cosentino S, Stern Y (2005) Metacognitive theory and assessment in dementia: do we recognize our areas of weakness? J Neuropsychol Soc 11(7):910–919
23. Cosentino S, Brickman A, Griffith EY, Habeck C, Cines S, Briner T, Stern Y (2014) Metamemory is associated with right insular volume in healthy aging and Alzheimer's disease. Poster to be presented at the annual meeting of the International Neuropsychological Society, Seattle, WA (in press)
24. Cosentino S, Metcalfe J, Butterfield B, Stern Y (2007) Objective metamemory testing captures awareness of deficit in Alzheimer's disease. Cortex 43(7):1004–1019
25. Cosentino S, Metcalfe J, Cary M, De Leon J, Karlawish JH (2011) Memory awareness influences everyday decision making capacity in alzheimer's disease. Int J Alzheimer's Dis (Article ID 483897)
26. Cosentino S, Metcalfe J, Holmes B, Steffener J, Stern Y (2011) Finding the self in metacognitive evaluations: metamemory and agency in nondemented elders. Neuropsychology 25(5):602–612. doi:10.1037/a0023972
27. Cosentino S, Zhu C, Metcalfe J, Cines S, Huey ED (submitted) Impaired Metamemory in Alzheimer's disease: a failure to attend, detect, or integrate?

28. Cotrell V, Wild K (1999) Longitudinal study of self-imposed driving restrictions and deficit awareness in patients with Alzheimer disease. Alzheimer Dis Assoc Disord 13(3):151–156
29. Craig, A.D. (2009). How do you feel - now? The anterior insula and human awareness. Nature Reviews Neuroscience 10(1):59–70.
30. Cummings JL (2000) Cognitive and behavioral heterogeneity in Alzheimer's disease: seeking the neurobiological basis. Neurobiol Aging 21(6):845–861
31. Dalla Barba G, Parlato V, Lavarone A, Boller F (1995) Anosognosia, intrusions and 'frontal' functions in Alzheimer's disease and depression. Neuropsychologia 33(2):247–259
32. DeBettignies BH, Mahurin RK, Pirozzolo FJ (1990) Insight for impairment in independent living skills in Alzheimer's disease and multi-infarct dementia. J Clin Exp Neuropsychol 12(2):355–363
33. Derouesne C, Thibault S, Lagha-Pierucci S, Baudouin-Madec V, Ancri D, Lacomblez L (1999) Decreased awareness of cognitive deficits inpatients with mild dementia of the Alzheimer type. Int J Geriatr Psychiatry 14:1019–1030
34. Duke LM, Seltzer B, Seltzer JE, Vasterling JJ (2002) Cognitive components of deficit awareness in Alzheimer's disease. Neuropsychology 16(3):359–369
35. Dymek MP, Atchison P, Harrell L, Marson DC (2001) Competency to consent to medical treatment in cognitively impaired patients with Parkinson's disease. Neurology 56(1):17–24
36. Farias ST, Mungas D, Jagust W (2005) Degree of discrepancy between self and other-reported everyday functioning by cognitive status: dementia, mild cognitive impairment, and healthy elders. Int J Geriatr Psychiatry 20(9):827–834. doi:10.1002/gps.1367
37. Farias ST, Mungas D, Reed BR, Harvey D, Cahn-Weiner D, Decarli C (2006) MCI is associated with deficits in everyday functioning. Alzheimer Dis Assoc Disord 20(4):217–223. doi:10.1097/01.wad.0000213849.51495.d9 ([pii] 00002093-200610000-00008)
38. Feinberg TE (2001) Altered egos: how the brain creates the self. Oxford University Press, New York
39. Fernandez-Duque D, Baird JA, Posner MI (2000) Executive attention and metacognitive regulation. Conscious Cogn 9(2):288–307
40. Gallo DA, Chen JM, Wiseman AL, Schacter DL, Budson AE (2007) Retrieval monitoring and anosognosia in Alzheimer's disease. (Research support, N.I.H., extramural research support, non-U.S. Government). Neuropsychology 21(5): 559–568. doi:10.1037/0894-4105.21.5.559
41. Gallo DA, Cramer SJ, Wong JT, Bennett DA (2012) Alzheimer's disease can spare local metacognition despite global anosognosia: revisiting the confidence-accuracy relationship in episodic memory. Neuropsychologia 50(9):2356–2364. doi:10.1016/j.neuropsychologia.2012.06.005
42. Galton CJ, Patterson K, Xuereb JH, Hodges JR (2000) Atypical and typical presentations of Alzheimer's disease: a clinical, neuropsychological, neuroimaging and pathological study of 13 cases. Brain 123(3):484–498
43. Glosser G, Goodglass H (1990) Disorders in executive control functions among aphasic and other brain-damaged patients. J Clin Exp Neuropsychol 12(4):485–501
44. Glosser G, Goodglass H, Biber C (1989) Assessing visual memory disorders. Psychol Assess 1:82–91
45. Glosser G, Cole L, Khatri U, DellaPietra L, Kaplan E (2002) Assessing nonverbal memory with the biber figure learning test–extended in temporal lobe epilepsy patients. Arch Clin Neuropsychol 17:25–35
46. Hannesdottir K, Morris RG (2007) Primary and secondary anosognosia for memory impairment in patients with Alzheimer's disease. Cortex 43(7):1020–1030
47. Hardy RM, Oyebode JR, Clare L (2006) Measuring awareness in people with mild to moderate Alzheimer's disease: development of the memory awareness rating scale–adjusted. Neuropsychol Rehabil 16(2):178–193. doi:10.1080/09602010500145646 (T5Q572T5H483LWU3)
48. Hunt L, Morris JC, Edwards D, Wilson BS (1993) Driving performance in persons with mild senile dementia of the Alzheimer type. J Am Geriatr Soc 41(7):747–752

49. Johnson JK, Head E, Kim R, Starr A, Cotman CW (1999) Clinical and pathological evidence for a frontal variant of Alzheimer disease. Arch Neurol 56(10):1233–1239
50. Johnson SC, Baxter LC, Wilder LS, Pipe JG, Heiserman JE, Prigatano GP (2002) Neural correlates of self-reflection. Brain 125(Pt 8):1808–1814
51. Johnson SC, Schmitz TW, Kawahara-Baccus TN, Rowley HA, Alexander AL, Lee J, Davidson RJ (2005) The cerebral response during subjective choice with and without self-reference. J Cogn Neurosci 17(12):1897–1906. doi:10.1162/089892905775008607
52. Johnson SC, Ries ML, Hess TM, Carlsson CM, Gleason CE, Alexander AL, Sager MA (2007) Effect of Alzheimer disease risk on brain function during self-appraisal in healthy middle-aged adults. Arch Gen Psychiatry 64(10):1163–1171. doi:10.1001/archpsyc.64.10.1163
53. Karlawish JH, Casarett DJ, James BD, Xie SX, Kim SY (2005) The ability of persons with Alzheimer disease (AD) to make a decision about taking an AD treatment. Neurology 64(9):1514–1519
54. Kaszniak AW, Zak M (1996) On the neuropsychology of metamemory: contributions from the study of amnesia and dementia. Learn Individ Differ 8(4):355–381
55. Koriat A, Levy-Sadot R (2001) The combined contributions of the cue-familiarity and accessibility heuristics to feelings of knowing. J Exp Psychol Learn Mem Cogn 27(1):34–53
56. Lawton MP, Brody EM (1969) Assessment of older people: self-maintaining and instrumental activities of daily living. Gerontologist 9(3):179–186
57. Lipinska B, Backman L (1996) Feeling-of-knowing in fact retrieval: further evidence for preservation in early Alzheimer's disease. J Int Neuropsychol Soc 2(4):350–358
58. Lopez OL, Becker JT, Somsak D, Dew MA, DeKosky ST (1994) Awareness of cognitive deficits and anosognosia in probable Alzheimer's disease. Eur Neurol 34(5):277–282
59. Mangone CA, Hier DB, Gorelick PB, Ganellen RJ, Langenberg P, Boarman R, Dollear WC (1991) Impaired insight in Alzheimer's disease. J Geriatr Psychiatry Neurol 4(4):189–193
60. Marson DC, Martin RC, Wadley V, Griffith HR, Snyder S, Goode PS, Harrell LE (2009) Clinical interview assessment of financial capacity in older adults with mild cognitive impairment and Alzheimer's disease. J Am Geriatr Soc 57(5):806–814. doi:10.1111/j.1532-5415.2009.02202.x (JGS2202)
61. Martyr A, Clare L, Nelis SM, Roberts JL, Robinson JU, Roth I, Morris RG (2011) Dissociation between implicit and explicit manifestations of awareness in early stage dementia: evidence from the emotional Stroop effect for dementia-related words. Int J Geriatr Psychiatry 26(1):92–99. doi: 10.1002/gps.2495 (Multicenter study research support, non-U.S. Government)
62. McCurdy LY, Maniscalco B, Metcalfe J, Liu KY, de Lange FP, Lau H (2013) Anatomical coupling between distinct metacognitive systems for memory and visual perception. J Neurosci 33(5):1897–1906. doi:10.1523/JNEUROSCI.1890-12.2013 (Research support, non-U.S. Government)
64. McDaniel KD, Edland SD, Heyman A (1995) Relationship between level of insight and severity of dementia in Alzheimer disease. CERAD (Consortium to Establish a Registry for Alzheimer's Disease) clinical investigators. Alzheimer Dis Assoc Disord 9(2):101–104
63. McGlynn SM, Kaszniak AW (1991) When metacognition fails: impaired awareness of deficit in Alzheimer's disease. J Cogn Neurosci 3:183–189
65. McKhann G, Drachman D, Folstein M, Katzman R, Price D, Stadlan EM (1984) Clinical diagnosis of Alzheimer's disease: report of the NINCDS-ADRDA work group under the auspices of department of health and human services task force on Alzheimer's disease. Neurology 34(7):939–944 (guideline practice guideline)
66. McKhann GM, Knopman DS, Chertkow H, Hyman BT, Jack CR Jr, Kawas CH, Phelps CH (2011) The diagnosis of dementia due to Alzheimer's disease: recommendations from the national institute on aging-Alzheimer's association workgroups on diagnostic guidelines for Alzheimer's disease. Alzheimers Dement 7(3):263–269. doi:10.1016/j.jalz.2011.03.005 (consensus development conference, NIH research support, non-U. S. Government)
67. Metcalfe J, Shimamura AP (eds) (1994) Metacognition: knowing about knowing. The MIT Press, London

68. Mez J, Cosentino S, Brickman AM, Huey ED, Manly JJ, Mayeux R (2012). Dysexecutive versus amnestic Alzheimer's disease subgroups: an analysis of demographic, genetic, and vascular factors in the national Alzheimer's coordinating center database. Alzheimer Dis Assoc Disord 8(4):365–366
69. Mez J, Cosentino S, Brickman AM, Huey ED, Mayeux R (2013) Different demographic, genetic and longitudinal traits in language versus memory Alzheimer's subgroups. J Alzheimer Dis 37(1):137–146
70. Michon A, Deweer B, Pillon B, Agid Y, Dubois B (1994) Relation of anosognosia to frontal lobe dysfunction in Alzheimer's disease. J Neurol Neurosurg Psychiatry 57(7):805–809
71. Mograbi DC, Brown RG, Morris RG (2009) Anosognosia in Alzheimer's disease—the petrified self. Conscious Cogn 18(4):989–1003
72. Mograbi DC, Brown RG, Salas C, Morris RG (2012). Emotional reactivity and awareness of task performance in Alzheimer's disease. Neuropsychologia 50(8):2075–2084. doi: 10.1016/j.neuropsychologia.2012.05.008 (Research support, non-U.S. Government)
73. Moulin CJ (2002) Sense and Sensitivity: metacognition in Alzheimer's Disease. In: Perfect TJ, Schwartz BL (eds) Applied metacognition. Cambridge University Press, Cambrdige
74. Moulin CJ, Perfect TJ, Jones RW (2000) Evidence for intact memory monitoring in Alzheimer's disease: metamemory sensitivity at encoding. Neuropsychologia 38(9):1242–1250
75. Moulin CJ, Perfect TJ, Jones RW (2000) The effects of repetition on allocation of study time and judgements of learning in Alzheimer's disease. Neuropsychologia 38(6):748–756
76. Neary D, Snowden JS, Bowen DM, Sims NR, Mann DM, Benton JS, Davison AN (1986) Neuropsychological syndromes in presenile dementia due to cerebral atrophy. J Neurol Neurosurg Psychiatry 49(2):163–174
77. Northoff G, Heinzel A, de Greck M, Bermpohl F, Dobrowolny H, Panksepp J (2006) Self-referential processing in our brain—a meta-analysis of imaging studies on the self. Neuroimage 31(1):440–457. doi: 10.1016/j.neuroimage.2005.12.002(meta-analysis research support, Government)
78. Okonkwo OC, Griffith HR, Copeland JN, Belue K, Lanza S, Zamrini EY, Marson DC (2008) Medical decision-making capacity in mild cognitive impairment: a 3-year longitudinal study. Neurology 71(19):1474–1480. doi:10.1212/01.wnl.0000334301.32358.48 (71/19/1474)
79. Okonkwo OC, Wadley VG, Griffith HR, Belue K, Lanza S, Zamrini EY, Marson DC (2008) Awareness of deficits in financial abilities in patients with mild cognitive impairment: going beyond self-informant discrepancy. Am J Geriatr Psychiatry 16(8):650–659
80. Okonkwo OC, Griffith HR, Vance DE, Marson DC, Ball KK, Wadley VG (2009) Awareness of functional difficulties in mild cognitive impairment: a multidomain assessment approach. J Am Geriatr Soc 57(6):978–984. doi:JGS2261 (10.1111/j.1532-5415.2009.02261.x)
81. Orfei MD, Varsi AE, Blundo C, Celia E, Casini AR, Caltagirone C, Spalletta G (2010) Anosognosia in mild cognitive impairment and mild Alzheimer's disease: frequency and neuropsychological correlates. Am J Geriatr Psychiatry 18(12):1133–1140. doi:10.1097/JGP.0b013e3181dd1c50
82. Ott BR, Lafleche G, Whelihan WM, Buongiorno GW, Albert MS, Fogel BS (1996) Impaired awareness of deficits in Alzheimer disease. Alzheimer Dis Assoc Disord 10(2):68–76
83. Pappas BA, Sunderland T, Weingartner HM, Vitello B, Martinson H, Putnam K (1992) Alzheimer's disease and feeling-of-knowing for knowledge and episodic memory. J Gerontol 47(3):159–164
84. Perrotin A, Belleville S, Isingrini M (2007) Metamemory monitoring in mild cognitive impairment: evidence of a less accurate episodic feeling-of-knowing. Neuropsychologia 45(12):2811–2826. doi:10.1016/j.neuropsychologia.2007.05.003 (S0028-3932(07)00184-4)
85. Perrotin, A., Mormino, E.C., Madison, C.M., Hayenga, A.O., Jagust, W.J. (2012). Subjective Cognition and Amyloid Deposition Imaging. JAMA Neurology, 69(2), 223–229
86. Pfeffer RI, Kurosaki TT, Harrah CH Jr, Chance JM, Filos S (1982) Measurement of functional activities in older adults in the community. J Gerontol 37(3):323–329

87. Prigatano GP (1991) Disturbances in self-awareness after traumatic brain injury. In: Prigatano GP, Schacter DL (eds) Awareness of deficit after brain injury. Oxford University Press, New York, pp 111–126
88. Reed BR, Jagust WJ, Coulter L (1993) Anosognosia in Alzheimer's disease: relationships to depression, cognitive function, and cerebral perfusion. J Clin Exp Neuropsychol 15(2):231–244
89. Ries ML, Jabbar BM, Schmitz TW, Trivedi MA, Gleason CE, Carlsson CM, Johnson SC (2007) Anosognosia in mild cognitive impairment: relationship to activation of cortical midline structures involved in self-appraisal. J Int Neuropsychol Soc 13(3):450–461. doi:10.1017/S1355617707070488 (S1355617707070488)
90. Ries ML, McLaren DG, Bendlin BB, Rowley Guofanxu HA, Birn R, Johnson SC (2012) Medial prefrontal functional connectivity-relation to memory self-appraisal accuracy in older adults with and without memory disorders. Neuropsychologia. doi:S0028-3932(11)00559-8
91. Roberts JL, Clare L, Woods RT (2009) Subjective memory complaints and awareness of memory functioning in mild cognitive impairment: a systematic review. Dement Geriatr Cogn Disord 28(2):95–109. doi:10.1159/000234911 (000234911)
92. Salmon E, Perani D, Herholz K, Marique P, Kalbe E, Holthoff V, Garraux G (2006) Neural correlates of anosognosia for cognitive impairment in Alzheimer's disease. Hum Brain Mapp 27(7):588–597. doi:10.1002/hbm.20203
93. Schmitter-Edgecombe M, Seelye AM (2011) Predictions of verbal episodic memory in persons with Alzheimer's disease. J Clin Exp Neuropsychol 33(2):218–225. doi:10.1080/13803395.2010.507184
94. Schmitz TW, Rowley HA, Kawahara TN, Johnson SC (2006) Neural correlates of self-evaluative accuracy after traumatic brain injury. Neuropsychologia 44(5):762–773. doi: 10.1016/j.neuropsychologia.2005.07.012 (Comparative study research support, N.I.H., extramural)
95. Shaked D, Farrell M, Cines S, Karlawish JH, Sullo E, Huey ED, Cosentino S (in press) Cognitive correlates of metamemory in ad: the role of memory, executive, and right hemisphere abilities
96. Shaked D, Karlawish J, Cines S, Sullo E, Devanand D, Cosentino S (2014) Memory awareness influences modification of everyday activities in cognitively impaired elders. Poster to be presented at the annual meeting of the International Neuropsychological Society, Seattle, WA (in press)
97. Shimamura AP (2000) Toward a cognitive neuroscience of metacognition. Conscious Cogn 9(2):313–323 (discussion 316–324)
98. Simmons WK, Reddish M, Bellgowan PS, Martin A (2010) The selectivity and functional connectivity of the anterior temporal lobes. Cereb Cortex 20(4):813–825. doi:10.1093/cercor/bhp149 (Research support, N.I.H., extramural)
99. Smith CA, Henderson VW, McCleary CA, Murdock GA, Buckwalter JG (2000) Anosognosia and Alzheimer's disease: the role of depressive symptoms in mediating impaired insight. J Clin Exp Neuropsychol 22(4):437–444
100. Souchay C (2007) Metamemory in Alzheimer's disease [Review]. Cortex 43(7):987–1003
101. Souchay C, Isingrini M, Gil R (2002) Alzheimer's disease and feeling-of-knowing in episodic memory. Neuropsychologia 40(13):2386–2396
102. Souchay C, Isingrini M, Pillon B, Gil R (2003) Metamemory accuracy in Alzheimer's disease and frontotemporal lobe dementia. Neurocase 9(6):482–492. doi:10.1076/neur.9.6.482.29376 (Clinical trial comparative study)
103. Souchay C, Isingrini M, Clarys D, Taconnat L, Eustache F (2004) Executive functioning and judgment-of-learning versus feeling-of-knowing in older adults. Exp Aging Res 30(1):47–62
104. Starkstein SE, Vazquez S, Migliorelli R, Teson A, Sabe L, Leiguarda R (1995) A single-photon emission computed tomographic study of anosognosia in Alzheimer's disease. Arch Neurol 52(4):415–420
105. Starkstein SE, Sabe L, Chemerinski E, Jason L, Leiguarda R (1996) Two domains of anosognosia in Alzheimer's disease. J Neurol Neurosurg Psychiatry 61(5):485–490

106. Starkstein SE, Chemerinski E, Sabe L, Kuzis G, Petracca G, Teson A, Leiguarda R (1997) Prospective longitudinal study of depression and anosognosia in Alzheimer's disease. Br J Psychiatry 171:47–52
107. Thies W, Bleiler L, Alzheimer's A (2013) 2013 Alzheimer's disease facts and figures. Alzheimers Dement 9(2):208–245. doi:10.1016/j.jalz.2013.02.003
108. Mars RB, Coles MG, Grol MJ, Holroyd CB, Nieuwenhuis S, Hulstijn W, Toni, I. (2005). Neural dynamics of error processing in medial frontal cortex. Neuroimage 28(4):1007–1013. doi: 10.1016/j.neuroimage.2005.06.041(Clinical trial research support, N.I.H., extramural research support, non-U.S. Government)
109. Tulving E (1985) How many memory systems are there? Am Psychol 40:385–398
110. Vilkki J, Servo A, Surma-aho O (1998) Word list learning and prediction of recall after frontal lobe lesions. Neuropsychology 12(2):268–277
111. Vogel A, Stokholm J, Gade A, Andersen BB, Hejl AM, Waldemar G (2004) Awareness of deficits in mild cognitive impairment and Alzheimer's disease: do MCI patients have impaired insight? Dement Geriatr Cogn Disord 17(3):181–187
112. Vogel A, Hasselbalch SG, Gade A, Ziebell M, Waldemar G (2005) Cognitive and functional neuroimaging correlate for anosognosia in mild cognitive impairment and Alzheimer's disease. Int J Geriatr Psychiatry 20(3):238–246
113. Zahn R, Moll J, Krueger F, Huey ED, Garrido G, Grafman J (2007) Social concepts are represented in the superior anterior temporal cortex. Proc Natl Acad Sci U S A 104(15):6430–6435. doi:10.1073/pnas.0607061104 (Comparative study research support, N.I.H., intramural research support, non-U.S. Government)
114. Zamboni G, Drazich E, McCulloch E, Filippini N, Mackay CE, Jenkinson M, Wilcock GK (2013) Neuroanatomy of impaired self-awareness in Alzheimer's disease and mild cognitive impairment. Cortex, 668–678. doi: 10.1016/j.cortex.2012.04.011 (S0010-9452(12)00138-4 [pii])
115. Zamboni G, Drazich E, McCulloch E, Filippini N, Mackay CE, Jenkinson M, Wilcock GK (2013) Neuroanatomy of impaired self-awareness in Alzheimer's disease and mild cognitive impairment. Cortex 49(3):668–678. doi: 10.1016/j.cortex.2012.04.011(Research support, non-U.S. Government)
116. Zanetti O, Vallotti B, Frisoni GB, Geroldi C, Bianchetti A, Pasqualetti P, Trabucchi M (1999) Insight in dementia: when does it occur? evidnce for a nonlinear relationship between insight and cognitive status. J Gerontology 54(2):100–106

Made in the USA
Columbia, SC
17 September 2026

60c90ccc-b88c-4e57-b91f-abd0b27bd02cR01